Optimization of Large Structural Systems

NATO ASI Series

Advanced Science Institutes Series

A Series presenting the results of activities sponsored by the NATO Science Committee, which aims at the dissemination of advanced scientific and technological knowledge, with a view to strengthening links between scientific communities.

The Series is published by an international board of publishers in conjunction with the NATO Scientific Affairs Division

A Life Sciences	Plenum Publishing Corporation
B Physics	London and New York
C Mathematical	Kluwer Academic Publishers
and Physical Sciences	Dordrecht, Boston and London
D Behavioural and Social Sciences	
E Applied Sciences	
F Computer and Systems Sciences	Springer-Verlag
G Ecological Sciences	Berlin, Heidelberg, New York, London,
H Cell Biology	Paris and Tokyo
I Global Environmental Change	

NATO-PCO-DATA BASE

The electronic index to the NATO ASI Series provides full bibliographical references (with keywords and/or abstracts) to more than 30000 contributions from international scientists published in all sections of the NATO ASI Series.
Access to the NATO-PCO-DATA BASE is possible in two ways:

– via online FILE 128 (NATO-PCO-DATA BASE) hosted by ESRIN,
Via Galileo Galilei, I-00044 Frascati, Italy.

– via CD-ROM "NATO-PCO-DATA BASE" with user-friendly retrieval software in English, French and German (© WTV GmbH and DATAWARE Technologies Inc. 1989).

The CD-ROM can be ordered through any member of the Board of Publishers or through NATO-PCO, Overijse, Belgium.

Optimization of Large Structural Systems

Volume II

edited by

G. I. N. Rozvany

**Professor of Structural Design,
Essen University,
Germany**

Springer-Science+Business Media, B.V.

Proceedings of the NATO/DFG Advanced Study Institute on
Optimization of Large Structural Systems
Berchtesgaden, Germany
23 September – 4 October 1991

Library of Congress Cataloging-in-Publication Data

NATO/DFG Advanced Study Institute (1991 . Berchtesgaden, Germany)
   Optimization of large structural systems : proceedings of the
NATO/DFG Advanced Study Institute, Berchtesgaden, Germany, 23
Sept.-4 Oct. 1991 / edited by G.I.N. Rozvany.
      p.   cm. -- (NATO ASI series. Series E, Applied sciences ; vol.
231)
   Includes index.
   ISBN 978-94-010-9579-2   ISBN 978-94-010-9577-8 (eBook)
   DOI 10.1007/978-94-010-9577-8
   1. Structural optimization--Congresses.   I. Rozvany, G. I. N.
II. North Atlantic Treaty Organization.   III. Deutsche
Forschungsgemeinschaft.   IV. Title.   V. Series: NATO ASI series.
Series E, Applied sciences ; no. 231.
TA658.8.N39 1991
624.1'771--dc20   92-43799

ISBN 978-94-010-9579-2

Printed on acid-free paper

TABLE OF CONTENTS

VOLUME II

Composite, Anisotropic and Nonlinear Materials

Z. Gürdal and R.T. Haftka
Optimization of Composite Laminates   623

P. Pedersen
Optimal Orientation of Anisotropic Materials
Optimal Distribution of Anisotropic Materials
Optimal Shape Design with Anisotropic Materials
Optimal Design for a Class of Non-Linear Elasticity   649

J.E. Taylor and J. Logo
Analysis and Design of Elastic/Softening Truss Structures Based on a
Mixed-Form Extremum Principle   683

M.P. Bendsøe, N. Olhoff and J.E. Taylor
A Unified Approach to the Analysis and Design of Elasto-Plastic
Structures with Mechanical Contact   697

C. Cinquini and M. Rovati
Optimal Orientation of Orthotropic Properties for Continuum Bodies and
Structural Elements   707

Neural Networks, Parallel Processing, Multicriteria and
Control Problems

L. Berke and P. Hajela
Application of Neural Nets in Structural Optimization   731

P. Hajela, B. Fu and L. Berke
Self-Organization in Neural Networks — Applications in Structural
Optimization   747

B.H.V. Topping and A.I. Khan
Parallel Computations for Structural Analysis, Re-Analysis and
Optimization   767

J. Koski
Multicriterion Structural Optimization — State of the Art 793

H.L. Thomas and G.N. Vanderplaats
Control-Augmented Structural Synthesis 811

N.S. Khot
Structural and Control Optimization 829

Applications

M.Z. Cohn
Theory and Practice of Structural Optimization 843

D.E. Grierson and C.-M. Chan
Design Optimization of Tall Steel Building Frameworks 863

D.E. Grierson and L. Xu
Design Optimization of Steel Frameworks Accounting for Semi-Rigid
Connections 873

D.E. Grierson and H. Moharrami
Design Optimization of Reinforced Concrete Building Frameworks 883

B.L. Karihaloo and S. Kanagasundaram
Optimum Design of Plane Structural Frames by Non-Linear Programming 897

B.L. Karihaloo and S. Kanagasundaram
Minimum-Cost Design of Reinforced Concrete Members by Non-Linear
Programming 927

W. Dobler, P. Erl and H. Rapp
Optimization of Sandwich Structures with Respect to Local Instabilities
with MBB-Lagrange 951

H.J. Baier
Optimization in Structural Dynamics with Applications 973

H.J. Baier
Optimization in Coupled Problems 987

H.J. Baier
Composite Laminate and Sandwich Optimization with Applications 997

H.A. Eschenauer, G. Schuhmacher and W. Hartzheim
Multidisciplinary Optimization of Fiber Composite Aircraft Structures 1011

Miscellaneous Topics

C.A. Mota Soares, J.I. Barbosa and C.M. Mota Soares
Shape Optimal Design of Axisymmetric Shell Structures   1023

N.V. Banichuk
Application of Analytical Models for the Optimization of Large Structural
Systems   1051

E. Schnack and G. Iancu
Optimization of Large Scale Systems in Elasticity   1073

W. Gutkowski, O. Mahrenholtz and M. Pyrz
Minimum-Weight Design of Structures under Non-Conservative Forces   1087

P. Morelle and V. Braibant
Multi-Model Optimization of Large Scale Structures   1101

M.B. Fuchs
The Analytic Solution of the Structural Analysis Problem and Its Use in
Structural Synthesis   1115

M.E.M. El-Sayed and T.S. Jang
Large Scale Structural Optimization with Nonlinear Goal Programming   1135

V.V. Kobelev
Isoperimetric Inequalities in Stability Problems   1155

B. Esping, P. Clarin and O. Romell
OCTOPUS — A Tool for Distributed Optimization of Multi-Disciplinary
Objectives   1165

TITLES OF CONTRIBUTED PAPERS   1195

SUBJECT INDEX   1199

NOTE: Short contributions to this NATO/DFG ASI whose text does not appear in full in these proceedings will be considered for publication in the international journal "**Structural Optimization**" (Springer-Verlag).

ILLUSTRATIONS — BERCHTESGADEN

Bird's-eye view of Berchtesgaden with Watzmann in the background   xviii

Church of Ramsau outside Berchtesgaden   xxii

Idyllic setting in the outskirts of Berchtesgaden with Watzmann   xxiii

Königssee with St. Bartholomä church and Watzmann   xxiv

VOLUME I

PREFACE   xi

LIST OF PARTICIPANTS   xiii

Opening Address by L. Berke   xix

Closing Address by B.L. Karihaloo   xxi

PRINCIPAL LECTURES

Optimality Criteria and Topology Optimization

G.I.N. Rozvany and M. Zhou
Continuum-Based Optimality Criteria (COC) Methods: An Introduction   1

M. Zhou and G.I.N. Rozvany
Iterative COC Methods
Part I. Iterative Methods Based on Continuum Formulation   27
Part II. Iterative COC Methods Formulated Directly for Discretized
Systems (DCOC Methods)   59

G.I.N. Rozvany, M. Zhou and W. Gollub
Layout Optimization by COC Methods: Analytical Solutions   77

G.I.N. Rozvany and M. Zhou
Layout and Generalized Shape Optimization by Iterative COC Methods   103

U. Kirsch and G.I.N. Rozvany
Design Considerations in the Optimization of Structural Topologies   121

M.P. Bendsøe and A. Ben-Tal
Truss Topology Optimization by a Displacements Based Optimality
Criterion Approach   139

K. Suzuki and N. Kikuchi
Layout Optimization Using the Homogenization Method   157

J. Fukushima, K. Suzuki and N. Kikuchi
Applications to Car Bodies: Generalized Layout Design of
Three-Dimensional Shells   177

Decomposition Methods and Approximation Concepts

J. Sobieszczanski-Sobieski
Optimization by Decomposition in Structural and
Multidisciplinary Applications   193

J.-F.M. Barthelemy and R.T. Haftka
Recent Advances in Approximation Concepts for Optimum Structural Design 235

H.L. Thomas and G.N. Vanderplaats
The State of the Art of Approximation Concepts in Structural Optimization 257

U. Kirsch
Approximate Models for the Optimization of Large Structural Systems   271

Sensitivity Analysis

R.T. Haftka and H.M. Adelman
Sensitivity of Discrete Systems   289

K.K. Choi and S. Wang
Design Sensitivity Analysis of Dynamics of Built-Up Structures Using
Ritz and Mode Acceleration Methods   313

K.K. Choi, I. Shim, J. Lee and H.T. Kulkarni
Design Sensitivity Analysis of Dynamic Frequency Responses of
Acousto-Elastic Built-Up Structures   329

J.S. Arora and T.H. Lee
Shape Design Sensitivity Analysis of Nonlinear Nonconservative Structural
Problems   345

G. Cheng and N. Olhoff
New Method of Error Analysis and Detection in Semi-Analytical Sensitivity
Analysis   361

N. Olhoff and J. Rasmussen
Method of Error Elimination for a Class of Semi-Analytical Sensitivity
Analysis Problems   385

B. Rousselet
Introduction to Shape Sensitivity — Three-Dimensional and Surface Systems 397

B. Rousselet
A Finite Strain Rod Model and Its Design Sensitivity 433

Z. Mróz
Design Sensitivity of Critical Loads and Vibration Frequencies of Nonlinear
Structures 455

K. Dems
Direct and Adjoint Approach to First- and Second-Order Shape Sensitivity
and Optimal Design of Structure 477

K. Dems
Shape Sensitivity Analysis and Optimal Design of Plates with Varying
External and Internal Boundaries 493

Mathematical Programming and Global Optima

C. Fleury
Dual Methods for Convex Separable Problems 509

C. Fleury
Sequential Convex Programming for Structural Optimization Problems 531

K. Svanberg
The Method of Moving Asymptotes (MMA) with Some Extensions 555

K. Svanberg
Some Second Order Methods for Structural Optimization 567

K. Svanberg
Local and Global Optima 579

J. Herskovits
An Interior Points Method for Nonlinear Constrained Optimization 589

S. Hernández
A New Mathematical Procedure for Global Optimization in Nonconvex
Problems 609

OPTIMIZATION OF COMPOSITE LAMINATES

Zafer Gürdal and Raphael T. Haftka
Virginia Polytechnic Institute and State University
Blacksburg, Virginia 24061

ABSTRACT. Design of composite structures can be viewed as a multi-faceted task, one which requires integration of issues related to composite mechanics, structural analysis, optimization, and manufacturing. The major coverage of the paper is on the issue of optimization, with special emphasis on the use of optimization for designing with discrete and integer valued variables required for the stacking-sequence optimization. Different techniques that can be used for stacking sequence optimization are introduced, and different aspects of their application are demonstrated for laminate buckling optimization.

1. Introduction

The design of laminated composite plates subject to various constraints has been a major topic of research in recent years (see, for example [1]). Typically the design variables are either the ply orientations of layers or the thicknesses of the layers that are assumed to have a given ply orientation. However, in many practical applications the ply orientations that may be used are limited to 0-deg, 90-deg and ±45-deg, and the thicknesses of the layers are limited to integer multiples of the lamina thickness. This means that the basic design problem is to determine the stacking sequence of the composite laminate—a problem which calls for integer programming techniques.

Integer programming techniques are often quite costly. For this reason, although numerous structural optimization algorithms have been applied to problems for composite laminates, relatively little has been done for optimum structural design using design variables with discrete values. So far the most common way of designing with discrete-valued design variables is to round-off the optimum values of the variables, obtained by assuming them to be continuous, to the nearest acceptable discrete value. For problems with n design variables there are 2^n possible rounded-off designs, and the problem of choosing the best one is formidable for large n. Furthermore, for some problems the optimum design may not even be one of these rounded-off designs.

More recently there has been more interest in applying integer programming techniques to structural optimization in general, and laminate design in particular. Schmit and Fleury [2] proposed a dual method combined with a convex approximation procedure that can be applied to discrete sizing problems, even though it is not guaranteed to converge to the optimum solution. Reinschmidt [3] used the Branch and Bound method for plastic design of frames by posing the problem as an integer linear programming problem. John et al. [4] merged the Branch and Bound algorithm with sequential linear programming for the discrete optimal design of trusses. A similar approach was used by Olsen and Vanderplaats [5] and applied to the design of a composite laminate and other problems. Shin et al. [6] applied a penalty function approach to the design of trusses. Mesquita and Kamat [7] solved the problem of laminate design using a nonlinear version of the branch and bound algorithm. Haftka and Walsh [8] showed that the problem of laminate design for maximum buckling load can be linearized by the use of ply identity design variables instead of ply thickness and ply angle variables. They then used the branch and bound method to solve this linear problem.

623

G. I. N. Rozvany (ed.), Optimization of Large Structural Systems, Vol. II, 623–648.
© 1993 Kluwer Academic Publishers.

With the recent interest in probabilistic search techniques such as simulated annealing and genetic algorithms (see Hajela [9]), these methods are also being applied to the composite laminate stacking sequence design problem. Lombardi [10] applied simulated annealing to the buckling maximization problem, while LeRiche [11] solved the same problem using a genetic algorithm.

The objective of the present paper is to review some of the integer programming techniques that have been used for laminate design and present some results of their application. These methods have been applied almost exclusively to the design of unstiffened laminates, and for this reason the paper is limited to that problem. Furthermore, in the interest of focusing the discussion, most of the presentation is limited to the design of laminate under buckling constraints.

2. Design Variable Definition

A composite laminate (see Fig. 1) is usually analyzed by classical lamination theory (CLT). With this theory, for symmetric laminates, properties are completely characterized by the A and D matrices. For example, for a balanced laminate the four major elements of the D matrix can be written in terms of material invariants and through the thickness integrals as

$$
\begin{aligned}
D_{11} &= U_1 W_0 + U_2 W_1 + U_3 W_3, & D_{12} &= U_4 W_0 - U_3 W_3, \\
D_{22} &= U_1 W_0 - U_2 W_1 + U_3 W_3, & D_{66} &= U_5 W_0 - U_3 W_3,
\end{aligned} \tag{1}
$$

where W_0, W_1 and W_3 are defined as

$$
W_0 = \int_{-h/2}^{h/2} z^2 dz, \quad W_1 = \int_{-h/2}^{h/2} z^2 \cos 2\theta dz, \quad \text{and} \quad W_3 = \int_{-h/2}^{h/2} z^2 \cos 4\theta dz, \tag{2}
$$

h is the total thickness of the laminate, z is the distance from the plane of symmetry (see Figure 1), and θ is the ply orientation angle. The material invariants in Eq. (1)

$$
U_1 = \frac{1}{8}(3Q_{11} + 3Q_{22} + 2Q_{12} + 4Q_{66}), \quad U_2 = \frac{1}{2}(Q_{11} - Q_{22}), \quad U_3 = \frac{1}{8}(Q_{11} + Q_{22} - 2Q_{12} - 4Q_{66}),
$$

$$
U_4 = \frac{1}{8}(Q_{11} + Q_{22} + 6Q_{12} - 4Q_{66}), \quad U_5 = \frac{1}{8}(Q_{11} + Q_{22} - 2Q_{12} + 4Q_{66}),
$$

$$\tag{3}$$

are related to the engineering properties by

$$
Q_{11} = \frac{E_1}{1 - \nu_{12}\nu_{21}}, \quad Q_{22} = \frac{E_2}{1 - \nu_{12}\nu_{21}},
$$

$$
Q_{12} = \frac{\nu_{12} E_2}{1 - \nu_{12}\nu_{21}} = \frac{\nu_{21} E_1}{1 - \nu_{12}\nu_{21}}, \quad \text{and} \quad Q_{66} = G_{12}. \tag{4}
$$

For the in-plane stiffnesses A_{11}, A_{12}, A_{22} and A_{66}, expressions similar to those given in Eq. (1) can be written by replacing the W's with V's where

$$V_0 = \int_{-h/2}^{h/2} dz, \quad V_1 = \int_{-h/2}^{h/2} \cos 2\theta dz, \quad \text{and} \quad V_3 = \int_{-h/2}^{h/2} \cos 4\theta dz. \tag{5}$$

This particular form of writing the A's and D's highlights the fact that the design problem is associated with the evaluation of the integrals. Of the three integrals given in Eq. (2) and (5), the first ones are independent of the stacking sequence and give the value of $h^3/12$ and h, respectively. Therefore, it can be concluded that the number of design variables needed to characterize a laminate for either in-plane conditions or out-of-plane conditions is really only two (corresponding to the last two integrals normalized by the values of their respective first integrals). As shown by Miki [12-14], this reduction to two design variables permits graphical solution for problems where we deal with only in-plane, or only out-of plane response. Because of the insight that such graphical solution techniques afford the designer, we demonstrate Miki's graphical technique in the next section.

For more general design constraints that involve a more complex combination of the in-plane and out-of-plane stiffnesses the use of normalized integrals as design variables loses its attractiveness. Still, the design variables need to define the stacking sequence so that the integrals can be calculated for the purpose of characterizing the laminate. Because of the history of using continuous design variables, ply-thickness and ply-orientation design variables are still popular. However, for the practical problem of designing a laminate from a fixed set of ply orientations, the most logical design variables are ply-identity design variables that define the identity of each ply (that is define its angle). For example, if we have four possible orientations and N plies, we can use N design variables that take the values of 1 to 4 to define the stacking sequence. Symmetry can be used to reduced this number to $N/2$. However, it is also possible to use zero-one ply identity design variables. In the present paper we consider laminates that are made from 0-deg, 90-deg, and ± 45-deg plies. The ply stacking sequence is defined in terms of four sets of ply-orientation-identity variables o_i, n_i, f_i^p and f_i^m, $i = 1, \cdots, N/2$, that are zero-one integer variables. The variable o_i, n_i, f_i^p or f_i^m is equal to one if there is a 0-deg, 90-deg, 45-deg or -45-deg ply, respectively, in the ith layer.

The advantage of these zero-one ply-identity variables is that the integrals, and therefore the A and D matrices are linear functions of these variables. For example, the integrals W_0, W_1 and W_3 are expressed as

$$W_0 = \frac{2t^3}{3} \sum_{k=1}^{N/2} p_k [(\frac{z_k}{t})^3 - (\frac{z_{k-1}}{t})^3] = \frac{2t^3}{3} \sum_{k=1}^{N/2} [k^3 - (k-1)^3](o_k + n_k + f_k^p + f_k^m),$$

$$W_1 = \frac{2t^3}{3} \sum_{k=1}^{N/2} p_k \cos 2\theta_k [(\frac{z_k}{t})^3 - (\frac{z_{k-1}}{t})^3] = \frac{2t^3}{3} \sum_{k=1}^{N/2} [k^3 - (k-1)^3](o_k - n_k), \tag{6}$$

$$W_3 = \frac{2t^3}{3} \sum_{k=1}^{N/2} p_k \cos 4\theta_k [(\frac{z_k}{t})^3 - (\frac{z_{k-1}}{t})^3] = \frac{2t^3}{3} \sum_{k=1}^{N/2} [k^3 - (k-1)^3](o_k + n_k - f_k^p - f_k^m),$$

where f_k^p and f_k^m do not appear in the expression for V_1 since the cosine of 90 degrees is equal to zero. The variable p_k in Eq (6) is equal to one if the kth ply is occupied and is equal to zero if the ply is empty. Constraints are applied during the optimization to ensure that p_k can be zero only for the outermost plies.

3. Graphical Optimization

A graphical procedure introduced by Miki [12-14] for the design of laminates with prescribed in-plane and flexural stiffnesses is a practical tool for design optimization. The procedure is suitable for multiple balanced angle-ply laminates of the type $[(\pm\theta_n)_{N_n}/(\pm\theta_{n-1})_{N_{n-1}}/\ldots/(\pm\theta_1)_{N_1}]_s$ where the total number of plies in the laminate is $N = 2\sum_{i=1}^{n} N_i$. In addition to the balanced angle-ply sub-laminates, one unidirectional lamina with principal material axes aligned with the axes of the laminate can be included into the stacking sequence.

The major effort of this design procedure is the construction of lamination parameter diagrams which describe the allowable region of in-plane, V_1^*, V_3^*, and flexural, W_1^*, W_3^*, lamination parameters. These parameters are obtained by normalizing the in-plane components, V_1 and V_3, by the total laminate thickness and the out-of-plane components, W_1 and W_3, by $h^3/12$. For a laminate of total thickness h, defining the volume fraction of the plies with $\pm\theta_i$ orientation angles as v_i, the in-plane lamination parameters are given as

$$V_1^* = \frac{V_1}{h} = \sum_{k=1}^{n} v_k \cos 2\theta_k, \quad \text{and} \quad V_3^* = \frac{V_3}{h} = \sum_{k=1}^{n} v_k \cos 4\theta_k, \tag{7}$$

where

$$v_i = \frac{2(z_i - z_{i-1})}{h}, \quad \text{and} \quad \sum_{i=1}^{n} v_i = 1. \tag{8}$$

Because of the normalization, the values of the lamination parameters are always bounded, $-1 \le V_1^*, V_3^* \le +1$. For a laminate with only one fiber orientation angle, the lamination parameters are

$$V_1^* = \cos 2\theta, \quad \text{and} \quad V_3^* = \cos 4\theta, \tag{9}$$

and the following relation, which defines the envelope of the region in Fig. 2, determines the allowable values of these parameters

$$V_3^* = 2V_1^{*2} - 1. \tag{10}$$

That is, values of all possible combinations of the lamination parameters are located along the curve ABC in Fig. 2 described by Eq. (10). Note that the points A, B, and C correspond to laminates with only [0], [$\pm$45]$_s$, and [90] ply orientations, respectively. If the laminate consists of two or more fiber orientations, then it is shown by Miki [12] that Eq. (10) becomes an inequality

$$V_3^* \ge 2V_1^{*2} - 1, \tag{11}$$

and the allowable region of the lamination parameters is the area bounded by the curve ABC in Fig. 2, independent of the number of different ply orientations. A single point inside the lamination parameter diagram, therefore, corresponds to laminates with two or more fiber orientations. For balanced angle-ply laminates with more than two orientations, there will be many combinations of the ply orientations that will produce the same lamination parameters and, therefore, the same stiffness properties. Each point inside the design space is called a lamination point, and corresponds to a laminate with a specific stiffness properties. It is also possible to put bounds on values of the various laminate in-plane stiffness properties, such as the engineering stiffnesses

$$E_x = \frac{1}{h}\left(\frac{A_{11}A_{22} - A_{12}^2}{A_{22}}\right), \quad E_y = \frac{1}{h}\left(\frac{A_{11}A_{22} - A_{12}^2}{A_{11}}\right), \quad G_{xy} = \frac{1}{h}A_{66}, \quad \text{and} \quad \nu_{xy} = \frac{A_{12}}{A_{22}}.$$

$$\tag{12}$$

This is achieved by drawing contours of constant effective engineering stiffnesses, obtained from Eqs. (12), for each of the engineering constants on the lamination parameter diagram as shown in Fig. 3. The figure indicates that, if no other constraints are specified, the maximum values of the E_x, E_y, and G_{xy} are all achieved for lamination points located on the boundary of the design space which require only one lamination angle. As expected, the maximal E_x and E_y are obtained for 0-deg and 90-deg laminates, respectively, and the maximal shear stiffness for $[\pm 45]_s$ laminate. The lamination angle $[\pm \theta]_s$ that maximizes the effective Poisson's ratio, on the other hand, is a function of material properties. For example, for T300/5208 graphite/epoxy and Scotchply 1002 glass/epoxy materials the laminates that maximize Poisson's ratio are $[\pm 25]_s$ and $[\pm 31]_s$, respectively.

Now, maximization of one of the engineering stiffness property with limits on the remaining properties can be achieved by first drawing the limiting contours on the diagram. This establishes the feasible domain on the diagram, and the contours of the property to be maximized can then be used to find the optimum design. Similarly, one can plot the contours of laminate strains for given loading or contours of failure criteria for imposing failure constraints.

An interesting application of the in-plane lamination parameter diagram is its possible use for designing laminates with predetermined ply orientation angles. It is shown by Miki [14] that the feasible region for laminates with fixed ply angles are polygons with vertices located on the envelope of the lamination parameter diagram. As mentioned earlier, if the design point is on the periphery of the diagram, the laminate can be constructed as an angle ply laminate with one fiber orientation. Therefore, vertices of the polygons correspond to laminates with a single ply orientation, which is one of the permissible values of the orientation angle. For example, for laminates constructed with 0-deg, ±45-deg, and 90-deg orientations, the design space is a triangle with vertices at A, B, and C in Fig. 2 corresponding to 0, ±45, and 90-deg laminates, respectively. If the available ply orientations are 0-deg, ±30-deg, ±60-deg, and 90-deg, then the design space is a trapezoidal. Interior points of the lamination diagram correspond to laminates with two or more ply orientations, and points along the edges of the polygons correspond to designs with various combinations of the vertex angles that are connected by that particular edge. The number of different laminates that can be constructed along a given edge depends on the number of plies considered for the design. If we consider having as many as n ply orientations for the design (for a total number of plies of $N = 2n$), then in addition to the vertices, we obtain $n - 1$ equally spaced design points along the edge. For example, in the case of an eight-ply (total) laminate with three available angles (triangular design space), there are five equally spaced design points with fiber orientations varying incrementally from one vertex to another as shown in Fig. 2. Note that the design points inside the triangular region also follow an incremental pattern, but are combinations of the three available angles.

4. Methods of Integer Programming

4.1. INTEGER LINEAR PROGRAMMING

For problems with discrete-valued design variables the standard form of a LP problem takes the form

$$
\begin{aligned}
\text{minimize} \quad & f(x) = c^T x \\
\text{such that} \quad & Ax = b, \\
& x_i \in X_i = \{d_{i1}, d_{i2}, \ldots, d_{il}\}, \quad i \in I_d,
\end{aligned}
\tag{13}
$$

where I_d is the set of design variables that can take only discrete values, and X_i is the set of allowable discrete values. Those problems with discrete-valued design variables are called discrete programming problems.

In general, a discrete programming problem can be converted to a form where design variables can assume only integer values. This conversion can be achieved by using the design variable x_i to represent the index j of the $d_{ij}, j = 1, \ldots, l$, Eq. (13). If the values in the discrete set are uniformly spaced, it is possible to scale the set to form a set of integer values only. The problem is then called an *integer linear programming* (ILP) problem,

$$
\begin{aligned}
&\text{minimize} &&f(x) = c_1^T x + c_2^T y \\
&\text{such that} &&A_1 x + A_2 y = b, \\
& &&x_i \geq 0 \text{ integer}, \\
& &&y_j \geq 0 .
\end{aligned}
\tag{14}
$$

This form, where certain design variables are allowed to be continuous, is referred to as *mixed integer linear programming* (MILP) problem. Problems where all variables are integer are called pure ILP problems or in short ILP problems. It is also common to have problems where design variables are used to indicate a *zero/one*-type decision-making situation. Such problems are referred to as *zero/one* or *binary* ILP problems. For example, a truss design problem where the presence of a particular member or its absence is represented by a binary variable falls into this category. Any ILP problem with an upper bound on the design variable x_i of $2^K - 1$ can be posed as binary ILP problem by replacing the variable with K binary variables $x_{i1}, \ldots, x_{iK}$ such that

$$
x_i = x_{i1} + 2x_{i2} + \ldots + 2^{K-1} x_{iK} .
\tag{15}
$$

It is also possible to convert the linear discrete programming problem to a binary ILP by using binary variables ($x_{ij} \in \{0, 1\}, j = 1, \ldots, l$) such that

$$
x_i = d_{i1} x_{i1} + d_{i2} x_{i2} + \cdots + d_{il} x_{il} ,
\tag{16}
$$

$$
\text{and} \qquad x_{i1} + x_{i2} + \cdots + x_{il} = 1 .
\tag{17}
$$

Most of the following discussion assumes problems to be pure ILP.

4.1.1. *Enumeration.* A practical approach to solving ILP problems is to try at random possible combinations of the integer values for the variables. A more systematic way of trying possible combinations of variables that will satisfy the requirements of a given problem can be explained by using the *enumeration tree* example of Garfinkel and Nemhauser [15]. Consider the binary ILP problem of choosing a combination of five variables such that the following summation is satisfied

$$
f = \sum_{i=1}^{5} i x_i = 5 .
\tag{18}
$$

A decision tree representing the progression of solution of this problem is composed of nodes and branches that represent the solutions and the combinations of variables that lead the those solutions, respectively (Figure 4). The top node of the tree corresponds to a solution which all the variables are turned off ($x_i = 0, i = 1, \ldots, 5$) with a function value of $f = 0$. Branching off from this solution are two paths corresponding to the two alternatives for the first variable. The branch which has $x_1 = 1$ has a function value of $f = 1$ and tolerates turning additional variables on without running into the risk of exceeding the required function value of 5. Of course the other branch is same as the initial solution, and can be branched further. Next, these two nodes are branched by considering the on and off alternatives for the second variable. The node arrived by taking $x_1 = x_2 = 1$ has $f = 3$ and is terminated as indicated by a vertical line. Such a vertex is said to be *fathomed*, because further branching would mean adding a number that would cause f to exceed its required value of 5. The other three vertices are said to be live, and can be branched further by considering the alternatives for the remaining variables in a sequential manner until either the created nodes are fathomed or the branches arrive at feasible solutions to the problem.

For this simple problem, after considering 19 possible combinations of variables, we identified 3 feasible solutions which are marked by an asterisk. This is a 40% reduction in the total number of possible trials, namely $2^5 = 32$, needed to identify all feasible solutions. For a structural design problem in which trials with different combinations of variables would possibly require expensive analysis an enumeration tree can yield substantial savings.

4.1.2. Branch-and-Bound Algorithm. The basic concept behind the enumeration technique forms the basis for this powerful algorithm suitable for MILP problems as well as nonlinear mixed integer problems [16,17]. The original algorithm developed by Land and Doig [18] relies on calculating upper and lower bounds on the objective function so that nodes that result in designs with objective functions outside the bounds can be fathomed and, therefore, the number of analyses required can be cut back. The first step of the algorithm is to solve the LP problem obtained from the MILP problem by assuming the variables to be continuous valued. If all the x variables for the resulting solution have integer values, there is no need to continue, the problem is solved. Suppose several of the variables assume noninteger values and the objective function value is f_1. The f_1 value will form a lower bound $f_L = f_1$ for the MILP since imposing conditions that require any of the noninteger valued variables to take integer values can only cause the objective function to increase. This initial problem with a solution denoted by x^1 is labeled as LP-1 and is placed in the top node of the enumeration tree.

The second step of the algorithm is to branch from the node into two new LP problems by adding a new constraint to the LP-1 that would involve only one of the noninteger variables, say x_k. One of the problems, LP-2, will require the value of the branched variable, x_k to be less than or equal to the largest integer smaller than x_k^1, and the other, LP-3, will have a constraint that x_k is larger than the smallest integer larger than x_k^1. These two problems actually do branch the feasible design space of the LP-1 into two segments. There are several possibilities for the solution of each of these two new problems. One of these possibilities is to have no feasible solution for the new problem. In that case the new node will be fathomed, and another variable will be branched. Another possibility is to reach an all integer feasible solution in which case the node will again be fathomed but the value of the objective function will become an upper bound f_U for the MILP problem. That is, beyond this solution point, any node that has an LP solution with a larger value of the objective function will be fathomed, and only those solutions that have the potential of producing an objective function between f_L and f_U will be pursued. If there are no solutions with an objective function smaller than f_U, then the node is an optimum solution. If there are other solutions with an objective function smaller than f_U, they may still include noninteger valued variables, and are labeled as *live* nodes. Live nodes are then branched again by considering one of the remaining noninteger values, and the resulting solutions are analyzed until all the nodes are fathomed.

The performance of the Branch-and-Bound algorithm relies heavily on the choice of noninteger variable to be used for branching, and the selection of node to be branched. If a selected node and branching variable leads to an upper bound close to the objective function of the LP-1 early in the enumeration scheme, then substantial computational savings can be obtained because of the elimination of branches that would not be capable of generating solutions lower than the upper bound. A rule of thumb for choosing the noninteger variable to be branched is to take the variable with the largest fraction. For the selection of the node to be branched, we choose, among all the live nodes, the LP problem which has the smallest value of the objective function; that node is most likely to generate a feasible design with a tighter upper bound.

Branch-and-Bound is only one of the algorithms for the solution of ILP or MILP problems. However, because of its simplicity it is incorporated into many commercially available computer programs [19,20]. An interesting application of the Branch-and-Bound algorithm (using the LINDO[20] package) to the stacking sequence design of laminated composite plates for improved buckling response is presented in section 5.3. There are a number of other techniques which are capable of handling general discrete-valued problems (see, for example, Ref. [21]). Some of these algorithms are good not only for ILP problems but also for NLP problems with integer variables. Particularly, methods based on probabilistic search algorithms are emerging for many applications, including structural design applications, that involve linear and nonlinear programming problems. Two of such techniques, namely simulated annealing and genetic algorithms, are discussed in sections 4.3.1 and 4.3.2, respectively.

4.2. PENALTY FUNCTION APPROACH

An extension of the penalty function approach which converts the constrained optimization problems to a sequence of unconstrained problems has recently been proposed [6] for problems with discrete valued design variables. We first present a brief description of the sequential unconstrained minimization technique (SUMT). SUMT transforms the constrained optimization problem into a sequence of unconstrained problems by creating an augmented objective function which combines the original objective function and penalty for constraint violation or near violation. Here we consider methods based on interior or extended interior penalty functions where the common form of the constraint penalty function is $1/g_j(x)$ with a linear or quadratic extension into the infeasible design space. With such techniques the constrained minimization problem,

$$
\begin{aligned}
\text{minimize} \quad & f(x)\,, \\
\text{subject to} \quad & g_j(x) \geq 0, \qquad j = 1, ..., n_g\,,
\end{aligned}
\tag{19}
$$

is replaced by the unconstrained problem

$$
\text{minimize} \quad \phi(x, r) = f(x) + r \sum_{j=1}^{n_g} y(g_j)\,,
\tag{20}
$$

where $x = (x_1, x_2, \ldots, x_n)^T$, n is the total number of design variables, n_g the number of constraints, r the penalty multiplier, and $y(g_j)$ is the penalty function. For a given value of the penalty multiplier, r, in Eq. (20) ϕ is often referred to as the response surface. As the penalty multiplier is decreased, the contours of the response surface conform with the original objective function and the constraints more closely. Therefore, minimization of the unconstrained problem is performed repeatedly as the value of r is decreased until the minimum value of the pseudo-objective function coincides with the value of the original objective function.

For problems with discrete-valued design variables additional penalty terms are included in the augmented-objective function $\phi(x, r)$ to reflect the requirement that the design variables take discrete values, $x_i \in \{d_{i1}, d_{i2}, \ldots, d_{il}\}$, $i = 1, 2, \cdots, n_d$, where n_d is the number of discrete design variables, d_{ik} is the k-th discrete value of the i-th design variable, and l is the number of discrete values for each design variable. In general, there can be more than one set of discrete values, with different design variables relating to different sets. Also, n_d can be smaller than n with remaining design variables being unrestricted with the exception of lower and upper bounds. The modified augmented objective function which includes the penalty terms due to constraints and the non-discrete values of the design variables is defined as

$$\text{minimize} \quad \phi(x, r, s) = f(x) + r \sum_{j=1}^{n_g} y(g_j) + s \sum_{i=1}^{n_d} \psi_d(x_i), \tag{21}$$

where s is a penalty multiplier for non-discrete values of the design variables, and $\psi_d(x_i)$ the penalty term for non-discrete values of the i-th design variable. Different forms for the discrete penalty function are also possible. Here, the penalty terms $\psi_d(x_i)$ are assumed to take the following sine-function form,

$$\psi_d(x_i) = \frac{1}{2} \left(\sin \frac{2\pi[x_i - \frac{1}{4}(d_{i(j+1)} + 3d_{ij})]}{d_{ij+1} - d_{ij}} + 1 \right), \qquad d_{ij} \leq x_i \leq d_{i(j+1)}. \tag{22}$$

While penalizing the non-discrete valued design variables, the functions $\psi_d(x_i)$ assure the continuity of the first derivatives of the augmented function at the discrete values of the design variables. The response surfaces generated by Eq. (21) are determined according to the values of the penalty multipliers r and s. In contrast to the multiplier r, which initially has a large value and decreased as we move from one response surface to another, the value of the multiplier s is initially zero and is increased gradually. One of the important factors in the application of this method is to determine when to activate s, and how fast to increase it to obtain the discrete optimum design. Clearly, if s is introduced too early in the design process, the design variables will be trapped away from the global minimum, resulting in a sub-optimal solution. To avoid this problem, the multiplier s has to be activated after several response surfaces which include only constraint penalty terms have been optimized. In fact, since usually the optimum design with discrete values is in the neighborhood of the continuous optimum, it may be desirable not to activate the penalty for the non-discrete design variables until reasonable convergence to the continuous solution is achieved. This is especially true for problems with a large number of design variables and/or if the intervals between discrete values are very close.

4.3. PROBABILISTIC SEARCH ALGORITHMS

A common disadvantage of most nonlinear minimization algorithms is their inability to distinguish local and global minima. Many structural design problems have more than one local minimum, and the identity of the local minimum obtained by the algorithm depends on the starting point.

Dealing with the problem of local minima becomes even worse if the design variables are required to take discrete values. First of all, for such problems the design space is discontinuous and disjointed, therefore derivative information is either useless or is not defined. Secondly, the use of discrete values for the design variables may introduce multiple minima corresponding to various combinations of the variables, even if the objective function for the problem has a single minimum for continuous variables. A methodical way of dealing with multiple minima for discrete optimization problems is to use either random search techniques that would sample the design space for a global minimum or to employ enumerative type algorithms. In either case, the efficiency of the solution process deteriorates dramatically as the number of variables is increased.

Two algorithms, *Simulated Annealing* and *Genetic Algorithms* (see, Laarhoven at al. [22] and Goldberg [23], respectively), have emerged more recently as tools ideally suited for optimization problems where a global minimum is sought. In addition to being able to locate near global solutions, these two algorithms are also powerful tools for problems with discrete-valued design variables.

Both algorithms rely on naturally observed phenomena and their implementation calls for the use of a random selection process which is guided by probabilistic decisions. In the following sections brief descriptions of the two algorithms are presented.

4.3.1. *Simulated Annealing.* The development of the simulated annealing algorithm was motivated by studies in statistical mechanics which deal with the equilibrium of large number of atoms in solids and liquids at a given temperature. During solidification of metals or formation of crystals, for example, a number of solid states with different internal atomic or crystalline structure that correspond to different energy levels can be achieved depending on the rate of cooling. If the system is cooled too rapidly, it is likely that the resulting solid state would have a small margin of stability because the atoms will assume relative positions in the lattice structure to reach an energy state which is only locally minimal. In order to reach a more stable, globally minimum energy state, the process of annealing is used in which the metal is reheated to a high temperature and cooled slowly, allowing the atoms enough time to find positions that minimize a steady state potential energy. It is observed in the natural annealing process that during the time spent at a given temperature it is possible to have the system jump to a higher energy state temporarily before the steady state is reached. As will be explained in the following paragraphs, it is this characteristic of the annealing process which makes it possible to achieve near global minimum energy states.

A computational algorithm that simulates the annealing process was proposed by Metropolis et al. [24], and is referred to as the *Metropolis algorithm*. At a given temperature, T, the algorithm perturbs the position of an atom randomly and computes the resulting change in the energy of the system, ΔE. If the new energy state is lower than the initial state, then the new configuration of the atoms is accepted. If, on the other hand $\Delta E \geq 0$, the perturbed state causes an increase in the energy, the new state might be accepted or rejected based on a random probabilistic decision. The probability of acceptance, $\mathcal{P}(\Delta E)$, of a higher energy state is computed as

$$\mathcal{P}(\Delta E) = e^{\left(\frac{-\Delta E}{k_B T}\right)}, \tag{23}$$

where k_B is the Boltzmann's constant. If the temperature of the system is high, then the probability of acceptance of a higher energy state is close to one. If, on the other hand, the temperature is close to zero, then the probability of acceptance becomes very small.

The decision to accept or reject is made by randomly selecting a number in an interval $(0, 1)$ and comparing it with $\mathcal{P}(\Delta E)$. If the number is less than $\mathcal{P}(\Delta E)$, then the perturbed state is accepted, if it is greater than $\mathcal{P}(\Delta E)$, the state is rejected. At each temperature, a pool of atomic structures would be generated by randomly perturbing positions until a steady state energy level is reached (commonly referred to as thermal equilibrium). Then the temperature is reduced to start the iterations again. These steps are repeated iteratively while reducing the temperature slowly to achieve the minimal energy state.

The analogy between the simulated annealing and the optimization of functions with many variables was established recently by Kirkpatrick et al. [25], and Cerny [26]. By replacing the energy state with an objective function f, and using variables x for the the configurations of the particles, we can apply the Metropolis algorithm to optimization problems. The method requires only function values. The moves in the design space from one point, x^i to another x^j causes a change in the objective function, Δf^{ij}. The temperature T now becomes a control parameter that regulates the convergence of the process. Important elements that affect the performance of the algorithm are the selection of the initial value of the "temperature", T_0, and how to update it. In addition, the number of iterations (or combinations of design variables) needed to achieve "thermal equilibrium" must be decided before the T can be reduced. These parameters are collectively referred to as the "cooling schedule".

The definition of the cooling schedule begins with the selection of the initial temperature. If a low value of T_0 is used, the algorithm would have a low probability of reaching a global minimum. The initial value of T_0 must be high enough to permit virtually all moves in the design space to be acceptable so that almost a random search is performed. Typically, T_0 is selected such that the acceptance ratio $\mathcal{X}$ (defined as the ratio of the number of accepted moves to total number of proposed moves) is approximately $\mathcal{X}_0 = 0.95$ [27]. Johnson et al. [28] determined T_0 by calculating

the average increase in the objective function, $\overline{\Delta f}^{(+)}$, over a predetermined number of moves and solved

$$\mathcal{X}_0 = e^{\left(\frac{-\overline{\Delta f}^{(+)}}{T_0} \right)}, \tag{24}$$

leading to

$$T_0 = \frac{\overline{\Delta f}^{(+)}}{\ln(\mathcal{X}_0^{-1})}. \tag{25}$$

Once the temperature is set, a number of moves in the variable space is performed by perturbing the design. The number of moves at a given temperature must be large enough to allow the solution to escape from a local minimum. One possibility is to move until the value of the objective function does not change for a specified number, M, of successive iterations. Another possibility suggested by Aarts and Korst [29] for discrete valued design variables is to make sure that every possible combination of design variables in the neighborhood of a steady state design is visited at least once with a probability of P. That is, if there are S neighboring designs, then

$$M = S \ln \left(\frac{1}{1-P} \right), \tag{26}$$

where Aarts and Korst [29] recommentd $P = 0.99$ for $S > 100$, and $P = 0.995$ for $S < 100$. For discrete valued variables there are often many options for defining the neighborhood of the design. One possibility is to define it as all the designs that can be obtained by changing a single design variable to its next higher or lower value. A broader immediate neighborhood can be defined by changing more than one design variables to their next higher or lower values. For an n variable problem, the immediate neighborhood has

$$S = 3^n - 1. \tag{27}$$

Once convergence is achieved at a given temperature, generally referred to as thermal equilibrium, the temperature is reduced and the process is repeated.

Many different schemes have been proposed for updating the temperature. A frequently used rule is a constant cooling update

$$T_{k+1} = \alpha T_k, \qquad k = 0, 1, 2, \ldots, K, \tag{28}$$

where $0.5 \leq \alpha \leq 0.95$. Nahar et al. [30] fix the number of decrement steps K, and suggests determination of the values of the T_k by numerical experiments. It is also possible to divide the interval $[0, T_0]$ into a fixed K number of steps and use

$$T_K = \frac{K-k}{K} T_0, \qquad k = 0, 1, 2, \ldots, K. \tag{29}$$

The number of intervals typically ranges from 5 to 20.

The use of simulated annealing for structural optimization has been quite recent. Elperin [31] applied the method to the design of a ten-bar truss problem where member cross-sectional dimensions were to be selected from a set of discrete values. Kincaid and Padula [32] used it for minimizing the distortion and internal forces in a truss structure. Optimal placement of active and passive members in a truss structure was investigated by Chen et al. [33] to maximize the finite-time energy dissipation to achieve increased damping properties. Application of the simulated annealing to laminate design problem was demonstrated by Lombardi [34].

4.3.2. *Genetic Algorithms.* Genetic algorithms use techniques derived from biology, and rely on Darwin's principle of survival of the fittest. When a population of biological creatures is allowed to evolve over generations, individual characteristics that are useful for survival tend to be passed on to future generations, because individuals carrying them get more chances to breed. Those individual characteristics in biological populations are stored in chromosomal strings. The mechanics of natural genetics are based on operations that result in structured yet randomized exchange of genetic information (i.e., useful traits) between the chromosomal strings of the reproducing parents, and consist of *reproduction, crossover,* occasional *mutation,* and *inversion* of the chromosomal strings.

Genetic algorithms, developed by Holland [35], simulate the mechanics of natural genetics for artificial systems based on operations which are the counterparts of the natural ones (even called by the same names), and are extensively used as multivariable search algorithms. As will be described in the following paragraphs, these operations involve simple, easy to program, random exchanges of location of numbers in a string, and, therefore, at the outset look like a completely random search of the extremum in the parameter space based on function values only. However, genetic algorithms are experimentally proven to be robust, and the reader is referred to Goldberg [23] for further discussion of the theoretical properties of genetic algorithms. Here we discuss the genetic representation of a minimization problem, and focus on the mechanics of three commonly used genetic operations, namely; reproduction, crossover, and mutation.

Application of the operators of the genetic algorithm to a search problem first requires the representation of the possible combinations of the variables in terms of bit strings that are counterparts of the chromosomes. Naturally, the fitness of a specific combination of genes is related in an artificial system to the objective function of the search problem. For example, if we have a minimization problem

$$\text{minimize} \quad f(x), \quad x = \{x_1, x_2, x_3, x_4\}, \tag{30}$$

a binary string representation of the variable space could be of the form

$$\underbrace{0\,1\,1\,0}_{x_1}\,\underbrace{1\,0\,1}_{x_2}\,\underbrace{1\,1}_{x_3}\,\underbrace{1\,0\,1\,1}_{x_4}, \tag{31}$$

where string equivalents of the individual variables are connected head-to-tail, and, in this example, base 10 values of the variables are $x_1 = 6, x_2 = 5, x_3 = 3, x_4 = 11$, and their ranges correspond to $\{15 \geq x_1, x_4 \geq 0\}, \{7 \geq x_2 \geq 0\},$ and $\{3 \geq x_3 \geq 0\}$. Because of the bit string representation of the variables, genetic algorithms are ideally suited for problems where the variables are required to take discrete or integer variables. For problems where the design variables are continuous values within a range $x_i^L \leq x_i \leq x_i^U$, one may need to use a large number of bits to represent the variables to high accuracy. The number of bits that are needed depends on the accuracy x^{incr} required for the final solution and can be calculated from

$$2^m \geq [(x_i^U - x_i^L)/x^{\text{incr}} + 1] , \tag{32}$$

where m is the number of digits. For example, if a variable is defined in a range $\{0.01 \leq x_i \leq 1.81\}$ and the accuracy needed for the final value is $x^{incr} = 0.001$, the smallest number of digits that satisfy the requirement would be $m = 11$, which actually produces increments of 0.00087 in the value of the variable, instead of the required value of 0.001.

Unlike the search algorithms discussed earlier that move from one point to another in the design variable space, genetic algorithms work with a population of strings (chromosomes). This aspect of the genetic algorithms is responsible for capturing near global solutions, by keeping many solution points that may have the potential of being close to minima (local or global) in the pool during the search process rather than singling out one point early in the process and running the risk of getting stuck at a local minimum. Working on a population of designs also suggests the possibility of implementation on parallel computers. However, the concept of parallelism is even more basic to genetic algorithms in that evolutionary selection can improve in parallel many different characteristics of the design. Also, the outcome of a genetic search is a population of good designs rather than a single design. This aspect can be very useful to the designer.

Initially the size of the population is chosen and the values of the variables in each string are decided by randomly assigning 0's and 1's to the bits. The next important step in the process is *reproduction*, in which individual strings with good objective function values are copied to form a new population, an artificial version of the survival of the fittest. The bias towards strings with better performance can be achieved by increasing the probability of their selection in relation to the rest of the population. One way to achieve this is to create a biased roulette wheel where individual strings occupy areas proportional to their function values in relation to the cumulative function value of the entire population. Therefore, the population resulting from the reproduction operation would have multiple copies of the highly fit individuals.

Once the new population is generated, the members are paired off randomly for crossover. The mating of the pair also involves a random process. A random integer k between 1 and $L - 1$, where L is the string length, is selected and two new strings are generated by exchanging the 0's and 1's that comes after the kth location in the first parent with the corresponding locations of the second parent. For example, the two strings of length $L = 9$

$$
\begin{aligned}
\textbf{parent 1:} \quad & 0\,1\,1\,0\,1\|0\,1\,1\,1 \\
\textbf{parent 2:} \quad & 0\,1\,0\,0\,1\|0\,0\,0\,1
\end{aligned}
\tag{33}
$$

are mated with a crossover point of $k = 5$, the offsprings will have the following composition,

$$
\begin{aligned}
\textbf{offspring 1:} \quad & 0\,1\,1\,0\,1\,0\,0\,0\,1 \\
\textbf{offspring 2:} \quad & 0\,1\,0\,0\,1\,0\,1\,1\,1
\end{aligned}
\tag{34}
$$

Multiple point crossovers in which information between the two parents are swapped among more string segments are also possible, but because of the mixing of the strings the crossover becomes a more random process and the performance of the algorithm might degrade, De Jong [36]. Exception to this is the two-point crossover. In fact, the one point crossover can be viewed as a special case of the two point crossover in which the end of the string is the second crossover point. Booker [37] showed that by choosing the end-point of the segment to be crossed randomly, the performance of the algorithm can actually be improved.

Mutation serves an important task of preventing premature loss of important genetic information by occasional introduction of random alteration of a string. As mentioned earlier, at the end of reproduction it is possible to have populations with multiple copies of the same string. In the worst scenario, it is possible to have the entire pool to be made of the same string. In such a case, the algorithm would be unable to explore the possibility of a better solution. Mutation prevents this uniformity, and is implemented by randomly selecting a string location and changing its value from 0 to 1 or vice versa. Based on small rate of occurrence in biological systems and on numerical experiments, the role of the mutation operation on the performance of a genetic algorithm is considered to be a secondary effect. Goldberg [23] suggests a rate of mutation of one in one thousand bit operations.

Application of genetic algorithms in optimal structural design has started only recently. The first application of the algorithm to a structural design was presented by Goldberg and Samtani [38] who used the 10-bar truss weight minimization problem. More recently, Hajela [39] used genetic search for several structural design problems for which the design space is known to be either nonconvex or disjoint. Rao et al. [40] address the optimal selection of discrete actuator locations in actively controlled structures via genetic algorithms.

5. Buckling Optimization of Plates

5.1. ANALYSIS FORMULATION

For thin rectangular laminated plates under in-plane compressive loads elastic stability and vibration are governed by the flexural rigidities of the plate. The governing differential equation for the buckling of an orthotropic laminated plate is given by

$$D_{11}\frac{\partial^4 w}{\partial x^4} + 2(D_{12} + 2D_{66})\frac{\partial^4 w}{\partial x^2 \partial y^2} + D_{22}\frac{\partial^4 w}{\partial y^4} = \lambda(N_x\frac{\partial^2 w}{\partial x^2} + N_y\frac{\partial^2 w}{\partial y^2} + 2N_{xy}\frac{\partial^2 w}{\partial x \partial y}), \qquad (35)$$

where flexural stiffnesses D_{11}, D_{12}, D_{22} and D_{66} are given in Eq. (1). For the case of a uniform thickness plate with free in-plane boundary conditions loaded by a uniform edge loads, N_x, N_y, and N_{xy} are equal to the applied edge loads.

For a balanced, symmetric laminate of dimensions a by b with simply supported boundaries under-in plane loads an assumed displacement function of the form

$$w(x,y) = \sum_{q=1}^{\infty}\sum_{p=1}^{\infty} W_{pq} \sin\frac{p\pi x}{a} \sin\frac{q\pi y}{b} \qquad (36)$$

satisfies the boundary conditions exactly. A truncated series with P, Q terms in the x, y directions, respectively, can be used to represent the PQ possible modes of buckling associated with the transverse displacement patterns. Substituting Eq. (36) into the equilibrium equation, Eq. (35), and applying Galerkin's method leads to an eigenvalue problem of the form

$$Kw = \lambda K_G w \qquad (37)$$

where the eigenvector w is composed of the unknown coefficients of the displacement function, $w = \{W_{11} \cdots W_{MN}\}^T$. Expressions for the K and K_G matrices can be found elsewhere in the literature.

In the case of a simply supported plate under biaxial compression loads per unit length of λN_x and λN_y acting in the x and y directions, respectively, as shown in Fig. 1, the plate buckles when the load amplitude parameter λ reaches a critical value λ_{cr} given as

$$\lambda_{cr}(p,q) = \frac{\pi^2 \left[D_{11} \left(\frac{p}{a} \right)^4 + 2(D_{12} + 2D_{66}) \left(\frac{p}{a} \right)^2 \left(\frac{q}{b} \right)^2 + D_{22} \left(\frac{q}{b} \right)^4 \right]}{\left(\frac{p}{a} \right)^2 N_x + \left(\frac{q}{b} \right)^2 N_y}, \tag{38}$$

where p and q are the number of half waves in the x and y directions, respectively, that minimize λ_{cr}.

5.2. GRAPHICAL BUCKLING MAXIMIZATION

Similar to the in-plane lamination diagram discussed earlier, a diagram can be constructed for the flexural lamination parameters [13]

$$W_1^* = \frac{12W_1}{h^3} = \sum_{k=1}^{n} s_k \cos 2\theta_k, \quad \text{and} \quad W_3^* = \frac{12W_3}{h^3} = \sum_{k=1}^{n} s_k \cos 4\theta_k, \tag{39}$$

where

$$s_i = \left(\frac{2z_i}{h} \right)^3 - \left(\frac{2z_{i-1}}{h} \right)^3. \tag{40}$$

The diagram for the flexural lamination parameters can be used for designing laminates for maximum buckling load under uniaxial and biaxial loads. Consider the case of prescribed values of the p and q corresponding to a given fixed ratio of applied transverse load to axial load, and, and panel aspect ratio r. Then, by manipulating Eq. (38) it can be shown that the contours of the critical load parameter λ_{cr} are straight lines in the flexural lamination diagram. A difficulty in using flexural lamination parameter in designing laminates with maximum buckling load is that p and q are seldom known apriory. Since these two numbers depends on the design variables, as well as the plate aspect ratio and the applied loads, it is not always possible to accurately predict them. Further discussion of the use of flexural lamination parameter diagram for buckling maximization is given in Ref. [13].

As in the case of in-plane lamination parameters, it is possible to construct the flexural lamination diagram for the case of a laminate with prescribed fiber orientations. The shape of the design space is same as the in-plane parameters with prescribed angles corresponding to vertices of polygons on the envelope of the diagram. However in this case the location of design points which are combinations of the given angles are not equally spaced (although the ones which are combinations of the angles corresponding to two vertices are still located along the edge that connect them) but are located through the use of Eq. (40).

5.3. LINEAR PROGRAMMING FORMULATION

Since the buckling load is a linear function of the flexural lamination parameters, which can be expressed in terms of the integer valued ply-identity variables, the problem can be posed as a linear integer programming problem as shown by Haftka and Walsh [8]. The laminate is again assumed to be symmetric and composed of 0-deg, 90-deg and ± 45-deg plies. Each ply has a constant thickness

t. For most of the examples in this paper, the laminate is also assumed to be balanced (i.e., the number of 45-deg plies is equal to the number of −45-deg plies). The laminate is assumed to have N plies with a total thickness of $h = Nt$. However, because in some situations the number of plies is unknown (it will be determined by the optimization process) the number of plies is assumed to be smaller than an upper limit N.

Two optimization problems are formulated. The first is the optimization of a laminate with a fixed thickness for maximum buckling load, and the second is the optimization of a laminate for minimum thickness for a given buckling load. The second optimization problem of minimizing the total thickness is somewhat more complex, and the reader is referred to [8] for details. For the first optimization problem the lowest (over values of p and q) buckling load λ^* is maximized. The objective λ^* is not a smooth function of the design variables, and the standard device for removing this problem is to add λ^* as a design variable and require it to be less than or equal to each $\lambda_{cr}(p, q)$. Thus, the optimization problem is formulated as

$$\text{find} \quad \lambda^*, \quad \text{and} \quad o_i, \ n_i, \ f_i^p, \ f_i^m, \quad i = 1, \cdots, N/2,$$
$$\text{to maximize} \quad \lambda^*$$
$$\text{such that} \quad \lambda^* \leq \lambda_{cr}(p, q), \quad p = 1, \cdots, p_f, \quad q = 1, \cdots, q_f,$$
$$o_i + n_i + f_i^p + f_i^m = 1, \quad i = 1, \cdots, N/2, \tag{41}$$

$$\text{and} \quad \sum_{i=1}^{N/2} f_i^p - f_i^m = 0.$$

The minimization over p and q is performed by checking for all values of p between 1 and p_f, and all values of q between 1 and q_f. The last constraint in Eq. (41) ensures that the number of 45-deg and -45-deg plies is the same, so that the laminate is balanced. Equations (1), (6) and (38) are used to calculate the buckling load which is clearly a linear function of the ply-identity design variables. Therefore, the optimization problem (41) is a linear integer programming problem.

In some cases it may be desirable to impose constraints on the stiffness of the plate. In the present study a limit on the in-plane stiffness in the x direction A_{11} was considered as an example of such constraints. A constraint requiring A_{11} to have a minimum value of A_{11}^0 can be written as

$$A_{11}/A_{11}^0 - 1 \geq 0. \tag{42}$$

As shown in [8] this constraint can be expressed as a linear function of the ply identity design variables similar to the buckling constraint.

6. Results

Stacking-sequences of rectangular plates designed for maximum buckling load under the action of biaxial compression are presented here based on the modified penalty function approach and the linear programming formulation. For all the results, typical graphite-epoxy ply properties of [$E_1 = 18.510^6$psi (128GPa), $E_2 = 1.8910^6$psi (13.0Gpa), $G_{12} = 0.9310^6$psi (6.4Gpa), $\nu_{12} = 0.3$, $t = 0.005$ in. (0.0127 cm)] were used.

6.1. PENALTY FUNCTION APPROACH

In order to establish results that can be used to compare with the results of designs with integer variables, first a series of results were generated for continuous problems. This was achieved through the use of modified penalty function approach by turning off the penalty terms for the non-discrete values of the design variables. The problems solved are for $a = 20$ in by $b = 10$ in (50.8 cm 25.4 cm) rectangular plates of specified number of plies with fiber orientation design variables. The critical eigenvalues are maximized for applied compressive load of $N_x = 1$ with varying N_y/N_x ratios.

Buckling loads of plates with four different thicknesses corresponding to 8, 12, 16, and 24 ply laminates are presented in Fig. 5. There is a substantial loss of load carrying capability as the ratio of the transverse compression to axial compression increases. For all the laminates, the presence of transverse compression with a magnitude equal to the axial compression causes a reduction of about 63% compared to uniaxial compression case. The optimal orientations of the surface layer fibers (indicated by dashed lines) and the layer adjacent to the mid-plane (solid lines) are shown in Fig. 6 for each of the four laminates. For the cases of uniaxial compression ($N_y = 0$) and large value of the transverse load ($N_y/N_x > 2.5$) the laminates tend to have the same fiber orientation at the mid-plane and at the surface layer (hence, through the entire thickness), which are ±45-deg and 90-deg, respectively. For intermediate load ratios, the fiber angles at the surface layer is larger than the mid-plane layers with difference being larger for thick 24-ply laminates. However, regardless of the laminate thickness the surface layers seem to assume the same fiber orientation for a given load ratio.

Next, the same design cases were repeated using two different sets of discrete fiber orientations; one for $0, \pm45,$ and 90-deg and another one with seven values with 15-deg intervals starting from 0-deg. Plots of the buckling load reduction as percentage of the buckling loads of the continuous design cases is shown in Fig. 7 for the four laminates with $0, \pm45,$ and 90-deg ply orientations. Discrete valued designs are accompanied with a substantial buckling load penalties for at least over a portion of the load ratio range considered. The largest penalty was for the load ratio of $N_y/N_x = 0.5$ (about 22% reduction), and was associated with the thin 8-ply and 12-ply laminates. However, buckling load penalties associated with different thicknesses appeared quite random without any obvious trends. A similar plot for the global optimum designs obtained through the use of the linear integer programming approach is shown in Fig. 8 for comparison. In general, there is a small amount of improvement in the buckling load reduction for most of the laminates. For example, the worst buckling load reduction (compared to the continuous designs) is still for the 8-ply laminate for a load ratio of $N_y/N_x = 0.5$, but it is only about 18% as compared to 22%. Also, there is a progression with increasing laminate thickness. The smallest and the largest buckling load reductions are associated with the 24-ply and the 8-ply laminates, respectively.

The lamination sequences obtained for the discrete valued designs are presented in Table 1 and 2 for the 8-ply and the 16-ply laminates. Included in the table are the lamination sequences for the continuous valued designs, the discrete designs obtained by using the modified penalty method, and the global optimal designs. If the design obtained by the penalty function approach is same as the global optimal design, the entry under the Global Opt. column is left blank. The penalty approach was unable to reach the global optimum in some cases, especially for laminates with large number of plies or for discrete sets with large number of choices. Notably, in every case the discrete designs obtained by the penalty function approach followed a pattern such that the orientation of the outer plies were larger than the plies close to the mid-plane, similar to the trend observed in the continuous designs. Global optimal designs, on the other hand had orientations that were quite random. Despite the difference in the lamination sequence, the differences in the buckling load were small. For the discrete set with seven angles, the designs obtained by the penalty approach were within 5% of the global designs. For the set with three choices, occasional large differences of as much as 14% were observed.

6.2. LINEAR PROGRAMMING FORMULATION

The computations were performed with the LINDO program [20] which employs the branch-and-bound algorithm. First uniaxial loading was considered, and the buckling load was maximized for various plate aspect ratios (a/b) for laminates with 16 plies. It is known (e.g. [1]) that for low aspect ratios the optimum ply angle is 0-deg, and for a/b larger than about 0.7 the optimum ply orientation is close to ± 45-deg. This can also be expected from Eq. (20) since for a/b larger than 0.7, D_{66} is the most important stiffness coefficient. A check was performed to see whether there was a transition range of a/b where the optimum stacking sequence would include both 0-deg and ± 45-deg plies. It was found that if such a transition range exists it is extremely narrow, since even changes in the fourth significant digit of the aspect ratio were not fine enough to locate it. When the number of plies (N) was not divisible by four, so that a balanced ± 45-deg laminate was precluded, the optimizer placed two 0-deg plies near the plane of symmetry of the laminate, as expected (because these less efficient plies have the smallest effect on D_{66} there).

Next the biaxial loading case was solved, and the results are presented in Fig. 9. It is known (e.g., [1]) that for aspect ratios less than 1.5 the optimum ply orientation is the same as for the uniaxial case, and for aspect ratios greater than 1.5, the value of the optimum ply angle increases rapidly as N_y/N_x increases, and that for large N_y/N_x, the optimum ply angle is 90 degrees. Therefore, the case of biaxial loading for a laminated plate with an aspect ratio of 2 was selected. The reference axial load N_x was fixed at 1 lb/in. (175 N/m), and the reference transverse load was increased from 0.1 to 3.0 lb/in. (17.5 to 525 N/m). The plate was specified to have 16 plies. Two transition ranges of N_y/N_x were found: one for N_y/N_x between 0.125 and 0.15 and the other for N_y/N_x between 2.4 and 2.45. The first range marked the transition from all ± 45-deg plies to a combination of 90-deg and ± 45-deg plies. The second range marked the transition from a combination of 90-deg and ± 45-deg plies to all 90-deg plies. As the ratio N_y/N_x increased to 0.15, first two 90-deg plies appeared, then four 90-deg plies ($N_y/N_x=0.25$), then six 90-deg plies ($N_y/N_x=1$), and finally all 90-deg plies ($N_y/N_x=2.45$). Also, as the transverse load N_y became larger than the axial load N_x, the ± 45-deg plies moved closer to the plane of symmetry until only 90-deg plies were present. This behavior is expected because when N_y dominates, the plate behaves like a plate of aspect ratio of 0.5 under uniaxial load, and for that case the optimum angle is in the direction of the loading.

When the number of contiguous plies in the same direction is large, composite laminates are known to experience matrix cracking. Therefore, it is desirable to limit the number of such contiguous plies. To demonstrate that such constraints can be easily added to the present formulation this constraint was imposed on the design obtained for $N_y/N_x = 2$ which had 5 contiguous 90-deg plies. This was implemented by adding the constraint

$$n_4 + n_5 + n_6 + n_7 + n_8 \leq 4. \tag{43}$$

The designs with and without this constraint are compared in Fig. 10. It is seen that the penalty for limiting the number of contiguous plies is quite small.

The case of $N_y/N_x = 2$ was used also for the purpose of checking on other aspects of the optimization. The first was the effect of the balanced laminate requirement. When this requirement was removed, the optimization selected a design with three 45-deg and five 90-deg plies. However, the buckling load changed by less than one hundredth of one percent. A second aspect was the effect of requiring that the design variables be integers. Noninteger design variables describe hybrid plies. For example, $o_1 = 0.5$, $n_1 = 0.5$ means that the first ply has properties which are the average of the elastic properties of 0-deg and 90-deg materials. When the requirement that the ply-identity variables be integers was removed the solution included two hybrid plies. For $i = 1$ the ply was 70 percent 45-deg and 30 percent 90-deg, and for $i = 4$ the ply was 70 percent -45-deg and 30 percent 90-deg, with the remaining plies being 90-deg. The effect on the buckling load was again quite small—less than five hundredths of one percent.

Another aspect of the optimization checked for $N_y/N_x = 2$ was the effect of introducing a minimum stiffness requirement. The optimum laminate for this case, was dominated by 90-deg plies, and has only 16 percent of the axial stiffness A_{11} of an all 0-deg laminate. A requirement that A_{11} be at least 50 percent of the unidirectional laminate was added, with and without the requirement of no more than four contiguous plies. The results are compared to the original design in Fig. 11. It is seen that the stiffness requirement is satisfied by putting 0-deg plies near the plane of symmetry where they have only a minimal effect on the bending stiffnesses, and hence on the buckling load. The reduction in the buckling load is about 8 percent. For this design the effect of adding the requirements of no more than 4 contiguous plies had a nontrivial effect (7 percent reduction) on the buckling load.

7. Acknowledgement

The work of the first and second authors was supported in part by NASA grant NAG-1-168 and NAG-1-643, respectively.

8. References

1. Haftka, R.T, and Gürdal, Z., Elements of Structural Optimization, 3rd Edition, Kluwer Academic Publishers, the Netherlands, 1992.
2. Schmit, L.A., and Fleury, C., "Discrete-Continuous Variable Structural Synthesis Using Dual Methods," AIAA Journal, 18, pp. 1515–1524, 1980.
3. Reinschmidt, K., "Discrete Structural Optimization," ASCE, J. Struct. Div., 97, 133-156, 1971.
4. John, K. V., Ramakrishnan, C. V. and Sharma, K. G., "Optimum Design of Trusses from Available Sections —Use of Sequential Linear Programming with Branch and Bound Algorithm," Eng. Opt., Vol. 13, pp. 119-145, 1988.
5. Olsen, G.R., and Vanderplaats, G.N., "Method for Nonlinear Optimization with Discrete Design Variables," AIAA Journal, 27, pp. 1584–1589, 1989.
6. Shin, D.K, Gürdal, Z., and Griffin, O. H., "A Penalty Approach for Nonlinear Optimization with Discrete Design Variables," Engineering Optimization, Vol. 16, pp. 29–42, 1990.
7. Mesquita, L., and Kamat, M.P., "Optimization of Stiffened Laminated Composite Plates with Frequency Constraints," Engineering Optimization, 11, pp. 77–88, 1987.
8. Haftka, R.T., and Walsh, J.L., "Stacking-Sequence Optimization for Buckling of Laminated Plates by Integer Programming," AIAA Journal (in Press).
9. Hajela, P., "Genetic Search–An Approach to the Nonconvex Optimization Problem," AIAA Journal, 28, 7, pp. 1205-1210, 1990.
10. Lombardi, M., "Ottimizzazione di Lastre in Materiale Composito con L'Uso di un Metodo di Annealing Simulato," Tesi di Laurea, Department of Structural Mechanics, University of Pavia, Pavia, Italy, 1990.
11. LeRiche, R., and Haftka, R.T., "Composite Plate Stacking Sequence Optimization for Buckling Load Maximization using a Genetic Algorithm," Paper submitted to the AIAA/ASME/ASCE/AHS/ASC 33rd, Structures, Structural Dynamics and Material Conference, Dallas Texas, April 13–15, 1992. "
12. Miki, M., "Material Design of Composite Laminates with Required In–plane Properties," Progress in Science and Engineering of Composites, Eds., T. Hayashi, K. Kawata, and S. Umekawa, ICCM–IV, Tokyo, pp. 1725–1731, 1982.
13. Miki, M., "Optimum Design of Fibrous Laminated Composite Plates Subject to Axial Compression," Proceedings, 3rd Japan-US Composite Materials Conference, Tokyo, pp. 673–680, 1986.
14. Miki, M., and Sugiyama, Yoshihiko "Optimum Design of Laminated Composite Plates Using Lamination Parameters," AIAA/ASME/ASCE/AHS 32th Structures, Structural Dynamics, and Materials Conference, Baltimore, MA., pp. 275–283, 1991.
15. Garfinkel, R. S., and Nemhauser, G. L., Integer Programming, John Wiley & Sons, Inc., New York, 1972.

16. Lawler, E. L., and Wood, D. E., "Branch-and-Bound Methods—A Survey," Operations research, 14, pp. 699–719, 1966.

17. Tomlin, J. A., "Branch-and-Bound Methods for Integer and Non-convex Programming," in Integer and Nonlinear Programming, J. Abadie (ed.), pp. 437–450, Elsevier Publishing Co., New York, 1970.

18. Land, A. H., and Doig, A. G., "An Automatic Method for Solving Discrete Programming Problems," Econometrica, 28, pp. 497–520, 1960.

19. Johnson, E. L., and Powell, S., "Integer Programming Codes," in Design and Implementation of Optimization Software, Greenberg, H. J. (ed.), pp. 225–240, 1978.

20. Schrage, L., Linear, Integer, and Quadratic Programming with LINDO, 4th Edition, The Scientific Press, Redwood City CA., 1989.

21. Kovács, L. B., Combinatorial Methods of Discrete Programming, Mathematical Methods of Operations Research Series, Vol. 2, Akadémiai Kiadó, Budapest, 1980.

22. Laarhoven, P. J. M. van., and Aarts, E., Simulated Annealing: Theory and Applications, D. Reidel Publishing, Dordrecht, The Netherlands, 1987.

23. Goldberg, D. E., Genetic Algorithms in Search, Optimization, and Machine Learning, Addison-Wesley Publishing Co. Inc., Reading, Massachusetts, 1989.

24. Metropolis, N., Rosenbluth, A.W., Rosenbluth, M.N., Teller, A.H., and Teller, E., "Equation of State Calculations by Fast Computing Machines," J. Chem. Physics, 21 (6), pp. 1087–1092, 1953.

25. Kirkpatrick, S., Gelatt, C. D., Jr., and Vecchi, M. P., "Optimization by Simulated Annealing," Science, 220 (4598), pp. 671–680, 1983.

26. Cerny, V., "Thermodynamical Approach to the Traveling Salesman Problem: An Efficient Simulation Algorithm," J. Opt. Theory Appl., 45, pp. 41–52, 1985.

27. Rutenbar, R. A., "Simulated Annealing Algorithms: An Overview," IEEE Circuits and Devices, January, pp. 19–26, 1989.

28. Johnson, D. S., Aragon, C. R., McGeoch, L. A., and Schevon, C., "Optimization by Simulated Annealing: An Experimental Evaluation. Part I. Graph Partitioning," Operations Research, 37, 1990, pp. 865–893.

29. Aarts, E., and Korst, J., Simulated Annealing and Boltzmann Machines, A Stochastic Approach to Combinatorial Optimization and Neural Computing, John Wiley & Sons, 1989.

30. Nahar, S., Sahni, S., and Shragowithz, E. V., in the Proceedings of 22nd Design Automation Conf., Las Vegas, June 1985, pp. 748-752.

31. Elperin, T, "Monte Carlo Structural Optimization in Discrete Variables with Annealing Algorithm," Int. J. Num. Meth. Eng., 26, 1988, pp. 815–821.

32. Kincaid, R. K., and Padula, S. L., "Minimizing Distortion and Internal Forces in Truss Structures by Simulated Annealing," Proceedings of the AIAA/ASME /ASCE/AHS/ASC 31st Structures, Structural Dynamics, and Materials Conference, Long Beach, CA., 1990, Part 1, pp. 327–333.

33. Chen, G.-S., Bruno, R. J., and Salama, M.,"Optimal Placement of Active/Passive Members in Structures Using Simulated Annealing," AIAA J., 29 (8), August 1991, pp. 1327–1334.

34. Lombardi, M., "Ottimizzazione di Lastre in Materiale Composito con l'uso di un Metodo di Annealing Simulato," Tesi di Laurea, Department of Structural Mechanics, University of Pavia, 1990.

35. Holland, J. H., Adaptation of Natural and Artificial Systems, The University of Michigan Press, Ann Arbor, MI, 1975.

36. De Jong, K. A., Analysis of the Behavior of a Class of Genetic Adaptive Systems (Doctoral Dissertation, The University of Michigan; University Microfilms No. 76-9381), Dissertation Abstracts International, 36 (10), 5140B, 1975.

37. Booker, L., "Improving Search in Genetic Algorithms," in Genetic Algorithms and Simulated Annealing, Ed. L. Davis, Morgan Kaufmann Publishers, Inc., Los Altos, CA. 1987, pp. 61–73.

38. Goldberg, D. E., and Samtani, M. P., "Engineering Optimization via Genetic Algorithm," Proceedings of the Ninth Conference on Electronic Computation, ASCE, February 1986, pp. 471–482.
39. Hajela, P., "Genetic Search—An Approach to the Nonconvex Optimization Problem," AIAA J., 28 (7), July 1990, pp. 1205–1210.
40. Rao, S. S., Pan, T.-S., and Venkayya, V. B., "Optimal Placement of Actuators in Actively Controlled Structures Using Genetic Algorithms," AIAA J., 29 (6), pp. 942–943, June 1991.

Table 1 : Optimum stacking sequence for 8-ply laminates under biaxial compression.

N_y/N_x	Continuous Optimum	±45–deg. discrete intervals		±15–deg. discrete intervals	
		Penalty Appr.	Global Opt.	Penalty Appr.	Global Opt.
0.0	$[\pm45]_{2s}$	$[\pm45]_{2s}$	-	$[\pm45]_{2s}$	-
0.25	$[\pm53.7/\pm49.8]_s$	$[\pm45]_{2s}$	$[\pm45/90_2]_s$	$[\pm60/\pm45]_s$	-
0.50	$[\pm64.3/\pm53.2]_s$	$[\pm45]_{2s}$	$[\pm45/90_2]_s$	$[\pm60]_{2s}$	$[\pm60/90_2]_s$
1.00	$[\pm73.5/\pm65.8]_s$	$[90_2/\pm45]_s$	-	$[\pm75/\pm60]_s$	-
1.50	$[\pm79.4/\pm70.5]_s$	$[90_2/\pm45]_s$	-	$[\pm75]_{2s}$	$[\pm75/90_2]_s$
2.00	$[\pm83.4/\pm78.1]_s$	$[90_2/\pm45]_s$	-	$[90_2/\pm75]_s$	$[90_2/\pm60]_s$
2.50	$[\pm89.2/\pm88.4]_s$	$[90_4]_{4s}$	-	$[90]_{4s}$	-

Table 2 : Optimum stacking sequence for 16-ply laminates under biaxial compression.

N_y/N_x	Continuous Optimum	±45–deg. discrete intervals		±15–deg. discrete intervals	
		Penalty Appr.	Global Opt.	Penalty Appr.	Global Opt.
0.0	$[\pm45]_{4s}$	$[\pm45]_{4s}$	-	$[\pm45]_{4s}$	-
0.25	$[\pm52.2/-/\pm46.5]_s$	$[\pm45]_{4s}$	$[\pm45_2/90_4]_s$	$[\pm60/\pm45_3]_s$	-
0.50	$[\pm65.3/-/\pm60.0]_s$	$[90_2\pm45_3]_s$	$[\pm45/90_6]_s$	$[\pm60]_{4s}$	$[\pm60_2/90_4]_s$
1.00	$[\pm74.9/-/\pm52.6]_s$	$[90_4/\pm45_2]_s$	$[90_2/\pm45/90_4]_s$	$[\pm75_2/\pm60_2]_s$	$[\pm75_2/\pm60/\pm45]_s$
1.50	$[\pm80.0/-/\pm64.1]_s$	$[90_6/\pm45]_s$	$[90_4/\pm45_2]_s$	$[\pm75_3/\pm60]_s$	$[\pm75/90_2/\pm75/90_2]_s$
2.00	$[\pm83.9/-/\pm71.8]_s$	$[90_6/\pm45]_s$	$[90_4/\pm45/90_2]_s$	$[90_2/\pm75_3]_s$	$[90_2/\pm75/90_4]_s$
2.50	$[\pm89.2/-/\pm87.9]_s$	$[90]_{8s}$	-	$[90_6/\pm75]_s$	-

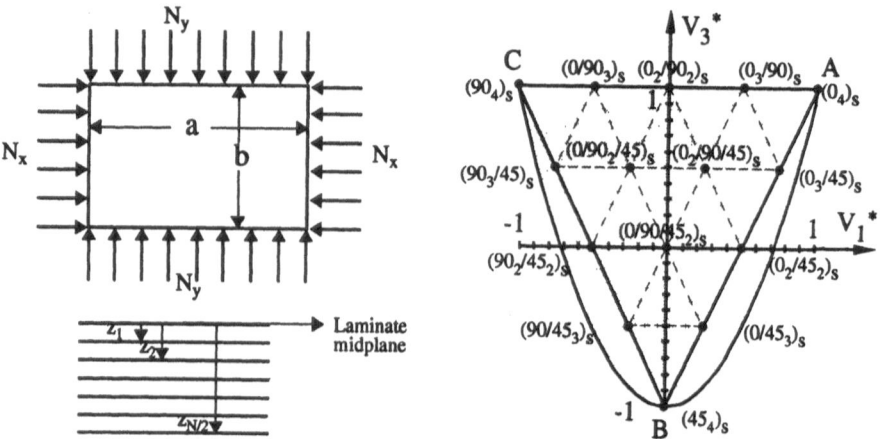

Figure 1: Laminated plate geometry and loading.   Figure 2: In-plane lamination diagram of
$[(\pm\theta_n)_{N_n}/\cdots/(\pm\theta_1)_{N_1}]_s$ laminates.

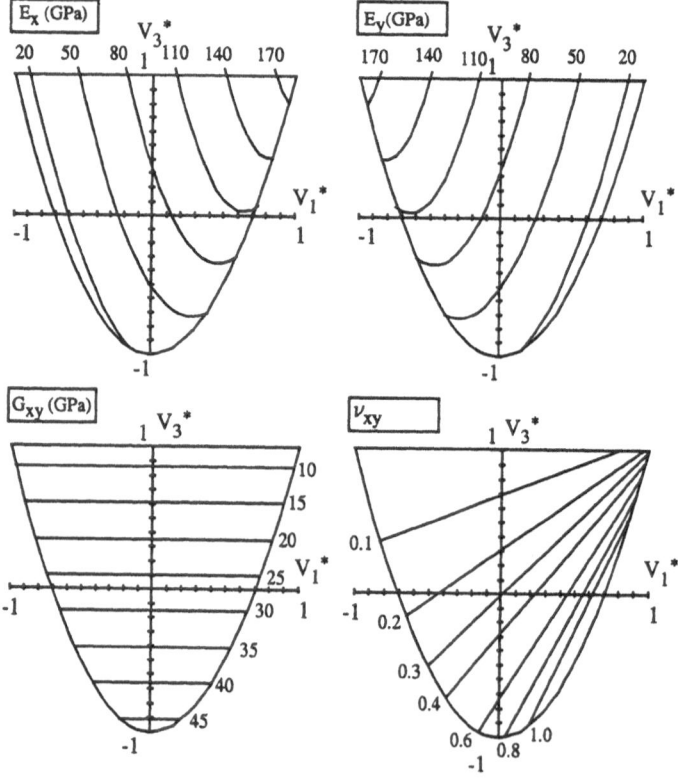

Figure 3: Contours of constant effective engineering constants.

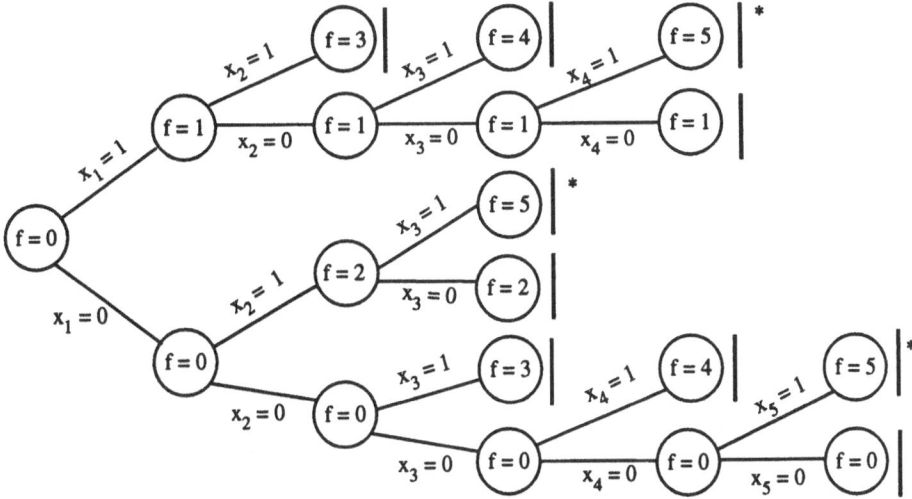

Figure 4: Enumeration tree

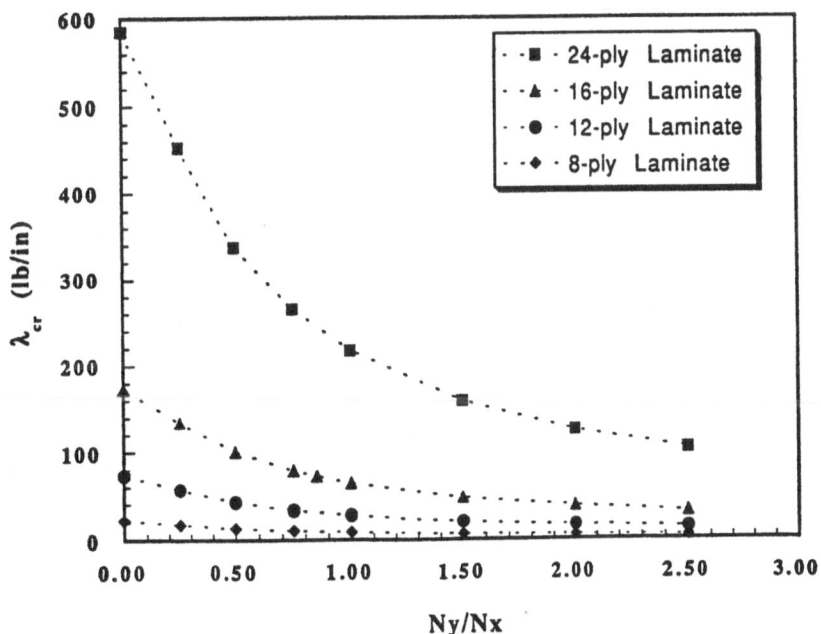

Figure 5: Maximum buckling loads of laminates with continuous fiber orientation variables $(a = 20\ in., b = 10\ in.)$.

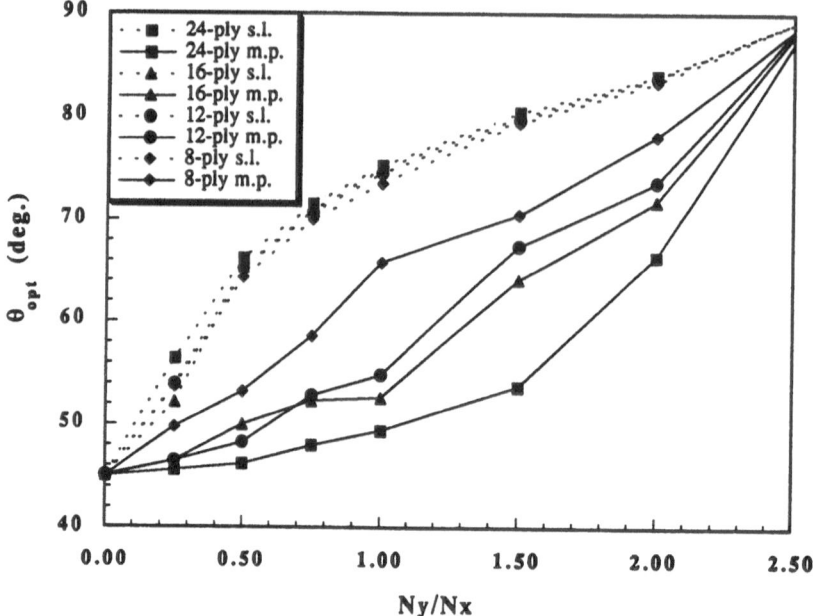

Figure 6: Optimum continuous fiber orientations for laminates with different thicknesses.

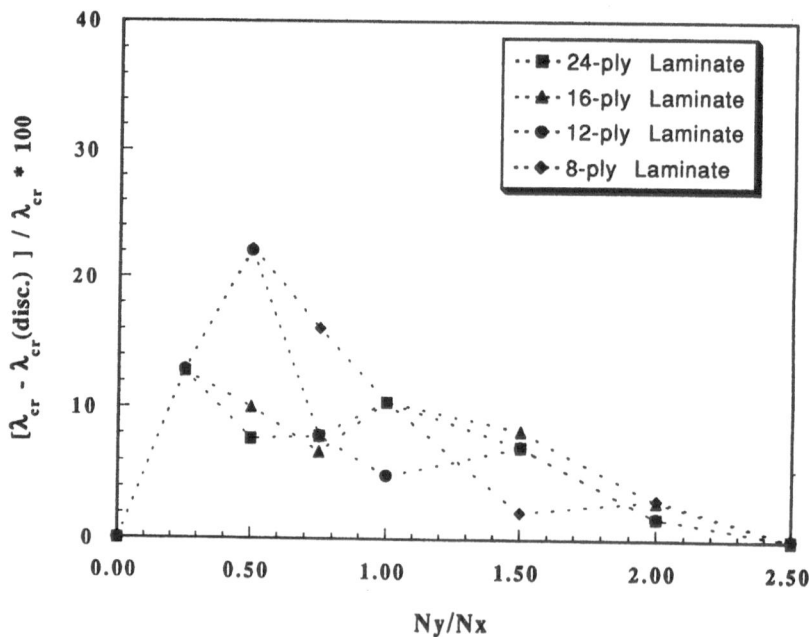

Figure 7: Buckling load reduction for laminates with discrete fiber orientations ($\theta_l = 0, \pm 45, 90$) obtained by the penalty function approach.

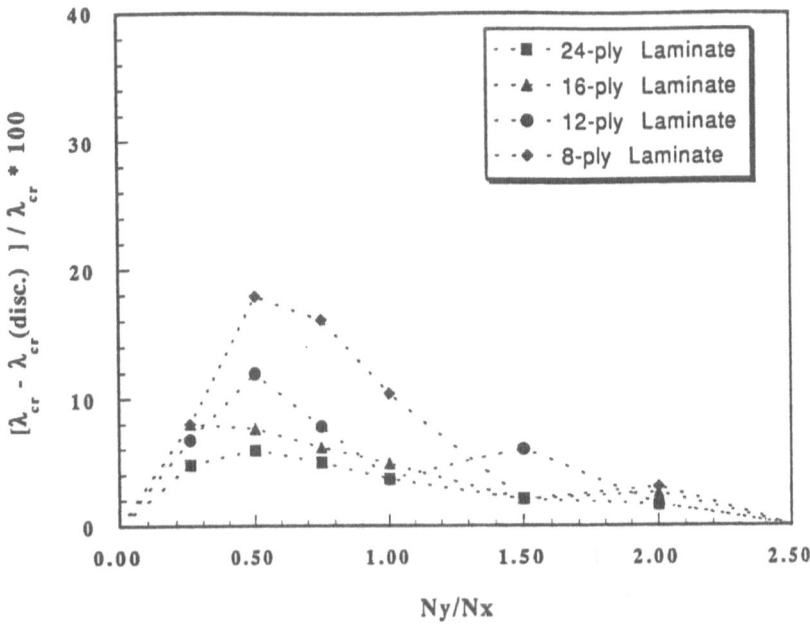

Figure 8: Buckling load reduction for global optimum designs with discrete fiber orientations $(\theta_l = 0, \pm 45, 90)$.

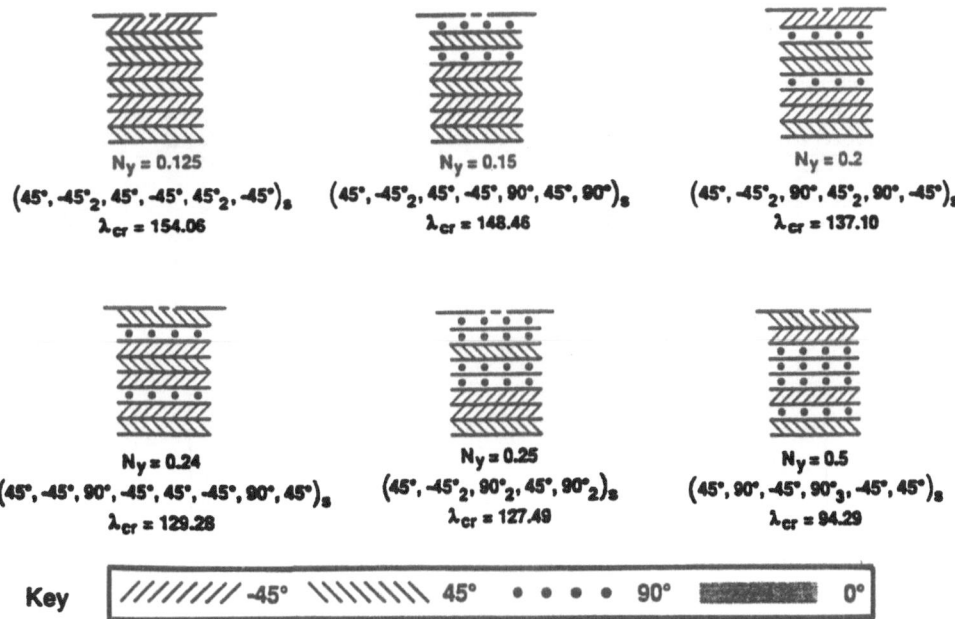

Figure 9: Maximum buckling load designs for 16-ply laminates ($a = 20\ in., b = 10\ in., N_x = 1\ lb/in.$).

648

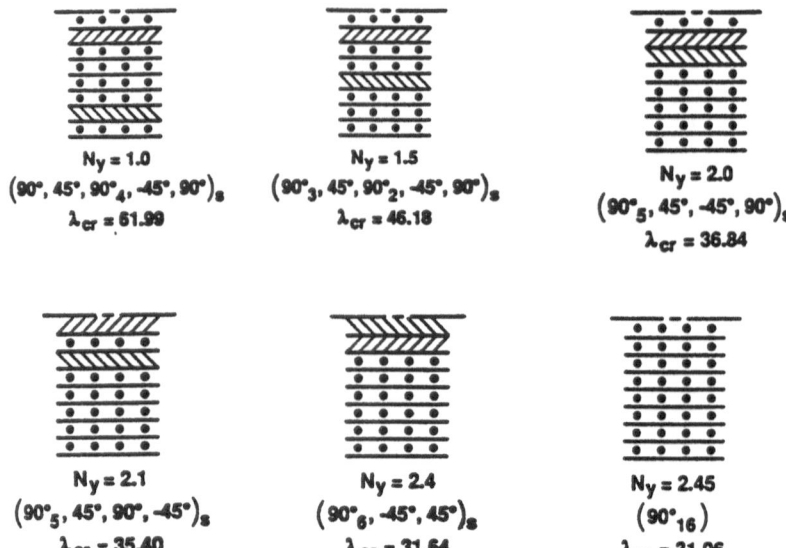

$N_y = 1.0$
$(90°, 45°, 90°_4, -45°, 90°)_s$
$\lambda_{cr} = 51.99$

$N_y = 1.5$
$(90°_3, 45°, 90°_2, -45°, 90°)_s$
$\lambda_{cr} = 46.18$

$N_y = 2.0$
$(90°_5, 45°, -45°, 90°)_s$
$\lambda_{cr} = 36.84$

$N_y = 2.1$
$(90°_5, 45°, 90°, -45°)_s$
$\lambda_{cr} = 35.40$

$N_y = 2.4$
$(90°_6, -45°, 45°)_s$
$\lambda_{cr} = 31.54$

$N_y = 2.45$
$(90°_{16})$
$\lambda_{cr} = 31.06$

Figure 9: (continued)

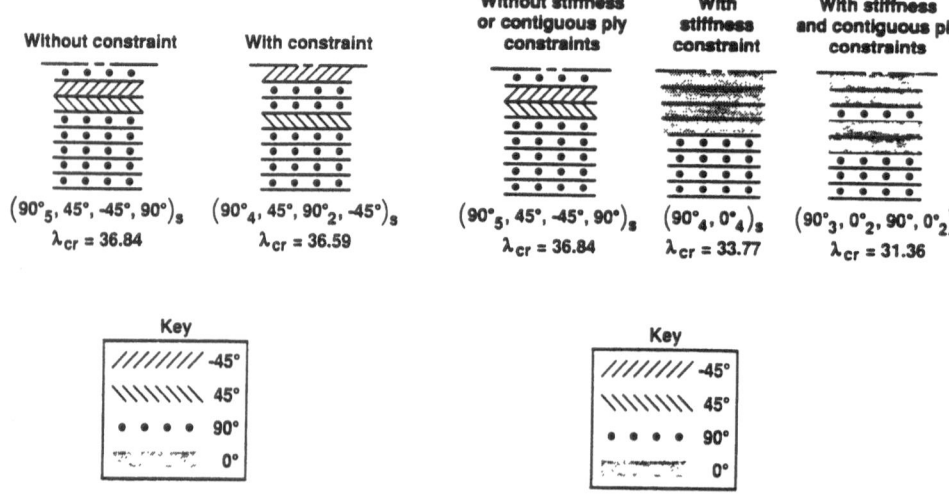

Without constraint

$(90°_5, 45°, -45°, 90°)_s$
$\lambda_{cr} = 36.84$

With constraint

$(90°_4, 45°, 90°_2, -45°)_s$
$\lambda_{cr} = 36.59$

Without stiffness or contiguous ply constraints

$(90°_5, 45°, -45°, 90°)_s$
$\lambda_{cr} = 36.84$

With stiffness constraint

$(90°_4, 0°_4)_s$
$\lambda_{cr} = 33.77$

With stiffness and contiguous ply constraints

$(90°_3, 0°_2, 90°, 0°_2)$
$\lambda_{cr} = 31.36$

Key

///////	-45°
\\\\\\\	45°
• • • •	90°
⁓ ⁓ ⁓	0°

Key

///////	-45°
\\\\\\\	45°
• • • •	90°
⁓ ⁓ ⁓	0°

Figure 10: Effect of constraint of no more than four contiguous plies in same direction on design for $N_y/N_x = 2$.

Figure 11. Effect of stiffness requirement on laminate design for $N_y/N_x = 2$.

OPTIMAL ORIENTATION OF ANISOTROPIC MATERIALS
OPTIMAL DISTRIBUTION OF ANISOTROPIC MATERIALS
OPTIMAL SHAPE DESIGN WITH ANISOTROPIC MATERIALS
OPTIMAL DESIGN FOR A CLASS OF NON–LINEAR ELASTICITY

Pauli Pedersen
Department of Solid Mechanics
The Technical University of Denmark
DK–2800 Lyngby, Denmark

ABSTRACT. Advanced materials are now used frequently in engineering design and that have opened for the possibility of material design. A general characteristic of these materials is that they are anisotropic, and this puts new demands on the analysis capabilities and optimization methods. In recent years a number of questions have been clarified, and the intention of the present paper is to distribute the knowledge gained. Active research areas are also commented on, and the concurrent design with a number of different design parameters are put forward.

1. Introduction

Often design parameters are classified under the headings of size, shape, and topology (layout); here mentioned in order of increasing importance. Design parameters related to materials to some extent fall outside these classes and are now extensively dealt with due to the rapid evolution of a number of new materials such as fibre–reinforced materials and ceramics. Two aspects of optimization are then of practical interest, viz. the influence on size, shape and topology design **with these materials**, and the more or less detailed design **of a material** for a specific purpose.

The two aspects are not uniquely defined; for example, we may argue that design of ply angles in a laminate is design **with plies**, but if the laminate is defined as a material, then it is design **of laminate material**. It is found valuable to recognize the different aspects. A general characteristic of the new materials is that they are anisotropic, and we shall start this paper with a **classification of constitutive matrices**. Although limited to two–dimensional problems, this classification includes information that is not commonly available. We shall restrict the possible materials only by the constraints of a positive definite, constitutive matrix.

Research related to material design has also focused on the very basic sensitivity analysis related to elastic strain energy. Many important results are not well known and as it is possible to prove these results at the energy level we shall do that. This analysis holds for one–, two– and three–dimensional problems, holds for different models, and holds for analytical as well as different numerical solutions. The most important result is the **localized determination of sensitivities** that we often encounter.

G. I. N. Rozvany (ed.), Optimization of Large Structural Systems, Vol. II, 649–681.
© 1993 Kluwer Academic Publishers.

With this basic knowledge we then, in section four, treat **optimal material orientation** for plates and discs. Here too, a number of rather general results are available, and we shall focus on these rather than on the specific problems. We may say that two– dimensional problems are to a large extent solved, but the three–dimensional problems are still a subject of intensive research. An interesting result is the **co–alignment of principal strains and stresses.**

Size optimization, such as distribution of plate thickness, is naturally influenced by the use of anisotropic material but in general the methods of isotropic design are applicable. Therefore, in section five, we focus on the **concurrent design of orientation and thickness.** Comments on non–linear elasticity and on the limitations of optimality criteria methods are given here.

A short account of the detailed material design, in terms of **design of constitutive matrices,** is given in section six. This very active research area already has a number of important results available, and we go through a specific example. The idea here is to see this as the third step after having first solved the orientational problem and then the total density (thickness) problem.

Finally, in section seven, the shape optimization based on anisotropic materials are discussed and recent solutions shown. The **influence from numerical modelling** is focused on, i.e. the number of elements and the element type in a finite element model. This naturally leads to the concurrent mesh design, a research area we only touch upon.

The paper is **not written as a review paper.** It concentrates mainly on the results of personal research. However, many groups are presently active in this area and their reports should be studied in addition to the present paper.

2. Classification of Constitutive Matrices for 2D Linear Elasticity

The reaction to the heading of this section might be that this information must be contained in the classical textbooks. However, material anisotropy is seldom treated in great depth and even in the well–known reference of LEKHNITSKII [1], we cannot find all the results needed for optimal design based on anisotropic behaviour. Hence, it is our intention to make this section self–contained.

First of all, **what is a constitutive law?** Is it described by material parameters alone?, and **what is a material?** The notions are often used as synonyms, and in fact we should add the notion of **model.** A two–dimensional constitutive law is not based on material behaviour alone, but also depends on the mathematical modelling, i.e. the reduction from three to two dimensions, as exemplified by the difference between plane strain and plane stress models. Furthermore, the definition of the concepts of strain and stress influences the constitutive law.

To be specific, let us define the constitutive law by the **constitutive matrix** $[C]$ (or by its inverse – the compliance matrix)

$$\{\sigma\}_x = [C]_x\{\epsilon\}_x \; ; \; \{\epsilon\}_x = [C]_x^{-1}\{\sigma\}_x$$

$$\begin{bmatrix} \sigma_{11} \\ \sigma_{22} \\ \sigma_{12} \end{bmatrix}_x = \begin{bmatrix} C_{11} & C_{12} & C_{13} \\ C_{12} & C_{22} & C_{23} \\ C_{13} & C_{23} & C_{33} \end{bmatrix}_x \begin{bmatrix} \epsilon_{11} \\ \epsilon_{22} \\ 2\epsilon_{12} \end{bmatrix}_x$$

$$(2.1)$$

As stated [C] is symmetric and it is also non–singular. Furthermore, we have indicated in eq. (2.1) that reference to a specific Cartesian x–coordinate system is assumed.

The concepts of **strain energy density** u and **stress energy density** u^C (complementary strain energy) are defined by their first variations as

$$\delta u := \{\sigma\}^T \{\delta \epsilon\} \ ; \ \delta u^C := \{\delta \sigma\}^T \{\epsilon\} \tag{2.2}$$

If linear elasticity is assumed, the definitions (2.2) result in

$$u = u^C = \tfrac{1}{2}\{\sigma\}^T\{\epsilon\} = \tfrac{1}{2}\{\epsilon\}^T[C]\{\epsilon\} = \tfrac{1}{2}\{\sigma\}^T[C]^{-1}\{\sigma\} \tag{2.3}$$

From the definition of energy density (2.2) it follows directly that [C] must be symmetric and positive definite, i.e. all its eigenvalues must be positive. So for **all 2D constitutive models**, we have

[C] symmetric, i.e. max. 6 parameters and

$$C_{11} > 0 \ ; \ C_{22} > 0 \ ; \ C_{33} > 0$$

$$C_{11}C_{22} - C_{12}^2 > 0 \ ; \ C_{22}C_{33} - C_{23}^2 > 0 \ ; \tag{2.4}$$

$$C_{11}C_{33} - C_{13}^2 > 0 \ ; \ |[C]| > 0$$

The conditions for being positive definite can be written in alternative forms.

To familiarize ourselves with these conditions, let us write the well–known constitutive matrix for **isotropic models with only 2 parameters**

$$[C] = \begin{bmatrix} C_{11} & ; & C_{12} & ; & 0 \\ C_{12} & ; & C_{11} & ; & 0 \\ 0 & ; & 0 & ; & \tfrac{1}{2}(C_{11}-C_{12}) \end{bmatrix} \tag{2.5}$$

for which the conditions (2.4) give $C_{11} > 0$; $C_{11}^2 - C_{12}^2 > 0$; $C_{11} - C_{12} > 0$ or, alternatively,

$$0 \le |C_{12}| < C_{11} \quad (\text{subset } 0 < C_{12} < C_{11}) \tag{2.6}$$

from which we see that $C_{12} < 0$ is possible, although often not the case. In plane stress $C_{11} = E/(1-\nu^2)$; $C_{12} = \nu C_{11}$ and in plane strain $C_{11} = E(1-\nu)/\big((1+\nu)(1-2\nu)\big)$; $C_{12} = \nu C_{11}/(1-\nu)$, so with $E > 0$ we get $-1 < \nu < 0.5$.

Let us then characterize the most important case of **orthotropic models with 4 parameters** and a specified direction. We note that only along the orthotropic directions are there no couplings between shear and normal stresses/strains

$$[C]_x = \begin{bmatrix} C_{11} & C_{12} & 0 \\ C_{12} & C_{22} & 0 \\ 0 & 0 & C_{33} \end{bmatrix}_x \tag{2.7}$$

To be more precise, we choose the larger modulus direction of the two orthogonal directions, i.e. $C_{11} > C_{22}$. Then the conditions (2.4) give additional relations $C_{22} > 0$; $C_{33} > 0$; $C_{11}C_{22} - C_{12}^2 > 0$, and from the last condition follows $\frac{1}{2}(C_{11}+C_{22}) - C_{12} > 0$. It is often assumed that $C_{12} > 0$, but this is not generally a real condition. Summing up, we have

$$|C_{12}| < \sqrt{C_{11}C_{22}} \; ; \; 0 < C_{22} < C_{11} \; ; \; 0 < C_{33} \tag{2.8}$$

$$\left(\text{subset } 0 \le C_{12} < \sqrt{C_{11}C_{22}} < \tfrac{1}{2}(C_{11}+C_{22})\right)$$

for the four parameters of the orthotropic models with **a uniquely chosen reference axis**.

Based on the quantities of the orthotropic constitutive matrix (2.7), we − in agreement with the literature on composite materials, (e.g. JONES [2]) − define **practical parameters** (often termed invariants, which is misleading)

$$C_1 := \tfrac{1}{8}(3(C_{11}+C_{22}) + 2(C_{12}+2C_{33}))$$

$$C_2 := \tfrac{1}{2}(C_{11}-C_{22})$$

$$C_3 := \tfrac{1}{8}((C_{11}+C_{22}) - 2(C_{12}+2C_{33})) \tag{2.9}$$

$$C_4 := \tfrac{1}{8}((C_{11}+C_{22}) + 2(3C_{12}-2C_{33}))$$

$$C_5 := \tfrac{1}{8}((C_{11}+C_{22}) - 2(C_{12}-2C_{33}))$$

Two of these are of major importance and have a distinct physical interpretation, i.e. C_2 and C_3. The **difference in modulus between the two orthotropic directions** is measured by C_2. The parameter C_3 is of the utmost importance for the optimal design. It has a physical interpretation as **relative shear modulus**, and according to its sign we classify the constitutive law as **low or high shear modulus**, respectively. From (2.8) we know that for "small" C_{33} we have $C_3 > 0$ and from (2.9) we see that for "large" C_{33} we get $C_3 < 0$. Thus

Low shear modulus:

$C_3 > 0$ i.e. $2C_{33} < \frac{1}{2}(C_{11}+C_{22}) - C_{12}$

High shear modulus:

$C_3 < 0$ i.e. $2C_{33} > \frac{1}{2}(C_{11}+C_{22}) - C_{12}$

(Isotropic $C_3 = 0$)

$$\tag{2.10}$$

This classification is not well known, and it is sometimes argued that $C_3 < 0$ is not possible. Most ordinary "materials" give $C_3 > 0$, but constructed materials like laminates can easily be designed to have $C_3 < 0$.

In a broader perspective, the notion of constitutive matrices is also used for **plate stiffnesses** – in the plane:

$$
\begin{bmatrix} N_{11} \\ N_{22} \\ N_{12} \end{bmatrix} = [A] \begin{Bmatrix} \epsilon_{11}^0 \\ \epsilon_{22}^0 \\ 2\epsilon_{12}^0 \end{Bmatrix}
\tag{2.11}
$$

as well as out of the plane:

$$
\begin{bmatrix} M_{11} \\ M_{22} \\ M_{12} \end{bmatrix} = [D] \begin{Bmatrix} \kappa_{11} \\ \kappa_{22} \\ 2\kappa_{12} \end{Bmatrix}
\tag{2.12}
$$

where $[A]$ is the matrix of extensional stiffnesses, $\{N\}$ is a vector of plate forces per unit length, $[D]$ is the matrix of bending stiffnesses, $\{M\}$ is a vector of plate moments per unit length, and $\{\epsilon^0\}$, $\{\kappa\}$ are midplane strains and curvatures, respectively.

In order to deal with **non–orthotropic constitutive matrices**, we define, in addition to the parameters $C_1 - C_5$ in (2.9),

$$
C_6 := \tfrac{1}{2}(C_{13} + C_{23})
\tag{2.13}
$$
$$
C_7 := \tfrac{1}{2}(C_{13} - C_{23})
$$

two parameters, that vanish for orthotropic laws. However, the orthotropic directions may be unknown and then we need a criterion to **test for existence of orthotropic directions**. This is derived in PEDERSEN [3] and is expressed using the parameters C_2, C_3, C_6 and C_7 (based on any reference axis) as

$$
C_7 C_2^2 - 4C_7 C_6^2 - 4C_6 C_3 C_2 = 0
\tag{2.14}
$$

If (2.14) is satisfied we choose the direction of orthotropy for which $C_2 > 0$. If (2.14) is not satisfied we have a non–orthotropic constitutive matrix, but still need a specific reference axis for comparison of equal constitutive matrices which may in reality just be mutually rotated. No clear tradition for selection of this direction seems to be available in the literature.

In our studies we have chosen the direction which maximizes C_{11}, i.e.

$$
\underset{\gamma}{\text{Max}} \left(C_{11} = C_{11}(\gamma) \right) \text{ for } 0 \leq \gamma \leq \pi
\tag{2.15}
$$

and in most (but not all) practical cases this is uniquely determined by

654

Fig. 2.1: Overview of the results.

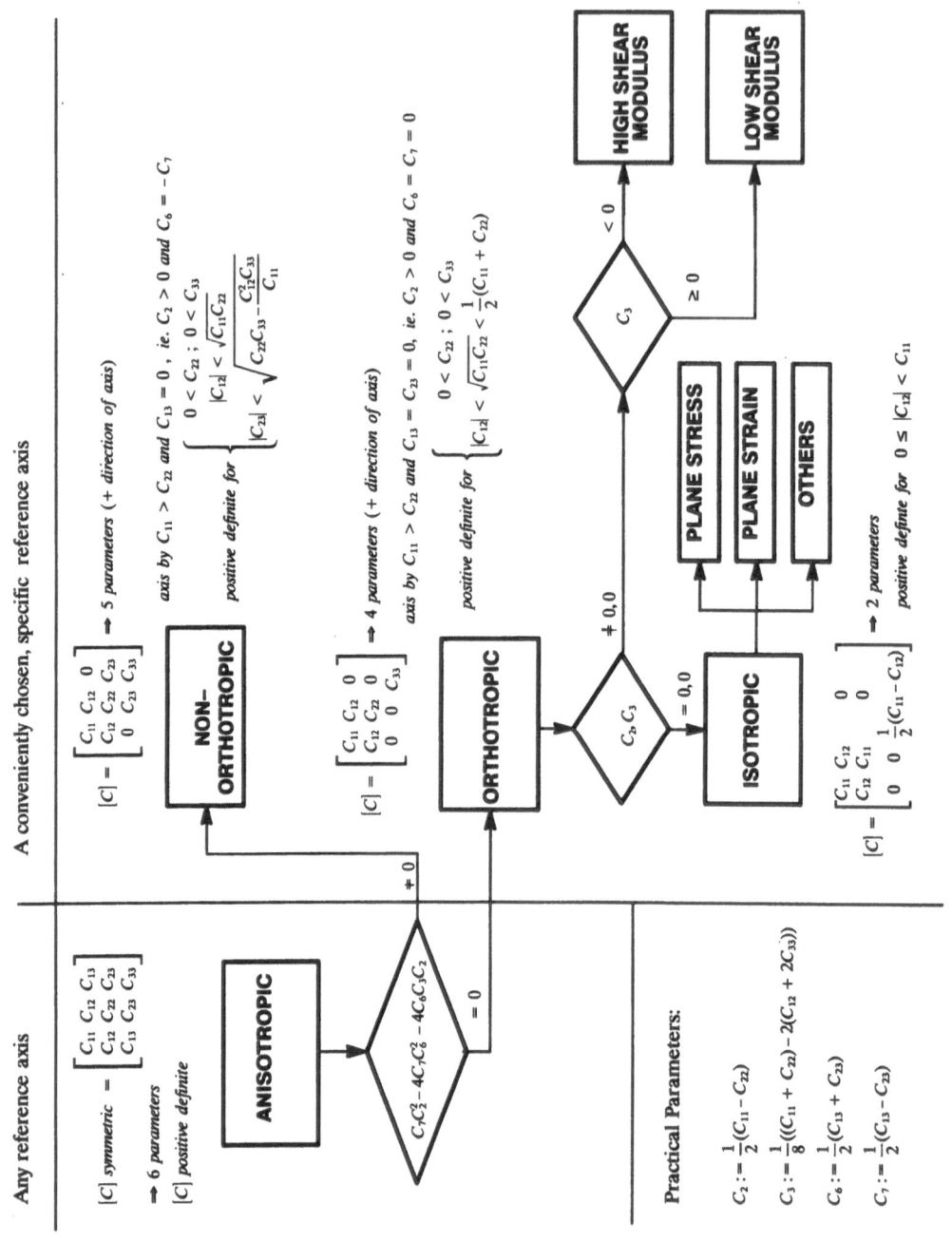

$$(C_{13})_x = 0 \ , \ \text{i.e.} \ (C_6)_x = -(C_7)_x \tag{2.16}$$

from which follows a general constitutive matrix

$$[C] = \begin{bmatrix} C_{11} & ; & C_{12} & ; & 0 \\ C_{12} & ; & C_{22} & ; & C_{23} \\ 0 & ; & C_{23} & ; & C_{33} \end{bmatrix} \ \text{with} \ C_2 = \tfrac{1}{2}(C_{11}-C_{22}) > 0 \tag{2.17}$$

This constitutive matrix is positive definite with the conditions

$$0 < C_{22} \ ; \ 0 < C_{33} \ ; \ |C_{12}| < \sqrt{C_{11}C_{22}}$$

$$|C_{23}| < \sqrt{C_{22}C_{33} - C_{12}^2 C_{33}/C_{11}} \tag{2.18}$$

to simplify (2.1) and (2.4).
The results of this section as a whole are shown in fig. 2.1.

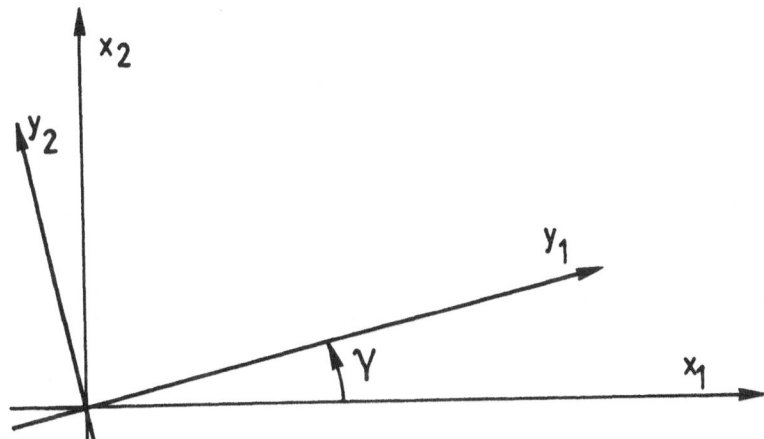

Fig. 2.2: The specific reference axis x_1 and a relatively rotated coordinate system $y_1;y_2$. Angle γ defined from x_1 to y_1, positive anti–clockwise.

Before ending the section we shall list the **transformation formulas** for constitutive laws from x–coordinates to y–coordinates as defined in fig. 2.2. Generally we have

$$(C_{11})_y = \frac{1}{2}(C_{11}+C_{22})_x + (C_2)_x\cos(2\gamma)$$
$$- (C_3)_x(1-\cos(4\gamma)) + (C_6)_x 2\sin(2\gamma) + (C_7)_x\sin(4\gamma)$$

$$(C_{22})_y = \frac{1}{2}(C_{11}+C_{22})_x - (C_2)_x\cos(2\gamma)$$
$$- (C_3)_x(1-\cos(4\gamma)) - (C_6)_x 2\sin(2\gamma) + (C_7)_x\sin(4\gamma)$$

$$(C_{12})_y = (C_{12})_x + (C_3)_x(1-\cos(4\gamma)) - (C_7)_x\sin(4\gamma)$$

$$(C_{33})_y = (C_{33})_x + (C_3)_x(1-\cos(4\gamma)) - (C_7)_x\sin(4\gamma) \qquad (2.19)$$

$$(C_{13})_y = -\frac{1}{2}(C_2)_x\sin(2\gamma) - (C_3)_x\sin(4\gamma)$$
$$+ \quad (C_6)_x\cos(2\gamma) + (C_7)_x\cos(4\gamma)$$

$$(C_{23})_y = -\frac{1}{2}(C_2)_x\sin(2\gamma) + (C_3)_x\sin(4\gamma)$$
$$+ \quad (C_6)_x\cos(2\gamma) - (C_7)_x\cos(4\gamma)$$

If the constitutive matrix is **orthotropic** in the x–coordinates it simplifies to

$$(C_{11})_y = (C_1)_x + (C_2)_x\cos(2\gamma) + (C_3)_x\cos(4\gamma)$$

$$(C_{22})_y = (C_1)_x - (C_2)_x\cos(2\gamma) + (C_3)_x\cos(4\gamma)$$

$$(C_{12})_y = (C_4)_x - (C_3)_x\cos(4\gamma) \qquad (2.20)$$

$$(C_{33})_y = (C_5)_x - (C_3)_x\cos(4\gamma)$$

$$(C_{13})_y = -\frac{1}{2}(C_2)_x\sin(2\gamma) - (C_3)_x\sin(4\gamma)$$

$$(C_{23})_y = -\frac{1}{2}(C_2)_x\sin(2\gamma) + (C_3)_x\sin(4\gamma)$$

Note that the definition is **from the orthotropic** direction and that the rotation angle γ is measured **positive anti–clockwise**. Alternative definitions are found in the literature.

Based on the transformations (2.19) we can prove the postulate (2.16) and see its limitations. The condition of $C_{11} > C_{22}$ gives

$$2C_2\cos2\gamma + 4C_6\sin2\gamma > 0 \; ; \; \text{i.e.} \; \tan2\gamma < -\frac{C_2}{2C_6}$$
$$(2.21)$$

with C_2, C_6 from the arbitrary coordinates. $\left(\text{From directions } x \text{ we have } \gamma = 0, \right.$ i.e. $\left. (C_2)_x > 0 \right)$. The condition of stationarity for C_{11} gives $\partial C_{11}/\partial\gamma = 0$, i.e.

$$- 2C_2\sin2\gamma - 4C_3\sin4\gamma + 4C_6\cos2\gamma + 4C_7\cos4\gamma = 0$$
$$(2.22)$$

and this also holds from the system x, for which $\gamma = 0$ results in

$$(C_6)_x = -(C_7)_x \; ; \; \text{i.e.} \; (C_{13})_x = 0 \tag{2.23}$$

The condition of maximum C_{11} with $\partial^2 C_{11}/\partial \gamma^2 < 0$ gives the relation

$$-4C_2 \cos 2\gamma - 16C_3 \cos 2\gamma - 8C_6 \sin 2\gamma - 16C_7 \sin 2\gamma < 0 \tag{2.24}$$

which in the x–coordinate system for $\gamma = 0$ reduces to

$$(C_2)_x + 4(C_3)_x > 0 \tag{2.25}$$

Mostly, there will be only one solution to (2.21), (2.22), (2.24), but this does not hold for all possible cases. A direct plot of $C_{11} = C_{11}(\gamma)$ will be most instructive.

3. Localized Determination of Sensitivity Results

The results from sensitivity analysis have strong relations to research on optimal design but are in reality of much wider importance and applicability. Often the results are derived for specific models, and the generality is lost or at least not visible.

When the quantity for which we seek the sensitivity is related to a global energy quantity, we have very **simplified results of localized nature**. These results will be derived here without reference to a specific model and are thus valid for one–, two– and three–dimensional models, for analytical calculation, and for numerical model and are valid independent of the numerical method chosen, say finite difference, finite element, or more global Galerkin approaches.

For dynamic problems the results are understandable intuitively and are well known. However, for static problems, the generality is not always appreciated. Thus although starting with the dynamic problems we shall focus on the static problems and also include important classes of non–linearity.

Now for the **dynamics**, let U be the elastic energy amplitude, T the specific kinetic energy, and ω^2 the squared eigenfrequency, such that $(\omega^2 T)$ is the kinetic energy amplitude based on the $e^{i\omega t}$ time dependence. For **conservative** problems we have constant total energy, i.e.

$$U - \omega^2 T = 0 \tag{3.1}$$

from which follows the Rayleigh quotient.

Our goal is to find the change in frequency as a function of change in design, here described by $\mathbf{h}$, which may be a scalar, a function or a number of such, i.e. a field quantity. The displacement field (the mode) is similarly characterized by $\mathbf{v}$, also a field quantity.

Stationarity of the Rayleigh quotient with respect to $\mathbf{v}$ gives

$$\partial U/\partial \mathbf{v} - \omega^2 \partial T/\partial \mathbf{v} = 0 \tag{3.2}$$

and the total differentials with respect to design $\mathbf{h}$ give

$$\frac{\partial U}{\partial \mathbf{h}} + \frac{\partial U}{\partial \mathbf{v}} \frac{\partial \mathbf{v}}{\partial \mathbf{h}} - \frac{d(\omega^2)}{d\mathbf{h}} T - \omega^2 \frac{\partial T}{\partial \mathbf{h}} - \omega^2 \frac{\partial T}{\partial \mathbf{v}} \frac{\partial \mathbf{v}}{\partial \mathbf{h}} = 0 \tag{3.3}$$

which by means of (3.2) simplifies to

$$\frac{d(\omega^2)}{dh} = \frac{1}{T}\left[\frac{\partial U}{\partial h} - \omega^2 \frac{\partial T}{\partial h}\right]_{\substack{fixed \\ mode}} \tag{3.4}$$

a well–known result from WITTRICK [4]. It should be appreciated that we can thus determine the change in eigenfrequency without involving the change in eigenmode. Eq. (3.4) is exact, not approximate.

This important observation also immediately implies **localization**. The elastic energy as well as the kinetic energy are accumulative, i.e.

$$U = \sum_e U_e \quad ; \quad T = \sum_e T_e \tag{3.5}$$

where U_e, T_e are the energies in domain e (element e in a finite element analysis) and the summations are extended over all the domains of our model. Furthermore, let the design parameter h_e be related only to domain e, i.e.

$$\frac{\partial U_i}{\partial h_e} = 0 \quad ; \quad \frac{\partial T_i}{\partial h_e} = 0 \quad \text{for} \quad e \neq i \tag{3.6}$$

but the total (physical) derivatives dU_i/dh_e and dT_i/dh_e may be different from zero.

Inserting (3.5) and (3.6) in (3.4) we get the localized determination of the sensitivity

$$\frac{\partial(\omega^2)}{\partial h_e} = \frac{1}{T}\left[\frac{\partial U_e}{\partial h_e} - \omega^2 \frac{\partial T_e}{\partial h_e}\right]_{\substack{fixed \\ mode}} \tag{3.7}$$

The local energies U_e and T_e are often simple functions of h_e and we often normalize to get $T = 1$. Then the result (3.7) can be written in a even simpler form and can often be interpreted in terms of the local energies instead of gradients of energies.

We shall now concentrate on static problems for which the sensitivity results are less intuitive. In general this is due to the different nature of the work of the external forces compared with the kinetic energy just treated. Let us start with the **work equation**

$$W + W^C = U + U^C \tag{3.8}$$

where W, W^C are physical and complementary work of the external forces, and U, U^C are physical and complementary elastic energy, also named strain and stress energy, respectively. The work equation (3.8) holds for any design h and therefore for the total differential quotient with respect to h

$$\frac{dW}{dh} + \frac{dW^C}{dh} = \frac{dU}{dh} + \frac{dU^C}{dh} \tag{3.9}$$

Now in the same way as h represents the design field generally, ϵ represents the strain field and σ represents the stress field. Remembering that as a function of h, ϵ we have W, U, while the complementary quantities W^C, U^C are functions of h, σ, we then get (3.9) in greater detail by means of

$$\frac{\partial W}{\partial h} + \frac{\partial W}{\partial \epsilon}\frac{\partial \epsilon}{\partial h} + \frac{\partial W^C}{\partial h} + \frac{\partial W^C}{\partial \sigma}\frac{\partial \sigma}{\partial h} = \frac{\partial U}{\partial h} + \frac{\partial U}{\partial \epsilon}\frac{\partial \epsilon}{\partial h} + \frac{\partial U^C}{\partial h} + \frac{\partial U^C}{\partial \sigma}\frac{\partial \sigma}{\partial h} \tag{3.10}$$

The **principles of virtual work**, which hold for solids/structures in equilibrium, are

$$\frac{\partial W}{\partial \epsilon} = \frac{\partial U}{\partial \epsilon} \tag{3.11}$$

for the physical quantities with strain variation, and for the complementary quantities with stress variation we have

$$\frac{\partial W^C}{\partial \sigma} = \frac{\partial U^C}{\partial \sigma} \tag{3.12}$$

Inserting (3.11) and (3.12) in (3.10) we get

$$\frac{\partial U^C}{\partial h} - \frac{\partial W^C}{\partial h} = -\left[\frac{\partial U}{\partial h} - \frac{\partial W}{\partial h}\right] \tag{3.13}$$

and for design–independent loads

$$\left[\frac{\partial U^C}{\partial h}\right]_{\substack{\text{fixed} \\ \text{stresses}}} = -\left[\frac{\partial U}{\partial h}\right]_{\substack{\text{fixed} \\ \text{strains}}} \tag{3.14}$$

as stated by MASUR [5]. Note that the only assumption behind this is the design–independent loads $\partial W/\partial h = 0$, $\partial W^C/\partial h = 0$.

To get further into a **physical interpretation** of $(\partial U/\partial h)_{\text{fixed strains}}$ (and by (3.14) of $(\partial U^C/\partial h)_{\text{fixed stresses}}$) we need the relation between external work W and strain energy U. Let us assume that this relation is given by the constant c

$$W = cU \tag{3.15}$$

For linear elasticity and dead loads we have $c = 2$, and in general we expect $c > 1$, i.e. we loose energy.

Parallel to the analysis from (3.8) to (3.10) and on the basis of (3.15), we obtain

$$\frac{\partial W}{\partial h} + \frac{\partial W}{\partial \epsilon}\frac{\partial \epsilon}{\partial h} = c\frac{\partial U}{\partial h} + c\frac{\partial U}{\partial \epsilon}\frac{\partial \epsilon}{\partial h} \tag{3.16}$$

which for design–independent loads $\partial W/\partial h = 0$ with virtual work (3.11), gives

$$\frac{\partial W}{\partial \epsilon}\frac{\partial \epsilon}{\partial h} = \frac{\partial U}{\partial \epsilon}\frac{\partial \epsilon}{\partial h} = \frac{c}{1-c}\frac{\partial U}{\partial h} \tag{3.17}$$

and thereby

$$\frac{dU}{dh} = \frac{\partial U}{\partial h} + \frac{\partial U}{\partial \epsilon}\frac{\partial \epsilon}{\partial h} = \frac{1}{1-c}\left[\frac{\partial U}{\partial h}\right]_{\substack{\text{fixed} \\ \text{strains}}} \tag{3.18}$$

Note in this important result that with $c > 1$ we have **different signs** for dU/dh and $(\partial U/\partial h)_{\text{fixed strains}}$.

For the case of **linear elasticity and dead loads** we have, with $c = 2$ and adding (3.14),

$$\frac{dU}{dh} = -\left[\frac{\partial U}{\partial h}\right]_{\substack{\text{fixed} \\ \text{strains}}} = \left[\frac{\partial U}{\partial h}\right]_{\substack{\text{fixed} \\ \text{stresses}}} \tag{3.19}$$

For the case of **non—linear elasticity** modelled one dimensionally by

$$\sigma = E\epsilon^n \tag{3.20}$$

and still "dead loads" $(W^C = 0)$ we get $c = 1+n$ and thereby

$$\frac{dU}{dh} = -\frac{1}{n}\left[\frac{\partial U}{\partial h}\right]_{\substack{\text{fixed} \\ \text{strains}}} = \frac{1}{n}\left[\frac{\partial U}{\partial h}\right]_{\substack{\text{fixed} \\ \text{stresses}}} \tag{3.21}$$

Now let us discuss the **localized determination** of sensitivities as given generally by (3.18). Again the strain energy is summed over all domains (3.5), and the design parameter h_e is local (3.6). Then we get

$$\frac{dU}{dh_e} = \frac{dU_e}{dh_e} + \sum_{i \neq e} \frac{dU_i}{dh_e}$$

$$= \frac{\partial U_e}{\partial h_e} + \frac{\partial U_e}{\partial \epsilon_e}\frac{\partial \epsilon_e}{\partial h_e} + \sum_{i \neq e} \frac{\partial U_i}{\partial \epsilon_i}\frac{\partial \epsilon_i}{\partial h_e} \tag{3.22}$$

where ϵ_e describes the strain field of domain e. Note that by (3.6) we have $\partial U_i/\partial h_e = 0$ for $i \neq e$. The results (3.18) tells that we need not calculate $\partial U_i/\partial \epsilon_e$, because

$$\frac{dU}{dh_e} = \frac{1}{1-c}\left[\frac{\partial U_e}{\partial h_e}\right] \tag{3.23}$$

and from (3.22)–(3.23) we can again determine the "indirect" effect

$$\frac{\partial U_e}{\partial \epsilon_e}\frac{\partial \epsilon_e}{\partial h_e} + \sum_{i \neq e} \frac{\partial U_i}{\partial \epsilon_i}\frac{\partial \epsilon_i}{\partial h_e} = \sum_{\text{all } i} \frac{\partial U_i}{\partial \epsilon_i}\frac{\partial \epsilon_i}{\partial h_e} = \frac{c}{1-c}\frac{\partial U_e}{\partial h_e} \tag{3.24}$$

Eq. (3.24) is the same as eq. (3.17) for the localized parameter h_e. For linear elasticity and dead loads we have $c = 2$, and the indirect effect is then twice the direct effect (with opposite signs).

We will often determine the strain energy by the strain energy density u_e and the domain volume V_e, i.e.

$$U_e = u_e V_e \tag{3.25}$$

We then naturally treat two groups of design parameters, i.e. the ones without influence on V_e and the ones without explicit influence on u_e. For the first group, say h_e is a parameter of the constitutive matrix, the sensitivity determination (3.23) simplifies to

$$\frac{dU}{dh_e} = \frac{V_e}{1-c}\left[\frac{\partial u_e}{\partial h_e}\right]_{\substack{\text{fixed} \\ \text{strains}}} \tag{3.26}$$

and for the second group, say h_e is a thickness or area parameter, the sensitivity determination (3.23) simplifies to

$$\frac{dU}{dh_e} = \frac{u_e}{1-c}\frac{\partial V_e}{\partial h_e} \tag{3.27}$$

In the following we shall apply these general sensitivity results to a number of specific problems.

4. Optimal Material Orientation for Plates and Discs

In this section our general design parameter h is taken to be the material orientation θ_e in the domain e. We assume linear elasticity and dead loads $(W = 2U)$ and thus have directly from (3.26)

$$\frac{dU}{d\theta_e} = -V_e\left[\frac{\partial u_e}{\partial \theta_e}\right]_{\substack{\text{fixed} \\ \text{strains}}} \tag{4.1}$$

For coupled plate/disc problems using traditional symbols from laminate analysis the energy density u_e is given by

$$u_e t_e = \frac{1}{2}\{\epsilon^0\}^T\{N\} + \frac{1}{2}\{\kappa\}^T\{M\} \ ;$$

$$\{N\} = [A]\{\epsilon^0\} + [B]\{\kappa\} \ ; \tag{4.2}$$

$$\{M\} = [B]\{\epsilon^0\} + [D]\{\kappa\} \ ,$$

and the combined result is

$$u_e t_e = \frac{1}{2}\{\epsilon^0\}^T[A]\{\epsilon^0\} + \{\epsilon^0\}^T[B]\{\kappa\} + \frac{1}{2}\{\kappa\}^T[D]\{\kappa\} \tag{4.3}$$

with t_e for plate thickness, $\{\epsilon^0\}$, $[A]$ for midsurface strains and extensional stiffnesses; and $\{\kappa\}$, $[D]$, $[B]$ for curvatures, bending stiffnesses and coupling stiffnesses. Applying the result (4.1) we get

$$\frac{dU}{d\theta_e} = -a_e\left[\frac{1}{2}\{\epsilon^0\}^T\left[\frac{\partial A}{\partial \theta_e}\right]\{\epsilon^0\} + \{\epsilon^0\}^T\left[\frac{\partial B}{\partial \theta_e}\right]\{\kappa\} + \frac{1}{2}\{\kappa\}^T\left[\frac{\partial D}{\partial \theta_e}\right]\{\kappa\}\right] \tag{4.4}$$

with a_e for domain area. Even for the **fully coupled problems** this result can be written

$$\frac{dU}{d\theta_e} = U_1\sin2\theta + U_2\cos2\theta + U_3\sin4\theta + U_4\cos4\theta \tag{4.5}$$

This follows from the fact that all the matrices [A], [B] and [D] originate from the constitutive matrix [C], which, as seen from (2.19), contains only the trigonometric functions of eq. (4.5).

Before treating the specific and simplified problems, it should be appreciated that according to (4.5) we, in the general case, can find **at most four different** solutions to $dU/d\theta_e = 0$. This follows from rewriting $dU/d\theta_e$ as a fourth order polynomial. However, analytical solutions are too complicated.

For **orthotropic materials** and models where only the cosine terms are involved, like $C_{13} = C_{23} \equiv 0$ in eq. (2.20), analytical solutions to $dU/d\theta_e$ are obtainable. Keeping in eq. (4.5) only the sine terms, we have for these (specially orthotropic/balanced) models

$$\frac{dU}{d\theta_e} = 2U_3\sin2\theta\left[\frac{U_1}{2U_3} + \cos2\theta\right] \tag{4.6}$$

Stationarity is then obtained for

$$\theta = 0 \;;\; \theta = \pi/2 \;;\; \theta = \pm\frac{1}{2}\arccos\left[-\frac{U_1}{2U_3}\right] \tag{4.7}$$

and, furthermore, supplementary angles will return the same energy density

$$u(\pi - \theta) = u(\theta) \tag{4.8}$$

when only $\cos2\theta$ and $\cos4\theta$ appear in the stiffness expressions. Thus, the orientational dependence is described completely by the interval $0 \le \theta \le \pi/2$.

Now let us discuss some specific results. First, **simply supported plates in pure bending**, for which the Navier analytical response solutions are available. With a load pressure by

$$p = p_{mn}\sin\frac{m\pi x_1}{a}\sin\frac{n\pi x_2}{b} \tag{4.9}$$

the transverse displacement is

$$w = p\,\frac{b^4}{\pi^4n^4}\frac{1}{\Phi_{mn}}$$

$$\Phi_{mn} := \eta_{mn}^4 D_{11} + D_{22} + \eta_{mn}^2\, 2(D_{12} + 2D_{33}) \tag{4.10}$$

$$\eta_{mn} := (mb)/(na)$$

with the **mode parameter** η_{mn} defined as the ratio between the two actual half-wavelengths. The same functional Φ give us eigenfrequencies by

$$\omega_{mn} = \frac{\pi^2 n^2}{b^2} \sqrt{\frac{\Phi}{\rho t}} \tag{4.11}$$

where ρ is the mass density of the plate. Also buckling load, say $(N_{x1})_{cr}$, is described by Φ as

$$(N_{x1})_{cr} = \frac{\pi^2}{b^2} \frac{\Phi}{\eta_{mn}^2} \tag{4.12}$$

Thus, maximizing Φ, we at the same time maximize buckling load, maximize frequency, and minimize displacements **for the same mode**, i.e. the same η_{mn}.

The solution which maximizing $\Phi = \Phi(\theta)$ gives as a specific case of (4.7)

$$\theta_{opt} = 0 \quad \text{or} \quad \theta_{opt} = \pm\,\pi/2 \quad \text{or} \quad \theta_{opt} = \pm \frac{1}{2} \arccos(-\delta)$$

$$\tag{4.13}$$

$$\delta := \frac{C_2}{4C_3} \frac{(\eta^4-1)}{(1-6\eta^2+\eta^4)} = \frac{(C_{11}-C_{22})}{\left(C_{11}+C_{22}-2(C_{12}+2C_{33})\right)} \frac{(\eta^4-1)}{(1-6\eta^2+\eta^4)}$$

where the global best extremum can be located by second order derivatives. The result (4.13) is derived in PEDERSEN [6] and can also be found in the paper by BERT [7] or the more recent paper by MUC [8]. We note from the definition of **the optimization parameter** δ that two inverse cases of η_1 and $\eta_2 = \eta_1^{-1}$ have opposite signs of δ and thereby give rise to complementary angles. The results are thus fully explained in fig. 4.1.

The result in fig. 4.1 can be directly explained in physical terms. When the deformation wavelength is large in a certain direction, then the fibredirection should be perpendicular to that direction, i.e. $\theta = 0^\circ$ or 90°, respectively. At a certain wavelength ratio, depending upon material but about $1:1.8$, use of skew fibre directions starts, and with a square deformation pattern, the fibredirection should be in the diagonal direction of the square.

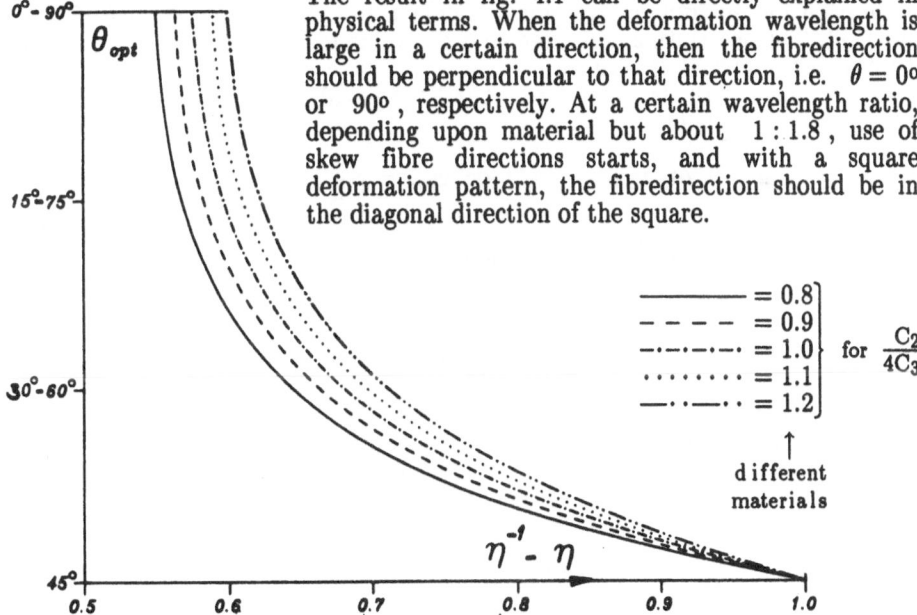

Fig. 4.1: Optimal orientation as a function of the mode parameter η for plate bending.

From an engineering point of view the main conclusions to be drawn from fig. 4.1 are:

· The optimal orientation depends mainly on the mode parameter η. Thus if the deformation pattern is known, the optimal fibredirection can be estimated directly.

· Cases of inverse mode parameters $\eta_1 = \eta_2^{-1}$ have complementary solutions $\theta_2 = \pi/2 - \theta_1$.

· For "extreme" values of η the optimal orientation is perpendicular to the long wavelength.

· The change of optimal fibredirection to a skew direction is very sensitive to the mode parameter.

· The optimal orientation is rather insensitive to the material parameters.

· The optimal orientation is independent of the position of the ply in the laminate, and thus the same for all plies.

· Local optima exist.

By simple numerical analysis problems with combined modes, as exemplified by adding terms (4.9), can also be optimized; PEDERSEN [9] shows such solutions.

We shall now move on to the disc problem and specifically treat optimal orientation for **maximum extensional stiffness**. For details see PEDERSEN [10], and [11]. The description is limited to **orthotropic material**, i.e. $C_{13} = C_{23} = 0$ in the principal material directions.

In agreement with the general result (4.7) the analytical results are very familiar with our plate result (4.13). We get

$$\psi_{\text{opt}} = 0 \ \text{ or } \ \psi_{\text{opt}} = \pm\, \pi/2 \ \text{ or } \ \psi_{\text{opt}} = \pm \tfrac{1}{2}\arccos(-\gamma)$$

(4.14)

$$\gamma := \frac{C_2}{4C_3}\frac{1+\eta}{1-\eta} = \frac{(C_{11}-C_{22})}{\left(C_{11}+C_{22}-2(C_{12}+2C_{33})\right)}\frac{1+\epsilon_{II}/\epsilon_I}{1-\epsilon_{II}/\epsilon_I}$$

where ψ is the angle from the larger principal strain direction $\left(|\epsilon_I| > |\epsilon_{II}|\right)$ to the larger modulus direction $(C_2 > 0$, i.e. $C_{11} > C_{22})$. The strain ratio is $\eta = \epsilon_{II}/\epsilon_I$. The fact that $|\eta| < 1$ and $C_2 > 0$ means that the **optimization parameter** γ has the same sign as the material parameter C_3. (Low shear modulus for $C_3 > 0$ and high shear modulus for $C_3 < 0$).

The second order derivate of the energy density is

$$\left[\frac{\partial^2 u_e}{\partial \psi_e^2}\right]_{\substack{\text{fixed} \\ \text{strains}}} = -8C_3(\epsilon_I-\epsilon_{II})^2\left[\cos2\psi(\gamma+\cos2\psi) - \sin^2 2\psi\right]$$

(4.15)

and with this we obtain the results given in Table 4.1 for solutions of global maximum or global minimum. For **non–orthotropic** materials, analytical solutions are difficult and extremum are found numerically, say with Newton–Raphson iterations. A more extended analysis than can be shown here, cf. POULSEN [12] offers information about appropriate starting points for such iterations.

Angle ψ of station- arity	Low shear modulus material $C_3 > 0$		High shear modulus material $C_3 < 0$	
	$0<\gamma<1$	$\gamma<1$	$\gamma<-1$	$-1<\gamma<0$
0°	Global min	Global min	Global min	Local max
$\pm 90^\circ$	Local min	Global max	Global max	Global max
$\cos 2\psi=-\gamma$	Global max			Global min

Table 4.1: Table for selection of optimality criterion wrt. global minimum or global maximum of total strain energy.

The numerical procedure for solving a specific problem is, briefly stated, as follows:

1) For a given design a finite element analysis gives the actual strain field, i.e. the principal strains with the ϵ_I direction in each element.

2) For each element the optimization parameter γ_e is evaluated by (4.14).

3) Based on Table 4.1 the new material angle (relative to the ϵ_I direction) is determined.

4) If actual changes are not within a given convergence criterion, return to 1) for a new analysis.

The total number of necessary iterations is normally about 5–10 , provided that an extreme convergence criterion is not specified. Convergence in terms of total energy is much faster than in terms of design variables.

It is shown in PEDERSEN [11] that for the optimal design the principal strain directions are aligned with the principal stress directions. Numerically it is found more efficient to redefine the material angle relative to the larger principal stress direction as an alternative to the larger principal strain direction.

As our first example we analyse the **short cantilever** in fig. 4.2, where the results for **uniform material orientation** throughout the model is shown. Two different materials are used as illustrative examples. If we extend the optimization with the possibility of local orientation in each of the 72 finite elements we can for the low shear modulus material further minimize the energy with a factor 0.51 or maximize with a factor 1.37 . For the high shear modulus material the corresponding numbers are 0.86 and 1.53 .

666

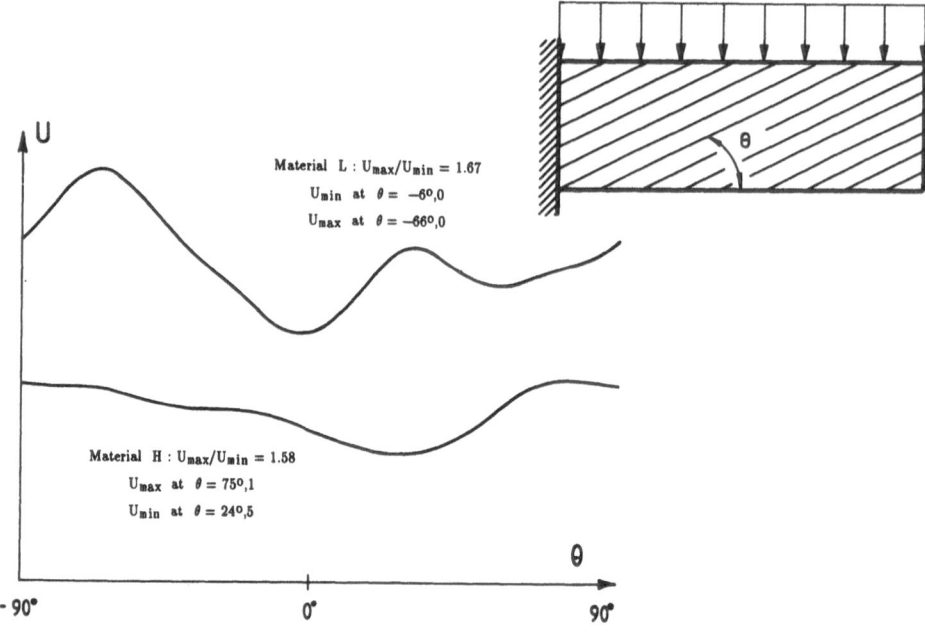

Fig. 4.2: Elastic energy for different material orientations. Upper curve for a low shear modulus material $(C_{11} = 8 ; C_{22} = 4 ; C_{12} = 1 ; C_{33} = 0.5)$ and lower curve for a high shear modulus material $(C_{11} = 8 ; C_{22} = 4 ; C_{12} = 3 ; C_{33} = 3.5)$.

As always, we **optimize the model**, and detailed models should also be analysed. Two such solutions are now shown. In fig. 4.3 we show the results of a **cantilever example** and in fig. 4.4 the results for a **bending loaded knee** example. For the 720 design parameters (cantilever), the stiffness is improved by a factor 1.8 , and for the 1408 design parameters (knee), by a factor 1.6 .

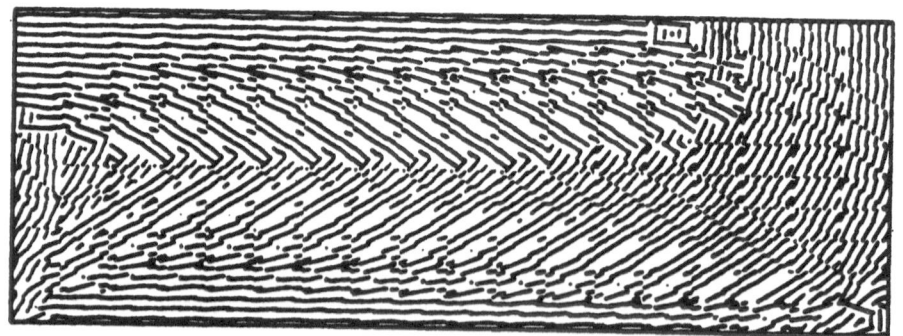

Fig. 4.3: Optimal orientation field for a uniformly loaded cantilever, based on a 720 element model. Relative material parameters: $(C_{11} = 80 ; C_{22} = 40 ; C_{12} = 10 ; C_{33} = 5)$.

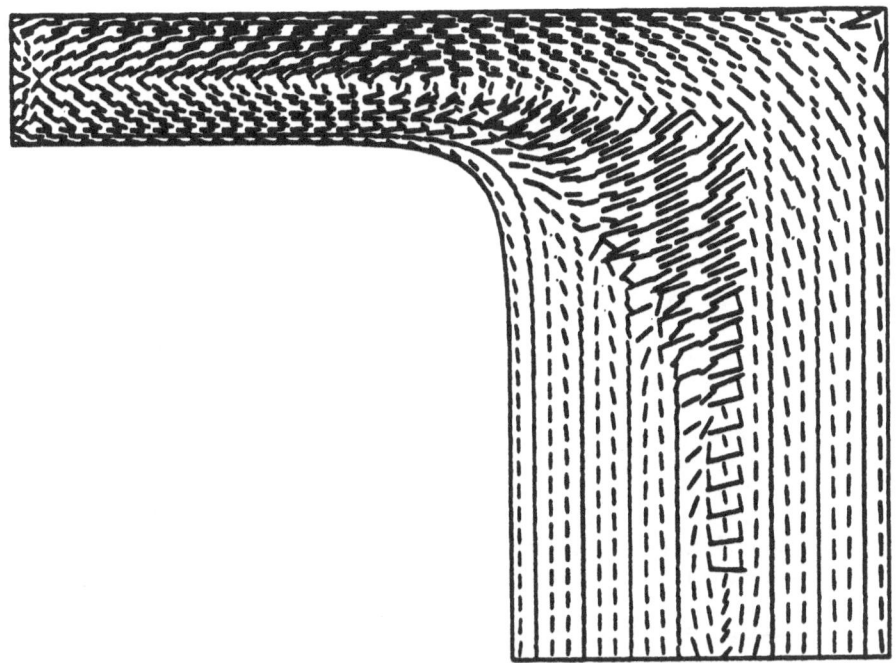

Fig. 4.4: Optimal orientation field for a bending loaded knee, based on a 1408 element model. Relative material parameters: $(C_{11} = 400 ; C_{22} = 100 ; C_{12} = 30 ; C_{33} = 75)$.

5. Thickness Distribution and Optimality Criterion

After the unconstrained optimization problem of orientation we now treat the also simple class of optimization problems, characterized by only a single constraint.

The actual constraint is a given volume $\bar{V}$, i.e.

$$V - \bar{V} = \sum_e V_e - \bar{V} = \sum_e a_e t_e - \bar{V} = 0$$

$$\frac{\partial(V - \bar{V})}{\partial t_e} = \frac{\partial V}{\partial t_e} = \frac{\partial V_e}{\partial t_e} = a_e = \frac{V_e}{t_e}$$

(5.1)

The domain volume, area and thickness are V_e, a_e and t_e, respectively.

The **necessary optimality criterion** with only a single constraint is **proportional gradients** for all design variables. When the objective is to minimize strain energy U by thickness design this criterion gives

$$\frac{\partial U}{\partial t_e} = A \frac{\partial V}{\partial t_e} \quad \text{for all } e$$

(5.2)

where A is the common factor of proportionality. From the general sensitivity results, say (3.27), we have

$$\frac{\partial U}{\partial t_e} = -\left[\frac{\partial U_e}{\partial t_e}\right]_{\substack{\text{fixed} \\ \text{strain}}} = -\frac{u_e V_e}{t_e} \tag{5.3}$$

and inserting (5.1) and (5.3) in (5.2) we get directly the well–known criterion of **constant energy density**, equal to the mean energy density $\bar{u}$

$$u_e = \bar{u} \quad \text{for all } e \tag{5.4}$$

See also the paper by MASUR [5] for this optimality criterion.

Solutions which satisfy (5.4) may correspond to maximum or minimum or just stationarity. Furthermore, the extremum may be local or global. Lastly, the existence of a design satisfying (5.4) is not proven, and a procedure for obtaining such a possible solution still has to be described.

How is a thickness distribution that fulfils (5.4) obtained? We shall discuss a practical procedure, cf. ROZVANY [13], which is based on a number of approximations. Firstly, we neglect the mutual influences from element to element, i.e. each element is redesigned independently (but simultaneously)

$$(t_e)_{\text{next}} = t_e + (\Delta t_e) \tag{5.5}$$

Secondly, the optimal mean energy density $\bar{u}$ is taken as the present mean energy density $\tilde{u}$. Thirdly, the element energy U_e is assumed to be constant through the change Δt_e and then, from (5.4), we get

$$\frac{U_e}{V_e(1 + \Delta t_e/t_e)} = \tilde{u} \quad , \text{i.e.}$$

$$\Delta t_e = t_e(u_e - \tilde{u})/\tilde{u} \quad \text{or} \quad (t_e)_{\text{next}} = t_e\, u_e/\tilde{u} \tag{5.6}$$

A relaxation power, say 0.8, is normally introduced in the formulation. It is natural to ask why the gradient of element energy is not taken into account

$$(U_e)_{\text{next}} = U_e + \frac{\partial U_e}{\partial t_e}\Delta t_e \tag{5.7}$$

but this is explained by the fact that although $\partial U/\partial t_e$ is known from (5.3), the gradient of the local energy (the element strain energy)

$$\frac{\partial U_e}{\partial t_e} = \left[\frac{\partial U_e}{\partial t_e}\right]_{\text{fixed strain}} + \left[\frac{\partial U_e}{\partial \epsilon}\right]\frac{\partial \epsilon}{\partial t_e} \tag{5.8}$$

is more difficult to determine. The two terms in (5.8) have different signs, and the other neglected terms $\partial U_e/\partial t_i$ for $e \neq i$ may also be of the same order.

Although the procedure (5.6) mostly works rather satisfactorily, we shall now extend our analysis to the coupled problem. The redesign procedure provided by

(5.6) neglects the mutual sensitivities, i.e. the change in element energy due to change in the thickness of the other elements. These sensitivities can be calculated by classical sensitivity analysis. Assume that the analysis is related to a finite element model

$$[S]\{D\} = \{A\} \tag{5.9}$$

where $\{A\}$ are the given nodal actions, $\{D\}$ the resulting nodal displacements and $[S] = \sum_e [S_e]$ the system stiffness matrix accumulated over the element stiffness matrices $[S_e]$ for $e = 1,2,...,N$.

Let t_e be without influence on $\{A\}$, we then get

$$[S] \frac{\partial \{D\}}{\partial t_e} = - \frac{\partial [S]}{\partial t_e} \{D\} = \{P_e\} \tag{5.10}$$

where the right–hand side $\{P_e\}$ is a pseudo load, equivalent to thickness design change. Knowing $\partial\{D\}/\partial t_e$ it is straightforward to calculate $\partial U_i/\partial t_e$. Generally, the computational effort involved corresponds to one additional load for each design parameter.

Then, with all the gradients $\partial U_i/\partial t_e$ available, we can formulate a procedure for simultaneous redesign of all element thicknesses

$$\{t\}_{next} = \{t\} + \{\Delta t\} \tag{5.11}$$

that takes the mutual sensitivities into account. In agreement with the optimality criterion (5.4) we change towards equal energy density $\tilde{u}$ in all elements. Formulated in terms of strain energy per area, we want

$$u_e t_e + \sum_{i=1}^{N} \frac{\partial (u_e t_e)}{\partial t_i} \Delta t_i = \tilde{u}(t_e + \Delta t_e) \quad \text{for } e=1,2,... \tag{5.12}$$

or in matrix notation

$$\{ut\} + [\nabla(ut)]\{\Delta t\} = \tilde{u}\left[\{t\} + \{\Delta t\}\right] \tag{5.13}$$

with solution

$$\{\Delta t\} = \left[[\nabla(ut)] - \tilde{u}[I] \right]^{-1} \left\{ (\tilde{u}-u)t \right\} \tag{5.14}$$

The gradient matrix $[\nabla(ut)]$ consists of the quantities $\partial(u_e t_e)/\partial t_i$. Note that with the assumption of fixed element energy, the strain energy per area is unchanged, i.e. $[\nabla(ut)] = [0]$ and we get the simple redesign formula (5.6).

An alternative formulation would be Newton–Raphson iterations directly on energy densities

$$(u_e - \tilde{u}) + \sum_{i=1}^{N} \frac{\partial(u_e - \tilde{u})}{\partial t_i} \Delta t_i = 0 \quad \text{for} \quad e=1,2,\ldots \tag{5.15}$$

or, in matrix notation,

$$[\nabla u]\{\Delta t\} = \tilde{u}\{1\} - \{u\} \tag{5.16}$$

Here, the gradient matrix $[\nabla u]$ constitutes $\partial u_e / \partial t_i$. An interesting formulation is obtained when we multiply every row e with area a_e and get

$$[\nabla(ua)]\{\Delta t\} = \{(\tilde{u} - u)a\} \tag{5.17}$$

The present matrix is now symmetric, which, to the author's knowledge, is not well known. Remembering that $u_e a_e = U_e / t_e$, we prove this directly from (5.3)

$$\frac{\partial^2 U}{\partial t_e \partial t_i} = -\frac{\partial(U_e / t_e)}{\partial t_i}$$

$$\frac{\partial^2 U}{\partial t_i \partial t_e} = -\frac{\partial(U_i / t_i)}{\partial t_e} \tag{5.18}$$

Therefore, as $\partial^2 U / (\partial t_e \partial t_i) = \partial^2 U / (\partial t_i \partial t_e)$, we have

$$[\nabla(ua)]^T = [\nabla(ua)] \tag{5.19}$$

The two specific examples optimized with respect to material orientation shall now be optimized, with respect to only thickness distribution, and then simultaneously with respect to thickness and orientation. These examples are taken from PEDERSEN [14]. In all the figures we have used the same way of visualizing of the results. The **design is characterized** by thickness and orientation, which are shown by hatching the triangular finite elements in the direction of the larger modulus direction and with the hatch density proportional to the thickness. Dark areas are therefore areas with large thicknesses. The **response is characterized** by the distribution of the strain energy density and by the direction of the larger principal stress. The technique of hatching is again used with hatch direction equal to principal stress direction and with hatch density proportional to strain energy density. Dark areas are therefore areas with energy concentration.

The **cantilever example** shown in fig. 5.1 is based on a 720 element model, with constant thickness and orientation in each element. For the uniform cantilever the mean and the maximum values of energy density are 787, 32919 (relative measures of stiffness and stress concentration). Only thickness optimization gives 414, 450, which means stiffness improved by a factor 1.9 and almost no stress concentration. With simultaneous thickness, orientational optimization we obtain 181, 199, i.e. a stiffness improved by a factor of 4.3.

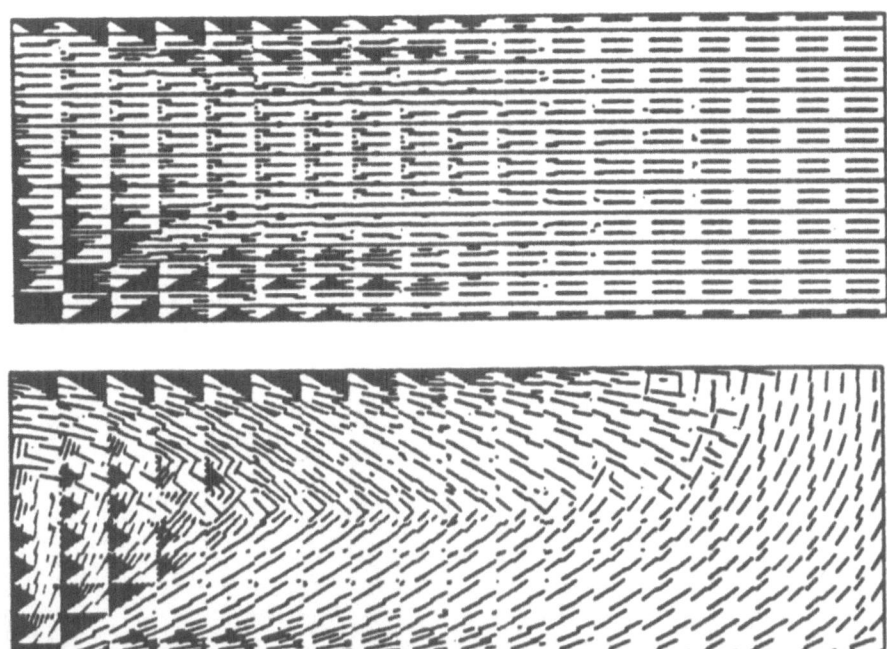

Fig. 5.1: Above: the optimal thickness distribution for the uniformly loaded cantilever, which in fig. 4.3 is shown with orientational optimization alone. Below: the combined thickness and orientational optimization.

The results for the **bending loaded knee** example are illustrated in fig. 5.2 and again we conclude that both thickness and orientational optimization need to be performed. The specific values for the uniform model are u_{mean}, $u_{max} = 726, 5002$; optimized only for thickness distribution we get 287 , 328 and simultaneous thickness; orientational optimization gives 151 , 161 . Totally, we can thus gain a factor of 4.8 for stiffness and almost eliminate stress concentration.

672

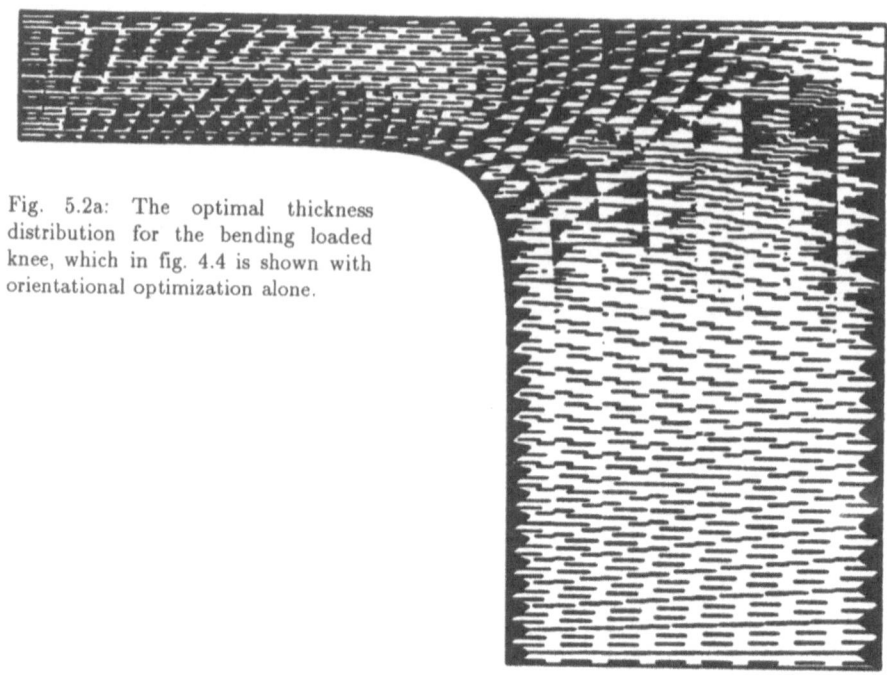

Fig. 5.2a: The optimal thickness
distribution for the bending loaded
knee, which in fig. 4.4 is shown with
orientational optimization alone.

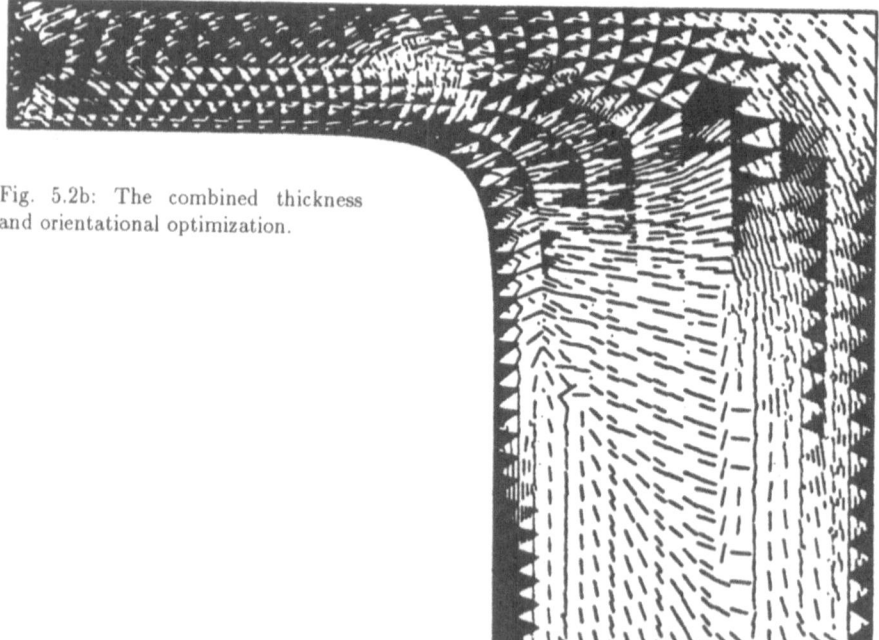

Fig. 5.2b: The combined thickness
and orientational optimization.

6. Detailed Material Design

The orientational design in section four and the thickness design in section five, in fact show how to use advanced materials. The detailed **design of the constitutive matrix** is also a very active research area, as documented by the works of LURIE, FEDOROV & CHEKAEV [15], KOHN [16], BENDSØE [17], DIAZ & BENDSØE [18], among others.

We shall here focus only on the objective of **maximize stiffness** (minimize strain energy) based on **orthotropic materials** classified as **low shear modulus** materials. In reality we shall go directly to the material model from the above–mentioned references, as illustrated in fig. 6.1.

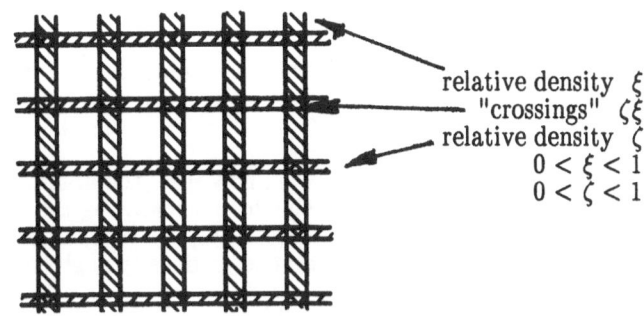

relative density ξ
"crossings" $\zeta\xi$
relative density ζ
$0 < \xi < 1$
$0 < \zeta < 1$

Fig. 6.1: A material model with two directional densities.

The total, relative volume densities ρ_e give the total volume V by

$$V = \sum_e V_e = \sum_e \tilde{V}_e \, \rho_e$$

$$\frac{\partial V}{\partial \rho_e} = \frac{\partial V_e}{\partial \rho_e} = \tilde{V}_e = \frac{V_e}{\rho_e}$$

(6.1)

where $\tilde{V}_e$ is the maximum volume of domain e. In principle, ρ_e acts like the thickness in section five. Often we shall omit the domain index e, and the relations of the directional relative densities ζ, ξ are

$$\rho = \zeta + \xi - \zeta\xi$$

$$\frac{\partial \rho}{\partial \zeta} = 1 - \xi \; ; \; \frac{\partial \rho}{\partial \xi} = 1 - \zeta$$

(6.2)

From BENDSØE [17] we take the orthotropic constitutive matrix, expressed in known modulus E and Poissons ratio ν and in the directional parameters ζ, ξ, as

$$[C] = \begin{bmatrix} [\tilde{C}] & ; & 0 \\ & & 0 \\ 0;0 & ; & C_{33} \end{bmatrix}$$

$$[\tilde{C}] = \frac{E\xi}{\alpha-\beta^2} \begin{bmatrix} 1 & ; & \beta \\ \beta & ; & \alpha \end{bmatrix} \quad ; \quad [\tilde{C}]^{-1} = \frac{1}{E\xi} \begin{bmatrix} \alpha & ; & -\beta \\ -\beta & ; & 1 \end{bmatrix}$$

$$\alpha = \xi^2 + \xi(1-\xi)/\zeta \quad ; \quad \beta = \xi\nu$$

$$(6.3)$$

and we assume the major principal axis chosen by

$$1 > \alpha \implies \zeta > \xi/(1+\xi) \tag{6.4}$$

The conditions of positive definite and limited $[\tilde{C}]$ are satisfied for

$$0 < \zeta < 1 \quad ; \quad 0 < \xi < 1 \quad ; \quad -1 < \nu < 1 \tag{6.5}$$

With the assumptions given, we know from the orientational results that the larger principal modulus direction should optimally be the same as the direction of the numerically larger strain ϵ_I $(|\epsilon_I| > |\epsilon_{II}|)$. Therefore with strain ratio $-1 < \eta := \epsilon_{II}/\epsilon_I < 1$ the strain energy density is given by

$$u = \frac{1}{2} \epsilon_I \{1;\eta\} [\tilde{C}] \begin{Bmatrix} 1 \\ \eta \end{Bmatrix} \epsilon_I \quad \text{or}$$

$$u = \frac{1}{2} \epsilon_I^2 \frac{E\xi}{(\alpha-\beta^2)} (1 + 2\beta\eta + \alpha\eta^2)$$

$$(6.6)$$

We know that principal stresses are generally also in the same direction, and for the present case it is proved that the numerically larger principal stress σ_I $(|\sigma_I| > |\sigma_{II}|)$ match with ϵ_I. Thus expressed in stresses and stress ratio $-1 < \mu := \sigma_{II}/\sigma_I < 1$ we have from (6.3)

$$u = u^C = \frac{1}{2} \sigma_I^2 \frac{1}{E\xi} (\alpha - 2\beta\mu + \mu^2) \tag{6.7}$$

This is the results from the orientational optimization. Next, in the same way as in the thickness optimization, we have

$$U = \sum_e U_e = \sum u_e \tilde{V}_e \rho_e$$

$$\frac{\partial U}{\partial \rho_e} = -\left[\frac{\partial U_e}{\partial \rho_e}\right]_{\substack{\text{fixed} \\ \text{strains}}} = -\frac{u_e V_e}{\rho_e}$$

$$(6.8)$$

and thus, with (6.1), again the result of uniform energy density equal to mean energy density $\bar{u}$

$$u_e = \bar{u} \tag{6.9}$$

Solution by optimality criterion iterations can thus give us ρ_e for $e = 1,2,...$

Lastly, we come to the detailed design. Knowing ρ_e, then how to optimize ζ_e, ξ_e?

We again have a single constraint problem (omitting index e)

$$\begin{array}{l} \text{Minimize } U = u\tilde{V}\rho \\ \text{over } \zeta, \xi \\[2mm] \text{with constraint } \rho = \zeta + \xi - \zeta\xi \end{array} \tag{6.10}$$

Derivatives with respect to the constraint are stated in (6.2) and derivatives with respect to u can be found from (6.6) to be:

$$\partial u/\partial\zeta = (1-\xi)(1+\xi\nu\eta)^2/N$$

$$N = \left(1 - \xi + \zeta\xi(1-\nu^2)\right)^2 \tag{6.11}$$

$$\partial u/\partial\xi = \left[\zeta\left(1 - \zeta(1-\nu^2)\right) + 2\zeta\nu\eta\right.$$
$$\left. + \eta^2\left[(1-\xi)^2 + \zeta\xi\left(2(1-\xi) + \xi\nu^2\right) + \zeta^2\xi^2(1-\nu^2)\right]\right]\Big/N$$

Then, by means of (6.11) and (6.2), the optimality criterion of constant ratios between objective and constraint gradients

$$\frac{\partial u/\partial\zeta}{\partial\rho/\partial\zeta} = \frac{\partial u/\partial\xi}{\partial\rho/\partial\xi} \tag{6.12}$$

gives the equation of optimality

$$(1-\zeta)(1+\xi\nu\eta)^2 = \zeta\left(1 - \zeta(1-\nu^2)\right) + 2\zeta\nu\eta$$
$$+ \eta^2\left[\left((1-\xi)^2 + \zeta\xi\left(2(1-\xi) + \xi\nu^2\right) + \zeta^2\xi^2(1-\nu^2)\right)\right] \tag{6.13}$$

to be solved together with (6.2) and (6.4)–(6.5).

In the stress formulation (6.7) the results are even more simple, and we get the final results after little algebra:

$$\zeta = \frac{\rho}{1+|\mu|} \quad ; \quad \xi = \frac{|\mu|\rho}{1+|\mu|-\rho} \tag{6.14}$$

in terms of the total relative density ρ and stress ratio $\mu := \sigma_{II}/\sigma_I$.

In the paper by THOMSEN [19] concurrent density and orientational design is solved with constraints on a cost defined functional.

7. Shape Design With Orthotropic Material

In this final section of the report we shall describe briefly the procedures for optimal shape design with anisotropic materials. In general we shall use the same methods, discussed in PEDERSEN [20], as for isotropic materials i.e. methods based on semi–analytical sensitivity analysis and linear programming.

This approach is very much in contrast to the other sections of the report. The overall problem formulation might be

Minimize Maximum Stress (energy)

by boundary design $\qquad$ (7.1)

possibly subject to geometry constraints

We immediately note two major differences compared with the problem previously treated. Firstly, the **objective is a local quantity**, so the sensitivity analysis is much more complicated. Secondly, the **design variables** are most effectively chosen **as global quantities**. In total, this points towards using a more numerically based approach. The fact that optimal shapes will often be uniformly stressed (uniform energy) should not be misunderstood. It is **not an optimality criterion**, but results when geometry constraints or multiple load cases do not avoid it.

We will first report the results of a recent M.Sc. thesis by TOBIESEN & JENSEN [21] and will then also subject these solutions to thickness and orientational optimization. The biaxially loaded hole for which an analytical solution exists with isotropic materials KRISTENSEN & MADSEN [22] will also be the subject of our study here.

Fig. 7.1: A course finite element model with pointers on the actual design elements, i.e. the elements influenced directly by boundary design.

With isotropic material the relevant reference stress in (7.1) will often be the von Mises stress or the maximum tensile stress. With anisotropy and perhaps laminated materials, a number of different directionally dependent reference stresses may be argued. We shall not take part in this discussion but simply **choose the energy density** u as the quantity to be minimized at the point(s) where it is maximum, which are mostly at the design boundary.

In fig. 7.1 we show a finite element model which by arguments of **symmetry** models the biaxially loaded hole. Analysis is based on **linear** displacement triangles (LDTR elements) or on **quadratic** displacement triangles (QDTR elements). The notion of **design elements** relates to the sensitivity analysis, but firstly we shall discuss the description of the hole boundary.

We choose a description with a super elliptic shape as the "basic" design with half axes a , b and power n . To this shape are accumulated a number of **orthogonal functions** that satisfy the required end–conditions. With the amplitudes of these functions c_i as design parameters in addition to a , b , n we have **maximum 8 design parameters**, and often 5 is enough. The description

$$c = c_0 + \sum_{i=1}^{5} c_i f_i(s) \; ; \; c_0 : (x_1/a)^n + (x_2/b)^n = 1$$

(7.2)

is illustrated in fig. 7.2.

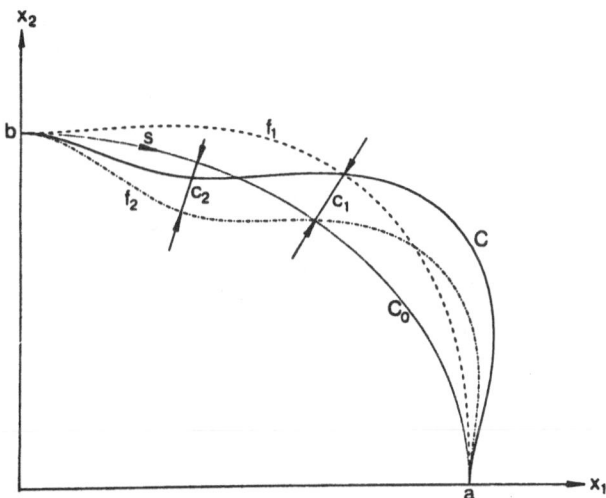

Fig. 7.2: Description of the shape of design, illustrated here with only two modification functions f_1 and f_2 .

Let us then see the resulting optimal shapes for increasing degree of orthotropy. The results in fig. 7.3 are taken directly from [21], and are based on the problem statement of

Minimize, Maximum energy density

for fixed model volume

(7.3)

with the load ratio being $3/2$ in x_1/x_2 directions respectively, in agreement with the shown isotropic solution.

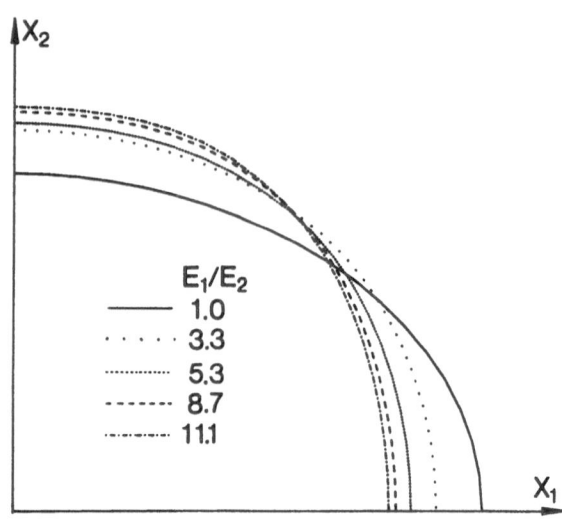

Fig. 7.3: Optimal boundary shapes for different "degrees" of orthotropy.

In TOBIESEN & JENSEN [21] the optimizations are carried through with increasing number of elements and results with LDTR elements are compared to results with QDTR elements. The conclusion is that **all models give satisfactory results**. Naturally, the QDTR modelling give more accurate results for the same number of elements. However, with CPU time as reference, the LDTR elements will be most favorable.

A few comments on the sensitivity analysis should be given. Let u_{cr} be a **critical energy density**, then from (2.3) we have at the actual point

$$\frac{du_{cr}}{dh} = \{\epsilon\}^T[C]\frac{d\{\epsilon\}}{dh} = \{\sigma\}^T[C]^{-1}\frac{d\{\sigma\}}{dh} \qquad (7.4)$$

with h being a, b, n, or c_i of description (7.2). From $d\{D\}/dh$ as by (5.10) we can easily determine $d\{\epsilon\}/dh$ or $d\{\sigma\}/dh$. It should be noted that the pseudo load which gives $d\{D\}/dh$ is calculated only from the change in the element stiffness matrices of the design elements shown in fig. 7.1. This should not be confused by the fact that after a redesign step the total model is changed, also outside the design element, but this is not related to the sensitivity analysis. New models are obtained by **automatic mesh design**, directly from the few design parameters.

After these pure shape design examples, we shall concurrently design also the thickness distribution and the orientational field. The classical example from [22] based on isotropic material is in fig. 7.4 shown with a thickness optimization, following the shape optimization. Without thickness optimization the (stiffness, stress concentration) is given relatively by $(u_{mean}, u_{max}) = (856, 3093)$ and after thickness optimization we get $(u_{mean}, u_{max}) = (789, 823)$. Almost uniform energy density is thus obtained, and a concurrent shape/thickness design is not expected to change these results.

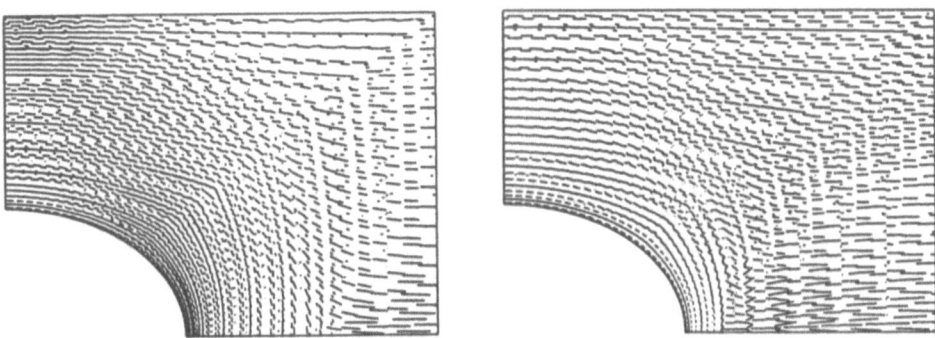

Fig. 7.4: Stress distributions for model with 1024 isotropic elements. To the left only shape design and then to the right shape/thickness design.

Finally, one of the examples from fig. 7.2 is optimized also for thickness distribution and finally for thickness as well as orientational orientation. This result is illustrated in fig. 7.5, and the relative numbers for stiffness and energy concentration are

only shape design:
$$(u_{mean}, u_{max}) = (3209, 12370)$$

shape followed by thickness design:
$$(u_{mean}, u_{max}) = (2900, 3320)$$

shape followed by thickness/orientational design:
$$(u_{mean}, u_{max}) = (1299, 1436)$$

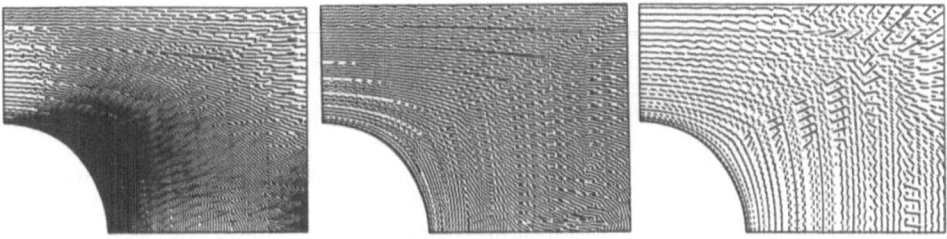

Fig. 7.5: Stress distributions for model with 1024 orthotropic elements $(C_{11}, C_{12}, C_{22}, C_{33}) = (108., 3.7, 12.4, 4.7)$. From left only shape design, then shape/thickness design and finally shape/thickness/orientational design.

8. Conclusion

In this tutorial introduction to research results from optimal design of and with anisotropic material, we have tried to focus on the more general aspects.

Our goal is to design concurrently structural shape, pointwise density or thickness, material orientation, and the detailed constitutive behaviour.

However, certain aspects, such as material orientation and material density, are given by optimality criteria that hold generally. Thus, advantageous decouplings are possible and should be utilized.

The description is restricted to two–dimensional problems, which to a large extent are now clarified. Interesting extensions to three–dimensional problems are the subject of active research, and connections to topology design are being pointed out.

It would be fair to say that the use of advanced materials has put optimal design into a higher level of application.

Lastly, once more, the present paper is a rather subjective point of view and the available time did not allow for appropriate reference to other research groups. Sorry.

References

[1] Lekhnitskii, S.G.: Theory of Elasticity of an Anisotropic Body, Mir Publishers, Moscow, 1981, p. 430.

[2] Jones, R.M.: Mechanics of Composite Materials, 2, McGraw–Hill, New York, N.Y., 1975, p. 355.

[3] Pedersen, P.: Combining Material and Element Rotation in One Formula, Comm. in Appl. Num. Meth., Vol. 6, 549–555, 1990.

[4] Wittrick, W.H.: Rates of Change of Eigenvalues, with Reference to Buckling and Vibration Problems, J.R. Aeronaut. Soc., Vol. 66, 590–591, 1962.

[5] Masur, E.F.: Optimum Stiffness and Strength of Elastic Structures, J. of the Engineering Mechanics Div., ASCE, EM5, 621–649, 1970.

[6] Pedersen, P.: On Sensitivity Analysis and Optimal Design of Specially Orthotropic Laminates, Eng. Opt., Vol. 11, 305–316, 1987.

[7] Bert, C.W.: Optimal Design of a Composite–Material Plate to Maximize its Fundamental Frequency, J. Sound and Vibration, Vol. 50, 229–239, 1977.

[8] Muc, A.: Optimal Fibre Orientation for Simply–Supported, Angle–Ply Plates under Biaxial Compression, Composite Structures, Vol. 9, 161–172, 1988.

[9] Pedersen, P.: Minimum Flexibility of Non–Harmonic Loaded Laminated Plates, In "Mechanical Characterization of Fibre Composite Materials" (ed. R. Pyrz), AUC, Denmark, 182–196, 1986.

[10] Pedersen, P.: On Optimal Orientation of Orthotropic Materials, Structural Optimization, Vol. 1, 101–106, 1989.

[11] Pedersen, P.: Bounds on Elastic Energy in Solids of Orthotropic Materials, Structural Optimization, Vol. 2, 55–63, 1990.

[12] Poulsen, H.H.: Optimal Material–Orientation, Thesis for M.Sc., Solid Mechanics, DTH, 1990 (in Danish).

[13] Rozvany, G.I.N.: Structural Design via Optimality Criteria, Kluwer, 1989, 463 p.

[14] Pedersen, P.: On Thickness and Orientational Design with Orthotropic Materials, Structural Optimization, to appear.

[15] Lurie, K.A., Fedorov, A.V. and Chekaev, A.V.: Regularization of Optimal Design Problems for Bars and Plates, J. Optim. Theory Appl., Vol. 37(4), 499–513, 1982.

[16] Kohn, R.: Composite Materials and Structural Optimization, to appear in Proc. Workshop on Smart/Intelligent Materials, Honolulu, Technomic Press, 1990.

[17] Bendsøe, M.P.: Optimal Topology Design and Homogenization, to appear in "Mechanics, Numerical Modelling and Dynamics of Materials" (ed. Blanc, Raous, Suquet), CNRS, 1991.

[18] Diaz, A.R. & Bendsøe, M.P.: Shape Optimization of Multipurpose Structures by a Homogenization Method, Report, Mech. Eng., Michigan State Univ., East Lansing, 1990.

[19] Thomsen, Jan: Optimization of Composite Discs, Structural Optimization, to appear soon.

[20] Pedersen, P.: Design for Minimum Stress Concentration – Some Practical Aspects, in "Structural Optimization" (eds. Rozvany & Karihaloo), Kluwer Academic Publishers, 225–232, 1988.

[21] Tobiesen, L. & Jensen, S.H.: Optimal Hulfacon i Ortotrop Skive (Optimal Hole Shape in Orthotropic Plate), Thesis for M.Sc., Solid Mechanics, DTH, 1990 (in Danish).

[22] Kristensen, E.S. & Madsen, N.F.: On the Optimum Shape of Fillets in Plates Subjected to Multiple In–Plane Loading Cases, Int. J. Num. Meth. Engn., Vol. 10, 1007–1019, 1976.

ANALYSIS AND DESIGN OF ELASTIC/SOFTENING TRUSS STRUCTURES BASED ON A MIXED-FORM EXTREMUM PRINCIPLE

J. E. Taylor*
University of Michigan
Ann Arbor, MI 48109
USA

and

Janos Logo**
Technical University of Budapest
H-1521 Budapest, Muegyetem rkp.3
HUNGARY

ABSTRACT   An extremum principle is presented for structural analysis problems expressed in mixed (stress and deformation) form. The structural model, exemplified for trussed structures, has each member composed of elements coupled in parallel. This type of formulation is convenient for the analysis of systems made up at least partially of materials with stress-limited capacity; element-wise elastic/softening constitutive properties are incorporated as an example. Certain classical extremum principles are recovered as special cases of the general mixed formulation. Also, the structural optimization problem of 'design for maximum strength' is formulated as a direct extension of the extremum principle for analysis. The effect on design of material degradation (softening) is observed through a set of numerical results for optimal truss configurations.

INTRODUCTION

A method is presented for the analysis and computational treatment of structures made up of an elastic/softening material. The non-simple material is modelled conveniently through application of a new form of mixed variational principle for the stress-deformation analysis [1]. This formulation for the mechanics comprises an extremum principle, in contrast to the familiar saddlepoint versions of mixed variational formulation. The elastic/softening material property is represented in a way that can be interpreted as a compound material, one where local elastic and yielding elements operate in parallel. With the characterization of the (local) material properties expressed implicitly in the (global) variational principle, the complexities often encountered with analysis involving materials with nonlinear and possibly nonsmooth properties is largely circumvented. Similar advantages were reported earlier [2,3] for a model of rate-independent elastoplasticity under proportional loading, and subsequently for contact problems [4] with stress-limited material properties.

*Professor, Department of Aerospace Engineering
**Visiting Research Fellow, Aerospace Engineering, University of Michigan

G. I. N. Rozvany (ed.), Optimization of Large Structural Systems, Vol. II, 683–695.
© 1993 *Kluwer Academic Publishers.*

The governing variational principle is stated in the form of a constrained nonlinear programming problem. A listing of the associated necessary conditions forms the basis for an interpretation of the mechanics analysis for this problem. Conventional extremum principles in structural mechanics are recovered as special cases within the general formulation, as a part of this analysis. A formulation for optimal structural design is obtained via a direct extension of the general extremum principle for analysis. Computational solutions for a number of example problems are obtained using a proprietary computer program. Example results are given for the analysis of general elastic/softening structures. Numerical results are presented as well for a set of solutions which depict the strongest/stiffest optimal structural design.

PROBLEM FORMULATION

The governing variational principle is expressed here in a form that is applicable for the treatment of discrete structures. The model provides for a bilinear elastic/softening character for each element of the structural system. Specifically, as it would appear for a trussed structure (1-D elements), the total element force is represented as the sum of a strictly elastic part and a second component that is elastic but with a limit on its magnitude. With symbol f_i representing this latter component, the total element force F_i is given by:

$$F_i = f_i + k_i e_i \tag{1}$$

Here e_i symbolizes element elongation and the constitutive properties of the element are given in part by constant coefficient k_i (stiffness of the elastic part) and the force limit $\bar{f}_i$ in the constraint

$$|f_i| - \bar{f}_i \le 0 \tag{2}$$

that applies to (1). A typical element force vs. elongation relation is pictured in Figure 1.

The variational model for global equilibrium analysis of a structural system made up of elements each with the stress-limited bilinear character given in (1,2) is stated as follows. With element component forces f_i and elongations e_i as independent unknowns and for system loads $\alpha \rho_\gamma$ acting on the structure, load factor α is to be maximized within the requirements (i) that the set F_i equilibrate the loads, and (ii), that the total system (mixed) energy is constrained from above. In other words, (α, f_i, e_i) are to be determined according to the extremum problem:

$$\text{max } \alpha$$
$$\alpha, f_i, e_i$$
$$\textbf{subject to}$$

$$(P) \qquad \sum_{i=1}^{N_m} - B_{\gamma i} (f_i + k_i e_i) + \alpha \rho_\gamma = 0 \qquad \gamma = 1,2,...,N \tag{C1}$$

$$|f_i| - \bar{f}_i \le 0 \qquad\qquad i = 1,2...,N_m \tag{C2}$$

$$\sum_{i=1}^{N_m} \frac{f_i^{\,2}}{2s_i} + \sum_{i=1}^{N_m} \frac{k_i e^2_i}{2} - \bar{\varepsilon} \leq 0 \qquad (C3)$$

The matrix $\mathbf{B}\gamma_i$ and load vector ρ_γ of equilibrium equations (C1), material limits $\bar{f}_i$, and constants k_i, s_i and E constitute data. Note that the *elastic property* s_i of component f_i of the total force first appears in the *energy* constraint (C3). An energy measure with the same structure as that in (C3) appeared in the study on sensitivity by Bendsøe and Sokolowski [5].

The classical *purely elastic* and ('deformation plasticity' version of) elastic/perfectly plastic models for structural analysis are imbedded within problem (P). These properties of the general formulation are verified below. Also the general character of the formulation is observed via an examination of the *necessary conditions* for the problem. In this direction, supposing a standard Lagrangian interpretation of problem (P) with multipliers λ_γ, μ_i and Λ associated with constraints (C1-C3) respectively, these conditions are stated as:

$$-1 + \sum_\gamma \lambda_\gamma \rho_\gamma = 0 \qquad (3)$$

$$-\sum_\gamma \mathbf{B}_{\gamma j}\, \lambda_\gamma + \mu^+_j - \mu^-_j + \Lambda \left(\frac{f_j}{s_j}\right) = 0 \qquad (4)$$

$$-\left(\sum_\gamma \mathbf{B}_{\gamma j}\, \lambda_\gamma\right) + \Lambda\, e_j = 0 \qquad (5)$$

$$\mu^+_j\, (f_j - \bar{f}_j) = 0 \quad ; \quad \mu^-_j\, (-f_j - \bar{f}_j) = 0 \qquad (6)$$
$$j = 1,2,...,N_m$$

$$\lambda_\gamma \left[\sum_{i=1}^{N_m} -\mathbf{B}_{\gamma i}\, (f_i + k_i\, e_i) + \alpha \rho_\gamma \right] = 0 \qquad (7)$$

$$\Lambda \left\{ \sum_{i=1}^{N_m} \left[\frac{f^2_i}{2s_i} + \frac{k_i e^2_i}{2} \right] - \bar{\varepsilon} \right\} = 0 \qquad (8)$$

and multipliers λ_γ, μ^+_j, μ^-_j, and Λ must be non-negative. Also it follows from (6) that $\mu^+ \mu^- = 0$.

Multipliers λ_γ are eliminated between equations (4) and (5) to obtain

$$-e_j + \mu^+_j - \mu^-_j + \frac{f_j}{s_j} = 0 \qquad (9)$$
$$j = 1,2,...,N_m$$

Thus the multipliers μ_j associated with constraints (C2), ie., with the limits $|f_i| \leq \bar{f}_i$, measure the difference between total elongation e_j and the quantity $\frac{f_j}{s_j}$ ($\frac{f_j}{s_j}$ is interpreted as a measure of elastic deformation). Accordingly, for the case $|f_j| - \bar{f}_j < 0$, from (6) the associated $\mu_j^{+,-}$ have value zero whereby (9) dictates compatibility between the component deformation e_j and $\frac{f_j}{s_j}$. On the other hand, for the case $(\mu_j^{+,-} \neq 0) \rightarrow$ $(|f_j| = \bar{f}_j$ the values

$$\mu^+_j - \mu^-_j = e_j - \frac{\bar{f}_j}{s_j} \quad ; \quad \mu^+_j \mu^-_j = 0 \tag{10}$$

$$j = 1,2,...,N_m$$

represent the (expected) measure of *plastic elongation*. [In a sense the $\mu_j^{+,-}$ might more aptly be identified as the kinematic relaxation; it is complementary to the force constraint (C2). Of course the strictly elastic component $k_j e_j$ of total force remains unaffected.]

To continue with the interpretation of necessary conditions (3-8), (5) reflects the relationship between member elongation e_j and (system displacement) measures λ_γ/Λ. (3) comprises a normalization of the set λ_γ weighted by load components ρ_γ. Also the results $\lambda_\gamma \neq 0$ all γ and $\Lambda \neq 0$ follow in general for this system, whereby from (7) and (8) the equilibrium constraint is enforced and the energy constraint is met via equality. It is remarked without elaboration that the set (3-8) together with the original constraints comprise necessary and sufficient conditions for the complete solution of the original analysis problem. Example computational solutions are presented in the next section.

As noted above, certain other 'extremum principle' forms for structural analysis are imbedded within problem (P) [or (P')]. For example, in the case $[\bar{f}_i \rightarrow 0$ all i $]$ $\rightarrow [f_i = 0$ all i $]$, (P) is reduced to the 'deformation only' form:

$$\max \alpha$$
$$\alpha, e_i$$
$$\text{subject to}$$

(P1) $$\sum_{i=1}^{N_m} - B\alpha_i k_i e_i + \alpha\rho_\gamma = 0$$

$$\alpha = 1,2,...,N$$

$$\sum_{i=1}^{N_m} k_i e^2_i/2 - \varepsilon \leq 0$$

As was the case with problem (P), for the solution to (P1) both equilibrium and energy constraints are tight. The result, that among equilibrium fields e_i the strain energy is maximized for the maximum load, is relatively uninteresting.

Problem statement (P) may be expressed alternately in the form:

$$\min \left[\sum_{i=1}^{N_m} \{ (f_i^2/2 \, s_i) + (k_i e_i^2/2) \} \right]$$
$$\alpha, f_i, e_i$$

(P')   **Subject to**

$$\sum_{i=1}^{N_m} - B_{\alpha i} (f_i + k_i e_i) + \alpha \rho_\gamma = 0$$

$$\alpha = 1,2,...,N$$

$$|f_i| - \overline{T}_i \leq 0$$

$$\underline{\alpha} - \alpha \leq 0$$

$$i = 1,2,...,N_m$$

The equivalence between problems (P) and (P') may be verified via a comparison of necessary conditions for the two problems. Form (P') might be judged more convenient for engineering practice, since the element $\underline{\alpha}$ of data in fact represents 'specified load'. A second 'purely elastic' model is realized from (P) if for specified energy bound ε the constraint values $\overline{T}_i$ are sufficiently large, ie. large enough so that constraints (C2) are nowhere active. This results in a simple mixed-form statement for elastic systems, where net member stiffness is given by $k_i + s_i$.

To consider yet another model, the form obtained by reduction from (P) according to $k_i = 0$, all i is

$$\max \alpha$$
$$\alpha, f_i$$
subject to

(P2)
$$\sum_{i=1}^{N_m} - B_{\alpha i} f_i + \alpha \rho_\gamma = 0$$

$$\gamma = 1,2,...,N$$

$$|f_i| - \overline{T}_i \leq 0$$

$$\sum_{i=1}^{N_m} f_i^2/2 \, s_i - E \leq 0$$

This 'force only' formulation models rate-independent elasto-plasticity, as reported in [2,3]. The further reduction from (P2) according to ε sufficiently large with specified (finite) values $\overline{T}_i$, namely

$$\max \alpha$$
$$\alpha, f_i$$
subject to

$$\sum_{i=1}^{N_m} - B_{\gamma i} f_i + \alpha \rho_\gamma = 0$$

$$\gamma = 1,2,...,N$$

$$|f_i| - \overline{T}_i \leq 0$$

recovers a statement of classical limit analysis. [Design in the context of idealized plastic response, a major subject within structural optimization (e.g., Prager [6], Rozvany, et al [7]) is not elaborated in this article.] Finally we note that the reduction from (P2) associated with $\bar{\tau}_i$ sufficiently large for all i and value ε finite is equivalent to the 'minimum complementary energy' principle.

DESIGN

The purpose in this section is to present a formalism that should be applicable for the design of structural systems made up of 'compound' members. This is accomplished via the specific example of a truss of given geometrical layout and specified proportioning of constituent properties in each member (the broader problem which includes the design of constituent properties as well is to be treated elsewhere). Thus design is characterized by a single vector of truss member sizes, say A_i. In order to expose these design variables, the truss equilibrium equations are rewritten as:

$$\sum_{i=1}^{N_m} - B\gamma_i A_i (\tau_i + E_i\varepsilon_i) + \alpha\rho\gamma = 0$$

$$\gamma = 1,2,...,N.$$

Also, the 'strength limit' for the degrading constituent of each member is represented in the form of constraints:

$$\ell_i A_i (\tau_i - \bar{\tau}_i) \le 0$$

$$\ell_i A_i (-\tau_i - \bar{\tau}_i) \le 0$$

$$i = 1,2,...,N_m$$

With these interpretations, the design problem is expressed:

$$\max_{A_i} \left[\max_{\alpha,\varepsilon_i,\tau_i} \alpha \right]$$

subject to

$$\sum_{i=1}^{N_m} - B\gamma_i A_i (\tau_i + E_i\varepsilon_i) + \alpha\rho\gamma = 0$$

$$\ell_i A_i (\tau_i - \bar{\tau}_i) \le 0$$

$$\gamma = 1,2,...,N$$

(D)

$$\ell_i A_i (-\tau_i - \bar{\tau}_i) \le 0$$

$$\sum_{i=1}^{N_m} (A_i \ell_{i/2}) \left[E_i\varepsilon^2_i + \tau^2_i/E_i \right] - \bar{\varepsilon} \le 0$$

$$i = 1,2,...,N_m$$

$$\sum_{i=1}^{N_m} \ell_i A_i - \mathfrak{R} \le 0$$

where the energy constraint is rewritten to be consistent in form, and the last constraint reflects a limit on structural resource. In this version, the quantities $B_{\gamma i}$, ρ_γ, E_i, $\tilde{E}_i$, ℓ_i, $\bar{\tau}_i$, ε and $\Re$ are data.

The optimality condition reflecting stationarity w.r.t. design A_i is (additional multiplier Λ_R associated with the resource constraint appears; all others are the same as those identified with problem P):

$$\sum_{\gamma=1}^{N} - \lambda\gamma\, B_{\gamma j}\, (\tau_j + E_j\epsilon_j) + \mu^+_j \ell_j\, (\tau_j - \bar{\tau}_j) \tag{11}$$
$$+ \mu^-_j\, \ell_j\, (-\tau_j - \bar{\tau}_j) + (\Lambda \ell_j/2)(E_j\epsilon_j^2 + \tau_j^2/\tilde{E}_j) + \Lambda_\Re \ell_j = 0$$

This equation is reduced using the remaining necessary conditions of (D) to the requirement that

$$\tau_j^2/2\tilde{E}_j + E_j\epsilon_j^2/2 \;=\; \Lambda_R/\Lambda - (\bar{\tau}_j/\Lambda)(\mu_j^+ - \mu_j^-) \tag{12}$$

$$\forall_j \in J^* = \{j;\, A_j > 0\}$$

Note that while (11) must be met uniformly, the specific form (12) is required only for members within the index set J^* defined in (12).

Also, for the same subset J^* of the member index set,

$$\mu_j^+ (\tau_j - \bar{\tau}_j) = 0$$
$$\mu_j^- (-\tau_j - \bar{\tau}_j) = 0 \qquad j \in J^*$$

For the special case where $\bar{\tau}_j$ are sufficiently large relative to (specified) ε and $\Re$ so that

$$|\tau_j| < \bar{\tau}_j \quad \forall_j,$$

the optimality condition can be simplified to the form:

$$\epsilon_j^2 = 2\Lambda_\Re/\Lambda(E_j + \tilde{E}_j) \tag{13}$$
$$j \in J^*$$

This corresponds to the (classical) result for strictly elastic optimal design, ie., the requirement that all design elements in the optimal structure are stressed to the same level (see e.g. Prager [8], Masur [9]).

A familiar consequence of (13) is that the optimal structure for linearly elastic response has at most N members, ie., it is statically determinate. Condition (12) for the more general situation representing the counterpart to (13) evidently provides for relaxation of this limitation on the optimal linearly elastic structure. Computational examples given below reflect a range of design results that include examples both for strictly elastic and for stress-limited structures.

COMPUTATIONAL EXAMPLES

As an alternative to treatment of the necessary conditions for their solution, or to the creation of a sensitivity-based model for the computations, the results presented in this section are obtained by application to an off-the-shelf program. This program [10] contains implementation of a number of methods for nonlinear, constrained minimization problems, notably for the SQP, and for the bundle method [11] to serve for nonsmooth problems.

Both the analysis problem (P) and design problem (D) were solved for the nine-bar truss of Figure 2. Bar forces and elongations for a set of energy levels are represented in Figures 3-5. The evolution of response is pictured in the 'load factor α' versus 'energy level' curve of Figure 6. Design results for specified load configuration ρ_γ, constituent stress limit $\bar{\tau}$, and Resource $\Re$, are given in the graphical sketches of Figure 7 for increasing values of energy (bound) $\bar{\varepsilon}$. Thicknesses of member lines in these sketches correspond to member size.

DISCUSSION

The structure predicted from optimal design formulation (D) is both strongest and stiffest among designs admissible within the resource constraint; this follows from the equivalence of problem statements (P) and (P'). Also, the characterization for degrading or softening material given as a part of (P) is readily extended. For example, the model for piecewise linear softening material (Besseling [12]) is obtained if the single 'stress-limited' constituent there is generalized to a set of such constituents. For T constituents, the term τ_i in the equilibrium constraint of (D) is replaced by $\tau_i^{(k)}$; $k = 1,2,...,T$, with independent constraints $|\tau_i^{(k)}| - \bar{\tau} \leq 0$.

Finally, models for substantially more interesting design problems are obtained if the properties of each constituent are represented via independent design variables.

REFERENCES

[1]   Taylor, J.E. (1991): "An Extremum Principle for a Mixed Form Model of Elastic/Softening Structures," (in manuscript).

[2]   Taylor, J.E. (1989): "Toward a Unified Model for Elastoplastic Structural Analysis," Mechs. Res. Comms., 6, 2, 125-131.

[3]   Ben-Tal, A. and Taylor, J.E. (1991): "A Unified Model for Elastoplastic Structural Analysis via Dual Variational Principles," Proc. SIAM-Dayton Conf. on Design Theory (V. Komkov, ed.), in press.

[4]   Bendsøe, M.P., Olhoff, N. and Taylor, J.E. (1990): "A Unified Approach to the Analysis and Design of Elasto-Plastic Structures with Mechanical Contact," 3rd NASA/AF Mtg. on Multidisciplinary Optimization, San Francisco, CA.

[5]   Bendsøe, M.P. and Sokolowski, J. (1988), "Design Sensitivity Analysis for Elastoplastic Analysis Problems," Mechs. of Structs. and Machines, 16, 81-102.

[6]   Rozvany, G.I.N. and Cohn, M.Z. (1970), "Lower Bound Optimal Design of Concrete Structures," ASCE (96), EMD, 1013-1030, December.

[7]   Prager, W. and Shield, R.T. (1967), "A General Theory of Optimal Plastic Design," J. Appl. Mech., 34, 1, 184-186.

[8]   Prager, W. (1974), "Introduction to Structural Optimization," course #CSM212, International Center for Mech. Sci., Udine.

[9]   Masur, E.F. (1970), "Optimum Stiffness and Strength of Elastic Structures," ASCE-EM5, 621-649.

[10]   Schittkowski, K. (1985/86): "A FORTRAN Subroutine Solving Constrained Nonlinear Programming Problems," Annals of Operation Research, 5, 485-500.

[11]   Lemarechal, C.M. and Bancora Imbert, C.(1985): "Le module M1FC1," INRIA-Report.

[12]   Besseling, J.F. (1984): "Models of Metal Plasticity: Theory and Experiment," in *Plasticity Today*, (A. Sawczuk an G. Bianchi, eds.), Elsevier-London & New York.

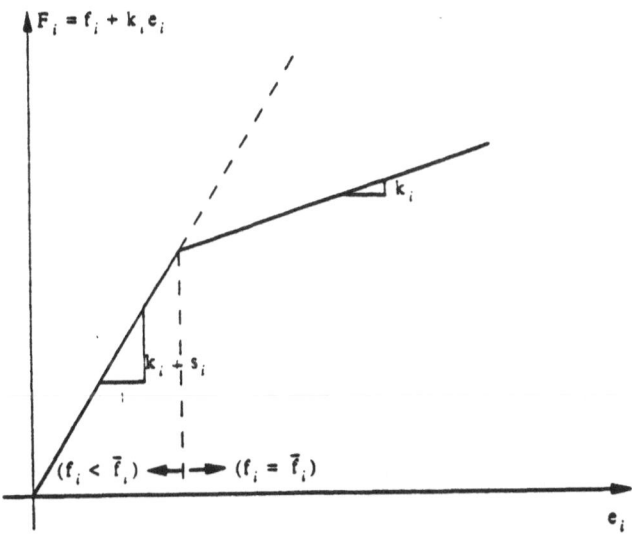

Figure 1:   Member 'Force vs. Elongation'

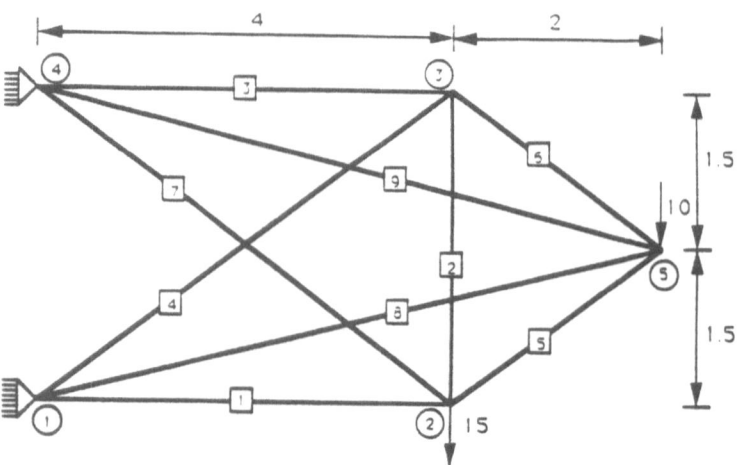

Figure 2:   Layout and Member Numbers for the Nine-Bar Truss

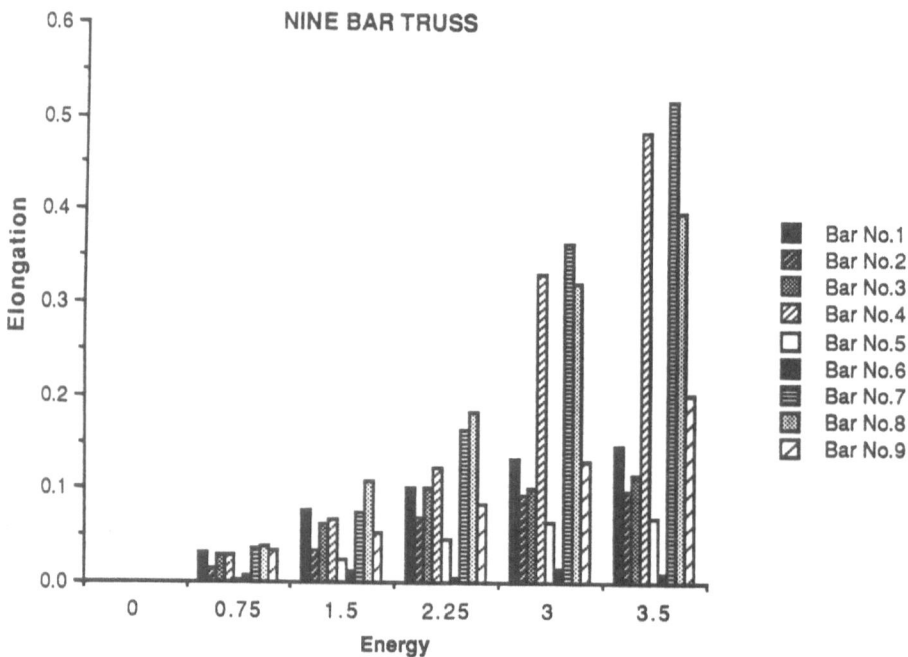

Figure 3:   Member Elongations for a Set of Energies

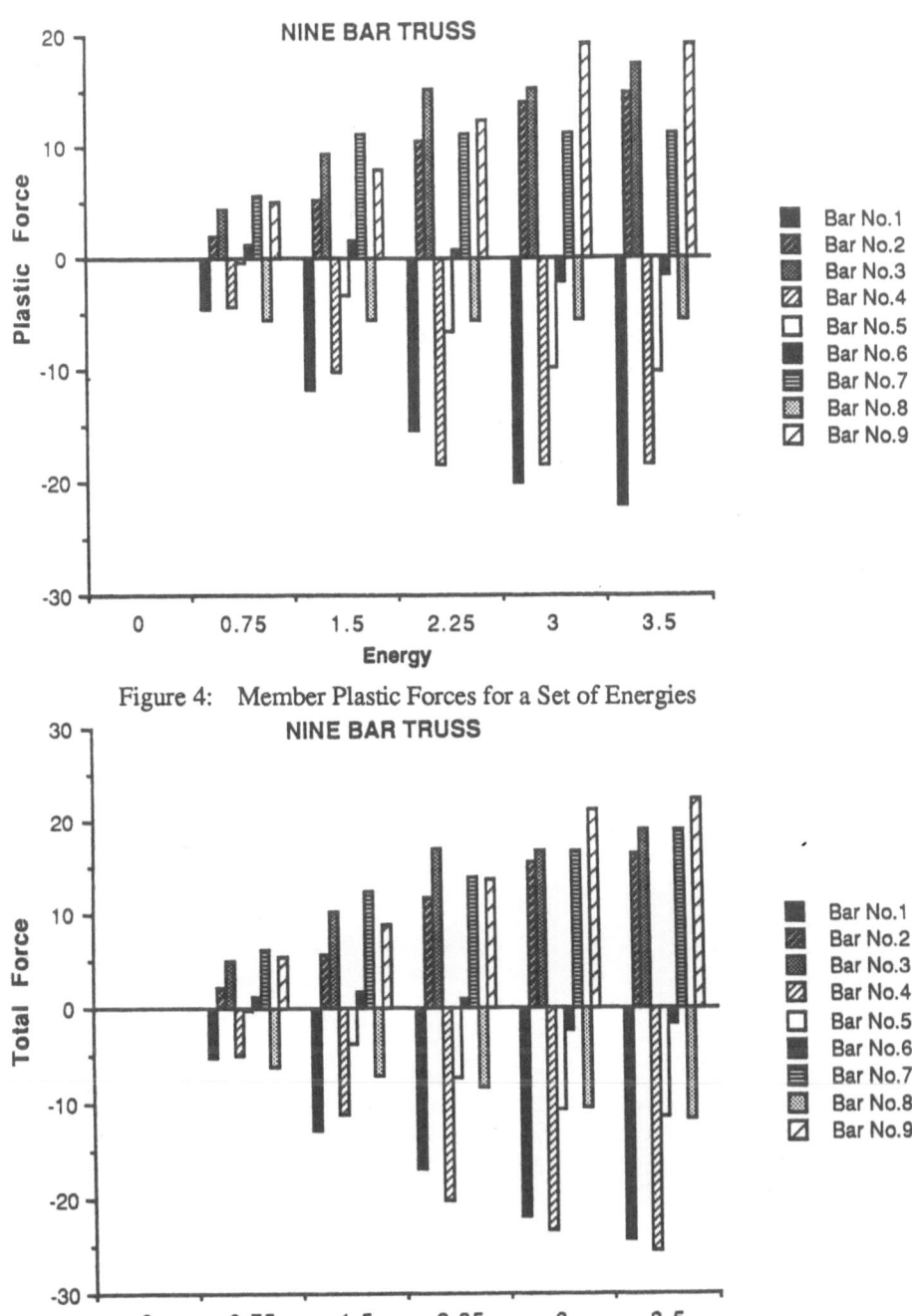

Figure 4:   Member Plastic Forces for a Set of Energies

Figure 5:   Member Total Forces for a Set of Energies

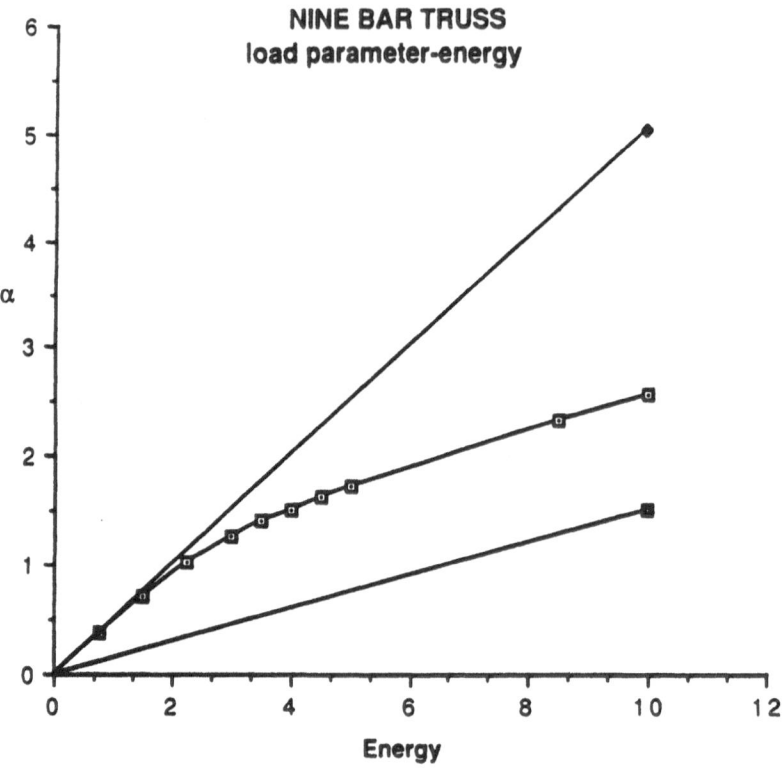

Figure 6: Evolution of Response: Load Factor α versus Energy

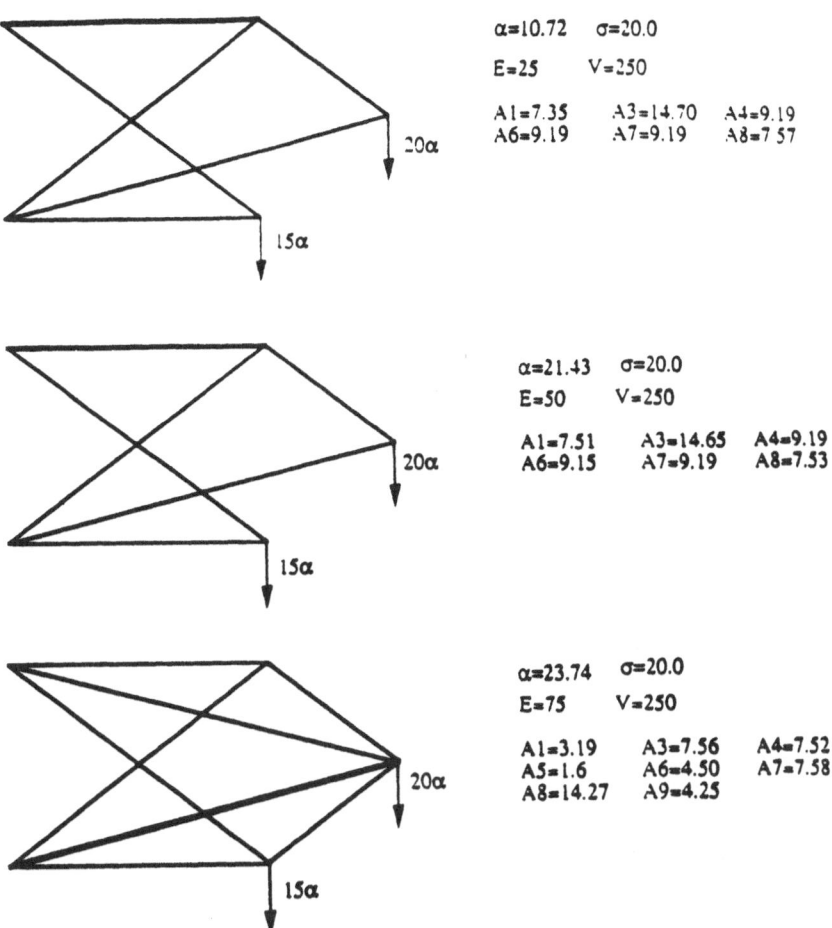

The following annotations appear beside each design:

α=10.72   σ=20.0

E=25   V=250

A1=7.35   A3=14.70   A4=9.19
A6=9.19   A7=9.19   A8=7.57

20α

15α

α=21.43   σ=20.0

E=50   V=250

A1=7.51   A3=14.65   A4=9.19
A6=9.15   A7=9.19   A8=7.53

20α

15α

α=23.74   σ=20.0

E=75   V=250

A1=3.19   A3=7.56   A4=7.52
A5=1.6   A6=4.50   A7=7.58
A8=14.27   A9=4.25

20α

15α

Figure 7:   Designs Under Given Load and for Three Values of Energy

A UNIFIED APPROACH TO THE ANALYSIS AND DESIGN OF ELASTO-PLASTIC STRUCTURES WITH MECHANICAL CONTACT

MARTIN P. BENDSØE
Mathematical Institute
The Technical University of Denmark
DK-2800 Lyngby, Denmark

NIELS OLHOFF
Institute of Mechnical Engineering
Aalborg University
DK-9220 Aalborg East, Denmark

JOHN E. TAYLOR
Department of Aerospace Engineering
The University of Michigan
Ann Arbor, MI 48109-2140, USA

ABSTRACT - With structural design in mind, a new unified variational model has been developed which represents the mechanics of deformation elasto-plasticity with unilateral contact conditions. For a design problem formulated as maximization of the load carrying capacity of a structure under certain constraints, the unified model allows for a simultaneous analysis and design synthesis for a whole range of mechanical behaviour.

1. INTRODUCTION

In order to express structural optimization, it is necessary to have an appropriate expression in variational form for the related structural mechanics analysis. With structural design in mind, a new unified variational model is presented which represents the mechanics of deformation elasto-plasticity with contact boundary conditions. The basic formulation of such problems is most natural in the form of a mixed variational model, with structural state expressed in terms of (independent) stress and displacement fields. However, a pure stress (force) method can also be obtained, and this constitutes the basis for the unified method developed for the formulation of a load carrying capacity design problem.

This paper constitutes an extension of earlier work on a unified model of elasto-plasticity that encompasses pure elasticity, elasto-plasticity and limit load analysis [1],[2]. The general model provides a monotone relation between the evolution of plastic deformation and contact and a global measure of system energy. For the optimization problems the variational formulation is usually added as so-called state equation constraints by including a set of necessary conditions for the variational statement, using both stress, displacement, plastic multipliers and

697

contact pressure as state variables (see e.g. Refs. [3]-[6] for sensitivity analysis and design for the case of contact, and Ref. [7]-[9] in the case of elasto-plastic behaviour). The approach advocated in this paper, however, treats the analysis problem in its original variational form. For certain design problems it is demonstrated that a more direct approach can be taken which combines the analysis and design goals into one broader variational statement. For a design problem formulated as maximization of the load carrying capacity of a structure under certain constraints, the resulting optimization problem becomes especially transparent and gives rise to very interesting relations between so-called design constraints and unilateral constraints arising from the plasticity and contact conditions.

It should be stressed that with a view to perform simultaneous analysis and design, a mixed variational model is less attractive because it implies a max-min formulation. Our goal has been to develop a *pure max* or *pure min* formulation such that *state- and design quantities* can be treated as *simultaneous variables* in the computational procedure. This is a potential advantage in terms of computational cost relative to, e.g., usual schemes where the equilibrium equations are solved "exactly" for each step of redesign, although, at the intermediate steps of redesign, this is not necessary because the design may be far from the optimal solution. Although the method proposed in this paper involves treatment of all the variables at the same time, the computertime may be reduced due to saving of a large number of iterations required for sequences of "exact" solutions of the equilibrium equations for intermediate designs.

2. VARIATIONAL MODELS

For the sake of notational simplicity we consider truss structures (FEM discretized continuum structures). For the analysis problem, a mixed variational model in bar forces q and nodal displacements x can be expressed as cf. [10]:

$$\max_{q} \min_{x} \left\{ x^T Bq - \frac{1}{2} q^T Qq - f^T x \right\} \tag{1}$$

where B denotes the compatibility matrix, Q the compliance matrix (i.e. $\frac{1}{2} q^T Qq$ is the complementary energy) and f the external nodal forces. This mixed formulation can be extended to elasto-plasticity with frictionless contact at nodal points by adding yield constraints of the form

$$| q_i / A_i | \leq \bar{\sigma}, \ i = 1, ..., NB \tag{2}$$

where $\bar{\sigma}$ is a given yield stress, A_i bar areas, and NB the number of bars, along with conditions

$$x_j \leq \bar{x}_j, \ i = 1, ... , NC \tag{3}$$

for the unilateral contact constraints. In (3), $\bar{x}_j$ are given initial gaps between nodal points and potential contact surfaces, and NC is the number of contact conditions. If no contact surfaces are specified the minimization over x in (1) implies $Bq = f$, and we have the well-known case of holonomic elasto-plastic analysis [11]:

$$\min_{q} \left\{ \frac{1}{2} q^T Q q \;\middle|\; Bq = f; \; |\, q_i/A_i \,| \;\le\; \bar{\sigma}, \, i = 1, \, \dots \, , \, NB \right\} \tag{4}$$

In the analysis of elasto-plastic structures, the limit load problem plays a key role [11] and is formulated as

$$\max_{\alpha, q} \left\{ \alpha \;\middle|\; Bq = \alpha f; \; |\, q_i/A_i \,| \;\le\; \bar{\sigma}, \, i = 1, \, \dots \, , \, NB \right\} \tag{5}$$

A mixed form of this problem is

$$\max_{q} \; \min_{x} \left\{ x^T Bq \;\middle|\; f^T x = 1; \; |\, q_i/A_i \,| \;\le\; \bar{\sigma}, \, i = 1, \, \dots \, , \, NB \right\} \tag{6}$$

The limit-load problem (5) and the elasto-plasticity problem (4) can be combined into one unified maximization problem, cf. Refs. [1], [2]:

$$\max_{\alpha, q} \left\{ \alpha \;\middle|\; Bq = \alpha f; \; \frac{1}{2} q^T Q q \le \bar{\epsilon}^{\,2}; \; |\, q_1/A_i \,| \;\le\; \bar{\sigma}, \, i = 1, \, \dots \, , \, NB \right\} \tag{7}$$

In this formulation the given constraint value $\bar{\epsilon}^{\,2}$ for the complementary energy controls the specific model problem that is handled by the variational statement. For small $\bar{\epsilon}$, (7) is an elasticity problem, and for large $\bar{\epsilon}$, where the constraint $\frac{1}{2} q^T Q q \le \bar{\epsilon}^{\,2}$ is inactive, we have the limit load problem. For intermediate values of $\bar{\epsilon}$, (7) is an elasto-plasticity model, and we have a monotone relation between the increase in the load carrying capacity α and the global measure $\bar{\epsilon}^{\,2}$ of system energy.

If we now include the contact conditions, the displacement variable x can be removed from (1) by use of conjugate duality (A. Ben-Tal: private communication). In the case of symmetric bounds on x, i.e. $|\, x_j \,| \;\le\; \bar{x}_j$, $j = 1, \, \dots \, , \, NC$, we get a formulation

$$\max_{\alpha, q} \left\{ \alpha - \Sigma \bar{x}_j \;\middle|\; (Bq - \alpha f)_j \;\middle|\; \frac{1}{2} q^T Q q \le \bar{\epsilon}^2; \; |\, q_i/A_i \,| \right.$$
$$\left. \le \bar{\sigma}, \, i = 1, \, \dots \, , \, NB \right\} \tag{8}$$

where we set $\bar{x}_j = \infty$ in the case of no contract conditions. If the constraint on complementary energy is not active we have defined a limit load problem for plasticity with possible contact.

3. FORMULATION OF UNIFIED ANALYSIS PROBLEM BY MEANS OF CASTIG-LIANO'S 2ND THEOREM

The pure minimum and maximum character, respectively, of the variational formulations of the analysis problems (4) and (7), may be preserved even if contact conditions are considered, if we make use of Castigliano's 2nd theorem to express the displacement conditions (3) in terms of forces. A similar advantage is achieved in the formulation (8), but we now consider one-sided bounds (3) on x_j, and entirely base our formulation on principles of mechanics.

Let us denote by r_j the possible external contact forces exhibited by frictionless plane surfaces that may constrain nodal displacements of the truss. If contact occurs, the equations of equilibrium changes to

$$Bq = f + ar \qquad (9)$$

where the matrix a projects the contact forces onto the directions of the external nodal forces f. The potential contact forces are orthogonal to the frictionless contact surfaces and taken to be nonnegative

$$r_j \geq 0 \qquad j = 1, \ldots , NC \qquad (10)$$

when directed along the outward normal. Non-zero contact forces r_j imply a change in the complementary energy $\frac{1}{2} q^T Qq$ through (9), and according to Castigliano's 2nd theorem the nodal point displacements x_j in the directions of the external forces r_j are simply given by

$$x_j = \frac{\partial (\frac{1}{2} q^T Qq)}{\partial r_j} = \left[\frac{\partial q}{\partial r_j} \right]^T Qq \, , j = 1, \ldots , NC \qquad (11)$$

Since r_j and hence x_j is directed away from the contact surface, the kinematic contact condition (3), where the initial gap $\bar{x}_j$ is given, is now expressed in terms of forces

$$- \left[\frac{\partial q}{\partial r_j} \right]^T Qq \leq \bar{x}_j \qquad j = 1, \ldots , NC \qquad (12)$$

In addition to (10) and (12), the unilateral contact problem is characterized by the fact that $r_j > 0$ and $r_j = 0$, respectively, depending on whether the condition (12) is satisfied as an equality or an inequality. Thus, as an additional governing condition for the contact problem, we have

$$r_j \left(\left[\frac{\partial q}{\partial r_j} \right]^T Qq + \bar{x}_j \right) = 0 \, , \, j = 1, \ldots , NC \qquad (13)$$

The variational formulation of the elasto-plastic analysis problem (4) may now be generalized as follows to allow for contact conditions

$$\min_{q,r} \left\{ \frac{1}{2} q^T Q q \; \middle| \; Bq = f + ar; \; |q_i/A_i| \le \bar{\sigma}, \; i = 1, \ldots, NB; \; r_j \ge 0 \right.$$

$$\left. -\left[\frac{\partial q}{\partial r_j}\right]^T Q q \le \bar{x}_j, \; r_j \cdot \left[\left[\frac{\partial q}{\partial r_j}\right]^T Q q + \bar{x}_j\right] = 0, \; j = 1, \ldots, NC \right\} \tag{14}$$

Similarly, we may extend the max formulation (7) to the following model that unifies the limit load problem with those of elasto-plasticity and unilateral contact

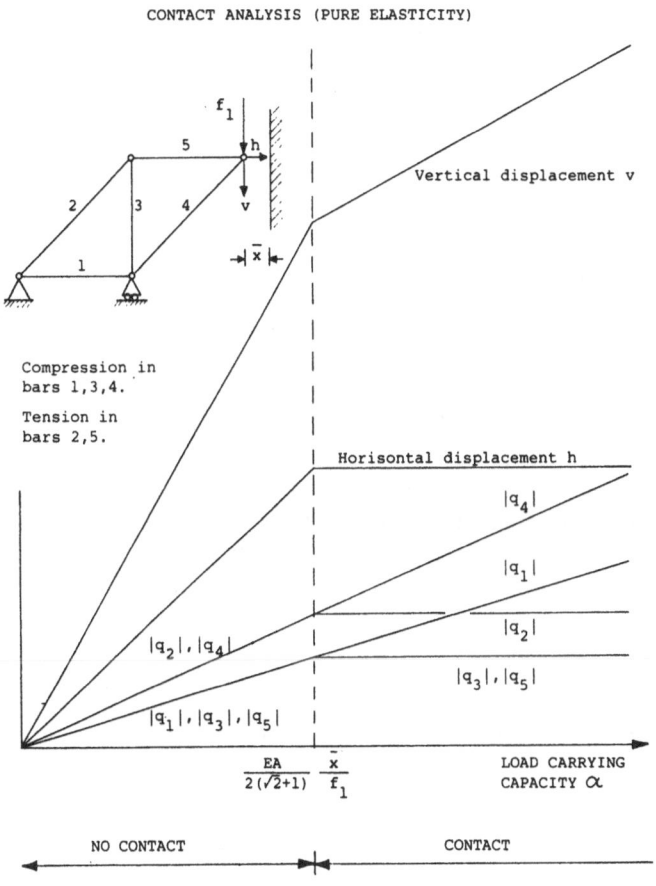

Fig. 1.

$$\max_{\alpha,q,r}\left\{ \alpha \;\middle|\; Bq = \alpha f + ar; \; \frac{1}{2}q^T Qq \leq \bar{\epsilon}^2; \; |q_i/A_i| \leq \bar{\sigma}, \right.$$

$$i = 1, \ldots, NB; \; r_j \geq 0, \; -\left[\frac{\partial q}{\partial r_j}\right]^T Qq \leq \bar{x}_j,$$

$$\left. r_j \cdot \left[\left[\frac{\partial q}{\partial r_j}\right]^T Qq + \bar{x}_j\right] = 0, \; j = 1, \ldots, NC \right\} \tag{15}$$

In the final statement of (14) and (15), the terms $\partial q/\partial r_j$ should be eliminated by suitable rearrangement of the equilibrium condition and separation of forces in nonredundant and redundant forces cf. [13].

To illustrate a solution to the analysis problem (15), consider the plane five-bar truss in Fig. 1 with specific bar stiffnesses EA, lengths $\ell_i = (1, \sqrt{2}, 1, \sqrt{2}, 1)\ell$, and large yield stress $\bar{\sigma}$ such that elasticity prevails. The truss is subjected to a given vertical external load f_1 at the upper, right hand nodal point, for which a vertical frictionless contact surface is defined along with the initial gap $\bar{x}$. The essence of the solution to the elastic contact problem is depicted in the figure through the dependence on the load parameter α of the bar forces q_i, $i = 1, \ldots, 5$, and the horizontal displacement h and vertical displacement v of the upper right hand nodal point.

4. DESIGN FORMULATION

The variational statements of maximization of the load carrying capacity constitutes a natural basis for the formulation of a unified analysis and design problem (see Ref. [12], Chapt. 10 for a discussion on the advantages of such a setting). Denoting by A_i the cross-sectional areas of the bars in a truss, a combined problem formulation, with A_i, $1 = 1, \ldots, NB$, as design variables, takes the form:

$$\underset{\alpha,q,r,A_i}{\text{maximize}} \quad \alpha$$

so: {Constraints of variational problem}

$$\left\{\begin{array}{l} \text{Constraints on design:} \\ \sum_{i=1}^{NB} l_i A_i = \bar{V}, \; \underline{A} \leq A_i \leq \bar{A} \end{array}\right\}$$

This problem maximizes the load carrying capacity of a given structure, for a given compliance and for a given volume. For a fixed volume, the compliance constraint controls whether the structure is in the elastic, elasto-plastic or limit load range, with or without contact, thus encompassing a broad class of model problems. Note that the problem includes displacement as well as stress constraints, but as *part of the analysis model*. Unlike in traditional design formulations an active displacement constraint gives rise to a contact force and an active stress constraint gives rise to a plastic deformation. *The problem thus finds the ultimate load carrying capacity, allowing the structure to yield and to explore the possibility of advantageous contact forces.* If displacement and stress constraints are to be included as part of the design model, minor modifications need to be included in the problem statement that has the effect of nullifying contact forces and plastic deformation from such constraints cf. [13].

Fig. 2 shows results for a case where contact is not present and illustrates the dependence of the optimal load carrying capacity α on the compliance constraint value $\bar{\epsilon}$ and the volume constraint value $\bar{V}$, respectively. Note that for an increase in volume the structure moves from the limit load range through an elasto-plastic range to the elastic range and through another elasto-plastic range before settling in the elastic range. The intermediate step is caused by one of the design constraints $A_i \leq A$ becoming active.

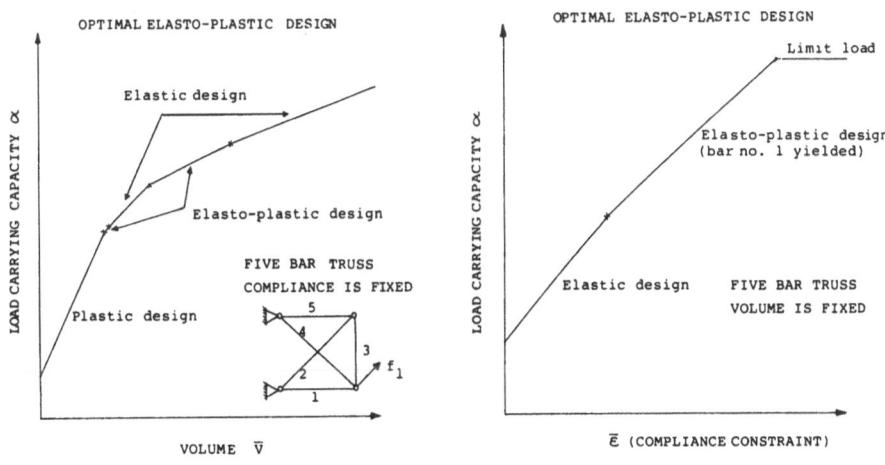

Fig. 2.

ACKNOWLEDGEMENT - The authors are grateful to Prof. A. Ben-Tal, Techion, Israel, for fruitful discussions on the subjects of this paper. This work was supported in part by the

Danish Technical Research Council through the Programme of Research on Computer Aided Design (MPB, NO). A guest professorship (JET) at the Mathematical Institute, The Technical University of Denmark, Sept. 15 to Dec. 15, 1989, is also gratefully acknowledged.

REFERENCES

[1]   *J.E. Taylor* (1989) Towards a Unified Model for Elasto-Plasticity Structural Analysis. Mechanics Research Communications, 16 (2), pp. 125-131.

[2]   *A. Ben-Tal, J.E. Taylor* (1989) Towards a Unified Model for the Elasto-Plastic Analysis of Structures. Report. Technion, Israel.

[3]   *M.P. Bendsøe, N. Olhoff, J. Sokolowski* (1985) Sensitivity Analysis of Problems of Elasticity with Unilateral Constraints. Journal of Structural Mechanics, 13, pp. 201-222.

[4]   *R.L. Benedict, J.E. Taylor* (1980) Optimal Design for Elastic Bodies in Contact. Proc. of NATO-ASI, Iowa City, Iowa. Optimization of Distributed Parameter Structures - Vol. 2, (Eds. E.J. Haug, J. Cea), Sijthoff and Nordhoff, Netherlands, pp. 1553-1589.

[5]   *P. Neittaanmaki* (1988) Optimal Shape Design in Contact Problems. In: Discretization Methods and Structural Optimization - Procedures and Applications, (eds. H.A. Eschenauer, E. Thierauf), Lecture Notes in Engineering, Vol. 42, Springer Verlag, pp. 248-254.

[6]   *Hlavacek, Bock, Lovisek* (1984-85) Optimal Control of Variational Inequality with Applications to Structural Analysis, Part 1, Part 2, Part 3, Applied Mathematics and Optimization. Vol. 11, pp. 111-143, and Vol. 13, pp. 117-136.

[7]   *M.P. Bendsøe, J. Sokolowski* (1989) Design Sensitivity Analysis of Elastic Plastic Analysis Problems. Mechanics of Structures and Machines, 16, pp. 81-102.

[8]   *I. Kaneko and G. Maier* (1981) Optimum design of plastic structures under displacement constraints, Comput. Methods Appl. Mech. Eng. 27, pp. 369-391.

[9]   *C. Cinquini, M. Rovati* (1989) Optimal Design of Elastic Plastic Structures. In: Computer Aided Optimal Structural Design, (eds. N. Olhoff, S. Kibsgaard), Lecture Notes, European Program for Advanced Education, Aalborg University.

[10]   *K. Washizu* (1983) Variational Methods in Elasticity and Plasticity, Pergamon, Oxford.

[11]   *M.Z. Cohn and G. Maier* (1977) Engineering Plasticity by Mathematical Programming. proceedings of the NATO Advanced Study Institute, University of Waterloo, Ontario, Canada. Pergamon Press, Oxford, England.

[12]   *R.T. Haftka, Z. Gürdal, M.P. Kamat* (1990) Elements of Structural Optimization. 2nd Edition, Kluwer, Dordrecht, The Netherlands.

[13]   *M.P. Bendsøe, N. Olhoff, J.E. Taylor* A Method for Simultaneous Analysis and design of Elasto-Plastic Structures with Contact. (In Preparation).

OPTIMAL ORIENTATION OF ORTHOTROPIC PROPERTIES FOR CONTINUUM BODIES AND STRUCTURAL ELEMENTS

Carlo Cinquini
Department of Structural Mechanics
University of Pavia
Via Abbiategrasso 211, I–27100 Pavia, Italy

Marco Rovati
Department of Structural Mechanics and Design Automation
University of Trento
Via Mesiano 77, I–38050 Trento, Italy

ABSTRACT

In this paper some notes on recent results obtained in the field of optimal design of orthotropic bodies and components are given. After a brief recall of the mechanical behaviour of orthotropic bodies and structural elements, a general formulation of the optimization problem is given: assuming the elastic energy as a meaningful measure of the global stiffness (or flexibility) of a three dimensional elastic body, the interest is focused on the determination of those local orientations of material symmetry axes in an orthotropic solid, which correspond to extreme values of the energy density. The stationarity conditions are computed, and a mechanical interpretation is given as well. Then, the two dimensional case is dealt with in details, both for the elastic and plastic cases, defining also a unified approach for approximate optimal solutions. Some simple examples of application are given as well.

INTRODUCTION

One of the earlier results on optimal orientation of material symmetry axes can be found in a work of Banichuk [5] where necessary conditions for optimal distribution of material properties in orthotropic bodies subjected to plane state of stress are given. Such results have found new impulses in very recent times, mainly due to the growing interest on fiber composite materials. In fact, the optimal orientation of reinforcing fibers seems to be an interesting and promising problem in mechanics of solids.

The aim of the present paper is to give some general results on optimal orientation of mechanical properties of orthotropic bodies and structural elements. Up to now, some papers have been published on the subject with reference to plane elastic problems for 2-D structures. After the pioneering work of Banichuk, the same problem was studied by Sacchi, Rovati [7] and Pedersen [8] which independently obtained similar results and mechanical interpretations of the optimality conditions. Mention has also to be made to Kartevelishvili, Kobelev [6] and Seregin, Troitskii [10]. Later Pedersen [3] performed a systematic study of the optimal solutions and obtained, for the plane stress problem, conditions for absolute maxima and minima. Numerical solutions of the problem can be found in Thomsen, Olhoff [15].

In this paper, at first an introduction to the mechanical behaviour of orthotropic solids and structural elements made of orthotropic materials is given. Then, the general formulation of an optimization problem is presented: in particular, the optimal orientation of the material symmetry axes in a three–dimensional orthotropic body are sought. The attention is focused on absolute maxima and minima, i.e., the determination of those orientations of orthotropy axes for which the body exhibits the stiffest (or the most flexible) response. The necessary optimality conditions are obtained, and in particular their mechanical interpreta-

707

tion is pointed out, namely the collinearity between principal directions of stress and strain at the optimum. Then two classes of solutions are given as well. The same optimization problem is then dealt with for the case of plane state of stress and for plates in bending. Optimization of orthotropic discs loaded in the plane is also performed at the limit state. A formal analogy between the optimality conditions for the elastic and the plastic cases is poited out, leading, in the last Section, to a unified approach to the optimization problems and to some approximate solutions.

The present paper is not written as a review paper on the subject, but is mainly based on the results of a joint research between the authors and Sacchi Landriani and Taliercio (Politecnico di Milano, Italy).

MECHANICAL BEHAVIOUR OF AN ORTHOTROPIC SOLID

Because of the physical nature of structural members made of composite materials, obtained by suitable compositions of several orthotropic plies, the theory of elasticity for orthotropic bodies is the keystone of analysis and design of composites. In this Section only the fundamentals on such a topic will be given, with particular reference to some aspects which will be useful later in the text. Further and deeper informations on the subject can be found, e.g., in Jones [2], Broutman and Krock [4] or Lekhnitskii [1].

A material possessing three mutually orthogonal planes of elastic symmetry, at each point, is said to be *orthogonally anisotropic*, or, more shortly and more commonly, *orthotropic*.

The coordinate axes normal to the plane of elastic symmetry are called the principal coordinate axes, or principal directions of orthotropy.

For an elastic and orthotropic material, in principal directions of orthotropy, stresses and strains are related by the generalized Hooke's law

$$\sigma_{ij} = C^0_{ijhk}\epsilon_{hk}, \quad i,j,h,k = 1,2,3 \tag{1}$$

where the tensor C^0_{ijhk} depends on 9 independent constants only. Rel. (1), defining the stress and strain vectors in a consistent way, can be rewritten in the following well known matrix form

$$\begin{Bmatrix} \sigma_{11} \\ \sigma_{22} \\ \sigma_{33} \\ \sigma_{23} \\ \sigma_{31} \\ \sigma_{12} \end{Bmatrix} = \begin{pmatrix} C^0_{1111} & C^0_{1122} & C^0_{1133} & 0 & 0 & 0 \\ C^0_{1122} & C^0_{2222} & C^0_{2233} & 0 & 0 & 0 \\ C^0_{1133} & C^0_{2233} & C^0_{3333} & 0 & 0 & 0 \\ 0 & 0 & 0 & C^0_{2323} & 0 & 0 \\ 0 & 0 & 0 & 0 & C^0_{1313} & 0 \\ 0 & 0 & 0 & 0 & 0 & C^0_{1212} \end{pmatrix} \begin{Bmatrix} \epsilon_{11} \\ \epsilon_{22} \\ \epsilon_{33} \\ 2\epsilon_{23} \\ 2\epsilon_{31} \\ 2\epsilon_{12} \end{Bmatrix}. \tag{2}$$

In the present paper the three–dimensional behaviour of orthotropic bodies, and related optimization problems, will be only discussed briefly as a general statement of an optimization problem, both for the sake of brevity and because of the fact that, up to now, in practical engineering composite materials are mainly adopted to build structural members which can be modelled as two–dimensional bodies. So that, we shall mainly confine our attention on two–dimensional elasticity problems. In such a way, the solid is assumed to possess a plane of elastic symmetry; in a general reference frame x_1-x_2-x_3 and taking the x_1-x_2 plane as a plane of material symmetry, the deformation is assumed to occour in this plane. In particular let us consider laminated composite theory and restrict ourselves to balanced laminates, made by stacking orthotropic layers (or plies) symmetrically about some flat surface, say, the x_1-x_2 plane.

All these assumptions allow us to assume the body built up as a two–dimensional orthotropic structural member, so that, in this Section, we focus our attention on the behaviour of orthotropic solids in plane elasticity.

If a plane state of stress is considered, one has

$$\sigma_{13} = \sigma_{23} = \sigma_{33} = 0 \tag{3}$$

from which it follows that $\epsilon_{13} = \epsilon_{23} = 0$. From the physical point of view, this is approximately the case of a thin plate loaded along its boundary by in–plane loads symmetrically distributed about the middle plane.

For these stress and strain distributions, the constitutive law (2) can be rewritten in a more suitable form, introducing the so called reduced (or plane) stiffness coefficients. On this purpose consider, in a general reference frame, the specific strain energy which is given by

$$\mathcal{E} = \frac{1}{2}\left(\sigma_{11}\epsilon_{11} + 2\sigma_{12}\epsilon_{12} + \sigma_{22}\epsilon_{22}\right). \tag{4}$$

Such an expression can be written in terms of strain components only, as

$$\mathcal{E} = \frac{1}{2}\left(C_{1111}\epsilon_{11}^2 + 4C_{1212}\epsilon_{12}^2 + C_{2222}\epsilon_{22}^2 + 2C_{1122}\epsilon_{11}\epsilon_{22} + C_{1133}\epsilon_{11}\epsilon_{33} + \right.$$
$$\left. + C_{2233}\epsilon_{22}\epsilon_{33} + 4C_{1112}\epsilon_{11}\epsilon_{12} + 4C_{1222}\epsilon_{12}\epsilon_{22} + 2C_{1233}\epsilon_{12}\epsilon_{33}\right). \tag{5}$$

Now, it is possible to express the component ϵ_{33} of the strain tensor in terms of the remaining nonvanishing components, by imposing

$$\sigma_{33} = C_{1133}\epsilon_{11} + 2C_{1233}\epsilon_{12} + C_{2233}\epsilon_{22} + C_{3333}\epsilon_{33} = 0 \tag{6}$$

from which

$$\epsilon_{33} = -\frac{C_{1133}\epsilon_{11} + 2C_{1233}\epsilon_{12} + C_{2233}\epsilon_{22}}{C_{3333}}. \tag{7}$$

Replacing Rel. (7) in the specific strain energy expression (4) we obtain

$$\mathcal{E} = \frac{1}{2}\left[\left(C_{1111} - \frac{C_{1133}^2}{C_{3333}}\right)\epsilon_{11}^2 + \left(C_{2222} - \frac{C_{2233}^2}{C_{3333}}\right)\epsilon_{22}^2 + \right.$$
$$+ 4\left(C_{1212} - \frac{C_{1233}^2}{C_{3333}}\right)\epsilon_{12}^2 + 2\left(C_{1212} - \frac{C_{1133}C_{2233}}{C_{3333}}\right)\epsilon_{11}\epsilon_{33} +$$
$$\left. + 4\left(C_{1112} - \frac{C_{1133}C_{1233}}{C_{3333}}\right)\epsilon_{11}\epsilon_{12} + 4\left(C_{1222} - \frac{C_{1233}C_{2233}}{C_{3333}}\right)\epsilon_{12}\epsilon_{22}\right] \tag{8}$$

or, in shorthand form

$$\mathcal{E} = \frac{1}{2}\left(Q_{11}\epsilon_{11}^2 + Q_{22}\epsilon_{22}^2 + 4Q_{66}\epsilon_{12}^2 + 2Q_{12}\epsilon_{11}\epsilon_{22} + 4Q_{16}\epsilon_{11}\epsilon_{12} + 4Q_{26}\epsilon_{12}\epsilon_{22}\right) \tag{9}$$

where the reduced stiffness coefficients have been defined in the following way (adopting a usual numbering for the indices)

$$Q_{11} = C_{1111} - \frac{C_{1133}^2}{C_{3333}} \qquad Q_{22} = C_{2222} - \frac{C_{2233}^2}{C_{3333}}$$

$$Q_{66} = C_{1212} - \frac{C_{1233}^2}{C_{3333}} \qquad Q_{12} = C_{1122} - \frac{C_{1133}C_{2233}}{C_{3333}} \tag{10}$$

$$Q_{16} = C_{1112} - \frac{C_{1133}C_{1233}}{C_{3333}} \qquad Q_{26} = C_{1222} - \frac{C_{1233}C_{2233}}{C_{3333}}.$$

In such a way, the stress–strain relationship, for a plane state of stress, can be written as

$$\begin{Bmatrix} \sigma_{11} \\ \sigma_{22} \\ \sigma_{12} \end{Bmatrix} = \begin{pmatrix} Q_{11} & Q_{12} & Q_{16} \\ Q_{12} & Q_{22} & Q_{26} \\ Q_{16} & Q_{26} & Q_{66} \end{pmatrix} \begin{Bmatrix} \epsilon_{11} \\ \epsilon_{22} \\ 2\epsilon_{12} \end{Bmatrix} \tag{11}$$

or, in pricipal directions of orthotropy, as

$$\left\{\begin{array}{c} \sigma_{11} \\ \sigma_{22} \\ \sigma_{12} \end{array}\right\} = \left(\begin{array}{ccc} Q_{11}^0 & Q_{12}^0 & 0 \\ Q_{12}^0 & Q_{22}^0 & 0 \\ 0 & 0 & Q_{66}^0 \end{array}\right) \left\{\begin{array}{c} \epsilon_{11} \\ \epsilon_{22} \\ 2\epsilon_{12} \end{array}\right\} \tag{12}$$

where the reduced stiffness coefficients, in principal directions, are defined as follows

$$Q_{11}^0 = C_{1111}^0 - \frac{\left(C_{1133}^0\right)^2}{C_{3333}^0} \qquad Q_{22}^0 = C_{2222}^0 - \frac{\left(C_{2233}^0\right)^2}{C_{3333}^0} \tag{13a}$$

$$Q_{66}^0 = C_{1212}^0 \qquad Q_{12}^0 = C_{1122}^0 - \frac{C_{1133}^0 C_{2233}^0}{C_{3333}^0}. \tag{13b}$$

The reduced stiffness coefficients Q_{ij}, in general coordinates, and Q_{ij}^0, in principal directions of orthotropy, are related each other through the following well known rotation formulas

$$Q_{11} = Q_{11}^0 m^4 + 2\left(Q_{12}^0 + 2Q_{66}^0\right) n^2 m^2 + Q_{22}^0 n^4 \tag{14a}$$

$$Q_{12} = \left(Q_{11}^0 + Q_{22}^0 - 4Q_{66}^0\right) n^2 m^2 + Q_{12}^0\left(n^4 + m^4\right) \tag{14b}$$

$$Q_{22} = Q_{11}^0 n^4 + 2\left(Q_{12}^0 + 2Q_{66}^0\right) n^2 m^2 + Q_{22}^0 m^4 \tag{14c}$$

$$Q_{16} = -Q_{22}^0 n^3 m + Q_{11}^0 n m^3 - \left(Q_{12}^0 + 2Q_{66}^0\right) n m \left(m^2 - n^2\right) \tag{14d}$$

$$Q_{26} = -Q_{22}^0 n m^3 + Q_{11}^0 n^3 m + \left(Q_{12}^0 - 2Q_{66}^0\right) n m \left(m^2 - n^2\right) \tag{14e}$$

$$Q_{66} = \left(Q_{11}^0 + Q_{22}^0 - 2Q_{12}^0\right) n^2 m^2 + Q_{66}^0 \left(m^2 - n^2\right)^2 \tag{14f}$$

where $n = \sin\theta$ and $m = \cos\theta$, and where θ represents the angle between the axis x_1 of the general reference frame and the axis z_1 of the frame oriented as the principal directions of orthotropy (see Fig. 1).

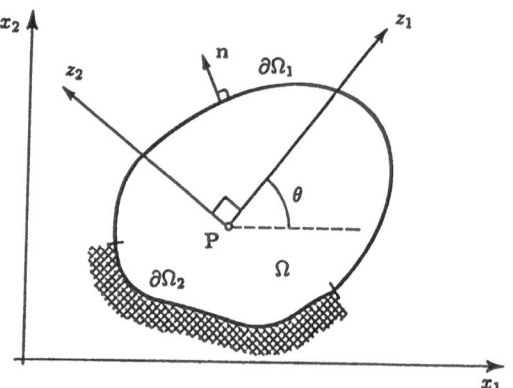

Figure 1. Definition of the angle θ.

The rotation formulas just shown will play a fundamental role in the optimization problems that will be presented later.

MECHANICAL RESPONSE OF COMPOSITE STRUCTURES

Although in the present paper our attention will be confined to optimal design of structural members subjected to plane states of stress only, in this Section a brief introduction to classical laminate theory is given, with the main objective to make clearer the notations that will be adopted later. Basic references on this topic can be indicated in Jones [2] or Vinson and Sierakowski [3].

As previously mentioned, most composite structures are plate type structures which are composed of thin layer, or laminae, of orthotropic material with, in general, different orientation of fibers in each layer, subjected to both in–plane forces and bending moments. In classical laminate analysis this structure is seen as a two–dimensional continuous body, with a total number (N) of orthotropic layers perfectly bonded together, under the hypothesis that the bondline is infinitely thin and non–shear–deformable. The usual assumption of Kirchhoff plate theory is made, i.e., the in–plane displacements follow a linear variation through the thickness of the plate. In the same reference frame introduced in the previous Section, such an hypothesis reads

$$u_1 = u_1^0 - x_3 \frac{\partial u_3^0}{\partial x_1} \qquad u_2 = u_2^0 - x_3 \frac{\partial u_3^0}{\partial x_2} \tag{15}$$

where u_i^0 ($i = 1, 2, 3$) are the midplane displacements.
The strain vector is then expressed as

$$\left\{ \begin{array}{c} \epsilon_{11} \\ \epsilon_{22} \\ 2\epsilon_{12} \end{array} \right\} = \left\{ \begin{array}{c} \epsilon_{11}^0 \\ \epsilon_{22}^0 \\ 2\epsilon_{12}^0 \end{array} \right\} + x_3 \left\{ \begin{array}{c} \chi_{11} \\ \chi_{22} \\ 2\chi_{12} \end{array} \right\} \tag{16}$$

where ϵ_{ij}^0 and χ_{ij} are the midplane strains and curvatures, respectively. Therefore, the stress distribution in the kth lamina is obtained by substituting Rel. (16) in the stress–strain relationship (11)

$$\left\{ \begin{array}{c} \sigma_{11} \\ \sigma_{22} \\ \sigma_{12} \end{array} \right\}_k = \left(\begin{array}{ccc} Q_{11} & Q_{12} & Q_{16} \\ Q_{12} & Q_{22} & Q_{26} \\ Q_{16} & Q_{26} & Q_{66} \end{array} \right)_k \left(\left\{ \begin{array}{c} \epsilon_{11}^0 \\ \epsilon_{22}^0 \\ 2\epsilon_{12}^0 \end{array} \right\} + x_3 \left\{ \begin{array}{c} \chi_{11} \\ \chi_{22} \\ 2\chi_{12} \end{array} \right\} \right). \tag{17}$$

In such a way, the resultant forces and moments (per unit length of the cross section) acting on the laminate can be obtained by integrating the stresses in each lamina over the total thickness (h) of the laminate

$$\left\{ \begin{array}{c} N_{11} \\ N_{22} \\ N_{12} \end{array} \right\} = \int_{-h/2}^{h/2} \left\{ \begin{array}{c} \sigma_{11} \\ \sigma_{22} \\ \sigma_{12} \end{array} \right\}_k dx_3 = \sum_{k=1}^{N} \int_{x_3^{k-1}}^{x_3^k} \left\{ \begin{array}{c} \sigma_{11} \\ \sigma_{22} \\ \sigma_{12} \end{array} \right\} dx_3 \tag{18}$$

$$\left\{ \begin{array}{c} M_{11} \\ M_{22} \\ M_{12} \end{array} \right\} = \int_{-h/2}^{h/2} \left\{ \begin{array}{c} \sigma_{11} \\ \sigma_{22} \\ \sigma_{12} \end{array} \right\} x_3 dx_3 = \sum_{k=1}^{N} \int_{x_3^{k-1}}^{x_3^k} \left\{ \begin{array}{c} \sigma_{11} \\ \sigma_{22} \\ \sigma_{12} \end{array} \right\} x_3 dx_3 \tag{19}$$

where N is the total number of laminae in the laminate. Then, substituting (17) in (18) and (19), it is possible to obtain the following constitutive equations for the laminate

$$\left\{ \begin{array}{c} N_{11} \\ N_{22} \\ N_{12} \end{array} \right\} = \left(\begin{array}{ccc} A_{11} & A_{12} & A_{16} \\ A_{12} & A_{22} & A_{16} \\ A_{16} & A_{26} & A_{66} \end{array} \right) \left\{ \begin{array}{c} \epsilon_{11}^0 \\ \epsilon_{22}^0 \\ 2\epsilon_{12}^0 \end{array} \right\} + \left(\begin{array}{ccc} B_{11} & B_{12} & B_{16} \\ B_{12} & B_{22} & B_{26} \\ B_{16} & B_{26} & B_{66} \end{array} \right) \left\{ \begin{array}{c} \chi_{11} \\ \chi_{22} \\ 2\chi_{12} \end{array} \right\} \tag{20}$$

$$\left\{ \begin{array}{c} M_{11} \\ M_{22} \\ M_{12} \end{array} \right\} = \left(\begin{array}{ccc} B_{11} & B_{12} & B_{16} \\ B_{12} & B_{22} & B_{26} \\ B_{16} & B_{26} & B_{66} \end{array} \right) \left\{ \begin{array}{c} \epsilon_{11}^0 \\ \epsilon_{22}^0 \\ 2\epsilon_{12}^0 \end{array} \right\} + \left(\begin{array}{ccc} D_{11} & D_{12} & D_{16} \\ D_{12} & D_{22} & D_{26} \\ D_{16} & D_{26} & D_{66} \end{array} \right) \left\{ \begin{array}{c} \chi_{11} \\ \chi_{22} \\ 2\chi_{12} \end{array} \right\} \tag{21}$$

with

$$A_{ij} = \sum_{k=1}^{N} (Q_{ij})_k \left(x_3^k - x_3^{k-1} \right) \tag{22}$$

$$B_{ij} = \frac{1}{2} \sum_{k=1}^{N} (Q_{ij})_k \left[\left(x_3^k \right)^2 - \left(x_3^{k-1} \right)^2 \right] \tag{23}$$

$$D_{ij} = \frac{1}{3} \sum_{k=1}^{N} (Q_{ij})_k \left[\left(x_3^k \right)^3 - \left(x_3^{k-1} \right)^3 \right]. \tag{24}$$

In the most general case, namely when both in-plane forces and bending moments are applied to the structure, the constitutive relationship in terms of characteristics can be written in compact form as

$$\left\{ \frac{N}{M} \right\} = \left(\frac{A}{B} \quad \frac{B}{D} \right) \left\{ \frac{\varepsilon^0}{\chi} \right\}. \tag{25}$$

Now, after this introductory part on general behaviour of orthotropic and composite materials, in next Sections, we shall deal with an optimization problem, starting with its most general formulation with reference to the three dimensional continuum case.

BOUNDS OF ELASTIC STRAIN ENERGY: GENERAL FORMULATION

Consider a linear elastic orthotropic body, defined on the open domain $\Omega \in \mathbf{R}^3$, with boundary $\partial\Omega \equiv \partial\Omega_1 \cup \partial\Omega_2$. Tractions p_i and displacements u_i ($i = 1, 2, 3$) are prescribed on $\partial\Omega_1$ and on $\partial\Omega_2$, respectively, while the body forces b_i are given in Ω.

Denote by x^i, $i = 1, 2, 3$, an orthogonal reference frame with axes aligned, at each point of the body, with the principal directions of orthotropy. Then, indicate with x^α, $\alpha = I, II, III$, the axes of a second reference frame coinciding, at each point, with the principal directions of strain. Let $g^{(\alpha)}$ be the unit vectors associated to the frame x^α, with components $g_i^{(\alpha)}$ referred to x^i axes. Now, the problem to be dealt with consists in determining the local mutual orientations of the frames x^i and x^α, in order to find extreme values (absolute maxima and minima) of the work performed by the external loads. Such optimal orientations can be specified through optimal values of the nine components $g_i^{(\alpha)}$ (which undergo the constraints $g_i^{(\alpha)} g_{(\beta)}^i = \delta_{(\beta)}^{(\alpha)}$, $\alpha, \beta = I, II, III$). On the other hand, for the sake of simplicity, it seems to be more convenient to characterize the mutual orientations of the two frames through a new set of design variables, namely the three Euler's angles θ_q ($q = 1, 2, 3$), as depicted in Fig. 2. Based on this choice of the Euler's angles, the relationship between the direction cosines of the frame x^α and the angles θ_q can be written as

$$\begin{pmatrix} g_1^{(I)} & g_2^{(I)} & g_3^{(I)} \\ g_1^{(II)} & g_2^{(II)} & g_3^{(II)} \\ g_1^{(III)} & g_2^{(III)} & \end{pmatrix} = \begin{pmatrix} c_{\theta_1} c_{\theta_2} & s_{\theta_1} c_{\theta_2} & s_{\theta_2} \\ -c_{\theta_1} s_{\theta_2} c_{\theta_3} - s_{\theta_1} s_{\theta_3} & -s_{\theta_1} s_{\theta_2} c_{\theta_3} + c_{\theta_1} s_{\theta_3} & c_{\theta_2} c_{\theta_3} \\ -c_{\theta_1} s_{\theta_2} s_{\theta_3} + s_{\theta_1} c_{\theta_3} & -s_{\theta_1} s_{\theta_2} s_{\theta_3} - c_{\theta_1} c_{\theta_3} & c_{\theta_2} s_{\theta_3} \end{pmatrix} \tag{26}$$

where, for the sake of brevity, the notations $s_{\theta_q} = \sin\theta_q$ and $c_{\theta_q} = \cos\theta_q$ were assumed. With these assumptions, the problem can be then stated as follows: find

$$\mathcal{F}^0 \equiv \min_{\theta_q} \mathcal{F}(\theta_q) = \min_{\theta_q} \frac{1}{2} \left(\int_{\partial\Omega_1} p^i \overline{u}_i \, dS + \int_\Omega b^i \overline{u}_i \, d\Omega \right) \quad i, q = 1, 2, 3 \tag{27}$$

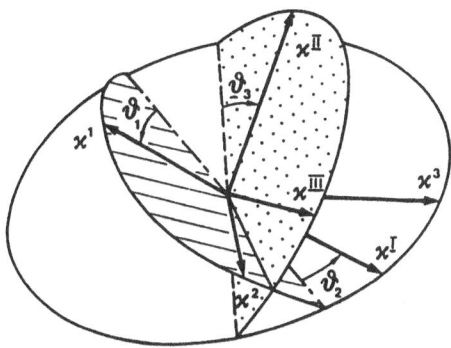

Figure 2. Euler's angles.

according to equilibrium, compatibility and stress-strain relationship. In Rel. (27), $\overline{u}_i$ are the components of the displacement field in equilibrium conditions. Now, it must be noted that the functional $\mathcal{F}$ to be minimized (or maximized) is equal, but opposite in sign, to $\mathcal{U}^0 = \min_{u_i} \mathcal{U}$, where $\mathcal{U}(u_i, \theta_q)$ is the total potential energy. In such a way, the constrained *min* problem (29) can be rewritten as the following unconstrained *min max* problem: *find*

$$\mathcal{F}^0 = \min_{\theta_q} \max_{u_i} \left(\int_{\partial\Omega_1} p^i u_i \, dS + \int_{\Omega} b^i u_i \, d\Omega - \int_{\Omega} \mathcal{E} \, d\Omega \right) \tag{28}$$

where the functions u_i belong to the class of the compatible displacement fields, and $\mathcal{E}$ is the specific strain energy

$$\mathcal{E} = \frac{1}{2} C^{ikrs} \epsilon_{ik} \epsilon_{rs}. \tag{29}$$

In the reference frame x^i, i.e., in principal directions of orthotropy, the specific strain energy (29) can be rewritten in explicit form as

$$2\,\mathcal{E} = C^{1111} \epsilon_{11} \epsilon_{11} + 2C^{1122} \epsilon_{11} \epsilon_{22} + 2C^{1133} \epsilon_{11} \epsilon_{33} + C^{2222} \epsilon_{22} \epsilon_{22} + 2C^{2233} \epsilon_{22} \epsilon_{33}$$
$$+ C^{3333} \epsilon_{33} \epsilon_{33} + 4C^{1212} \epsilon_{12} \epsilon_{12} + 4C^{2323} \epsilon_{23} \epsilon_{23} + 4C^{3131} \epsilon_{31} \epsilon_{31}. \tag{30}$$

Now, the stationarity conditions with respect to the Euler's angles θ_q ($q = 1, 2, 3$), involve only the term $\mathcal{E}$ of the functional $\mathcal{F}$, and for this reason the specific strain energy can be assumed as a meaningful parameter of the global stiffness/flexibility of the body. Then, the solution of the problem can be found as the solution of the following set of three nonlinear algebraic simultaneous equations

$$\frac{\partial \mathcal{E}}{\partial \theta_q} = 0 \qquad q = 1, 2, 3 \tag{31}$$

for the three unknowns θ_q. More explicitly, denoting $e_{(\beta)}$, $\beta = I, II, III$, the principal strains Eq.s (31) can be rewritten as

$$\frac{\partial \mathcal{E}}{\partial \theta_q} = \sigma^{ik} \frac{\partial \epsilon_{ik}}{\partial \theta_q} = 2 \sum_{\beta=I}^{III} e_{(\beta)} \frac{\partial g_i^{(\beta)}}{\partial \theta_q} g_k^{(\beta)} = 0. \tag{32}$$

Now, Rel. (32) allows us to write explicitly the stationarity conditions as

$$\epsilon_{3j}^k \sigma^{ij} \epsilon_{ik} = 0 \tag{33a}$$

$$\left(\epsilon_{1j}^k s_{\theta_1} - \epsilon_{2j}^k c_{\theta_1} \right) \sigma^{ij} \epsilon_{ik} = 0 \tag{33b}$$

$$2 \left(e_{(III)} - e_{(II)} \right) \sigma^{ik} g_i^{(II)} g_k^{(III)} = 0 \tag{33c}$$

where the symbol ϵ_{ij}^{k} denotes Ricci's tensor. By means of suitable linear combinations of Eq.s (33), it is possible to rewrite such optimality conditions in the following way

$$\left(e_{(I)} - e_{(II)}\right) \sigma^{ik} g_i^{(I)} g_k^{(II)} = 0 \tag{34a}$$

$$\left(e_{(II)} - e_{(III)}\right) \sigma^{ik} g_i^{(II)} g_k^{(III)} = 0 \tag{34b}$$

$$\left(e_{(III)} - e_{(I)}\right) \sigma^{ik} g_i^{(III)} g_k^{(I)} = 0. \tag{34c}$$

In the most general case, i.e., when $e_{(I)} \neq e_{(II)} \neq e_{(III)}$, the above conditions imply that

$$\sigma^{I\,II} = \sigma^{II\,III} = \sigma^{III\,I} = 0. \tag{35}$$

This means that, at the optimum, principal directions of stress are collinear with principal directions of strain. Collinearity between principal directions of stress and strain can be achieved, in particular, when the principal directions of orthotropy are aligned with those of stress and strain (*trivial* solutions). Such a condition is satisfied when

$$i)\ \ \theta_1 = \theta_2 = \theta_3 = 0 \qquad \Longrightarrow\quad x^1 \equiv x^I,\ x^2 \equiv x^{II},\ x^3 \equiv x^{III} \tag{36a}$$

$$ii)\ \ \theta_1 = \theta_2 = 0,\ \theta_3 = \pi/2 \quad \Longrightarrow\quad x^1 \equiv x^I,\ x^2 \equiv x^{III},\ x^3 \equiv -x^{II} \tag{36b}$$

$$iii)\ \ \theta_1 = \pi/2,\ \theta_2 = 0,\ \theta_3 = \pi/2 \Longrightarrow\ x^1 \equiv -x^{II},\ x^2 \equiv x^I,\ x^3 \equiv x^{III} \tag{36c}$$

$$iv)\ \ \theta_1 = \pi/2,\ \theta_2 = \theta_3 = 0 \qquad \Longrightarrow\quad x^1 \equiv x^{III},\ x^2 \equiv x^I,\ x^3 \equiv x^{II}. \tag{36d}$$

The remainder two trivial solutions (i.e., with $x^3 \equiv x^I$) have to be seen as limit cases for $x^I \to x^3$, otherwise the Euler's angles degenerate and cannot be longer defined. This yields

$$v)\ \ \theta_1 = \theta_2 = \theta_3 = \pi/2,\ \text{or}$$
$$\theta_1 = 0,\ \theta_2 = \pi/2,\ \theta_3 = 0 \ \Longrightarrow\ x^1 \equiv x^{II},\ x^2 \equiv x^{III}, x^3 \equiv x^I \tag{36e}$$

$$vi)\ \ \theta_1 = \theta_2 = \pi/2,\ \theta_3 = 0,\ \text{or}$$
$$\theta_1 = 0,\ \theta_2 = \theta_3 = \pi/2 \ \Longrightarrow\ x^1 \equiv -x^{III},\ x^2 \equiv x^{II},\ x^3 \equiv x^I. \tag{36f}$$

Although the complete analytical solution of the general case seems to be extremely involved, solutions other than the trivial ones can be easily obtained by imposing collinearity between only one axis of orthotropy and one principal direction of strain. For instance, by subsequently aligning x^I with x^1, x^2 and x^3 respectively, one gets the three *semi-trivial* solutions

$$x^I \equiv x^1:\quad \theta_1 = \theta_2 = 0$$
$$\cos 2\theta_3 = \frac{2\left(C^{1122} - C^{3311}\right) e_{(I)} + \left(C^{2222} - C^{3333}\right)\left(e_{(II)} + e_{(III)}\right)}{\left[C^{2222} + C^{3333} - 2\left(C^{2233} + 2C^{2323}\right)\right]\left(e_{(II)} - e_{(III)}\right)} \tag{37a}$$

$$x^I \equiv x^2:\quad \theta_1 = \pi/2\ \theta_2 = 0$$
$$\cos 2\theta_3 = \frac{2\left(C^{1122} - C^{2233}\right) e_{(I)} + \left(C^{1111} - C^{3333}\right)\left(e_{(II)} + e_{(III)}\right)}{\left[C^{1111} + C^{3333} - 2\left(C^{3311} + 2C^{3131}\right)\right]\left(e_{(II)} - e_{(III)}\right)} \tag{37b}$$

$$x^I \equiv x^3:\quad \theta_1 = \theta_2 = \pi/2\ \text{or}\ \theta_1 = 0,\ \theta_2 = \pi/2$$
$$\cos 2\theta_3 = \pm\frac{2\left(C^{3311} - C^{2233}\right) e_{(I)} + \left(C^{1111} - C^{2222}\right)\left(e_{(II)} + e_{(III)}\right)}{\left[C^{1111} + C^{2222} - 2\left(C^{1122} + 2C^{1212}\right)\right]\left(e_{(II)} - e_{(III)}\right)}. \tag{37c}$$

Analogous solutions can be obtained by aligning x^{II} or x^{III} with one of the principal directions of orthotropy. The general solution for the three dimensional case is still unknown; the complete solution for the simpler cubic case has beeen found by Rovati and Taliercio [11].

In the next Sections we shall focus our attention to the two dimensional case, for which complete solutions are available and mechanical interpretations and applications more easily found.

THE OPTIMIZATION PROBLEM IN PLANE ELASTICITY

In this Section, with reference to plane elasticity, the optimal orientation of fibers or, more in general, of principal directions of orthotropy, will be sought, with the aim of highligh some particular mechanical features of optimal solutions, which easily suggest possible practical applications. As reference, see, for instance, Banichuk [5], Pedersen [8] and [9] and Sacchi and Rovati [7].

Finally, an optimization problem, in which both fiber orientation and fiber density are adopted as design variables, will be presented and discussed (see also Thomsen and Olhoff [15]).

In an orthogonal reference frame x_1–x_2–x_3, let us consider a solid body defined on the open domain $\Omega \in \mathbb{R}^2$ (in the plane x_1–x_2) with boundary $\partial\Omega \equiv \partial\Omega_1 \cup \partial\Omega_2$, sufficiently smooth. Moreover, let the body be subjected to a plane state of stress, with tractions and displacements prescribed on $\partial\Omega_1$ and $\partial\Omega_2$ respectively. Assume also x_3 to be a principal direction for the strain tensor. Moreover, the body is regarded as locally orthotropic in the plane x_1–x_2 and the principal directions of orthotropy z_1, z_2 are specified through the angle $\theta(x_1, x_2)$, made by the z_1 and x_1 axes (see Fig. 1).

So, with the mechanical properties formerly shown, the behaviour of the solid is governed by the stress–strain relationship and by the following mixed boundary value problem: the equilibrium equations (with $i, j = 1, 2$), expressed in terms of generalized variables (n_i are the components of the outward normal unit vector, defined on $\partial\Omega$, and overbars indicate the closure of the set), and in the absence of body forces, are

$$N_{ji,j} = 0 \quad \text{in } \Omega \tag{38}$$

$$N_{ij} = N_{ji} \quad \text{in } \overline{\Omega} \tag{39}$$

$$N_{ij}n_j = p_i \quad \text{on } \partial\Omega_1 \tag{40}$$

and the compatibility equations with the displacement boundary conditions are

$$\epsilon_{ij} = \frac{1}{2}(u_{i,j} + u_{j,i}) \quad \text{in } \overline{\Omega} \tag{41}$$

$$u_i = \hat{u}_i \quad \text{on } \partial\Omega_2 \tag{42}$$

where $\hat{u}_i$ ($i = 1, 2$) are prescribed displacement functions.

For given orthotropic properties, i.e. for given values of the coefficients A_{ij}^0, and assuming as design variable the angle θ, the problem consists in finding the optimal function $\theta(x_1, x_2)$ in order to minimize the work performed by the prescribed external loads p_i (compliance problem), (see Banichuk [5]). In other words, as already seen in the previous Section for the three dimensional continuum body, the problem can be stated as follows: *find, in agreement with equilibrium and compatibility*

$$\mathcal{F}^0 \equiv \min_\theta \mathcal{F}(\theta) = \min_\theta \frac{1}{2} \left(\int_{\partial\Omega_1} p_i \overline{u}_i dS \right) \quad i, j = 1, 2. \tag{43}$$

Again, the functional to be minimized is equal, but opposite in sign, to $\mathcal{U}^0 = \min_{u_i} \mathcal{U}$, being $\mathcal{U}$ the total potential energy. So, the optimization problem can be also written as: *find*

$$\mathcal{F}^0 \equiv \min_\theta \mathcal{F}(\theta) = \min_\theta \left(-\mathcal{U}^0 \right). \tag{44}$$

Moreover, it is well known that, in the class of the compatible displacement fields, the solution of the above mentioned boundary value problem minimizes the functional $\mathcal{U}$. In such a way it is possible to rewrite again the problem in the following form: *find*

$$\mathcal{F}^0 \equiv \mathcal{F}(\theta) = \min_\theta \left(-\min_{u_i} \mathcal{U}\right) = \min_\theta \max_{u_i} (-\mathcal{U})$$

$$= \min_\theta \max_{u_i} \left(\int_{\partial\Omega_1} p_i u_i dS - \iint_\Omega \mathcal{E} d\Omega\right) \tag{45}$$

where $\mathcal{E}$ represents the specific elastic strain energy that, in our case, can be written explicitly as

$$\mathcal{E} = \frac{1}{2}\left(A_{11}\epsilon_{11}^2 + A_{22}\epsilon_{22}^2 + 4A_{66}\epsilon_{12}^2 + 2A_{12}\epsilon_{11}\epsilon_{22} + 4A_{16}\epsilon_{11}\epsilon_{12} + 4A_{26}\epsilon_{12}\epsilon_{22}\right). \tag{46}$$

In such a way, the constrained *min* problem (43) has been transformed in a unconstrained *min max* problem. Then the stationarity conditions due to a compatible variation of the state variables u_i return the equilibrium equations (38) and (40).

In order to compute the optimality condition with respect to the design variable θ it appears convenient to introduce a new local reference frame y_1-y_2 coinciding, in each point of the body, with the principal direction of strain. Let us define such reference system through the angle $\phi(x_1, x_2)$ made by the z_1 and y_1 axes.

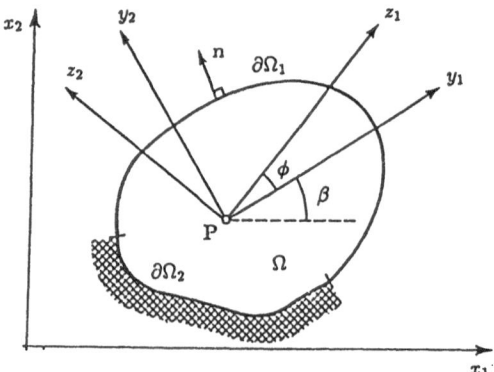

Figure 3. Definition of the design variable.

From Fig. 3 it appears that the design variable is given by $\theta = \beta + \phi$; nevertheless the angle $\beta(x_1, x_2)$ is univocally defined when the strain state is known. Thus it is possible to assume as new design variable the angle $\phi(x_1, x_2)$. By virtue of this assumption, and in the local frame y_1-y_2, one has

$$2\mathcal{E} = A_{11}\epsilon_I^2 + A_{22}\epsilon_{II}^2 + 2A_{12}\epsilon_I\epsilon_{II} \tag{47}$$

where ϵ_I and ϵ_{II} are the principal components of the strain tensor.

By means of the rotation formulas (14), and defining

$$A = \left(A_{11}^0 + A_{22}^0 - 2A_{12}^0 - 4A_{66}^0\right)(\epsilon_I - \epsilon_{II})^2 \tag{48}$$

$$B = 2\left[\left(-A_{22}^0 + A_{12}^0 + 2A_{66}^0\right)\epsilon_I^2 + \left(-A_{11}^0 + A_{12}^0 + 2A_{66}^0\right)\epsilon_{II}^2 + \right.$$
$$\left. + \left(A_{11}^0 + A_{22}^0 - 2A_{12}^0 - 4A_{66}^0\right)\epsilon_I\epsilon_{II}\right] \tag{49}$$

$$C = A_{22}^0\epsilon_I^2 + A_{11}^0\epsilon_{II}^2 + 2A_{12}^0\epsilon_I\epsilon_{II} \tag{50}$$

Rel. (47) becomes

$$2\mathcal{E} = A\cos^4\phi + B\cos^2\phi + C. \tag{51}$$

Now, the optimality condition (see Rel. (45)) is given by

$$\frac{\partial(-\mathcal{E})}{\partial\phi} = 0 \quad \text{in } \Omega \tag{52}$$

that, in explicit form, furnishes

$$\sin\phi\cos\phi\left(2A\cos^2\phi + B\right) = 0 \quad \text{in } \Omega. \tag{53}$$

By virtue of such a condition, it clearly appears that the orthotropy axes can be oriented, with respect to the principal axes of strain, in three different ways (see Banichuk [5])

$$\sin\phi = 0 \qquad \text{in } \Omega \tag{54}$$
$$\cos\phi = 0 \qquad \text{in } \Omega \tag{55}$$
$$\cos^2\phi = -\frac{B}{2A} \qquad \text{in } \Omega \text{ with } 0 < -\frac{B}{2A} < 1. \tag{56}$$

This result has also a mechanical interpretation; in fact it must be noticed that the condition (56) can be explicitly written as

$$\cos^2\phi = -\frac{c_2\epsilon_I^2 + c_1\epsilon_{II}^2 - (c_1 + c_2)\,\epsilon_I\epsilon_{II}}{(c_1 + c_2)\,(\epsilon_I - \epsilon_{II})^2} \tag{57}$$

where, for the sake of brevity, the following coefficients have been introduced as functions of the elastic stiffnesses only

$$c_1 = A_{11}^0 - A_{12}^0 - 2A_{66}^0 \tag{58}$$
$$c_2 = A_{22}^0 - A_{12}^0 - 2A_{66}^0. \tag{59}$$

Finally, Rel. (57) becomes

$$\cos^2\phi = \frac{c_2\epsilon_I - c_1\epsilon_{II}}{(c_1 + c_2)\,(\epsilon_I - \epsilon_{II})} = -\frac{B}{2A}. \tag{60}$$

As in the three dimensional case, at this point it is possible to show how the necessary stationarity condition (53), and consequently conditions (54), (55) and (56), are conditions of collinearity between the principal directions of stress and strain.
Indeed, in that case, the stress–strain relationship is written as

$$\left\{\begin{array}{c} \sigma_I \\ \sigma_{II} \\ 0 \end{array}\right\} = \begin{pmatrix} A_{11} & A_{12} & A_{16} \\ A_{12} & A_{22} & A_{26} \\ A_{16} & A_{26} & A_{66} \end{pmatrix} \left\{\begin{array}{c} \epsilon_I \\ \epsilon_{II} \\ 0 \end{array}\right\} \tag{61}$$

where the third equation represents the condition of collinearity between the principal directions

$$A_{16}\epsilon_I + A_{26}\epsilon_{II} = 0. \tag{62}$$

Now, making use again of Rel.s (58) and (59), it is possible to rewrite (14d) and (14e) as

$$A_{16} = \sin\phi\cos\phi\left(c_2\sin^2\phi - c_1\cos^2\phi\right) \tag{63}$$
$$A_{26} = \sin\phi\cos\phi\left(c_2\cos^2\phi - c_1\sin^2\phi\right) \tag{64}$$

and consequently Rel. (62) becomes

$$\sin\phi\cos\phi\left[(c_1 + c_2)(\epsilon_I - \epsilon_{II})\cos^2\phi - (c_2\epsilon_I - c_1\epsilon_{II})\right] = 0 \tag{65}$$

which, by virtue of (60), allows us to obtain again

$$\sin\phi\cos\phi\left(2A\cos^2\phi + B\right) = 0. \tag{66}$$

This means that the principal directions of stress and strain are collinear if one at least of the conditions (54) to (56) is fullfilled.

It is worth noting that Rel.s (54) and (55) have as a consequence the simultaneous collinearity between principal directions of strain, stress and orthotropy. Instead, Rel. (56) furnishes a third possible orientation of the orthotropy axes, not coinciding with the principal directions of stress and strain tensors, which, on the other hand, are still collinear.

The question that now arises is if all the conditions (54) to (56) represent conditions for local minima. An answer can be given by checking the second variation of the functional, with respect to the design variable. In other words we have to check the sign of the following expression

$$\frac{\partial^2\left(-\mathcal{E}\right)}{\partial\phi^2} = \left(\cos^2\phi - \sin^2\phi\right)\left(2A\cos^2\phi + B\right) - 4A\sin^2\phi\cos^2\phi. \tag{67}$$

Such an expression provides conditions for the three possible orientations of orthotropy axes, as determined by the sign of

$$\left.\frac{\partial^2\left(-\mathcal{E}\right)}{\partial\phi^2}\right|_{\sin\phi=0} = 2A + B \tag{68}$$

$$\left.\frac{\partial^2\left(-\mathcal{E}\right)}{\partial\phi^2}\right|_{\cos\phi=0} = -B \tag{69}$$

$$\left.\frac{\partial^2\left(-\mathcal{E}\right)}{\partial\phi^2}\right|_{\cos^2\phi=\frac{-B}{2A}} = \frac{B}{A\left(B + 2A\right)}. \tag{70}$$

If, for instance, we assume that the solutions $\sin\phi = 0$ and $\cos\phi = 0$ correspond to local minima, then

$$2A + B \geq 0 \tag{71}$$

$$-B \geq 0. \tag{72}$$

These inequalities, after substitution in (70), furnish

$$\frac{B}{A\left(B + 2A\right)} \leq 0. \tag{73}$$

This result means that if the solution showing collinearity between principal axes of orthotropy and strain is a local minimum, then the third possible orientation (70) correspond to a local maximum. On the other hand, it is not difficult to verify, by replacing (48) and (49) in (71) and (72), that

$$A_{11}^0 - A_{12}^0 - 2A_{66}^0 \geq 0 \tag{74a}$$

$$A_{22}^0 - A_{12}^0 - 2A_{66}^0 \geq 0. \tag{74b}$$

These inequalities, from a physical point of view, can be assumed true. This means that the orientation of the orthotropy axes stated by Rel. (56) cannot correspond to an optimal solution. It must be noticed

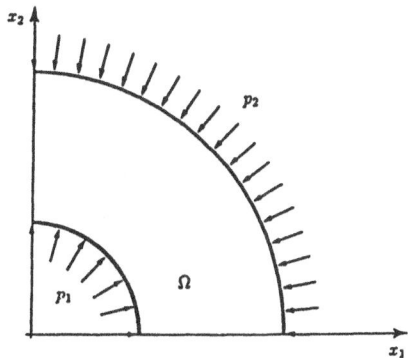

Figure 4. Example of application.

that, although these considerations can be applied to most of the materials, Pedersen (see [8] and [9]) has pointed out that in some cases (namely for materials which show a *relatively high* shear stiffness) the optimal solution is characterized by the intermediate orientation of orthotropic axes (i.e., by Rel (56)).

Now, as an example of application, consider the simple case of an anular disc subjected to internal and external pressures p_1 and p_2 (see Fig. 4); if R_1 and R_2 are the inner and external radii respectively, the principal directions of strain are in radial and circumferential directions and the equilibrium differential equation written in polar coordinates is

$$\frac{1}{r}\frac{\partial(r\sigma_{rr})}{\partial r} - \frac{\sigma_{\theta\theta}}{r} = 0 \tag{75}$$

with the boundary conditions $\sigma_{rr}(R_1) = -p_1$, $\sigma_{rr}(R_2) = -p_2$.

Now it is necessary to distinguish two cases.

CASE 1: $\sin\phi = 0$. Then the stiffness coefficients are $A_{11} = A_{11}^0$, $A_{12} = A_{12}^0$ and $A_{22} = A_{22}^0$, and the equilibrium equation becomes

$$u_{rr} + \frac{u_r}{r} - \frac{A_{22}^0}{A_{11}^0}\frac{u}{r^2} = 0 \tag{76}$$

the solution of which, with the above mentioned boundary conditions and defining $k^2 = A_{22}^0/A_{11}^0$, is

$$u(r) = \frac{\left(\frac{p_2 R_2^{-k-1} - p_1 R_2^{-k-1}}{A_{12}^0 + kA_{11}^0}r^k + \frac{p_1 R_1^{k-1} - p_2 R_1^{k-1}}{A_{12}^0 - kA_{11}^0}r^{-k}\right)}{R_1^{k-1}R_2^{-k-1} - R_1^{-k-1}R_2^{k-1}}. \tag{77}$$

CASE 2: $\cos\phi = 0$. In this case the stiffness coefficients are given by $A_{11} = A_{22}^0$, $A_{22} = A_{11}^0$ and $A_{12} = A_{12}^0$. The differential equation is then analogous to (76) and the solution can be obtained by replacing k with q in (77), where $q^2 = A_{11}^0/A_{22}^0$.

Now, the solution of the optimization problem can be obtained by comparing the values of the compliance corresponding to the displacement fields of Case 1 and Case 2. The compliance is given by

$$C = \pi p_1 R_1 u(R_1) + \pi p_2 R_2 u(R_2) \tag{78}$$

and it is not difficult to verify that the value of the compliance depends on the ratio A_{11}^0/A_{22}^0. In particular, if $A_{11}^0 = A_{22}^0$, the cases $\sin\phi = 0$ and $\cos\phi = 0$ give rise to the same solution.

The result here obtained seems to suggest the possibility to deal with optimization problems making use of the isotropic solution. In other words, to determine, as a first step, the principal directions of stress (or strain) for the structural member seen as isotropic, then to orient the orthotropy axes along these directions; if displacement boundary conditions are absent, then the orthotropic body is able to assume a compatible configuration which shows collinearity between stress, strain and principal directions of orthotropy. If A_{11}^0 is not equal to A_{22}^0, this practical tool furnishes a local minimum corresponding to solution (54) or (55). For instance, if we consider a thin tube made of an orthotropic material subjected to torsion, it is well known that the isotropic solution is given by the Bredt's formula and the principal directions of stress are directed at $\pi/4$ with respect to the axis of the cylinder. So, a solution with the orthotropy axes in this direction can be regarded as an optimal solution.

Now it is possible to show that the results obtained for a plane state of stress can be easily extended to flexural problems, and in particular it can be shown that Eq. (53) represents the optimality condition also for the minimum compliance problem of a solid elastic plate in bending. In such a case, and for the same body formerly defined, the constitutive law is given by Rel. (21) in absence of midplane strains

$$\begin{Bmatrix} M_{11} \\ M_{22} \\ M_{12} \end{Bmatrix} = \begin{pmatrix} D_{11} & D_{12} & D_{16} \\ D_{12} & D_{22} & D_{26} \\ D_{16} & D_{26} & D_{66} \end{pmatrix} \begin{Bmatrix} \chi_{11} \\ \chi_{22} \\ 2\chi_{12} \end{Bmatrix}. \tag{79}$$

It is necessary to point out that if, for the sake of simplicity, the plate is assumed orthotropic but not a laminate, the plane coefficients A_{ij} do not depend on x_3, so Rel. (24) furnishes:

$$D_{ij} = \frac{h^3}{12} Q_{ij}. \tag{80}$$

Now, the minimum compliance problem can be stated as follows: find, *in agreement with equilibrium and compatibility*

$$\mathcal{F}^0 \equiv \min_\theta \mathcal{F}(\theta) = \min_\theta \frac{1}{2} \left(\iint_\Omega p_3 \, \bar{u}_3 \, d\Omega \right) \tag{81}$$

where $p_3(x_1, x_2)$ and $\bar{u}_3(x_1, x_2)$ are the load distribution and the displacement function at equilibrium respectively, both in direction of x_3.

Once again, the *min* problem (81) can be transformed in the following *min max* problem: *find*

$$\mathcal{F}^0 \equiv \min_\theta \mathcal{F}(\theta) = \min_\theta (-\mathcal{U}^0) = \min_\theta \max_{u_3} \left(\frac{1}{2} \iint_\Omega p_3 \, u_3 \, d\Omega - \iint_\Omega \mathcal{E} \, d\Omega \right) \tag{82}$$

where $\mathcal{U}$ is the total potential energy and $\mathcal{E}$ the specific strain energy, given, in principal directions of curvature, by

$$\mathcal{E} = \frac{h^3}{24} \left(A_{11}\chi_I^2 + A_{22}\chi_{II}^2 + 2A_{12}\chi_I\chi_{II} \right). \tag{83}$$

Proceeding as in plane stress case, we adopt as a new design variable the angle $\phi(x_1, x_2)$ made by the principal directions of curvature and the principal directions of orthotropy. By virtue of Rels. (48) to (50), the strain energy (83) can be rewritten as

$$\mathcal{E} = \frac{h^3}{24} \left(A\cos^4\phi + B\cos^2\phi + C \right). \tag{84}$$

Finally, the stationarity condition with respect to the displacement u_3 furnishes the equilibrium equations in Ω and on the boundary, whereas the stationarity condition with respect to the angle ϕ (optimality condition) reads again

$$\sin\phi\cos\phi \left(2A\cos^2\phi + B \right) = 0. \tag{85}$$

So, it clearly appears that the mechanical interpretation of the optimality condition formerly shown can be repeated without any change in the case of plates in bending.

As a conclusion, an extension of the plane stress optimization problem can be given minimizing the work done by the external loads, with an upper bound on the total amount of the structural cost or, in other words, with a bound on the density of reinforcement, which can be regarded as responsible for the cost of the structure. If we assume as meaningful parameters the stiffnesses A_{11}^0 and A_{22}^0, no longer assumed as given, the problem, taking as design variables both the angle $\theta(x_1, x_2)$ and the functions $A_{11}^0(x_1, x_2)$ and $A_{22}^0(x_1, x_2)$, can be stated in the following way: find

$$\mathcal{F}^0 \equiv \min_{\theta, A_{ii}^0} \mathcal{F}(\theta, A_{ii}^0) = \min_{\theta, A_{ii}^0 \geq A^0} \left(\frac{1}{2} \iint_{\Omega} p_i \overline{u}_i \, ds \right) \tag{86}$$

subject to the equilibrium and compatibility conditions and to the global constraint

$$\iint_{\Omega} (A_{11}^0 + A_{22}^0) \, d\Omega \leq K^0 \tag{87}$$

with A^0 and K^0 prescribed values. Here A^0 can be seen as the contribution to the stiffness due to the matrix in a composite. Making use of the Lagrange multiplier method, problem (86), (87) can be stated in the form: find

$$\mathcal{F}^0 \equiv \min_{\theta} \mathcal{F}(\theta, A_{ii}^0, \alpha) = \inf_{\theta, A_{ii}^0 \geq A^0} \sup_{u_i, \alpha \geq 0} \left(\int_{\partial\Omega_1} p_i u_i \, ds - \iint_{\Omega} \mathcal{E} \, d\Omega \right.$$
$$\left. + \alpha \iint_{\Omega} (A_{11}^0 + A_{22}^0) \, d\Omega - \alpha K^0 \right) \tag{88}$$

where α is a Lagrange multiplier.

Now, the stationarity conditions with respect to displacements u_i return the equilibrium equations, in Ω and on the boundary. Once again, the stationarity with respect to the orientation angle of orthotropy axes furnishes

$$\sin\phi \cos\phi \, (2A \cos^2\phi + B) = 0. \tag{89}$$

Finally, as a consequence of a variation of the stiffness coefficients A_{ii}^0, we obtain the following inequalities that must be fulfilled by the strain field at the optimum

$$(\epsilon_I - \epsilon_{II})^2 \cos^4\phi + 2(\epsilon_I \epsilon_{II} - \epsilon_{II}^2)\cos^2\phi + \epsilon_{II} \leq \frac{2}{\alpha} \tag{90}$$

$$(\epsilon_I - \epsilon_{II})^2 \cos^4\phi + 2(\epsilon_I \epsilon_{II} - \epsilon_I^2)\cos^2\phi + \epsilon_I^2 \leq \frac{2}{\alpha}. \tag{91}$$

For the two cases $\sin\phi = 0$ and $\cos\phi = 0$, corresponding as shown before to local minima, both Rels. (90) and (91) give

$$\epsilon_{II}^2 \leq \frac{2}{\alpha} \quad \text{and} \quad \epsilon_I^2 \leq \frac{2}{\alpha}. \tag{92}$$

If the stiffnesses A_{ii}^0 do not attain their lower value A^0, then inequalities (92), at the optimum, become equalities and the following relation holds

$$|\epsilon_I| = |\epsilon_{II}| = \text{const.} \quad \text{in } \Omega. \tag{93}$$

This strong constraint on the strain field can be, on the other hand, avoided if A_{11}^0 (or A_{22}^0) is equal to A^0 (i.e. if the reinforcement of the composite in the z_1 (resp. z_2) direction vanishes), and one of the inequalities (94) can be recovered. A related problem has been dealt with by Thomsen and Olhoff [15], where some numerical solution are also given.

So, it has been shown that in the minimum compliance problem, with a constraint on the total structural cost, the condition of collinearity between principal directions of stress, strain and orthotropy still holds, but an optimal design can be obtained only for particular values of the components of strain.

OPTIMAL LIMIT DESIGN IN THE PLANE CASE

As in the previous Section, consider again an orthogonal reference frame $x_1 - x_2$, where a solid body of domain $\Omega \in \mathbf{R}^2$ is defined, with boundary $\partial\Omega \equiv \partial\Omega_1 \cup \partial\Omega_2$ on which tractions and displacements are prescribed, respectively. Now we assume that the body is made by a non structural matrix component and reinforced by a system of fibers orthogonal each others. In such a way the solid is regarded as locally orthotropic, and the local orthogonal frame $z_1 - z_2$ denotes the principal directions of orthotropy, (refer again to Fig. 1). Moreover, the stresses in the material are supposed to be entirely carried by the fibers, both in traction and in compression.

For a prescribed load distribution p on $\partial\Omega_1$, the stress field in the body, expressed in terms of characteristics, is specified through the tensor $\underline{N}$, which satisfies the equilibrium equation (38) to (40), again for vanishing body forces. The fibers are supposed to have cross sectional area variable along the fiber length. In principal directions of orthotropy, at each point of the body, the strength of the fibers, and therefore of the material, is defined as

$$R^+{}_{I,II} = H_{I,II}\,\sigma^+ \tag{94}$$

$$R^-{}_{I,II} = H_{I,II}\,\sigma^- \tag{95}$$

where H_I and H_{II} represent the cross sectional areas of the fibers in direction z_1 and z_2 respectively, while σ^+ and σ^- are the ultimate stresses in traction and compression. The strength of the material is assumed to be defined through the components of the tensors $\underline{\underline{R}}^+$ and $\underline{\underline{R}}^-$ (see the symbolic representations of Fig.s 5a and 5b).

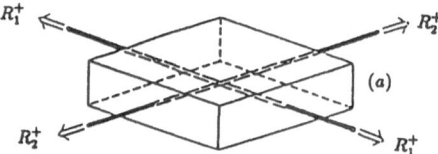

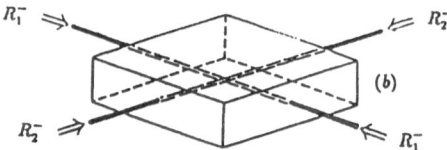

Figure 5. Symbolic representation of the strength criterion.

Finally, the strength criterion, in orthogonal coordinates, can be stated as follows (see Fig. 6)

$$R^+{}_{nn} \geq N_{nn} \geq -R^-{}_{nn} \quad , \forall n \tag{96}$$

under the conditions of non–negativeness

$$R^+{}_{nn} \geq 0 \quad , \forall n \tag{97}$$

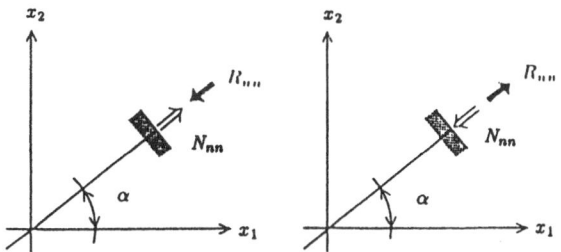

Figure 6. Strength criterion.

$$R^-_{nn} \geq 0 \quad , \forall n \tag{98}$$

Instead of a representation given in terms of components, the strength criterion can be written in tensor form

$$\mathrm{tr}\,(\underline{R}^+ - \underline{N}) \geq 0 \quad \text{and} \quad \det(\underline{R}^+ - \underline{N}) \geq 0 \tag{99}$$

$$\mathrm{tr}\,(\underline{R}^- + \underline{N}) \geq 0 \quad \text{and} \quad \det(\underline{R}^- + \underline{N}) \geq 0 \tag{100}$$

In the generalized stress space, the ultimate strength domain defined by (99) and (100) can be represented as in Fig. 7. Or, analogously, in the space N_I, N_{II} of the principal stresses, as in Fig. 8.

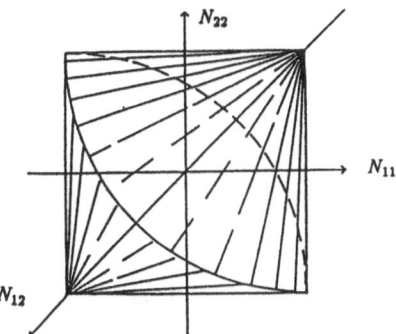

Figure 7. Ultimate strength domain in the generalized stress space.

From Fig. 8 it clearly appears that the ultimate strength domain, for given values of the limit stresses σ^+ and σ^-, depends on the cross sectional areas of the fibers.

By virtue of such a remark, in the present Section the total amount of structural material (fibers), i.e., the size of the strength domain, will be minimized, for prescribed values of the external loads, according to equilibrium and strength criterion. In such a way, assuming as design variables the local values of the cross sectional areas of the fibers, the optimal design problem can be stated as follows: find

$$\mathcal{F}^0 \equiv \min_{H_{I,II}} \mathcal{F}(H_{I,II}) = \min_{H_{I,II} \geq 0} \iint (H_I + H_{II}) d\Omega \tag{101}$$

724

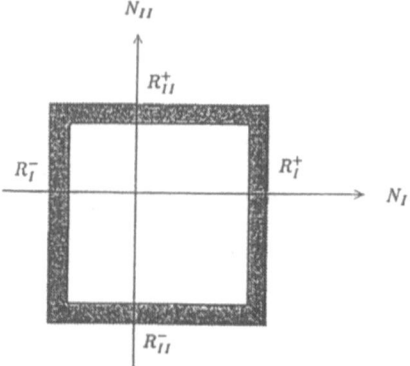

Figure 8. Ultimate strength domain in the space of principal stresses.

according to the equilibrium conditions (38) to (40) and to the inequality constraints

$$\mathrm{tr}\left(\underline{\underline{R}}^+ - \underline{N}\right) \geq 0 \qquad \mathrm{tr}\left(\underline{\underline{R}}^- + \underline{N}\right) \geq 0 \tag{102}$$

$$\det\left(\underline{\underline{R}}^+ - \underline{N}\right) \geq 0 \qquad \det\left(\underline{\underline{R}}^- + \underline{N}\right) \geq 0 \tag{103}$$

By virtue of (94) and (95) the design variables H_I and H_{II} are replaced by R_I^+ and R_{II}^+, taking into account that

$$R_I^+ R_I^- = R_{II}^+ R_{II}^- = 0 \tag{104}$$

and consequently (101) reads

$$\mathcal{F}^0 \equiv \min_{H_{I,II}} \mathcal{F}(H_{I,II}) = \min_{\underline{\underline{R}}^\pm \geq 0} \iint_\Omega \left[\mathrm{tr}(\underline{\underline{R}}^+) + \mathrm{tr}(\underline{\underline{R}}^-)\right] \, d\Omega \tag{105}$$

The minimization problem (105) subject to the constraints (38) to (40), (102) and (103) can be transformed in the search of the stationarity of the unconstrained Lagrangian functional $\mathcal{L}$, defined as

$$\mathcal{L} = \iint_\Omega \left(\mathrm{tr}\underline{\underline{R}}^+ + \mathrm{tr}\underline{\underline{R}}^-\right) \, d\Omega + \iint_\Omega \phi^T \mathrm{div}\underline{\underline{N}} \, d\Omega + \int_{\partial\Omega_1} \phi^T \left(\underline{\underline{N}}\underline{n} - \underline{p}\right) \, dS +$$

$$+ \iint_\Omega \mu_1 \mathrm{tr}\left(\underline{\underline{R}}^+ - \underline{N}\right) \, d\Omega + \iint_\Omega \alpha_1 \det\left(\underline{\underline{R}}^+ - \underline{N}\right) \, d\Omega +$$

$$+ \iint_\Omega \mu_2 \mathrm{tr}\left(\underline{\underline{R}}^- + \underline{N}\right) \, d\Omega + \iint_\Omega \alpha_2 \det\left(\underline{\underline{R}}^- + \underline{N}\right) \, d\Omega \tag{106}$$

obtained by appending to the functional to be minimized the constraints through the Lagrangian multipliers ϕ, μ_1, μ_2, α_1 and α_2.
In such a way, the optimization problem can be rewritten in the following way: *find*

$$\mathcal{V} = \inf \sup \mathcal{L} \tag{107}$$

where the *infimum* must be computed with respect to the state variables $\underline{\underline{R}}^+ \geq 0$, $\underline{\underline{R}}^- \geq 0$, $\underline{N}$, and the *supremum* with respect to Lagrangian multipliers ϕ, $\mu_1 \leq 0$, $\mu_2 \leq 0$, $\alpha_1 \leq 0$ and $\alpha_2 \leq 0$.

The following orthogonality constraints hold

$$\mu_1 \operatorname{tr} \left(\underline{\underline{R}}^+ - \underline{\underline{N}} \right) = 0 \tag{108}$$

$$\mu_2 \operatorname{tr} \left(\underline{\underline{R}}^- + \underline{\underline{N}} \right) = 0 \tag{109}$$

$$\alpha_1 \det \left(\underline{\underline{R}}^+ - \underline{\underline{N}} \right) = 0 \tag{110}$$

$$\alpha_2 \det \left(\underline{\underline{R}}^- + \underline{\underline{N}} \right) = 0 \tag{111}$$

$$\mu_1 \mu_2 = 0 \tag{112}$$

$$\alpha_1 \alpha_2 = 0 \tag{113}$$

The necessary stationarity conditions with respect to a variation of the state variables N_{11}, N_{22} and N_{12} are respectively

$$- \phi_{1,1} - \mu_1 - \alpha_1 \left(R_{22}^+ - N_{22} \right) + \mu_2 + \alpha_2 \left(R_{22}^- + N_{22} \right) = 0 \tag{114}$$

$$- \phi_{2,2} - \mu_1 - \alpha_1 \left(R_{11}^+ - N_{11} \right) + \mu_2 + \alpha_2 \left(R_{11}^- + N_{11} \right) = 0 \tag{115}$$

$$- \phi_{1,2} + \alpha_1 \left(R_{12}^+ - N_{12} \right) - \alpha_2 \left(R_{12}^- + N_{12} \right) = 0 \tag{116}$$

Then, the stationarity conditions with respect to R_{11}^+, R_{22}^+ and R_{12}^+ read respectively

$$1 + \mu_1 + \alpha_1 \left(R_{22}^+ - N_{22} \right) \geq 0 \tag{117}$$

$$1 + \mu_1 + \alpha_1 \left(R_{11}^+ - N_{11} \right) \geq 0 \tag{118}$$

$$\alpha_1 \left(R_{12}^+ - N_{12} \right) = 0 \tag{119}$$

and, analogously, the stationarity with respect to R_{11}^-, R_{22}^- and R_{12}^- implies

$$1 + \mu_2 + \alpha_2 \left(R_{22}^- + N_{22} \right) \geq 0 \tag{120}$$

$$1 + \mu_2 + \alpha_2 \left(R_{11}^- + N_{11} \right) \geq 0 \tag{121}$$

$$\alpha_2 \left(R_{12}^- + T_{12} \right) = 0 \tag{122}$$

Such necessary conditions suggest some phisical considerations, concerning the optimal solution of the problem.

Remark 1. When the cross sectional areas of the reinforcing fibers are not vanishing (i.e., when $R_{11}^+ > 0$ and $R_{22}^+ > 0$), then the body, at the optimum, shows a fully stress and a corner stress behaviour, simultaneously. In fact, if $R_{11}^+ > 0$ and $R_{22}^+ > 0$, Rel.s (117) and (118) read

$$1 + \mu_1 + \alpha_1 \left(R_{22}^+ - N_{22} \right) = 0 \tag{123}$$

$$1 + \mu_1 + \alpha_1 \left(R_{11}^+ - N_{11} \right) = 0 \tag{124}$$

from which one has

$$\alpha_1 \left(R_{22}^+ - N_{22} \right) = \alpha_1 \left(R_{11}^+ - N_{11} \right) \tag{125}$$

and multiplying both the sides for $\left(R_{11}^+ - N_{11} \right)$

$$\alpha_1 \left(R_{11}^+ - N_{11} \right) \left(R_{22}^+ - N_{22} \right) = \alpha_1 \left(R_{11}^+ - N_{11} \right)^2 \tag{126}$$

Now, it is worth noting that the left hand side of Rel. (126), by virtue of the orthogonality condition (110), vanishes, for any value of $\alpha_1 \leq 0$. This means that

$$\left(R_{11}^+ - N_{11} \right)^2 = 0 \tag{127}$$

or

$$R_{11}^+ = N_{11} \tag{128}$$

Now, by virtue of equality (128), Rel. (125), which holds again for any $\alpha_1 \leq 0$, shows that also the following condition holds

$$R_{22}^+ = N_{22} \tag{129}$$

In such a way, Rels. (128) and (129) are simultaneously verified, that is, in other words, both the fully stress and the corner stress conditions hold at the optimum.

Remark 2. The optimal solution is characterized by local collinearity between stress, strength and strain rate. In fact, in principal directions of stress, i.e., when $N_{12} = 0$, and for $\alpha_1 \leq 0$, Rel. (119) implies also $R_{12} = 0$. Moreover, such a result, substituted in Rel. (116) and taking into account the orthogonality condition (113), returns $\phi_{1,2} = 0$.

By virtue of its own nature, the Lagrangian multiplier $\underline{\phi} = \{\phi_1, \phi_2\}^T$ can be seen as a displacement rate vector, and consequently its derivatives as strain rates. Hence, the principal directions of stress, strength and strain rate must be collinear in the optimal solution.

Remark 3. The stationarity conditions allow also for some kinematical considerations on the mechanism of the optimal solution. In fact, Rel. (117) furnishes

$$1 + \mu_1 + \alpha_1 \left(R_{22}^+ - N_{22} \right) \geq 0 \tag{130}$$

or

$$\mu_1 \geq -1 - \alpha_1 \left(R_{22}^+ - N_{22} \right) \tag{131}$$

The last inequality, after substitution in (114), gives

$$-\phi_{1,1} - \alpha_1 \left(R_{22}^+ - N_{22} \right) + \mu_2 + \alpha_2 \left(R_{22}^- + N_{22} \right) \geq -1 - \alpha_1 \left(R_{22}^+ - N_{22} \right) \tag{132}$$

which, by virtue of the fully stress condition (129) and of the orthogonalities (112) and (113), furnishes

$$1 - \phi_{1,1} \geq 0. \tag{133}$$

Following the same considerations also for the other analogous equations, one obtains the following set of inequalities

$$1 + \phi_{1,1} \geq 0 \tag{134}$$

$$1 - \phi_{2,2} \geq 0 \tag{135}$$

$$1 + \phi_{2,2} \geq 0 \tag{136}$$

By virtue of the fact that in Eq.s (133) and (135) the equality holds if $R_{11}^+ > 0$ and $R_{22}^+ > 0$ (or, analogously, if in Eq.s (134) and (136) $R_{11}^- > 0$ and $R_{22}^- > 0$) and, conversely, the inequality is true if the fibers vanish. This means that the optimal solution shows a mechanism characterized by a constant deformation in the fiber direction and by a sign–restricted deformation if the fibers vanish.

A UNIFIED APPROACH FOR APPROXIMATE OPTIMAL SOLUTIONS

As shown in the former Sections, optimal elastic solutions for orthotropic bodies are characterized, in most cases, by collinearity between principal directions of stress, strain and orthotropy. Analogously, in optimal design at the limit state, collinearity again appears between principal directions of stress, strain rate and strength. In the last case, the optimal solution is also characterized by a constant strain rate field in the fiber directions, over the body, with the possibility to have unconstrained strain rates if the fibers vanish.

In the elastic case, an analogous behaviour of optimal solutions is shown, if not only fiber orientation is assumed as design variable, but also the cross sectional areas of fibers (i.e., the local density of fibers) enter into the design. These common features of the optimal solution in the elastic and plastic cases suggest the possibility of a unified method of solution for both the problems.

At first, it must be noticed that the conditions, in terms of strains, or strain rates, ϵ (see Rel.s (93) and (134) to (136))

$$|\epsilon| = \text{const.} \quad \text{in fiber directions} \tag{137a}$$

$$|\epsilon| \leq \text{const.} \quad \text{if the fibers vanish} \tag{137b}$$

represent a common feature for the two problems.

On the other hand, the possibility of generating solutions with constant strain (or strain rate) fields seems to be possible only for few trivial cases. In such a way, a first possibility to avoid these difficulties can be seen in a re–formulation of the problem, allowing for some lines of discontinuity inside the body, i.e., taking into account the possibility to find an optimal solution characterized by constant strain (or strain rate) over subsets of the domain Ω of the body. Of course, such an hypotesis leads to possible discontinuities of the state variables and of the fiber paths, along the boundaries of discontinuity. Moreover the problem becomes in this way much more difficult, because the number and the shapes of the internal interfaces are both unknowns, and the optimal design problem becomes a free boundary problem.

Nevertheless, a simpler way to proceed to optimal solutions can be found if one observes that the conditions (137) are the same conditions fulfilled by a Michell truss, or, more precisely, by a truss–like continuum, i.e., by a double infinity of bars of infinitesimal length that follow the lines of a dense net formed by two orthogonal families of curves. It is worth noting that this classical result was derived for a problem concerning minimum weight design of isotropic structural elements under plane states of stress, at the limit state. On the other hand, the results here obtained show how the same optimality conditions are necessary conditions also for minimum weight plastic design and minimum elastic compliance problems, *for orthotropic bodies*.

The idea to deal with this kind of problems through truss–like solutions is mainly suggested from the peculiar meaning of the optimality conditions, but it also finds confirmation in the numerical solutions shown by Olhoff and Thomsen [15] and Bendsøe [18] and [19]. Such solutions, obtained through a direct optimization procedure, without using optimality criteria, show exactly a truss like behaviour of the optimal solution, when both orientation and fiber density are adopted as design variables.

As an example of application, consider a cantilever plate subject to a concentrated load P at its end. If we leave the formulation in the 2–D continuum and consider a cantilever truss, then the problem becomes a topological problem, with locations of the joint as unknowns, and the solution will result as a consequence of the adopted choices. For instance, if we approximate the plate with a five–bar truss, then the optimal solutions (for minimum elastic compliance and for minimum weight at the limit state) are shown in Fig. 9, where it must be observed that both the solutions are characterized by the same topology.

In Fig. 9, and in the following examples, the lower bound for the structural volume in the elastic case has been assumed to be $V_{min} = kAL$. The influence of the topology can be shown better by increasing the number of bars in the truss; in Fig. 10 the only two possible solutions characterized by constant strain (or strain rates) are illustrated for a ten–bar truss.

In this case it is well clear how the layout of the structure influences the solution. These two examples, on the other hand, constrain the solution to locate the joints in prescribed positions of the design domain. If some of such constraints are relaxed then, as shown in Fig. 11, the optimal solution can be improved.

In this way, a first approximation of the optimal structural layout has been found, and further improvements can be obtained, for instance, by increasing the number of bars. In other words, when an approximately optimal topology has been found, then starting from this tentative design it is possible to find better solutions transforming the initial structure in a more complex one by suitable increase of bars and joints.

In fact, referring to the structure of Fig. 12 and for the minimum volume problem at the limit state (but the consideration can be easily repeated in analogous way for the elastic case), it can be shown that the

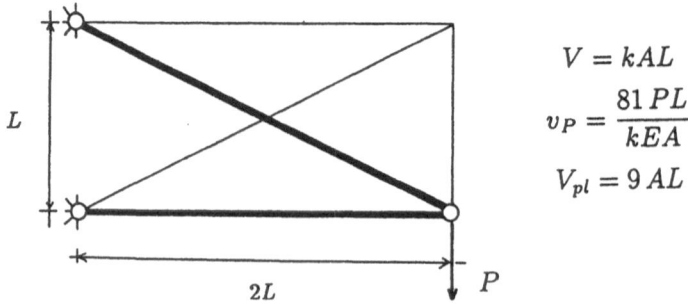

$$V = kAL$$
$$v_P = \frac{81\,PL}{kEA}$$
$$V_{pl} = 9\,AL$$

Figure 9. Example problem for optimization. Starting structure: five–bar truss.

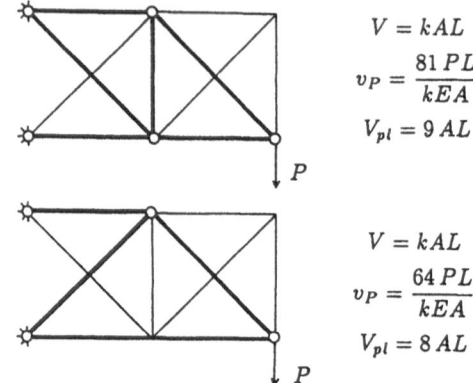

$$V = kAL$$
$$v_P = \frac{81\,PL}{kEA}$$
$$V_{pl} = 9\,AL$$

$$V = kAL$$
$$v_P = \frac{64\,PL}{kEA}$$
$$V_{pl} = 8\,AL$$

Figure 10. Example problem for optimization. Starting structure: ten–bar truss.

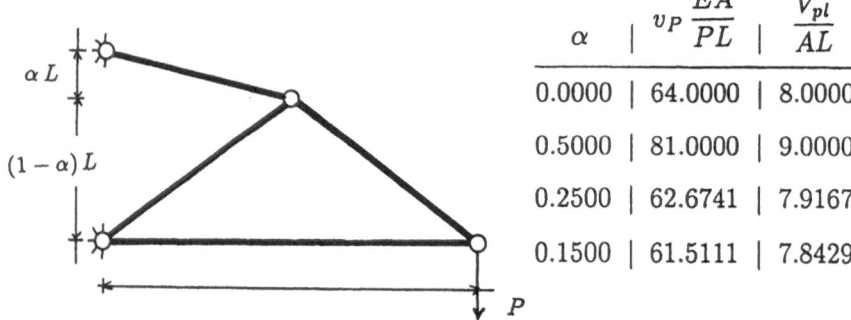

α	$v_P \dfrac{EA}{PL}$	$\dfrac{V_{pl}}{AL}$
0.0000	64.0000	8.0000
0.5000	81.0000	9.0000
0.2500	62.6741	7.9167
0.1500	61.5111	7.8429

Figure 11. Optimal solution when also the joint locations are design variables.

minimum weight of the truss $ABCDE$ (5 joints, 6 bars) is less or equal to the minimum weigh of the truss

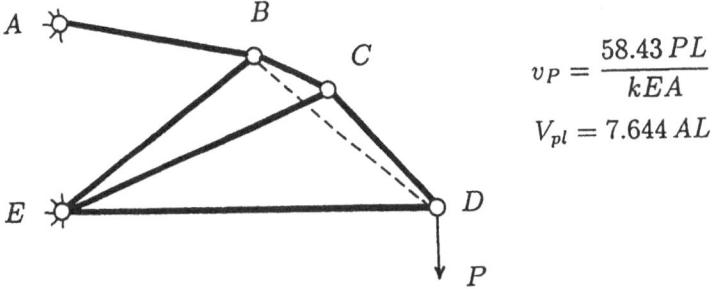

$$v_P = \frac{58.43\,PL}{kEA}$$

$$V_{pl} = 7.644\,AL$$

Figure 12. Improved solution.

$ABDE$ (4 joints, 4 bars), i.e.,

$$\inf \sum_{k=1}^{6} A_k L_k \leq \inf \sum_{k=1}^{4} A_k L_k \tag{138}$$

where the *infimum* are computed with respect to the joint locations. The inequality (138) can be immediatly verified if one considers that the right hand side corresponds to the minimum weight of the truss $ABCDE$ when the joint C belongs to the line connecting the joints B and E. It is well clear that if this geometric constraint is assumed, then the space of feasible solutions for the *inf* problem at the left hand side of inequality (138) is restricted, and therefore a lower value for the objective function can be found if the unconstrained problem is considered.

The simple examples presented here clearly show how the optimal topology problem leads to practical solutions, easy to handle, but also show that the *optimal structure* can be obtained only increasing indefinitely the number of bars and joints. Such solution, on the other hand, has only a theoretical meaning, while practical applications can be found with a finite number of elements in the truss.

ACKNOWLEDGEMENTS

The present work has been made possible by financial supports of Italian Ministry of University and Scientific and Technological Research (M.U.R.S.T.), which is here gratefully acknowledged. The authors are also indebted to Prof. N. Olhoff (University of Aalborg, DK) and Prof. P. Pedersen (The Technical University of Denmark) for the useful discussions on the subject.

REFERENCES

1. S.G. Lekhnitskii, *Theory of Elasticity of an Anisotropic Body*, Mir Publisher, Moscow, 1981.
2. R.M. Jones, *Mechanics of Composite Materials*, McGraw Hill, New York, 1975.
3. J.R. Vinson, R.L. Sierakowski, *The Behaviour of Structures Composed of Composed Materials*, Martinus Nijhoff Publ., Dordrecht, 1986.
4. L.J. Broutman, R.H. Krok, *Composite Materials*, Vol. 7, *Structural Design and Analysis, Part 1*, (C.C. Chamis Ed.), Academic Press, New York, 1975.
5. N.V. Banichuk, *Optimization Problems for Elastic Anisotropic Bodies*, Arch. Mech., Vol. 33, N. 6, pp. 347–363, 1981.

6. V.M. Kartevelishvili, V.V. Kobelev, *Rational Schemes for Reinforcing Laminar Plates from Composite Materials*, P.M.M., Vol. 48, N. 1, pp. 68–88, 1984.

7. G. Sacchi Landriani, M. Rovati, *Optimal Design for Two-Dimensional Structures made of Composite Materials*, Journal of Engineering Materials and Technology, Vol. 113, pp. 88–92, 1991.

8. P. Pedersen, *On Optimal Orientation of Orthotropic Materials*, Structural Optimization, Vol. 1, pp. 101–106, 1989.

9. P. Pedersen, *Bounds on Elastic Energy in Solids of Orthotropic Materials*, Structural Optimization, Vol. 2, pp. 55–63, 1990.

10. G.A. Seregin, V.A. Troitskii, *On the Best Position of Elastic Symmetry Planes in an Orthotropic Body*, P.M.M., Vol. 45, pp. 139–142, 1981.

11. M. Rovati, A. Taliercio, *Optimal Orientation of the Symmetry Axes of Orthotropic 3-D Materials*, Proc. Int. Conf. on *Engineering Optimization in Design Processes* (H.A. Eschenauer, C. Matthek, N. Olhoff Eds.), Karlsruhe Nuclear Research Center, Germany, September 3–4, 1990, pp. 127–134, Springer–Verlag, Berlin, 1991.

12. N.V. Banichuk, V.V. Kobelev, *On Optimal Plastic Anisotropy*, P.M.M., Vol. 51, N. 3, pp. 489–495, 1987.

13. G. Sacchi Landriani, M. Rovati, *Optimal Limit Design of Fiber Reinforced Orthotropic Bodies*, Proc. COMETT Course on *Computer Aided Optimal Design of Structures*, University of Pavia, Italy, 4–8 September 1989.

14. P. Pedersen, *Combining Material and Element Rotation in One Formula*, Communications in Applied Numerical Methods, Vol. 6, pp. 549–555, 1990.

15. J. Thomsen, N. Olhoff, *Optimization of Fiber Orientation and Concentration in Composites*, Proc. CEEC Course on *Design and Analysis of Structures made of Composite Materials*, Politecnico di Milano, Italy, 28 May – 1 June 1990.

16. J. Thomsen, *Optimization of Composite Discs*, Report N. 21 of the Institute of Mechanical Engineering, Aalborg University, Denmark, 1990.

17. N. Olhoff, M.Ph. Bendsøe, J. Rasmussen, *On CAD-Integrated Structural Topology and Design Optimization*, Report N. 27 of the Institute of Mechanical Engineering, Aalborg University, Denmark, 1990.

18. M.Ph. Bendsøe, *Optimal Shape Design as a Material Distribution Problem*, Proc. COMETT Course on *Computer Aided Optimal Design of Structures*, University of Pavia, Italy, 4–8 September 1989.

19. M.Ph. Bendsøe, *Composites as a Basis of for Topology Optimization*, Proc. CEEC Course on *Design and Analysis of Structures made of Composite Materials*, Politecnico di Milano, Italy, 28 May – 1 June 1990.

APPLICATION OF NEURAL NETS IN STRUCTURAL OPTIMIZATION

LASZLO BERKE
NASA Lewis Research Center
Cleveland, Ohio

PRABHAT HAJELA
Rensselaer Polytechnic Institute
Troy, New York

ABSTRACT. The biological motivation for Artificial Neural Net developments is briefly discussed, and the most popular paradigm, the feedforward supervised learning net with error back propagation training algorithm, is introduced. Possible approaches for utilization in structural optimization is illustrated through simple examples. Other currently ongoing developments for application in structural mechanics are also mentioned.

1. Introduction

It is uncomfortable for most humans to contemplate the proposition that our conscious self appears to exist in a massively parallel distributed processor, a three pound wet personal computer. Certainly no evidence currently exists to the contrary. This wet PC is made up of hundreds of billions of threshold logic processing elements called neurons or brain cells, each incessantly receiving and sending information from and to thousands of other neurons.

Most of this self aware information processing machinery is located in the cerebral cortex that is the size of a thin newspaper crumpled up and wrapped over other brain matter for careful packaging inside its hard cover. This small volume of highly organized conscious living tissue represents the ultimate technology in the known Universe. The rest of the human body serves mainly as its life support system to aid it in its pleasure seeking, pain avoiding, self preservation and self replication pursuits. Some researchers view it as an existence proof for the possibility of a conscious computer somewhat like birds were the motivating existence proof for heavier than air flight.

The thin sheet of human cerebral cortex is also hard at work since ancient times to understand itself. It set out a few decades ago to extend its computational powers and addressable memory capacity, its weakest natural capabilities. In this it was eminently successful and already created supercomputers that do more calculations than all of the almost five billion existing human brains could perform in the same amount of time. In an almost parallel effort it made attempts to

731

G. I. N. Rozvany (ed.), Optimization of Large Structural Systems, Vol. II, 731–745.
© 1993 Kluwer Academic Publishers.

similarly extend its cognitive powers and named that enterprise Artificial Intelligence or AI. For that end two approaches are being pursued.

One approach concentrates on developing artificial reasoning capabilities referred to as Expert Systems. Their development and application became an industry during the past decade. It is a historical trivia now that one of the first areas of early feasibility studies [1] was in structures. The other approach is to develop biologically motivated computational models taking clues from the much studied processes that are ongoing in all biological neural networks and their sensors from the nervous system of slugs to that of humans. These models are usually referred to as Artificial Neural Nets or ANN to indicate their original motivation. This name also sells well.

With the help of rapidly advancing computer science more and more sophisticated ANN models of selected intelligent brain functions are being introduced both in hardware and software implementations. Vision, perception, natural language understanding, classification, associative memory, learning and accumulation of expertise are targets. ANN research is truly multidisciplinary, the participants come from neurobiology, psychology, physics, the medical, mathematical, computer science and engineering disciplines. Human brain cells can now be grown in a laboratory dish. Let us hope genetic engineers will not join in and create further uncomfortable questions. One such question would be is there a "critical net complexity", biological or artificial, at which self awareness occurs instead of a comatose reactive model, like our current computers?

The work reported here is among the first attempts to investigate the applicability of ANN to problems in structural mechanics. There appears to be now an upsurge of interest as signified by a few ANN sessions at engineering conferences that are being organized.

2. Basic Concepts of Artificial Neural Nets

It is more productive to view ANN technology in much simpler terms than neurobiology. "Parallel Distributed Processing" (PDP), the title of the two volumes [2,3] that helped perhaps the most in the recent upsurge of interest in neural nets, indicates a computer science view. When a biological entity provides the early motivation to develop its artificial counterpart, and the counterpart becomes successful, it takes on an evolutionary path of its own departing from its biological counterpart. Airplanes are not covered with feathers and they provide little information about how birds fly or what is their nesting habit. Neural nets are also rapidly losing their feathers except when the study of brain functions, instead of the development of a new class of powerful information processing approach, is at the focus.

As stated earlier only a brief and incomplete introduction is given here and the three volumes of References 2, 3 and 4 are suggested as introductory material. Reference 4 contains classroom exercises for the material in references 2 and 3, and also diskettes with codes covering most popular ANN paradigms.

As a simplified representation Figure 1 shows a sketch of a generic brain cell, or neuron, with the following components: the cell body, the "axon" that transmits stimulus from one neuron to the other, the "dendrites" and their "synapses" that serve as the connections. In a human brain the number of neurons is in the high billions, each connected to thousands of other neurons forming an immense network. Figure 1 also shows a minute portion of this network in the thin sheet of the cerebral cortex.

Of the brain functions studied by various ANN models only learning is of interest here. Learned information is thought to be represented in a biological neural net by a pattern of synoptic connection strengths that modify the incoming stimuli, strengthening or inhibiting them. When the accumulation of the modified stimuli in a cell reaches a certain threshold the cell "fires" sending out its stimulus to the thousands of cells downstream in a massively parallel fashion. Learning in turn is thought to be associated with the development and retention of a pattern of the synoptic connection strengths in various regions of this immense network. The learning rule appears to be to strengthen those synoptic connections at each feedback cycle that transmit information that cause other neurons to fire, and weaken those that transmit information that do not cause any cell to fire. Hebb [5] introduced the concept in qualitative terms in 1949 and now a number of variations [2] exist of the Hebbian learning rule. The quantitative training rules in use today have the same flavor but rely on mathematical concepts of optimization in their derivation.

The basic component of a neural net is an artificial neuron indicated in Figure 2. It uses a simple approach to mimic the complex operation of a biological neuron. A jth artificial neuron in a net receives information, labeled x_i, from the incoming n connections from other neurons of the net. The values x_i are modified by connection strengths w_{ij} associated with the jth neuron which performs a summation resulting in the value r. The jth neuron processes this value through an activation function producing its output z_i. This output is then sent to all the connecting neurons stimulating their "firing" governed by an "activation function". There are a number of activation functions from simple threshold to the most commonly used sigmoid function shown on Figure 3.

Artificial neurons can be assembled in principle into any architecture. One classification is whether the information travels only in one direction or neurons are also allowed to send information backwards creating interactive activations and competition to fire. These to fundamental cases are indicated in Figures 4 and 5 respectively. The first class, referred to as feedforward nets with supervised learning, was utilized in this study for the accumulation of expertise through exposure to known cases. The second class has a number of important variations used among other tasks for classification, associative memory and optimization. Figure 5 shows a small Hopfield net that can be used for associative memory and certain class of optimization problems. An other recurrent net is the Adaptive Resonance Theory net or ART net and is utilized in the lecture that follows.

A feedforward net consist of a layer of input nodes and a layer of output nodes defined by the number of I/O variables of the problem at hand. It has been shown by Kolmagorov and Cybenko that with an additional "hidden" layer any nonlinearity can be represented. In practice it is often found that more than one hidden layer may improve the net's training and modeling performance. How many nodes to use in the hidden layers is an art developed by practice.

"Training" of feedforward nets involves the establishment or evolution of the connection strengths w_{ij} everywhere in the net. Once that is established by a learning algorithm the trained net responds to a new input within the domain of its training by "propagating" it through the net and producing an output. This output is an estimate within certain error of the output that the actual computational, physical or some other process would have produced.

Variations of the "generalized delta error back propagation" algorithm is used usually to train the net. It is essentially a special purpose steepest descent algorithm to adjust the w_{ij} connection strengths, and other additional internal parameters that are sometimes added to increase the flexibility of the net. In principle other optimization methods can also be used and the development of efficient learning algorithms is an active area of research. The thing to remember is that like in any optimization problem one should try to "get away" with the smallest number of training, that is optimization variables.

Most currently available neural net capabilities are simulations of the distributed parallel processing aspect on serial machines, and such simulations were used also in this study. Neural nets represent premier applications for parallel machines or for the developments of special purpose hardware. These approaches are all happening and neural nets enjoy vigorous funding and developments worldwide.

To start out with an application one requires a set of known input and output pairs that one has to generate by the "real" process one is planning to simulate. The number of training pairs, and how should they span the intended domain of training requires experimentation and experience. The same statement is also valid for the architecture of the neural net one intends to use. The examples given latter will provide some idea of what is required for a successful application. Nonlinear regression is essentially what is being performed in the training process. It can be mentioned here that ANN is only one of many possibilities for the broader class of trainable networks. Polynomial nets with simple polynomials on a chip in their hard wired nodes have a long history including early studies [6] for structures.

The representation of the input variables can be improved if known functional relations among them are also represented. To include reciprocal variables effecting structural stiffness is a simple example. The Functional Link concept [7] is a generalization of this idea where functions of the variables used in functional expansions are also included as input. This artificial introduction of additional nonlinearities can be used to simplify the net architecture, eliminate hidden layers, and reduce the training effort.

The small net of Figure 4 has five input nodes in its input layer, seven nodes in its single hidden layer, and two output nodes in its output layer. In further discussions this net would be designated a (5-7-2) net signifying the number of nodes in its three layers. A net can provide an n-to-m mapping which in this case is a 5-to-2 mapping. Feedforward nets are currently limited to moderate I/O configuration. Combined use of unsupervised/supervised learning systems can cluster a large problem into smaller ones eliminating the limitation in most cases.

For the "design expert" examples discussed later a version of the publicly available NASA/JSC developed capability NETS was used. It is available from COSMIC and its user manual provides a good introduction for someone who would like to experiment with neural nets. Many commercial capabilities are now available to suit any price range.

3. Neural Nets in Structural Design Optimization.

The exploitation of computer technology by computational structural mechanics falls in three categories of higher and higher levels of abstraction. Procedural codes, expert systems and neural nets represent "number crunching", "expert judgments" and finally a "feel" for a problem domain respectively. These three levels simulate increasing intellectual content with less and less participation needed by the human user. If the eventual goal is the development of automated expert design capability the various neural net paradigms will certainly participate in any such future capabilities.

The major advantage of a trained neural net over the original (computational) process is that results can be produced in orders of magnitude less computational effort than the original process. This effort, once the net is trained, is also insensitive to the effort it takes to generate an output by the original process. Consequently benefits can be higher for those problem areas that are computationally very intensive, such as optimization, especially in multidisciplinary settings. There is of course a catch, namely that in those cases to generate the sufficient training set is also costlier.

One of the basic concepts of applying neural nets presented here is to train an appropriate net to replace analyses of given structural configurations during optimization iterations with neural net estimates. This approach could provide an extremely fast sensitivity analysis capability. Another application is to train a neural net to provide estimates of the actual optimum structures directly, circumventing the usual analysis-optimization iterations. These two approaches will be illustrated with a few examples. References 8 and 9 are earlier discussions of ANN applications in Structural mechanics.

Multidisciplinary design optimization provides particularly intriguing possibilities. For example, nets could be trained for each of the disciplines and connected to represent appropriate coupling or to use an additional net that learns the important coupling effects. Optimization problems that involve expensive iterative analyses due material nonlinearities could perhaps also benefit by neural net simulations. Feasibility studies were recently conducted to capture nonlinear material behavior with ANN to be then coupled with conventional analysis. Training data were obtained both by executing procedural codes for a set of cases covering the intended domain and also by using material test data directly.

4. Neural Net Assisted Optimization.

This feasibility study to simulate analysis with the quick response of neural nets was motivated by the approximation concepts in structural optimization. The idea here was to train a neural net to provide computationally inexpensive quick estimates of analysis output needed for sensitivity evaluations that in turn are needed by most optimization codes. Figure 6 is a diagram of the fundamental idea. First an initial design would be created that is to be then refined and modified during the design effort. The selected design variables are to be varied to span their expected domain. Analysis is performed for each set of design variables, and the neural net training pairs are constructed using the sets of design variables and the associated analysis results as the input-output pairs. These input-output pairs are then used to train an appropriate net which once trained can respond to any new variations of the design variables, within the selected domain, with instant estimate of the output a conventional analysis would produce.

The familiar five bar and ten bar truss "toy" problems, given in Figures 7 and 8 respectively, were used for this initial feasibility study. First various sets of input-output training pairs, and net configurations were examined to find the combination that produced trained nets with good results in terms of the accuracy of its estimates and in terms of reducing the efforts needed for the training. Once an acceptable trained neural net was obtained it was attached to an optimizer that would call it instead of calling a conventional analysis capability.

The optimization involved constraints on the nodal displacements, consequently the input-output training pairs consisted of the bar areas as inputs and nodal displacements as output variables respectively. How many pairs in a set, and within what range of variations, is required as a minimum, is itself a research question.

At this point one has to mention that the number of iterations required during training to obtain desired levels of accuracy ranges from a few hundred to tens of thousands. These numbers are routinely accepted in neural net applications even for small nets as in this study. For this level of experimentation one often initiates a run on a PC or a workstation and lets it run to a large number overnight.

A number of net configurations were examined starting with a 5-to-4 mapping with a (5-4) net with no hidden layer. The four output variables were the four nodal displacements indicated in Figure 7. Because the active constraints were essentially d2 and d4 the rest of the nets considered only these two displacements as output. First a (5-2) net was used, then to capture essential nonlinearities present in displacement calculations the reciprocals and products were also added as independent input variables up to (20-2) net configurations. The intent was not to use hidden layers in order to minimize the training efforts. The results were acceptable with better than five percent prediction accuracy. Finally (5-7-4) and (5-7-2) nets were trained and attached to a conventional optimizer in the spirit of Figure 6. The results are presented in Table. 1. The results are not very good when all four displacements were considered as constraints but improved dramatically when only the two larger displacements were constrained needing a neural net only for their prediction. Similar experimentations were conducted for the ten bar truss. The results shown in Table 2 were obtained by training a (10-6-6-2) neural net with two hidden layers.

These were initial experimentation to prove the feasibility of the concept. The tools used were "plain vanilla backprop" (as they say in ANN circles) research codes. Performance can be approved with learning more about what training pairs and what net architectures are the best.

5. Neural Nets as Expert Designers

The next set of experiments were conducted to explore the idea of training a neural net to estimate optimum designs directly for given design conditions and bypass all the analyses and optimization iterations of the conventional approach. It is conceivable in practice that successful similar designs could be collected within some domain of design conditions, input-output pairs defined, and then a neural net trained to serve as "intelligent corporate memory" that can provide a new design for new design requirements. The preliminary design scenario discussed earlier, and the approach described can be carried a step further. Instead of training a neural net to simulate analyses for a conventional optimization algorithm, one can create sufficient number of optimum designs to span a desired domain either by the approach discussed above or by conventional optimization. Training pairs can then be assembled from the design conditions and the associated optimum designs to train a net. Figure 9 indicates the flow of events.

5.1 TEN BAR TRUSS DESIGN EXPERT

Now let us suppose that we work in a company that markets equipment that is mounted in all cases on ten bar trusses as shown in Figure 8. These trusses have to carry the equipment weight (2 X 100 K) at the two lower free nodes which can not deflect more than 2 in. for proper operation of the equipment. The sizes of the dimensions L1, L2 and H of the trusses can vary between 300 and 400 inches depending on the particular installation. The engineer who was designing the trusses for the past 30 years and could simply tell the optimum bar areas for any combination of those dimensions has just retired. Can we create an accurate ANN simulation of this departed expert? Yes we can, and rather simply!

As a cautious first experiment only H was varied in 5 inch increments between 300 and 400 inches and the resulting 21 trusses were optimized by conventional methods. Eleven designs with H ending in 10 inches were put into the training set and the designs with H ending in 5 inches were retained for testing the trained net. This was a very benign case with very smooth changes in all the variables. Consequently only about 200 training iterations, costing less than a minute on a SUN 386i work station, produced better than one percent prediction accuracy.

One has to mention that for this single input variable case one could have simply plot the curves of the optimum areas and weight against H on a graph paper and draw a curve across them with a conventional regression analysis using a polynomial or other suitable family of functions and be just as accurate. The neural net essentially learned those plots in terms of its sigmoid functions. In case of multivariable input it learns a multidimensional sigmoid surface fit again equivalent to one based on conventional regression analysis methods.

After the above limited learning exercise the 3-to-11 mappings were performed between L1, L2, H, the ten bar areas A1,...,A10 and the optimum weight Wt, creating the "ten bar optimum design expert" to replace our retired expert designer. A rather limited training set was created obtaining optimum designs for only ten random sets of L1, L2, and H. It is interesting to note that ten training pairs did quite well in this case versus the hundreds of training pairs used for the neural net assisted optimization study. Of course here only three variables are varied to cover a domain. The resulting ten optimum designs are given in Table 3. Optimum designs were also obtained for an other seven random sets of L1, L2, and H as checks on the estimates of the net once trained. Table 3 shows the seven design conditions and the optimum designs.

The (3-14-11) net given in Figure 10 was used with the number of nodes in the hidden layer coming from 3 + 11 = 14 as the formula advised by experts for maximum number possibly needed. As an other practical detail it was found during experimentation with various options that it is beneficial to code the .1 minimum sizes as .5 instead. The active midpoint of the sigmoid activation function is probably the explanation.

NETS worked rather well, and 1% RMS accuracy was obtainable with 200 iterations and using only the default values for the learning parameters provided by NETS. Because of this good performance exercises were conducted to overtrain the net. Letting it run for 5000 iterations an RMS accuracy of .0062% was obtained. Overtraining is to be avoided because the neural net at that point becomes a memory with lessened ability to generalize. The overtrained net actually reproduced the training results exactly, but it did a little worse if anything against the seven check conditions than the net trained only for 1% RMS accuracy. Figure 14 is a bar chart comparing for one of the check cases the exact optimum and the estimates of the optimum design by the net trained to 1% and to 0.0062% RMS accuracy. The reader will be the judge.

Table 5 shows the net predictions and the percent errors for the seven check cases of Table 4 and for the net trained for 1% RMS accuracy. As can be seen the results are quite satisfactory and certainly would be good enough information to produce the ten bar trusses for these new design conditions.

The mental activities our retired expert designer employed to come up with his optimum designs have been replaced by a trained (3-14-11) neural net of similar capability for this limited task. The expert knowledge is captured in the extremely compressed form of a 3x14 and a 14x11 matrix of connection weights. Think of what it would take to come up with verbalized expert system rules to achieve the same with a rule based or any other expert system.

5.2 WING DESIGN EXAMPLE

The well known intermediate complexity wing was used next as a more practical problem. The 158 elements were linked to obtain a reduced set of 57 size variables that together with the weight were considered as the 58 output variables. A family of 15 variations of the wing was chosen within 25 percent of the original wing geometry. The variations were governed by a single master variable. Figure 12a shows the wing and 12b the family within the selected domain. These "instances" were optimized first by conventional methods and thirteen were used to train a (1-60-58) net. The remaining two designs were used to check the training accuracy. Figure 12c is a histogram of both the training and test accuracy.

5.3 TRUSS RING EXAMPLE

Figure 13a shows a truss ring with 60 truss elements and Figure 13b shows the expected "design domain" spanned by the family of rings produced by variations of the inner and outer radii. Aluminum material properties were used in the finite element model. Three static load conditions and stress, displacement and frequency constraints were considered in the optimization. The 60 bars were linked to obtain a reduced set of 25 size variables. Five cases of inner and outer radii and five cases of frequency constraints were combined into 125 cases. Each case was optimized by a SUMT procedure and also by a feasible direction and a quadratic programming technique as a check on the optimums obtained. Of the 125 cases 120 were selected to train a (3-30-28) net and 5 were used as test cases. The inner and outer radii and the frequency constraints were selected as three training input variables and the 25 bar areas and the optimum weight as the output variables. Figure 13c is a histogram showing the prediction error of the trained neural net expert truss-ring designer. Considering the complexity of the problem and the nonsmooth nature of the optimized design variations from case to case the accuracy is acceptable. Probably better accuracy would have been obtained with clustering the 120 cases into similar groups by an unsupervised training paradigm, or by using a hybrid capability that clusters the training set before training the feedforward net. When a reduced set of 34 designs clustered within a preselected optimum weight were used the maximum error was reduced to 5 percent.

6. Closing Remarks

What has been presented here has to be considered as initial feasibility studies. A few researchers are starting to investigate artificial neural net applications in structural mechanics and structural design. In cases where a design is to be modified and refined many times, the creation of a trained net could become less costly than the many analysis or optimization runs that would otherwise be performed. The low cost instant answers to what-if questions would also encourage broader search for the optimum design solution.

One has to mention that in some cases the same capability can also be obtained by creating polynomial or other conventional curve or surface fits over the same training data by well known methods. The Functional Link concept [7] departs significantly from the original biological motivation and brings in powerful mathematical functionalities. Neural nets currently have the advantage of the massive Government and private research and development efforts that are already making powerful capabilities available and some with user seductive features. Special purpose neural net machines with programmable hard wired arrays will eventually become available and will offer an entirely new way to attack problems.

7. References

1. Melosh, R. J., Berke, L., Marcal, P. V., (1978) Knowledge-Based Consultant for Structural Analysis Strategy, NASA Conference Publication 2059.

2. Rummelhart, R. J., McClelland, J. L. (1986) Parallel Distributed Processing, Vol 1: Foundations, The MIT Press, Cambridge, Massachusetts

3. McClelland, J. L., Rummelhart, R. J. (1986) Parallel Distributed Processing, Vol 2: Psychological and Biological Models, The MIT Press, Cambridge, Massachusetts

4. McClelland, J. L., Rummelhart, D. E. (1988) Explorations in Parallel Distributed Processing, The MIT Press, Cambridge, Massachussets

5. Hebb, D. O., (1949) The Organization of Behavior, Wiley, New York.

6. Bogner, F.K., Berke, L. (1977) Structural Applications of Automated Learning Networks, University of Dayton Research Institute, UDRI-TM-77-06, Dayton, Ohio.

7. Pao, Yo-Han, (1989) Adaptive Pattern Recognition and Neural Networks, Addison-Wesley Publishing Co.

8. Berke, L., Hajela, P. (1990) Application of Artificial Neural Nets in Structural Mechanics, NASA TM 102420

9. Hajela, P., Berke, L. (1990) Neurobiological Computational Models in Structural Analysis and Design, 31st Structures, Structural Dynamics and Materials Conference, Part 1, AIAA, 1990, pp. 345-355.

740

8. FIGURES

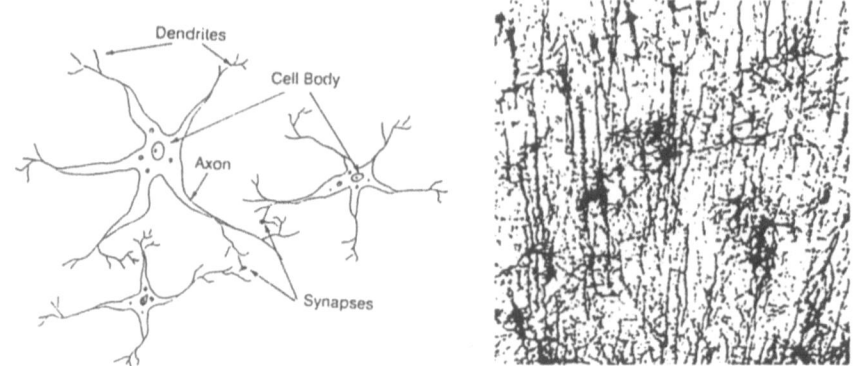

Figure 1. Biological neurons and neural net

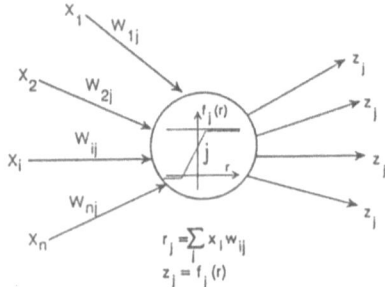

Figure 2. Artificial neuron

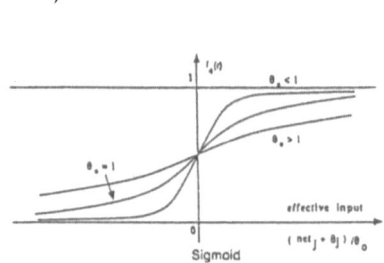

Figure 3. Sigmoid activation function

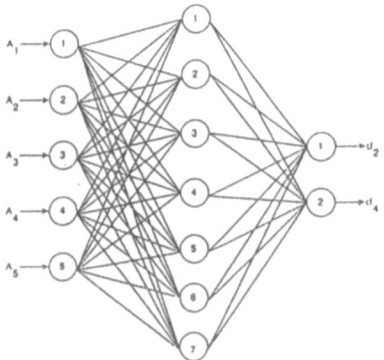

Figure 4. Feedforward net

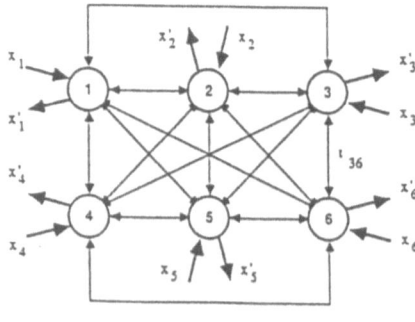

Figure 5. Interactive activation net

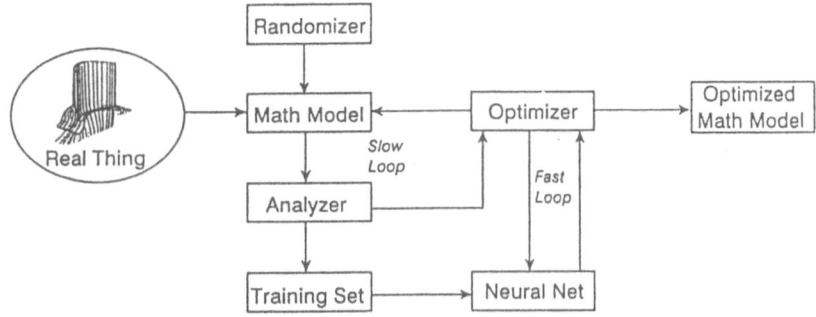

Figure 6. Neural net assisted optimization

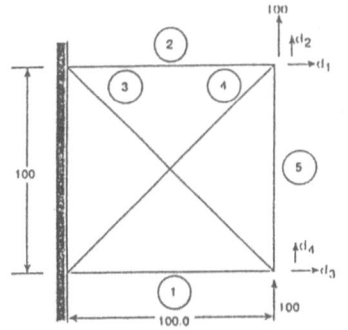

Figure 7. Five-bar-truss

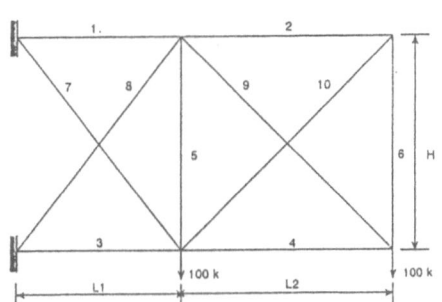

Figure 8. Ten-bar-truss

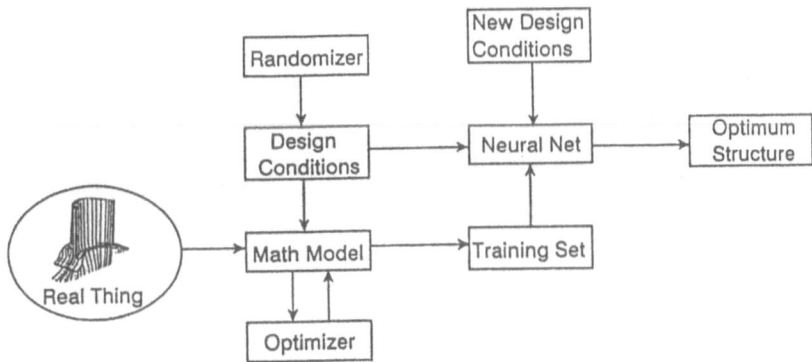

Figure 9. Neural net as design expert

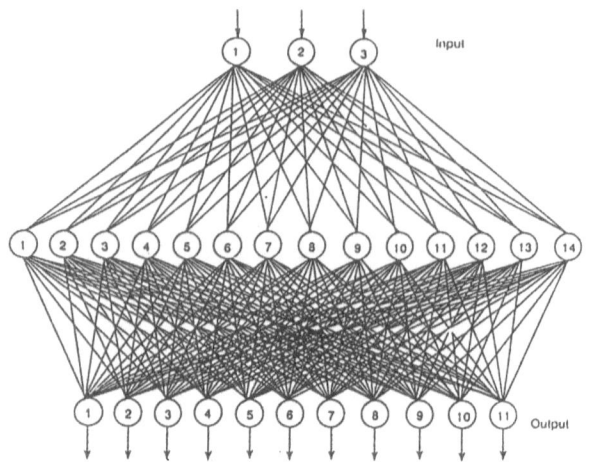

Figure 10. 3-14-11 neural net trained as ten-bar-truss design expert

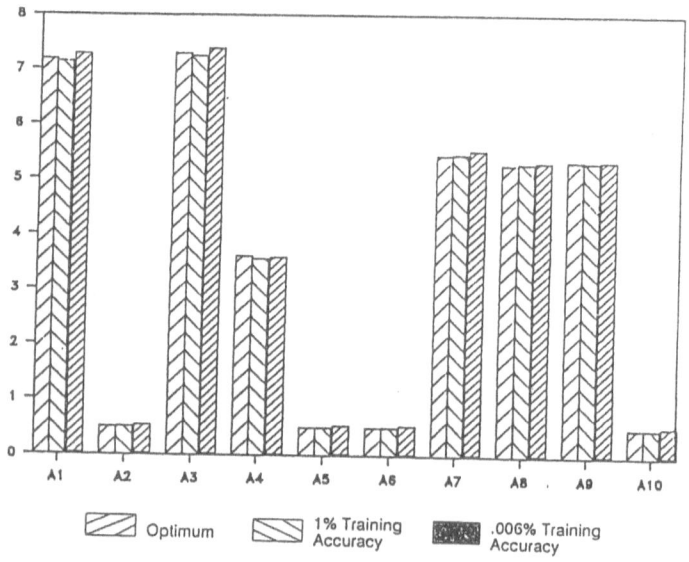

Figure 11.   Prediction accuracy of ten-bar-truss neural net design expert
trained for two levels of RMS accuracy

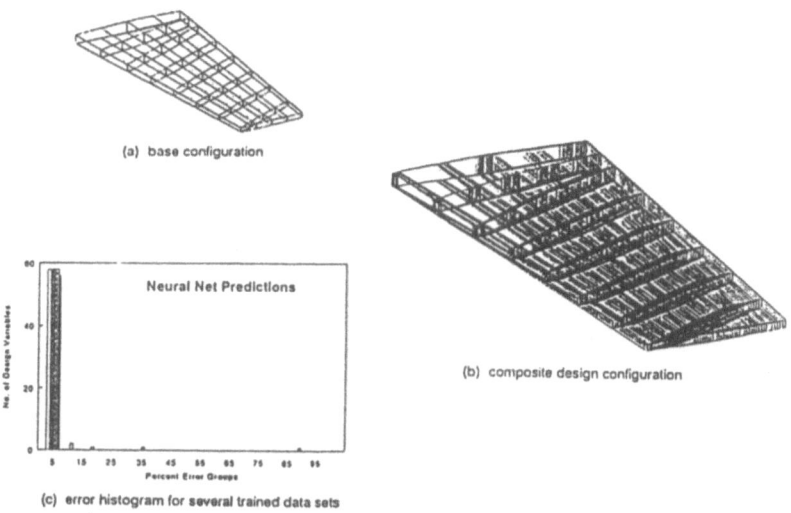

Figure 12. Intermediate complexity wing example

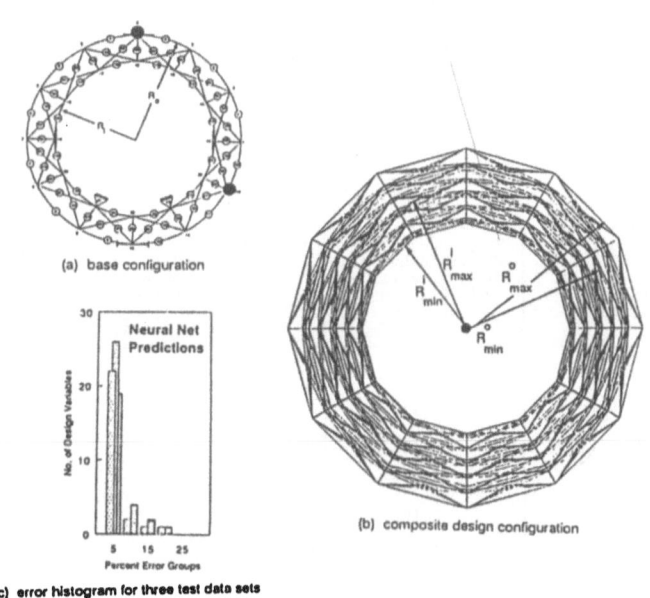

Figure 13. Truss-ring example

9. TABLES

Table 1. Optimal designs for five-bar-truss example using trained neural nets for analysis

Network Description		Design Variables					Objective Functions
		x_1	x_2	x_3	x_4	x_5	
(5-7-4) 200 training sets, all four output displacements mapped.	Initial	1.0	1.0	1.0	1.0	1.0	58.28
	Final	2.131	2.032	2.679	2.766	1.0	128.626
(5-7-4) 500 training sets, all four output displacements mapped.	Initial	1.0	1.0	1.0	1.0	1.0	58.28
	Final	1.952	2.013	2.763	2.760	1.0	127.759
(5-7-2) 100 training sets, two vertical displacements mapped.	Initial	1.0	1.0	1.0	1.0	1.0	58.28
	Final	1.535	1.778	2.29	2.265	1.0	107.56
(5-7-2) 100 training sets, two vertical displacements scaled as constraints and mapped.	Initial	1.0	1.0	1.0	1.0	1.0	58.28
	Final	1.505	1.584	2.138	2.211	1.0	102.399
	Exact Solution	1.5	1.5	2.121	2.121	1.0	100.0

(xx-yy-zz) denotes a three layer architecture with xx input layer nodes, yy hidden layer nodes, and zz output layer nodes.

Table 2. Optimal design for ten-bar-truss using trained neural nets for analysis

Design Variables	Network Description			
	(10-6-6-2)* 100 Training Sets Used in a Range of ±25% About Optimum	(10-6-6-2)* 400 Training Sets Used in a Range of ±25% About Optimum	(10-6-6-2) 100 Training Sets Used in a Range of 0.01-55.0 in^2 Output Scaled to Reduce Range of Variation	Solution From Exact Analysis
x_1	30.774	30.967	30.508	30.688
x_2	0.112	0.100	0.100	0.100
x_3	17.40	19.136	26.277	23.952
x_4	11.425	14.279	11.415	15.461
x_5	0.108	0.100	0.100	0.100
x_6	0.487	0.434	0.413	0.552
x_7	5.593	5.593	5.593	8.421
x_8	22.953	20.031	21.434	20.606
x_9	20.886	19.966	22.623	20.554
x_{10}	0.100	0.100	0.100	0.100
Objective Function	4692.49	4666.71	5010.22	5063.81

* Lower bound of design variables used as initial design - was infeasible.

Table 3. Ten-bar-truss optimum designs for training

| Input | | | Output | | | | | | | | | | |
L1	L2	H	A1	A2	A3	A4	A5	A6	A7	A8	A9	A10	WT
310	350	380	6.89	.1	7.00	3.62	.1	.1	5.24	5.08	5.35	.1	1356.7
345	326	360	7.40	.1	7.50	3.56	.1	.1	5.62	5.45	5.31	.1	1456.4
371	329	310	8.95	.1	9.10	4.17	.1	.1	6.33	6.14	5.74	.1	1684.5
360	300	340	7.69	.1	7.83	3.47	.1	.1	5.91	5.73	5.25	.1	1492.6
315	340	340	7.64	.1	7.76	3.93	.1	.1	5.53	5.36	5.56	.1	1407.6
380	355	390	7.47	.1	7.60	3.58	.1	.1	5.67	5.50	5.32	.1	1605.7
322	319	400	6.35	.1	6.46	3.13	.1	.1	5.21	5.05	5.03	.1	1314.2
400	300	400	9.25	.1	9.41	3.93	.1	.1	6.77	6.56	5.56	.1	1780.9
300	400	300	9.27	.1	9.39	5.25	.1	.1	5.73	5.57	6.57	.1	1593.8
311	350	315	8.33	.1	8.45	4.36 .	.1	.1	5.70	5.53	5.88	.1	1464.6

Table 4. Ten-bar-truss optimum designs for testing trained neural net

| Input | | | Optimum Solutions | | | | | | | | | | |
L1	L2	H	A1	A2	A3	A4	A5	A6	A7	A8	A9	A10	WT
342	351	383	7.18	.1	7.29	3.60	.1	.1	5.44	5.27	5.34	.1	1466
360	360	360	7.93	.1	8.06	3.93	.1	.1	5.74	5.56	5.56	.1	1593
320	350	360	7.38	.1	7.50	3.82	.1	.1	5.43	5.26	5.49	.1	1417
340	370	340	8.29	.1	8.41	4.28	.1	.1	5.74	5.56	5.82	.1	1578
310	350	380	6.89	.1	6.99	3.62	.1	.1	5.24	5.08	5.35	.1	1356
345	326	360	7.39	.1	7.51	3.56	.1	.1	5.62	5.45	5.30	.1	1456
371	329	310	8.95	.1	9.16	4.17	.1	.1	6.33	6.14	5.74	.1	1684

Table 5. Trained neural net estimates for the seven test cases

A1	7.138	8.000	7.365	8.398	6.908	7.310	8.916
%	0.580	0.780	0.200	1.300	0.260	1.080	0.380
A2	0.503	0.505	0.502	0.507	0.502	0.501	0.504
A3	7.245	8.160	7.485	8.558	7.013	7.448	9.089
%	0.620	1.240	0.200	1.760	0.330	0.820	0.770
A4	3.554	3.957	3.777	4.354	3.551	3.533	4.177
%	1.230	0.690	1.130	1.730	1.900	0.760	0.170
A5	0.499	0.499	0.500	0.499	0.500	0.500	0.500
A6	0.501	0.501	0.500	0.501	0.500	0.500	0.500
A7	5.456	5.783	5.461	5.781	5.309	5.586	6.359
%	0.290	0.750	0.570	0.710	1.320	0.610	0.460
A8	5.293	5.594	5.301	5.598	5.164	5.416	6.157
%	0.470	0.610	0.780	0.680	1.650	0.620	0.280
A9	5.326	5.574	5.477	5.863	5.336	5.311	5.740
%	0.260	0.250	0.240	0.740	0.260	0.210	0.000
A10	0.499	0.500	0.499	0.499	0.497	0.503	0.500
WT	1466.000	1598.000	1417.000	1585.000	1362.000	1451.000	1693.000
%	0.000	0.310	0.000	0.440	0.440	0.340	0.530

SELF-ORGANIZATION IN NEURAL NETWORKS - APPLICATIONS IN STRUCTURAL OPTIMIZATION

P. HAJELA & B. FU
Mechanical Engineering, Aeronautical Engineering & Mechanics
Rensselaer Polytechnic Institute, Troy, New York

L. BERKE
Structural Mechanics Branch
NASA Lewis Research Center, Cleveland, Ohio

ABSTRACT. The present paper discusses the applicability of ART (Adaptive Resonance Theory) networks, and the Hopfield and Elastic networks, in problems of structural analysis and design. A characteristic of these network architectures is the ability to classify patterns presented as inputs into specific categories. The categories may themselves represent distinct procedural solution strategies. The paper shows how this property can be adapted in the structural analysis and design problem. A second application is the use of Hopfield and Elastic networks in optimization problems. Of particular interest are problems characterized by the presence of discrete and integer design variables. The parallel computing architecture that is typical of neural networks is shown to be effective in such problems. Results of preliminary implementations in structural design problems are also included in the paper.

Introduction

A neural network is a computational model that has its philosophical basis in biological processes. At the most fundamental level, interest in such an approach is inspired by an observation that even less advanced biological life forms are equipped with a level of neuronal activity that allows them to perform basic life preserving functions. The basic computational processor in such neural networks is an artificial neuron, modelled to simulate the behavior of its biological counterpart. A very large number of such interconnected neurons comprise an artificial neural network. Neurobiological methods of computation are distinctly different from traditional numerical computing, and the more recently emergent strategies of symbolic processing and computation.

There has been significant recent activity in adapting this computational model in various fields of engineering. Applications have included image processing and pattern recognition [1-2], fault detection [3], diagnostic systems [4], and formulations which allow the use of neural networks as numerical optimization algorithms [5-6]. To a large extent, these applications are based on neural networks that have an architecture defined by one or another model of learning and/or association that has originated in psychology. In essence, these architectures all involve the basic neuronal computing element; only the

747

G. I. N. Rozvany (ed.), Optimization of Large Structural Systems, Vol. II, 747–765.
© 1993 Kluwer Academic Publishers.

pattern of connectivity is changed to accommodate differences in the models. The network may either have feed-forward characteristics only, or may have feedback loops. Likewise, networks may be fully connected in that each neuron is linked to every other neuron in the network, or the processing elements may be sparsely connected. While the feedforward and feedback architectures influence network training, the property of connectivity is intimately related to the parallelism in the system.

Subsequent sections of this paper examine the application of self-organizing neural networks in problems of structural analysis and design. An overview of three neural network models, the ART network, the Hopfield network, and the Elastic net is presented. This overview includes a discussion of the network architecture, the approach adopted for network training, and the suitability of the architecture for parallel processing. The adaptation of these models in structural design are then presented. The ART network has been primarily used for vector classification, and draws extensively on the associative reasoning characteristic of these networks. Specific applications pertaining to the conceptual design of structural systems are discussed in this context. Although the Hopfield network and the Elastic net may also be used in vector classification, the present paper only examines their applications in direct optimization problems. The approach is shown to be applicable in combinatorial optimization problems that are an immediate consequence of the presence of discrete variables. Illustrative applications in structural optimization are presented, along with general limitations of the approach.

Self-Organization in Neural Networks

The neural network architecture that has been most widely studied in the context of structural analysis and design is the multilayer perceptron model [7-11]. The training of these networks is described as supervised, wherein a set of input-output parameters are presented repetitively to the network, and the network weights adjusted such that the network produced output is as close as possible to the known output. Once trained, such networks can be used to obtain estimates of outputs for input patterns not included in the training set. The multilayer perceptron networks are characterized by the absence of any feedback from the output to the input neurons, and are therefore unconditionally stable. This stability is not without a price, as such networks have a limited repertoire in their ability to model the functioning of the brain.

Networks that do have a feedback path between the output and input neurons are described as recurrent networks. Although unconditional stability is not assured in these architectures, a more realistic representation of the memory process can be attained. Furthermore, a form of learning that is of special interest in the present work is one referred to as self-organization, such as may be required if no known output is available to train the network. The ART network, and the Hopfield and Elastic nets have the recurrence property, and are also classified as belonging to the general category of self-organizing networks.

ART Networks

The representation of human memory, albeit in a limited form, by an artificial neural network, has been the subject of extensive research [12-14]. A special feature that must be incorporated in this model of memory is that it must allow for recognition of previously encountered patterns, and at the same time accommodate and store new input without destroying the existing contents. This stability-plasticity requirement has contributed to the proposition of adaptive resonance theory, or ART, networks.

ART networks belong to the general category of self-organizing or unsupervised-learning systems. ART networks are essentially distinguished on the basis of the form of input information they can accept (binary or continuous), and on the approach used to process this information. The essence of the approach can be explained most simply by considering a network that can only process binary input patterns, and discussions in this paper will be limited to this model. As stated earlier, ART networks attempt to model the process by which humans learn and recognize invariant properties of a given problem domain. A special characteristic of such networks is the plasticity that allows the system to learn new concepts, and at the same time retain a stability that prevents destruction of previously learned information. ART networks accommodate these requirements through interactions between different subsystems, designed to process previously encountered and unfamiliar events, respectively. Before discussing the application of these networks in structural design, a brief description of the network will be presented.

A schematic representation of an ART network architecture is shown in Figure 1. As stated earlier, these networks function as vector classifiers, determining if an input vector presented to the network has features that are similar to those of one of the stored patterns. There are essentially five interacting subsystems that are required to model the vector classification process. As shown in the figure, these interacting subsystems are a set of two gain controls, an attentional subsystem consisting of two layers of neurons referred to as the comparison and the recognition layers, respectively, and a reset control, also termed as the orienting subsystem. The attentional subsystem is designed to process familiar events. By itself, however, it is unable to process new patterns and at the same time retain stability of existing patterns. In an isolated mode, it is either rigid to new patterns or exhibits extreme instability with regard to stored patterns. This shortcoming is addressed by the orienting subsystem or the reset control. The function of each component subsystem is summarized in the following sections.

Attentional Subsystem

This subsystem consists of two layers of artificial neurons - the comparison and the recognition layer. In the present paper, these layers will be referred to as F1 and F2 , respectively. The input pattern is first processed by the F1 layer neurons, as shown in Figure 2, and the output presented to the F2 layer neurons through a filter of stored weights as shown in Figure 3. The F2 layer neurons are subjected to a lateral inhibition mechanism, in which the output of each neuron is fed back to itself with a positive weight (enforcement), and is presented to all other neurons in the layer with a negative weight (inhibition). This allows only one neuron in the F2 layer to fire - a "winner-take-all" strategy. The output of this activated neuron then generates a template pattern for comparison with the input pattern. A vigilance or tolerance level is assigned to this comparison, and if the two patterns are similar, the input pattern is automatically classified under the same category as the flag of the fired F2 neuron. Otherwise, a reset wave is generated from the orienting subsystem.

Orienting Subsystem

The role of the orienting subsystem is primarily one of processing unfamiliar patterns. If the pattern generated by the fired F2 layer neuron is different from the input pattern, this subsystem generates a reset signal, whereby the fired neuron is disabled for a second round of presentation of the input patterns. The next dominant neuron in the F2 layer fires as a result, and the process of similarity assessment is repeated until a match is located. If no similar category can be found, as would be true for an unfamiliar pattern, a new category is established.

Gain Controls

The gain controls Gain 1 and Gain 2 are vital for the process of training and classification. They play a central role in regulating the firing of F1 layer neurons. In the binary system described here, an F1 layer neuron fires if it satisfies the "two-thirds rule" proposed by Carpenter and Grossberg [12]. This rule states that an F1 layer neuron will fire only if at least two of the three input signals that it receives are equal to unity.

Vector Classification

The vector to be classified is presented to the comparison layer. As shown in Figure 2, each neuron in the F1 layer receives three binary inputs - from the input, from Gain 1, and a feedback from the F2 layer. The process is initiated by setting Gain 1 to unity and Gain 2 to zero. As a result of the latter, the feedback from the recognition layer is exactly zero. Further, since Gain 1 is unity, the two-thirds rule dictates that at the first presentation, the output of the F1 neurons will be exactly the same as the input. Associated with the j-th neuron in the F2 layer, are a set of weights b_{ij}, $i=1 \rightarrow n$ and $j=1 \rightarrow m$, where n is the dimensionality of the input vector, and m is the total number of neurons in the F2 layer. As shown in Figure 3, each F2 neuron receives as an input, the dot product of the vector output from the comparison layer and the weights b_{ij}. The neuron with the largest input is declared as the winner with an output unity, while outputs of all other F2 neurons is set to zero. Now that the recognition layer neurons are active, Gain 1 is set to zero. The output from the winning F2 neuron is multiplied by a set of stored binary weights t_{ji} (zero or one) as shown in Figure 2, and passed back to each of the n neurons in the F1 layer. The two-thirds rule is invoked once again, and the F1 layer neurons with at least two unity inputs would fire creating a new output signal. Clearly, if the weights t_{ji} are similar to the input pattern, a match is obtained, and the input vector is classified under the current dominant neuron in the F2 layer. If there is significant dissimilarity, the previously winning F2 neuron is disabled and the process is repeated. If no match is found, a new F2 neuron weights is trained to recognize the new input pattern.

Issues of Implementation

The training process ensures that the t_{ji} weights are binary valued, and that a scaled version of these weights is available as the bottom-up weights b_{ij}. These weights are representative of the different training patterns that the system has learnt to recognize. This learning is unsupervised in that there are no target patterns to emulate. Rather, the weights are developed as a result of successive reinforcement with similar patterns. Details of the network training are beyond the scope of this paper. For the present development, it is sufficient to know that the dynamics of network learning are based on models found in psychology. This subject is discussed in greater detail in Reference 15. It is also important to indicate that ART networks are indeed amenable to parallel processor implementation. Since the stored patterns must be compared to the input pattern in sequence, the bottom-up filtering which is required can be performed simultaneously on a large number of processors. With minor modifications in the network architecture, this parallel processing would also allow for determining more than one stored pattern that possesses similar features to the input pattern.

Applications in Structural Design

Applications of the binary ART network in the conceptual design of structural systems have been considered in Reference 15. There are two distinct design processes that were considered in the present work. The first encompasses a class of structures where the structural layout was generally known and the load and support conditions were allowed to vary. The second class of problems is one where the loads and support points were assumed to be given, and the object of the design was to generate a near optimal structural topology. In either case, significant design experience with similar problems was assumed to be available, and dependent upon the identifiable critical features of the problem, a distinct procedural design process associated with the problem could be formulated. The problem may therefore be best understood as a use of ART networks to provide a memory capacity or a knowledge base for design, from which information can be recovered upon presentation of relevant features. The approach, therefore, draws upon the notion that human memory operates according to associative principles.

As a simple illustration of this idea, consider a problem in which the optimal cross section of a beam is to be determined for an allowable stress σ_{al} and for minimum weight. For simplicity, it may be assumed that the beam has a rectangular cross section, and further, either its cross-sectional depth or width are assumed fixed. In such a situation, the optimal cross section along the span of the beam is determined as follows:

$$h = \sqrt{\frac{6M}{b\sigma_{al}}} \quad \text{if b remains constant} \tag{1a}$$

$$b = \sqrt{\frac{6M}{h^2\sigma_{al}}} \quad \text{if h remains constant} \tag{1b}$$

This determination of optimal section properties is based entirely on the value of the bending moment at that section, where the latter is determined by the end supports and the type of loading that is applied. The type of loads and supports can thus be considered as critical features of the problem. Combination of these features result in a specific moment distribution, and hence a distinct sizing of the beam section for uniform strength. As a simple illustration, the problem features can be represented by a 4-digit binary string; the relationship between the binary numbers and the physical features is summarized in Table 1. A total of 16 distinct combinations can be obtained in this representation. These combinations and their associated procedural processes are shown in Table 2.

For the comparison layer in the attentional subsystem, this problem requires a total of four nodes (one node per digit of the input pattern). Note that this discussion is confined to binary coded ART networks only. The recognition layer in this subsystem would typically have as many nodes as the number of distinct patterns presented for categorization. Reference 15 describes the results on numerical experiments with this simple problem, designed to study the influence of the vigilance parameter on the network's ability to learn and to classify the given patterns. As is to be expected, higher values of the vigilance parameter (0.65-0.75) result in the recognition of minor differences in the input pattern. This finer distinction is of increased significance as the number of problem features is increased. As an illustration of this concept, the beam was divided into a number of smaller segments, and the presence or absence of a load on that segment was described by a binary variable. This subdivision may be necessary as the problem definition becomes more elaborate. Another level of refinement would be to include the magnitude of loads and moments as part of the problem features. The ART network, therefore, functions as an efficient feature classifier. Once classification of a problem is complete, a predefined procedural process takes over. Present studies are directed at looking at a continuous variable implementation of ART, and combining it with the multilayer perceptron network to include procedural processing in the distributed computing environment.

Hopfield Networks

Another class of networks that also exhibits the self-organizing behavior is the widely studied Hopfield network. Figure 4 shows a schematic sketch of a recurrent network that is representative of this architecture. There is a single layer of processing neurons. The output of each is returned through a set of weighting coefficients to the input of all neurons in the network. Each neuron also receives as an input, a bias value denoted as I_i in the figure.

The state of the network is defined by the values of all neurons in the network. If one considers a network where the output of each neuron is binary, then for n neurons, there are 2^n states that can be achieved. The weights w_{ij} and the bias inputs must be determined for each of these states; this would then represent a trained network. When a new input vector is presented for classification, the various stored states would be examined to determine the one closest to the input. This search is conducted on the basis of determining the minimum of an energy functional, where the latter is defined in terms of the state of each neuron, the interconnection weights, and the external or bias input.

The input to the i-th neuron Z_i is calculated as follows,

$$Z_i = \sum_j w_{ij} V_j + I_i \qquad (2)$$

where Y_j is the output of the j-th neuron, and is obtained as follows:

$$Y_j = f(Z_j) \qquad (3)$$

Here, f is defined as the activation function; threshold and sigmoid functions are typically used in neural computing. In order to obtain a stable network, the energy functional E is defined in the Lyapunov form [16] as,

$$E = -\frac{1}{2} \sum_i \sum_j w_{ij} V_i V_j - \sum_i V_i I_i \qquad (4)$$

and the change in energy δE due to the change in the state of a neuron δV_j is written as follows:

$$\delta E = -\left(\sum_i w_{ij} V_i + I_j \right) \delta V_j \qquad (5)$$

The network is said to have assumed a stable state when a stationary value of the energy functional is realized.

As stated earlier, there are two ways in which this paradigm can be used in the structural design problem.

a. The network is used as an associative memory, wherein the network recovers the class to which an input pattern belongs, even if the input pattern is incomplete. This is similar to the purported use of the ART network.

b. The use of this network in direct optimization problems. This aspect will be explored in somewhat greater depth in this paper.

Associative Memory

Human memory works in an associative manner, where partial recollection of some features can trigger the recall of larger amounts of related features. The Hopfield network can be used to simulate this behavior, and the basic steps to achieve this are as follows. First, the memory itself is constructed and stored in the form of the interconnection weights. This weight matrix is obtained as follows,

$$W = \sum_k V_{(k)} V_{(k)}^T \qquad (6)$$

where W is the weight matrix and $V_{(k)}$ is a vector defining the k-th desired state of all neurons. The partially correct or incomplete input pattern is momentarily applied as the output of the network, and is then removed. The network then relaxes along the energy contours, and settles to a state that is most similar to the presented input. Since the network follows the local contours of the energy function, there is a possibility of getting trapped in a local minimum. The Boltzmann machine, a stochastic variant of the Hopfield paradigm, and one based on the simulated annealing concept, has a better chance of locating the global optimum of the energy function.

Optimization Applications

A second area of Hopfield network applications in structural design is in the direct optimization problem. In contrast to the associative memory modeling, the present application requires that the optimization problem be mapped on to the network domain. This is accomplished by formulating an energy functional for the network in such a manner that its minimization also results in an optimal solution to the given problem. The essence of the approach is to adjust the strength of the neuron interconnection weights, and hence also the output of each neuron such that a stationary value of the energy function is realized. The output of the neurons, also referred to as the state of the system, are then translated into the desired optimal solution. The present discussion is directed towards illustrating the mapping of an optimal design problem in the framework of the Hopfield networks, and is restricted to a discrete variable, optimal assignment problem.

As an example problem, consider a truss assembly that is made up of a few groups of equal length members. In each group of elements, there may exist errors in lengths such as those introduced during manufacture; these errors are assumed to be small in relation to nominal values of member lengths. Further, it is assumed that the member lengths can be determined precisely. The problem then, is one of determining an optimal arrangement of available members to minimize the surface distortion and the pre-load in the structural members. For the truss assembly shown in Figure 5, the problem is simplified further in that the nominal lengths of all structural elements are the same.

This problem is essentially one of optimal assignment, where a particular member is assigned to a special position so as to reduce the overall shape distortion and to minimize the member preloads. If a variable V_{Xi} is defined to be unity when member X is assigned to position i, and zero otherwise, a mathematical statement of the optimal placement problem may be obtained as follows:

$$\text{Minimize} \quad \sum_{r=1}^{ND} (u_r)^2 \tag{7}$$

Subject to:

$$\sum_{X} V_{Xi} = 1 \tag{8a}$$

$$\sum_i V_{Xi} = 1 \tag{8b}$$

$$\sum_X \sum_i V_{Xi} = n \tag{8c}$$

where u_r is the displacement in the r-th dof. Here, constraints given by eqns. (8a) and (8b) require that only one member be assigned to each position, and the constraint of eqn. (8c) is introduced to ensure that all 'n' available positions are assigned an element. To solve this problem using a Hopfield network requires that it be mapped into the network domain in such a manner that the optimal solution be interpreted from the output of the network neurons.

As discussed in the previous section, an energy functional must first be generated for the assignment problem. The functional used in the present work was of the following form:

$$E = \frac{A}{2} \sum_X \sum_i \sum_{j \neq i} V_{Xi} V_{Xj} + \frac{B}{2} \sum_i \sum_X \sum_{Y \neq X} V_{Xi} V_{Yi} +$$

$$+ \frac{C}{2} \left(\sum_X \sum_i V_{Xi} - n \right)^2 + \frac{D}{2} \sum_{r=1}^{ND} (u_r)^2 \tag{9}$$

Note that V_{Xi} can be zero or unity, indicating the absence/presence of a member X at location i. In the above expression, the first three terms are required to account for the constraints given by eqns. (8a-8c). The fourth term simply represents the objective function given in eqn. (7), and can be written explicitly as,

$$\sum_{r=1}^{ND} (u_r)^2 = \sum_{r=1}^{ND} \left(\sum_{s=1}^{ND} k_{rs}^{inv} f_s \right)^2$$

$$= \sum_{r=1}^{ND} \left(\sum_{s=1}^{ND} (k_{rs}^{inv} f_s)^2 + \sum_{s=1}^{ND} \sum_{t=1}^{ND} k_{rs}^{inv} k_{rt}^{inv} f_s f_t \right) \tag{10}$$

where, k_{rs}^{inv} is the rs term in the inverse of the stiffness matrix, ND is the number of degrees of freedom, and f_s is the s-th term in residual force vector.

The network weights and the state of the system must evolve so as to generate the minimum energy state. This would correspond to an optimal assignment of the available members. The evolution of the neurons is defined by the following equation:

$$\frac{du_{Xi}}{dt} = -u_{Xi} - \frac{\partial E}{\partial V_{Xi}} \tag{11}$$

Using a first order finite difference approximation for the time derivative on the left hand side of eqn. (11), and setting $\Delta t = 1$, one obtains an expression for the total input to neuron Xi at time t+1 as follows:

$$u_{Xi} = -\frac{\partial E}{\partial V_{Xi}}$$

$$= -A\sum_{j\neq i} V_{Xj} - B\sum_{Y\neq X} V_{Yi} - C\left(\sum_Y \sum_j V_{Yj} - n\right) -$$

$$-D\sum_{r=1}^{ND}\left(\sum_{s=1}^{ND} k_s^{inv} f_s \frac{df_s}{dV_{Xi}} + \sum_{r=1}^{ND}\sum_{s=1}^{ND} k_s^{inv} k_t^{inv} f_s \frac{df_t}{dV_{Xi}}\right) \tag{12}$$

Comparing this with the expression for the input to neuron Xi given as,

$$u_{Xi} = \sum_Y \sum_j w_{Xi,Yj} V_{Yj} + I_{Xi} \tag{13}$$

one obtains an expression for the updated weights w as follows:

$$w_{Xi,Yj}^{(t+1)} = -A\delta_{XY}(1-\delta_{ij}) - B\delta_{ij}(1-\delta_{XY}) - C -$$

$$-De_X e_Y \sum_{r=1}^{ND}\left(\sum_{s=1}^{ND} k_s^{inv} C_{sj} C_{si} + \sum_{s=1}^{ND}\sum_{t=1}^{ND} k_s^{inv} k_t^{inv} C_{sj} C_{ti}\right) \tag{14a}$$

$$I_{Xi} = Cn \tag{14b}$$

These weights are then used with the old V_{Yj}'s and I_{Xi} to determine u_{Xi} from eqn. (13); these u_{Xi} are then processed through the gain function,

$$V_{Xi} = \frac{1}{2}\left(1 + \tanh\left(\frac{u_{Xi}}{u_0}\right)\right) \tag{15}$$

to obtain new values of V_{Xi} for a second round of network evolution. For a stable network, successive iterations produce smaller and smaller changes in the output, with the system settling to a stable state. Additional details on the implementation of this approach, and numerical results, are presented in Reference 16.

Elastic Nets

This network is also an illustration of the self-organizing behavior in neural networks, and has an architecture that is very similar to the recurrent Hopfield network. The approach belongs to a general category of deformable models that use an energy minimization process to attain an optimal solution. It is generally applicable to a class of NP complete (nonpolynomial time) problems, and is illustrated here in reference to the quadratic assignment problem (QAP).

This problem, first proposed by Koopmans and Beckmann [17], deals with the optimal assignment of a set of factories to another set of given locations, and can be stated as follows:

Minimize

$$\sum_i C_{is(i)} + \sum_i \sum_j f_{ij} d_{s(i)t(j)} \tag{16}$$

where, $C_{is(i)}$ is the cost of assigning factory i to location s, f_{ij} represents the flow of goods between factories i and j, $d_{s(i)t(j)}$ is the distance between locations s and t to which factories i and j have been assigned. The nature of this problem makes it difficult to obtain exact solutions whenever a modest number of assignment decisions are at stake. Hence, heuristic strategies that can generate near-optimal solutions in a reasonable time must be considered. The elastic net paradigm is a heuristic approach with a strong geometrical flavor. It was proposed by Durbin and Willshaw [18] as a solution to the travelling salesman problem (TSP). An elastic "rubber band" was allowed to deform and touch all cities. The dynamic variables were position coordinates on the band that varied according to a gradient descent prescription obtained from an appropriately selected energy function.

A similar idea is adapted into the solution of the QAP and discussed in detail in Reference 19. The essence of the approach is outlined here for completeness. At the very outset, an energy function E is defined as follows:

$$E(y_i, V_{is}) = \mu \sum_i \sum_s C_{is} V_{is} (y_i - x_s)^2 + \nu \sum_i \sum_j f_{ij} (y_i - y_j)^2 \tag{17}$$

Here, V_{is} is a binary variable with value unity when factory i is matched to location s, and is zero otherwise; x_s, $s=1 \to n$ denotes the set of locations, and y_i, $i=1 \to n$ denotes the factories. The constants μ and ν determine the relative strengths of attraction forces felt by a factory from the locations and from its neighbors.

If the probability density function for the states of the system V_{is} and y_i follows the Gibbs distribution, then the problem is reduced to determining the minimum of an effective energy function as follows [19]:

$$E_{eff}(y_i) = -T \sum_s \log \sum_i e^{-\frac{1}{T} \mu C_{is} (y_i - x_s)^2} + \nu \sum_i \sum_j f_{ij} (y_i - y_j)^2 \tag{18}$$

Note that the variable V_{is} has been eliminated in this manipulation. Such an approach enhances the possibility of locating the global minimum of E_{eff}, as the process can be done in stages, with the temperature T lowered gradually for each stage. The change in the state Δy_i can be obtained from a steepest descent strategy as follows,

$$\Delta y_i = -\frac{\partial E_{eff}(y_i)}{\partial y_i}$$

$$= 2\mu \sum_s w_{is}(x_s - y_i) + 2\nu \sum_k f_{ki}(y_k - y_i) \tag{19}$$

where,

$$w_{is} = \frac{e^{-\frac{1}{T}\mu C_{s}(y_i - x_s)^2}}{\sum_k e^{-\frac{1}{T}\mu C_{ks}(y_k - x_s)^2}} \qquad (20)$$

Note that the expression for Δy_i is similar to eqn. (13) obtained for the Hopfield network. The elastic net can therefore be represented by an architecture similar to the Hopfield net, with the exception that in the elastic net, the input to a neuron is not subjected to processing through an activation function. The adjustment in the state Δy_1 is performed at gradually decreasing values of T till there is no further change. This results in a minimum of the effective energy function, and is also the desired optimal assignment.

As an illustration of this approach, consider a five factory assignment problem. Since the flow and distance data are symmetric, i.e. $f_{ij} = f_{ji}$ and $d_{ij} = d_{ji}$, the data may be represented in a compact form as shown in Table 3. Note that C_{is} represents the linear term in the cost function, and may be addressed by a linear programming strategy. The focus of this approach resides in the quadratic term of the cost function. Hence, we will assume that $C_{is(i)} = 1$. For the five factory problem, the data in Table 4 was used. The locations where factories can be placed are shown in Figure 6. This figure also shows the initial distribution of the factories along a small circle around the centroid of desired factory locations. The trajectories traced by the factories in moving to their optimal assignments is shown in Figure 7. Details of this implementation including the dependence of the approach on problem parameters, is discussed in Reference 19. That publication also presents similar studies with larger scale problems.

Closing Remarks

The present paper provides a broad overview of three neural computing models that exhibit self-organizing behavior. Applications of these computational models in problems similar to those encountered in structural design, are also discussed. The ART and Hopfield networks can both be used as efficient vector classification procedures. A simple beam design problem is presented to illustrate the use of vector classification in structural design. Applications of the Hopfield network and Elastic net in direct optimization problems are also presented. Both examples discussed are representative of combinatorial optimization problems, a class of generically difficult problems encountered in structural design. Although these neural computing models are applicable to such problems, they do not necessarily provide a general solution procedure. The mapping of an optimization problem to the network domain is not trivial, and is different for each problem. To this extent, the approach has drawbacks similar to the optimality criteria methods. However, the computational advantages available from a distributed processing implementation of these methods, cannot be ignored in light of the computational requirements of combinatorial optimization problems.

Acknowledgements

This work was supported under research grant NAG 3-1196 from the NASA Lewis Research Center.

References

1. Psarrou, A. and Buxton, H., "A Neural Network Approach to the Computation of Vision Algorithm", *First IEE International Conference on Artificial Neural Networks*, pp67, 1989.

2. Costa, L.D.F. and Sandler, M.B., "Neural Networks and Hough Transform for Pattern Recognition", *First IEE International Conference on Artificial Neural Networks* , pp81, 1989.

3. Ramirez, M.R. and Arghya, D., "A Faster Learning Algorithm for Back-Propagation Neural Networks in NDE Applications", *2nd International Conference on AI in Civil and Structural Engineering* , in AI&CE, ed. B.H.V. Topping, pp275, Civil Comp Publications, Edinburgh, 1991.

4. Cheu, R.L., Ritchie, S.G., Recker, W.W., and Bavarian, B., "Investigation of a Neural Network Model for Freeway Incident Detection", *2nd International Conference on AI in Civil and Structural Engineering* , in AI&CE, ed. B.H.V. Topping, pp267, Civil Comp Publications, Edinburgh, 1991.

5. Tank, D.W. and Hopfield, J.J., "Simple Neural Optimization Networks: An A/D Converter, Signal Decision Circuit, and a Linear Programming Circuit", *IEEE Transactions on Circuits and Systems* , Vol. **CAS-33**, May 1986.

6. Ramanujam, J. and Sadayappar, P., "Optimization by Neural Networks," *IEEE International Conference on Neural Networks*, **Vol. 2**, pp. 325, 1988.

7. P. Hajela and L. Berke, "Neurobiological Computational Models in Structural Analysis and Design", *proceedings of the 31st AIAA/ASME/ASCE/AHS/ASC SDM Meeting* , pp345, Long Beach, California, April 1990, to be published in *Computers and Structures*, 1991.

8. P. Hajela and L. Berke, "Neural Network Based Decomposition in Optimal Structural Synthesis", *Computing Systems in Engineering* , Vol. 2, No. 5, 1991.

9. Swift, R., and Batill, S., "Application of Neural Networks to Preliminary Structural Design", AIAA Paper No. 91-1038, *Proceedings of the 32nd AIAA/ASME/ASCE/AHS/ASC SDM Meeting* , Baltimore, Maryland, April 1991.

10. Rehak, D.R., Thewalt, C.R., and Doo, L.B., "Neural Network Approaches in Structural Mechanics Computations", *Computer Utilization in Structural Engineering* , ed. J.K. Nelson, Jr., ASCE proceedings from Structures Congress, 1989.

11. Z. Szewczyk, P. Hajela and B. Fu, "Neural Network Approximations in a Simulated Annealing Based Optimal Structural Design, *Proceedings of the Society of Engineering Science Meeting* , Gainesville, Florida, 1991.

12. Carpenter, G.A. and Grossberg, S., "A Massively Parallel Architecture for a Self-Organizing Neural Pattern Recognition Machine. *Computer Visision, Graphics, and Image Processing* , Vol. 37, pp54, 1987.

13. Grossberg, S. (ed.), *The adaptive brain , Vol. I and II* , Arsterdam, North-Holland: Elsevier, 1987.

14. Grossberg, S. (ed.), *Neural networks and neural intelligence* , Cambridge, MA: MIT Press, 1988.

15. Hajela, P., Fu, B. and Berke, L., "ART Networks in Automated Conceptual Design of Structural Systems", *2nd International Conference on AI in Civil and Structural Engineering* , in AI&SE, ed. B.H.V. Topping, pp263, Civil Comp Publications, Edinburgh, 1991.

16. Fu, B. and Hajela, P., "Minimizing Distortion in Truss Structures - A Hopfield Network Solution", in review for *the 33rd AIAA/ASME/ASCE/AHS/ASC SDM meeting* , Dallas, Texas, April 1992.

17. Koopmans, J.C., and Beckmann, M.J., "Assignment Problems and the Location of Economic Activities". *Econometrica* , Vol. 25, pp53, 1957.

18. Durbin, R., and Willshaw, G., "An Analogue Approach to the Travelling Salesman Problem Using an Elastic Net Method". *Nature (London)* , Vol. **326**, pp689, 1987.

19. Fu, B. and Hajela, P., "An Elastic Net Solution to the Quadratic Problem ", in review for *Journal of Engineering Optimization* , 1991.

Table 1   Binary representation of problem features

A	0	The left end is a simple support.
	1	The left end is a clamped support.
B	0	The right end is a simple support.
	1	The right end is a clamped support.
C	0	The load is a force.
	1	The load is a moment.
D	0	The load is concentrated.
	1	The load is uniformly distributed.

Table 2. ART Codes and related procedural processes

[0010]

$$M = \begin{cases} \dfrac{M_0}{L}x & 0 < x < a \\ \dfrac{M_0}{L}x - M_0 & a < x < L \end{cases}$$

[0000]

$$M = \begin{cases} W\dfrac{b}{L}x & 0 < x < a \\ W\dfrac{b}{L}x - W(x - a) & a < x < L \end{cases}$$

[0011]

$$M = 0$$

[0001]

$$M = \dfrac{1}{2}qLx - \dfrac{1}{2}qx^2$$

[1110]

$$M = \begin{cases} \dfrac{6M_0 a}{L^2}\left(1 - \dfrac{a}{L}\right)x - M_0\left(-1 + 4\dfrac{a}{L} - \dfrac{3a^2}{L^2}\right) & 0 < x < a \\ \dfrac{6M_0 a}{L^2}\left(1 - \dfrac{a}{L}\right)x - M_0\left(-1 + 4\dfrac{a}{L} - \dfrac{3a^2}{L^2}\right) - M_0 & a < x < L \end{cases}$$

[1100]

$$M = \begin{cases} \dfrac{Wb^2}{L^3}(3a + b) - \dfrac{Wab^2}{L^2} & 0 < x < a \\ \dfrac{Wb^2}{L^3}(3a + b) - \dfrac{Wab^2}{L^2} - W(x - a) & a < x < L \end{cases}$$

[1111]

$$M = 0$$

[1101]

$$M = \dfrac{qL}{2}x - \dfrac{qL^2}{12} - \dfrac{1}{2}qx^2$$

Table 2. Continued

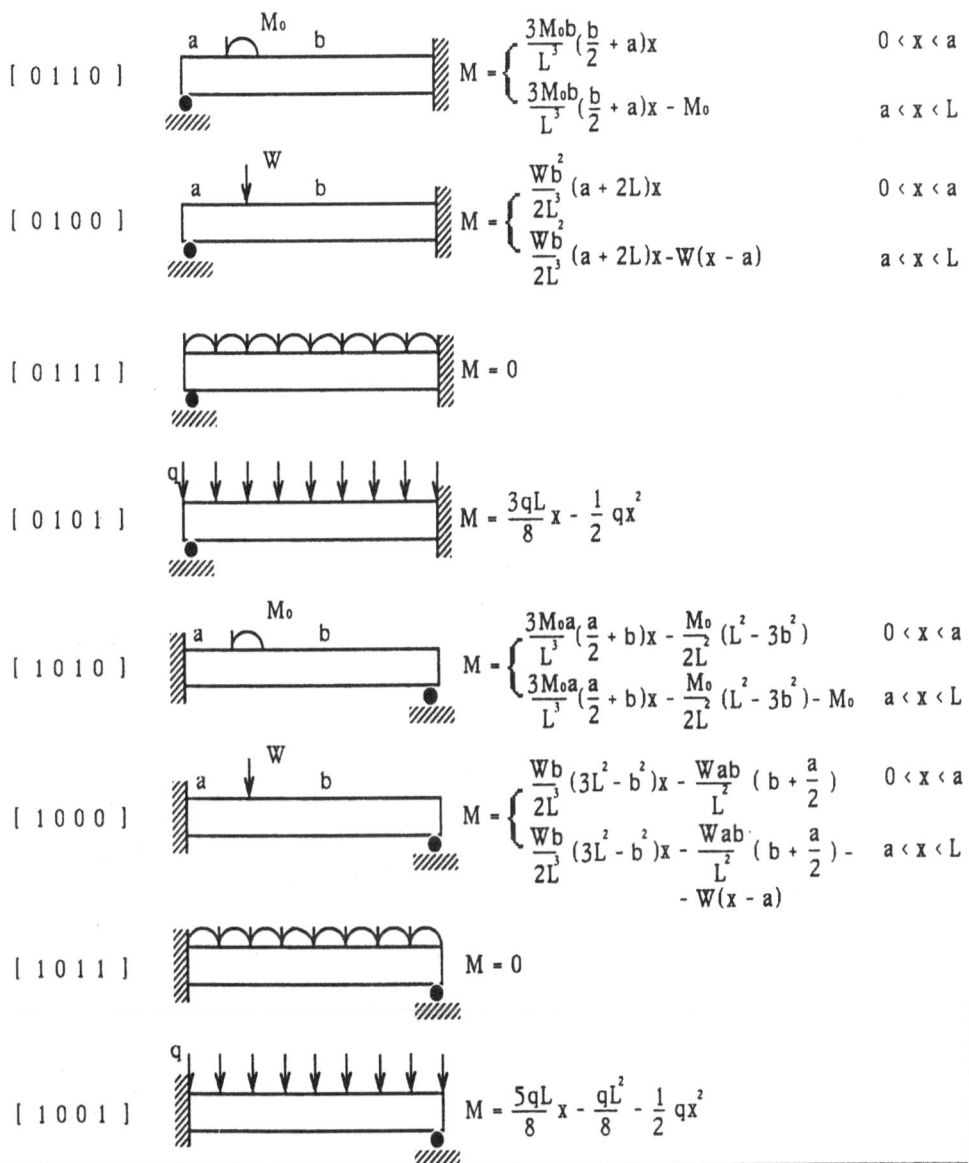

[0 1 1 0] $M = \begin{cases} \dfrac{3M_0b}{L^3}(\dfrac{b}{2}+a)x & 0 < x < a \\[3mm] \dfrac{3M_0b}{L^3}(\dfrac{b}{2}+a)x - M_0 & a < x < L \end{cases}$

[0 1 0 0] $M = \begin{cases} \dfrac{Wb^2}{2L^3}(a+2L)x & 0 < x < a \\[3mm] \dfrac{Wb^2}{2L^3}(a+2L)x - W(x-a) & a < x < L \end{cases}$

[0 1 1 1] $M = 0$

[0 1 0 1] $M = \dfrac{3qL}{8}x - \dfrac{1}{2}qx^2$

[1 0 1 0] $M = \begin{cases} \dfrac{3M_0a}{L^3}(\dfrac{a}{2}+b)x - \dfrac{M_0}{2L^2}(L^2-3b^2) & 0 < x < a \\[3mm] \dfrac{3M_0a}{L^3}(\dfrac{a}{2}+b)x - \dfrac{M_0}{2L^2}(L^2-3b^2) - M_0 & a < x < L \end{cases}$

[1 0 0 0] $M = \begin{cases} \dfrac{Wb}{2L^3}(3L^2-b^2)x - \dfrac{Wab}{L^2}(b+\dfrac{a}{2}) & 0 < x < a \\[3mm] \dfrac{Wb}{2L^3}(3L^2-b^2)x - \dfrac{Wab}{L^2}(b+\dfrac{a}{2}) - & a < x < L \\ \qquad - W(x-a) \end{cases}$

[1 0 1 1] $M = 0$

[1 0 0 1] $M = \dfrac{5qL}{8}x - \dfrac{qL^2}{8} - \dfrac{1}{2}qx^2$

Table 3   Data representation for the QAP

-	d_{12}	d_{13}	...	d_{1n}
f_{21}	-	d_{23}	...	d_{2n}
f_{31}	f_{32}	-		
...	...			
f_{n1}	f_{n2}			

Table 4   Data used in QAP with five factories

-	1	1	2	3
5	-	2	1	2
2	3	-	1	2
4	0	0	-	1
1	2	0	5	-

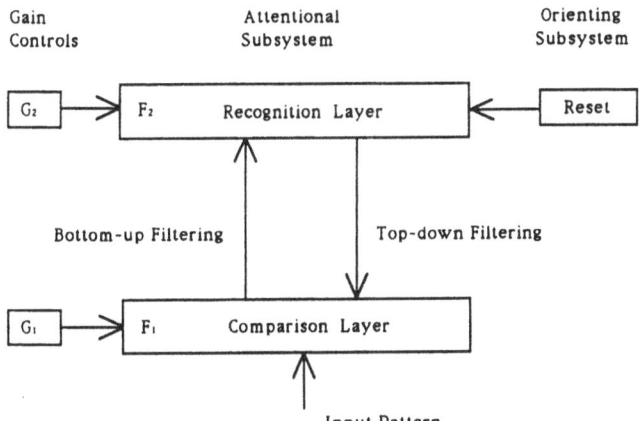

Figure 1   Schematic arrangement of an ART network

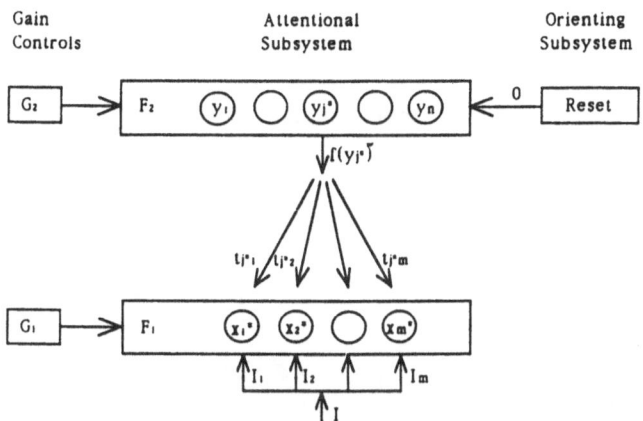

Figure 2   The attentional subsystem (top-down filtering)

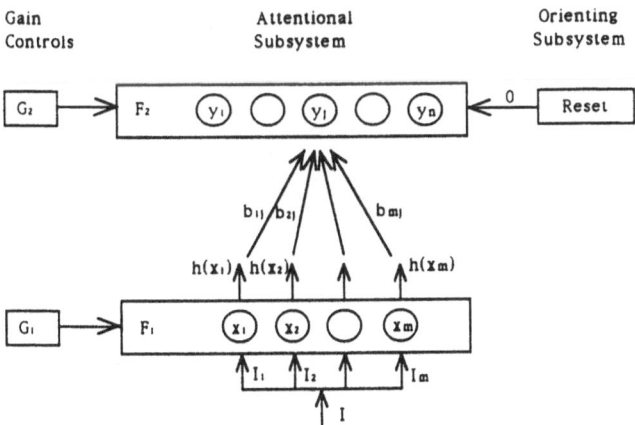

Figure 3   The attentional subsystem (bottom-down filtering)

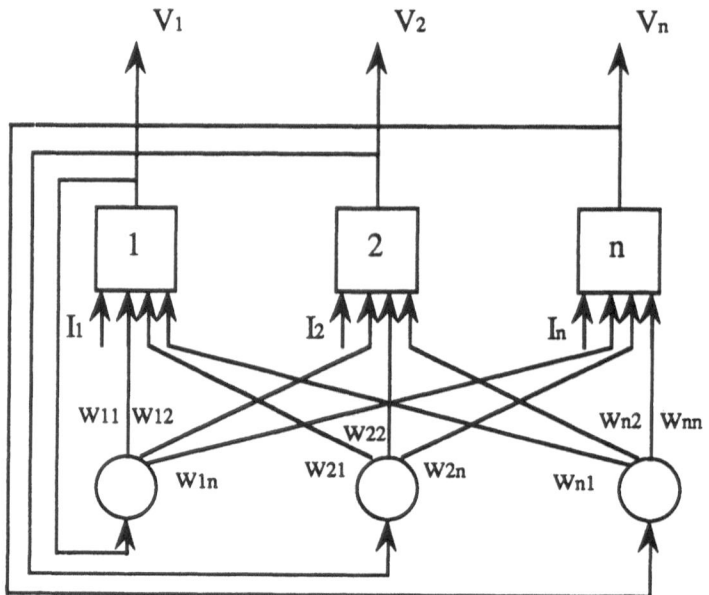

Figure 4   Architecture of Hopfield network

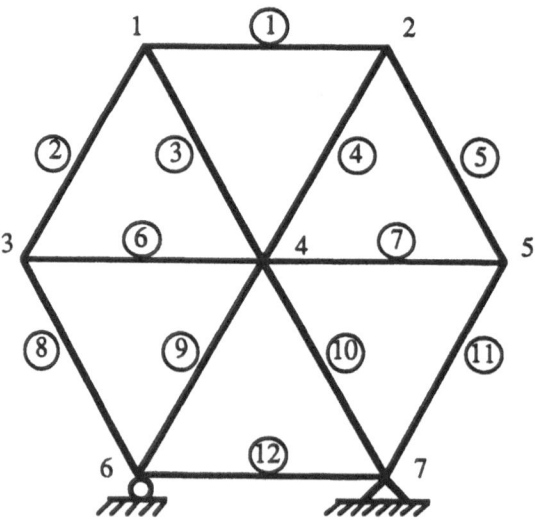

Figure 5   A 12-member planar truss

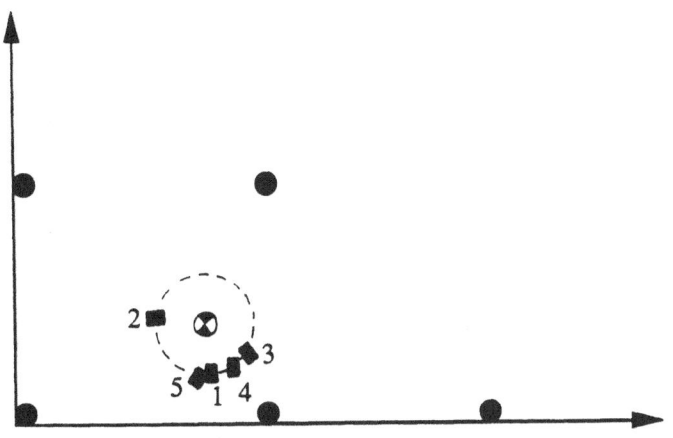

Figure 6   Initial and desired locations of factories

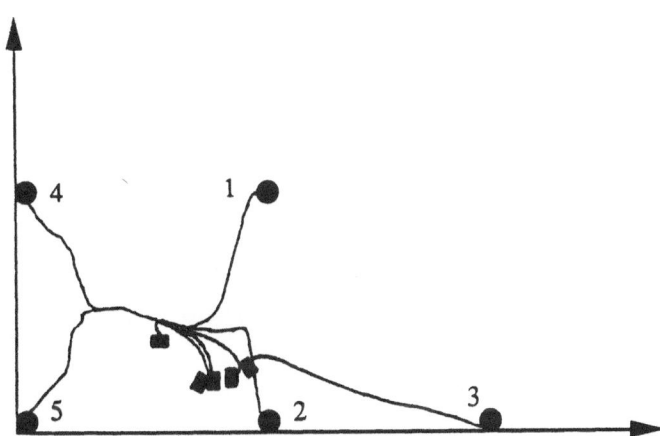

Figure 7   Trajectories of factory movements as the elastic net relaxes into an optimal configuration

Parallel Computations for Structural Analysis, Re-Analysis and Optimization

B.H.V. Topping[†] and A.I. Khan[‡]
†*Professor of Structural Engineering*
‡*Research Associate*
Heriot-Watt University,
Riccarton,
Edinburgh, EH14 4AS,
United Kingdom

Abstract This paper reviews recent developments in parallel processing technology. Developments in transputer systems and associated processors are described. Structural Optimization is shown to be a series of sub-problems. Each of these sub-problems is reviewed and the scope for parallelisation discussed.

1 Introduction

Recent developments in parallel computing has enabled rapid linear, non-linear and dynamic analysis of complex structures [5, 33, 56, 57, 58, 60, 62, 63]. This same computational power may be utilised for the efficient structural optimization of structures. Relatively little effort has, however, been made in this direction to date [53, 61, 74]. The process of optimization consists of a number of sub-problems including:

- formulation of the design model and idealization of the analysis model;

- structural analysis;

- sensitivity analysis;

- design of the structure using optimization techniques; and

- error analysis and re-idealization of the structure using adaptive procedures.

767

G. I. N. Rozvany (ed.), Optimization of Large Structural Systems, Vol. II, 767–792.

Hinton *et al* [35, 36] have demonstrated how the adaptive procedure may be effectively used in an integrated structural optimization approach to design. In this paper it is suggested that all the sub-problems of this sequential integrated approach may be parallelised. Figure 1 shows the parallel integrated approach to structural optimization. Before reviewing each of the sub-problems shown in Figure 1 parallel computing systems will be reviewed.

2 Parallel Computing Systems

The hardware system for parallel computers may generally be classified on the following basis:

SISD - single instruction single data. This is the traditional sequential computer consisting of one processor, memory and one communication channel as shown in Figure 2.

Obviously this computer architecture will always be restricted by the part of the three elements with the least capacity. With the present technology providing almost unlimited memory and extremely fast central processing units, the bottleneck is the communication channel.

SIMD - single instruction multiple data. This architecture may be available in distributed and shared memory systems. The shared memory SIMD architecture is illustrated in Figure 3 and the distributed memory SIMD architecture in Figure 4. This architecture overcomes the above mentioned bottleneck with the SISD systems through implementing multiple processors and memory module/modules working on a single instruction sets. Many large parallel computers use this principle taking advantage of the simplified programming task as the number of data sets are program independent and the same instruction is performed on all processors at any given time.

Host communication may however prove to be troublesome for this architecture since each processor must share the same communication path through the pipeline.

MIMD - multiple instructions multiple data. This architecture may again be subdivided into shared memory systems and distributed memory systems. The MIMD shared memory network is illustrated in Figure 5. This implementation has the disadvantage that processors may have to queue to access the same area of memory.

The distributed memory system applies to systems such as transputer networks where each processor has its own local memory and communicates with other processors through channels, as shown in the Figure 6.

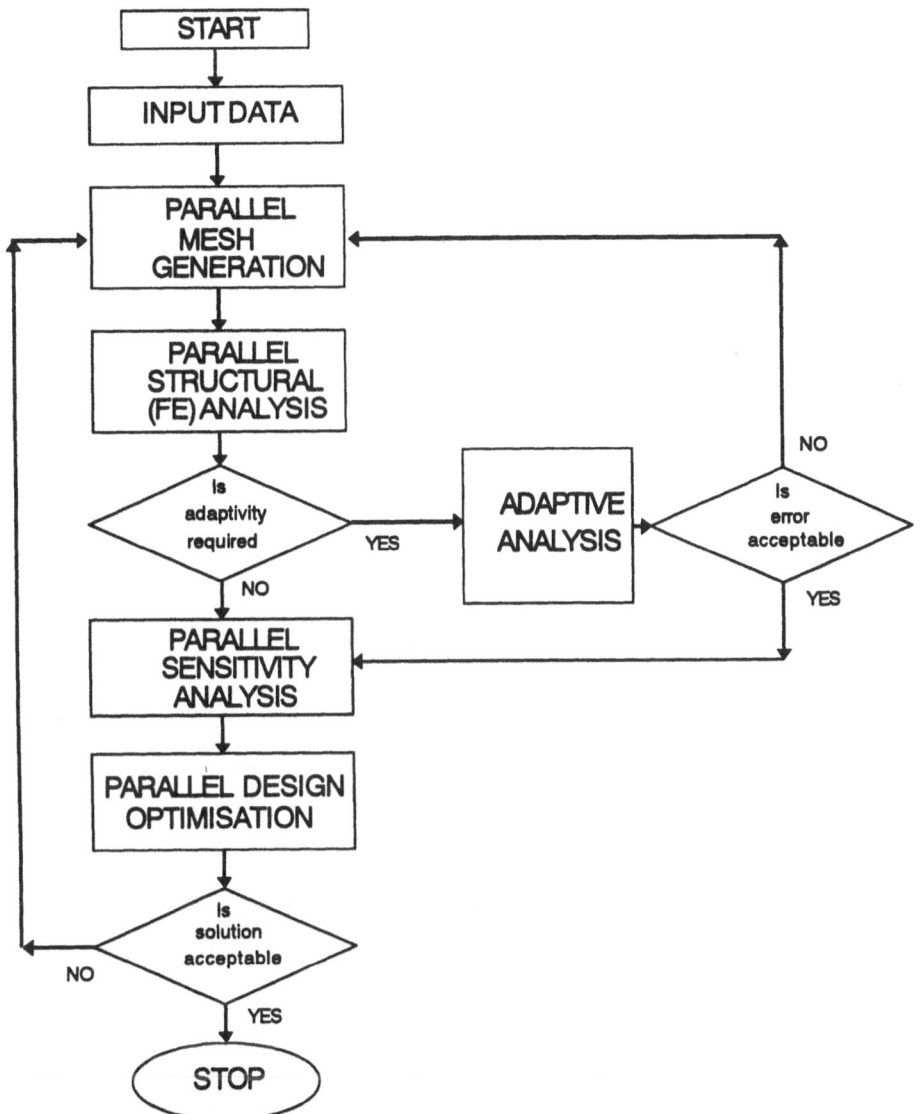

Figure 1. *Parallel Integrated Approach to Structural Optimization
after the Sequential Approach due to Hinton et al [35]*

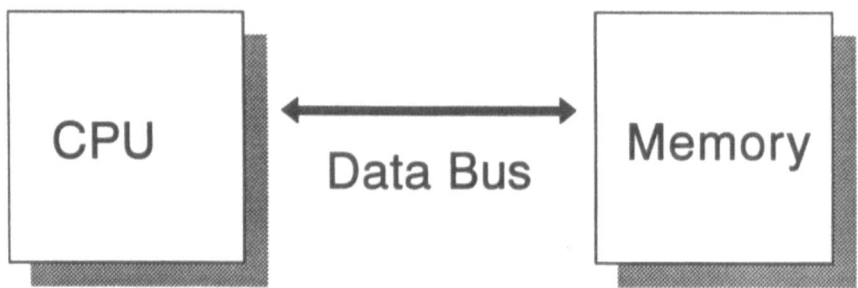

Figure 2. *The traditional sequential computer*

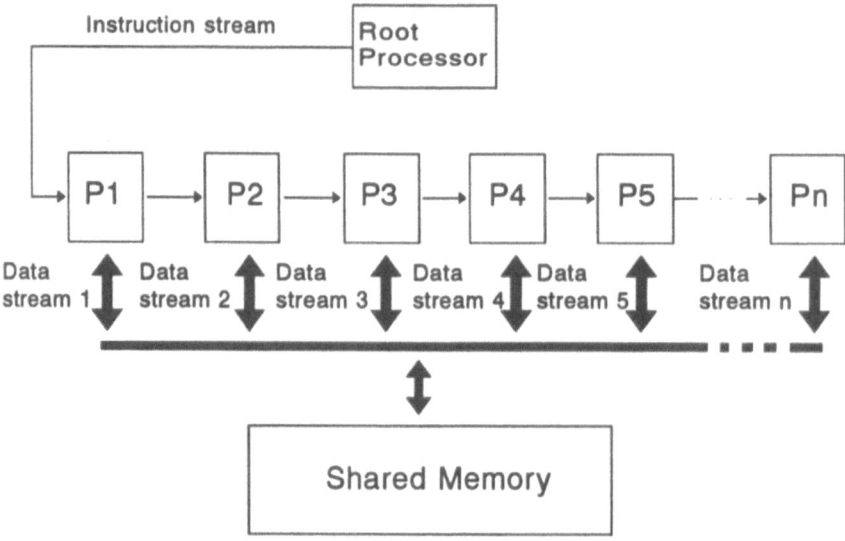

Figure 3. *Single Instruction Multiple Data (SIMD) architecture with shared memory*

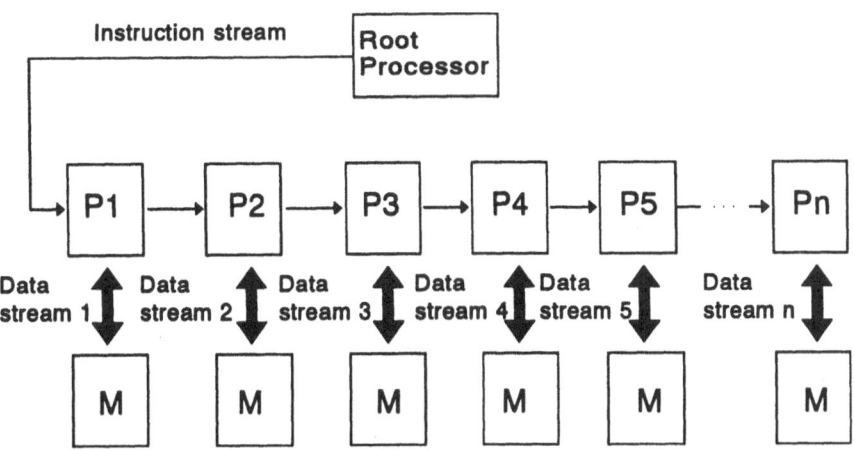

Figure 4. *Single Instruction Multiple Data (SIMD) architecture with distributed memory*

This implementation, Figure 6, defeats both the host communication problem (SIMD) and the bottleneck created by several processors queuing up to access the same data (shared memory MIMD).

MISD - multiple instructions single data. This architecture has yet to be implemented.

There is no obvious limitation to the number of processors in a MIMD architecture, but practical restrictions will be imposed by computational inefficiencies, communication requirements and hardware constraints. The distributed memory systems may be further divided according to the nature of the intercommunication. *Not switched* specifies broadcast or point to point communication, whereas *switched* implements a crossbar switch set according to a specified configuration. The last option is more expensive and complex to implement but provides the ultimate flexibility. A system with point to point communications connecting all processors has yet to be built.

As already indicated the computer architectures above may be further classified as:

- **Local memory systems** : where each processor carries its own on-board memory which is not available to other processors in the network. Data transfer takes place through interprocessor communication which may be bus-based or through serial links.

- **Shared memory** : where common memory area is made available to all the processors in the network and data transfer occurs through this shared memory.

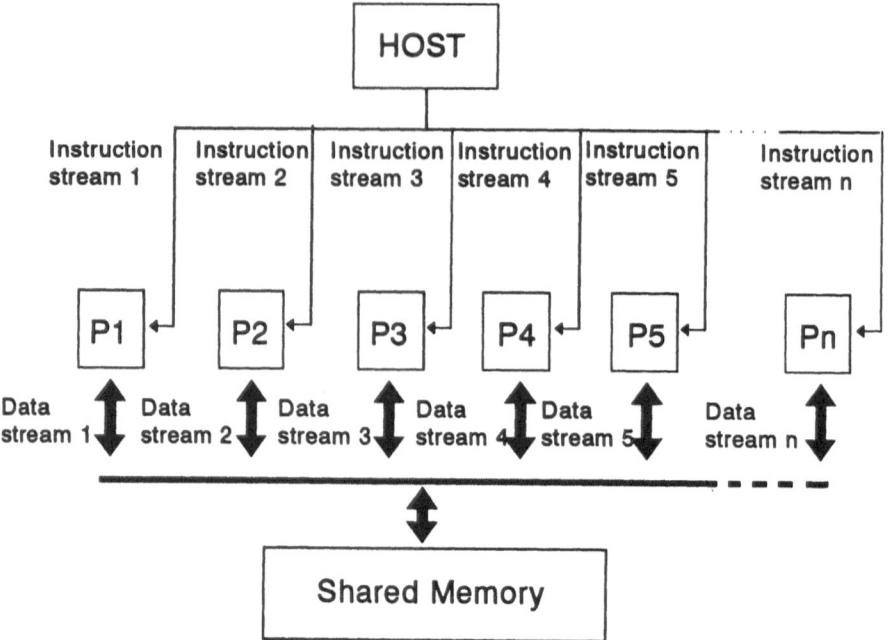

Figure 5. *Multiple Instruction Multiple Data (MIMD) network using shared memory*

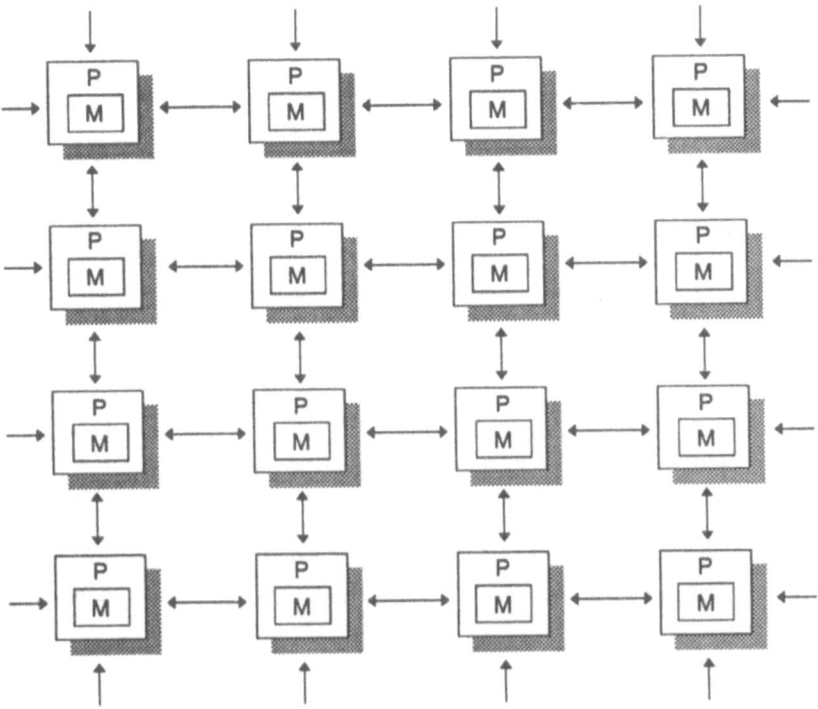

Figure 6. *Multiple Instruction Multiple Data (MIMD) network with distributed memory*

2.1   Granularity

Parallel computing systems may also be defined as coarse, medium or fine grained. If a few high performance processors are linked together to form a network it is termed a coarse grained system e.g. CRAY-2, CRAY X-MP and ETA-10. A fine grain system results from the coupling of a large number of processor which individually may or may not possess high performance capabilities. The performance of the processor is generally evaluated on the basis of the number of instructions it can execute per second or the number of floating point operations it can perform per second. Examples of these fine grained machines are Connection Machines of Thinking Machine Inc. with up to 65,536 processors and the Floating Point System T-series hypercube with up to 16,384 processors.

Medium grained machines fall somewhere between the first two categories. Sequent Balance 8000, Encore Multimax and FLEX/32 may be quoted as examples of medium grained machines. The number of processors present within a parallel computing system determines the systems granularity. Generally speaking systems with less than a hundred processors are termed as coarse grained systems. Systems lying within a range of $100 < no\_of\_processors < 1000$ are called medium grained systems. Systems having the number of processors in thousands are termed as fine grained systems.

2.2   Speed-up and Efficiency

Theoretically a program should run N times faster on a network containing N processors when compared to a single processor. In reality the *"Speed-ups"* achieved are less than the theoretical values owing to the increase in the communication time with the increase in the number of processors.

The *"Speed-up"* for a parallel system is defined as: $S = \frac{T_{seq}}{T_N}$; where the T_{seq} is the time taken by the code to execute on a single processor and T_N is the time taken for the same code to execute on a parallel system having N processors. The efficiency of the parallel system for a given algorithm is defined as: $E = \frac{S}{N}$

3   Transputer based parallel systems

3.1   Introduction

In 1985 the T414 transputer was launched by INMOS to form modular parallel processing **MIMD** systems. The transputer comprised a CPU, some memory and four fast DMA (Direct Memory Access) serial-communication links. The provision of serial communication links instead of a conventional bus type communication interface ensured that the communication bandwidth remained in balance with the computational power as the number of processors is increased in the network. The

Processor type	No. of Links	FPU	Mflops	Max. Link speed
T414-20	4	not present	0.09	20 Mbits/sec
T800-30	4	present	2.25	20 Mbits/sec
T9000-50	4	present	25	100 Mbits/sec
TMS320C40	6	present	50	160 Mbits/sec
i860	bus system	present	80	-
Sparc 2	bus system	present	4.2	-
VAX 11/780	-	present	0.54	-

Table 1. *Comparison of some of the major features of processors*

conventional bus-based communication inevitably led to a communication bottleneck as the processors contended for control of the bus.

The T414 transputer was then followed by T800 transputer which carried an on-board FPU (floating point unit) to speed up floating point operations. The T800 is a 32 bit CMOS (Complementary Metal-Oxide Semiconductor) microcomputer with a 64 bit floating point unit and graphics support [38]. The T800 may directly access 4Gbytes of linear address space. For a 30 MHz device the data rate for the accessible memory is up to 40 Mbytes/sec. The standard INMOS links cater for interprocessor communication. Each link can transfer data bi-directionally at up to 2.35Mbytes/sec.

The T800 carries 4Kbyte of on-chip RAM, a configurable memory interface and four standard INMOS communication links. A T800 running at 30MHz achieves a throughput of 15 MIPS (million instructions per second) and can sustain a rate of 2.25 Mflops (million floating point operations per second).

The transputer has two 32 bit timers. One timer is used for high priority tasks and is incremented every microsecond. The other timer is for low priority processes and is incremented every 64 microseconds. These timers are used to schedule the processes running on the transputer. A process may be made to wait until the timer has reached a certain value or interprocessor communication may be made time dependent meaning that if a process is unable to send or receive a message within a given number of timer ticks then it may return without carrying out the communication and become active again. The transputer can host concurrent processes running at the same or different priorities. The sharing of the processor time by the concurrent processes is controlled by the relevant timers which allocate the processor time slices according to the priority status of the processes. The processing capacities of some processor are given in table 1

Transputer based parallel systems provide attractive performance/hardware cost ratios. The present day supercomputers have the performance/hardware cost ratio of about 10 Flops/dollar while a T800 transputer system may have a ratio higher than 200 [60]. However transputer systems usually require more programming effort than many of the SIMD and MIMD supercomputers.

3.2   Classification of parallel algorithms

In order to understand the degree of parallelism offered by a transputer based parallel systems a broad classification of parallel algorithms may be established. The subject of parallel algorithms being in its infancy, general classifications are difficult to form. However successfully implemented parallel algorithms on Multiple Instructions Multiple Data machines such as transputers, may be classified in the following categories:

- **Processor Farming:** in which each processor in the network carries the same code and executes it in isolation from all the other processors. This is also called Independent Task or Event Parallelism.

- **Geometric Parallelism:** in which each processor executes the same code on data corresponding to a sub-region of the system being simulated and communicates boundary data to neighbouring processors handling neighbouring sub-regions.

- **Algorithmic Parallelism:** in which each processor is responsible for part of the algorithm, and all the data passes through each processor.

3.3   Classification of transputer based systems

The transputer based systems can be broadly classified into two groups, namely PC based systems and systems controlled by microcomputers such as a Sun or VAX.

INMOS [38] through such firms such as Transtech [71, 72] offer PC based transputer systems. They usually comprise a mother board fitted into the expansion slot of PC with the option of adding on other mother boards in cascaded form. Details of such a system can be found in [41, 71, 72]. Up to 10 size one transputer modules may be fitted on a PC based motherboard, the transputer links 1 and 2 are connected in a default pipeline and links 0 and 3 are passed through a linking switch which can connect links 0 and 3 of one transputer on the mother board to the link 0 and 3 of any other transputer in the network. This link switch is controlled by a dedicated transputer on the mother board. The first transputer on the mother board, called the ROOT transputer, is connected to the PC host for downloading the code and performing the input/output operations. The communication between the processors has to be managed through user written software for the application. The present day compilers for such systems only provide primitives for communications over specified channels. At any one instance each transputer cannot be connected to more than four other transputers thus extra code has to be written by the user to manage communication between processors which are not directly connected.

The Computing Surface (CS) produced by Meiko is an example of a microcomputer controlled transputer system. The CS comes with a global back plane bus, called the System Supervisor, to supplement the transputer point to point serial communication links. This bus carries control signal and hardware diagnostics information around the machine. The bus is now equipped with a VLSI switching chip and a software facility called the Computing Surface Network (CSN). The advantage of this hardware-software based facility is twofold. It enables the programmer to specify point to point communication over the network without getting involved into the low level network topology of the Computing Surface and heterogeneous networks can be built up comprising for example the i860 chips, SPARC processors and T800 transputers. The i860 processor delivers 80 Mflops which out runs the present T800 by a factor of thirty-five. The SPARC processor enables the Computing Surface to adopt the SunOS kernel and other facilities providing full Sun workstation emulation.

The new TMS320C40 processor from Texas Instruments offers six links of 160 Mbits/sec and a processing capability of 50 MFlops. The processors may be assembled in transputer like networks using the six links. These processors have the capability of communicating with similar processors through memory interface enabling further communication ability above the six links.

4   A critical overview of the transputer based systems

The transputer based systems are seen as occupying a mid-point between the coarse and fine grained parallel systems, however from the original concepts of the transputer design it appears they were visualized as flexible scalable systems leading to fine grained systems owing to their low cost and ability to keep the communication bandwidth in balance with the increase in the number of transputers. The use of transputers is more inclined toward coarse grained systems owing to large communication times as compared to computation times. The communication time overheads for the transputers result from the following:

- The bi-directional data transfer rate over the serial links for T800 transputers is (40/2.25) times slower than the data rate for the internal memory thus an application waiting for data from a serial link has to wait (40/2.25) times longer than if the same data was available on the internal memory of the processor.

- Additional instructions have to be provided in the code for routing the messages over the processor which are not connected directly, adding to communication overheads.

- The number of messages passed in the network.

- The individual length of the messages in bytes.

The above stated factors hinder the transputer based systems to operate in fine grain mode. The T800 transputer also faces stiff challenge from the bus-based Intel chip, i860, which can deliver 80 Mflops as compared to T800 which has a peak rate of 2.25 Mflops. However INMOS have announced their next transputer, the T9000, is claimed to posses 200 MIPS and 25 Mflops computational rates. The most significant feature of this transputer would be its capability to communicate at a much higher speed as compared with the T800. Infact it has been claimed that the T9000 can route messages over networks containing 1000 or more processors with communication delays of only a few microseconds where the T800 software exhibits delays of hundreds of milliseconds in a similar situation. This is accomplished due to T9000's packet switched "virtual" communication system. The bi-directional data rate for the T9000 is 100 Mbits/sec which is five time faster than the bi-directional data rate for the T800 (20 Mbits/sec). Any number of virtual channels may be defined, limited only by the maximum bandwidth of the links and the programmers own performance criteria, thus eliminating the the constraint of having only four bi-directional links per transputer. This allows for the optimal use of the physical links, while any virtual channel is idle the physical link may be used by another process for communication. The T9000 will also be provided with a hardware based router chip which would facilitate and expedite the message passing over distant processors.

5 Structural Analysis

Parallel processing provides increased computational power for structural analysis by using a number of strategies:

- Since multiple processors are available the analysis problem may be sub-divided by geometrically dividing the idealization into a number of sub-domains. Since each of these domains will be subject to the same instructions this may be accommodated through Processor Farming or Geometric Parallelisation. This division into subdomains is referred to as domain decomposition.

- Alternatively the system of equations for the whole structure may be assembled and solved in parallel without recourse to a physical domain decomposition of the problem.

5.1 Domain Decomposition

Domain Decomposition Techniques may be described as "Divide-and-Conquer" algorithms since their object is to divide the larger problem into a series of smaller subproblems. There are numerous techniques for domain decomposition [14, 15, 77] however two basic approaches are common:

- **Iterative Techniques** are used to solve the whole problem iteratively and information concerning nodes common to two or more domains must be communicated between processors after each iteration. Figure 7 shows an 'L' shaped domain which has been decomposed into seven sub-domains for solution using a parallel dynamic relaxation scheme [73].

The concept of domain decomposition methods dates back to Schwarz alternating procedure [15, 77]. This method is based on overlapping sub-domains. Recently the work on the algorithm has been re-examined and substantially extended [14, 46].

- **Sub-structuring** techniques have been established for over thirty years [65]. In this approach the sub-domains are treated as complex structural elements and the formulation for each sub-domain is carried out by only taking into account the degrees of freedom of the boundary nodes. Once the displacements of the boundary nodes have been determined the displacements within each substructure may be determined independently.

5.1.1 Discretization of a finite element mesh into subdomains For the implementation of a parallel algorithm it is necessary that the given domain of the problem is descretized into finite number of subregions or subdomains. In this regard it is important that the domain decomposition algorithm should be able:

- to handle irregular mesh geometries of arbitrarily shaped domains to make the method completely general;

- to minimize the interface problem by providing partitioning interfaces which deliver minimum boundary node connectivities. This aids in the reduction of the magnitude of the boundary problem and in physical terms reduces the communication overheads in the parallel computing system; while

- the subdomains should carry approximately equal amounts of computational load, this again is to satisfy the physical requirement of the system ensuring that all the processors finish with their work at about the same time and that the controlling processor is not forced to wait on account of a lagging processor.

Mesh decomposing algorithms using the above requirements as guide lines have been presented by Farhat [24] and Al-Nasra and Nguyen[4].

The domain-decomposition technique has been applied to the solution of nonlinear problems using a Dynamic Relaxation (DR) based finite element technique. The discretization of a structure into seven subdomains is shown in Figure 7. The code was parallelized using a one-dimensional array of transputers. The *"speed-ups"* were found to be a function of:

- The number of interface boundary nodes per sub-domain.

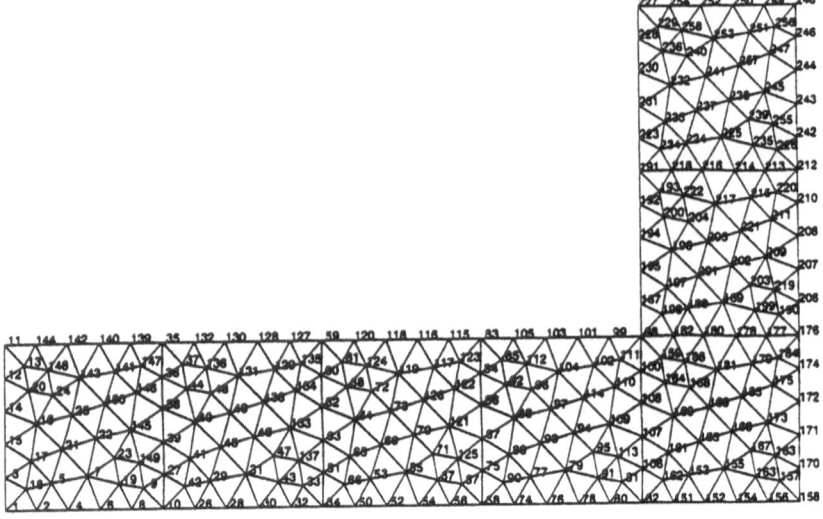

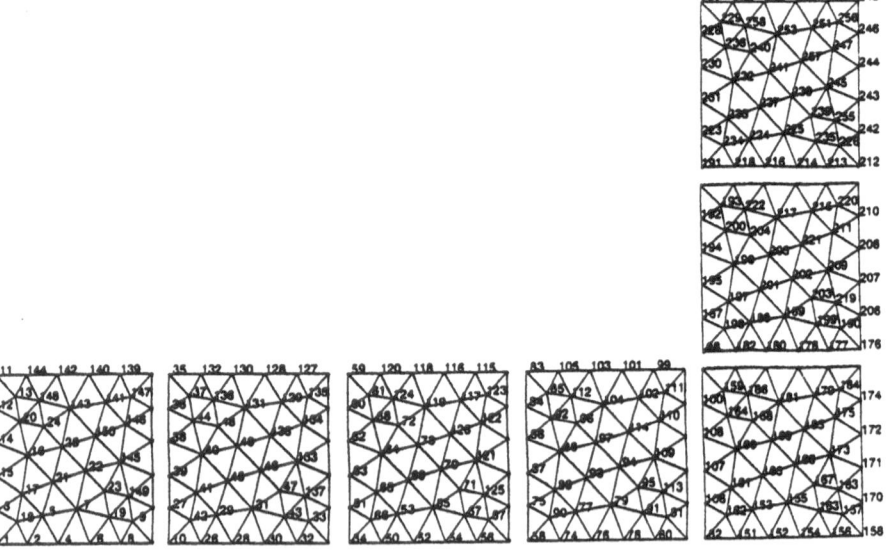

Figure 7. *Domain Decomposition of an 'L' shaped domain for Dynamic Relaxation Solution from reference [41]*

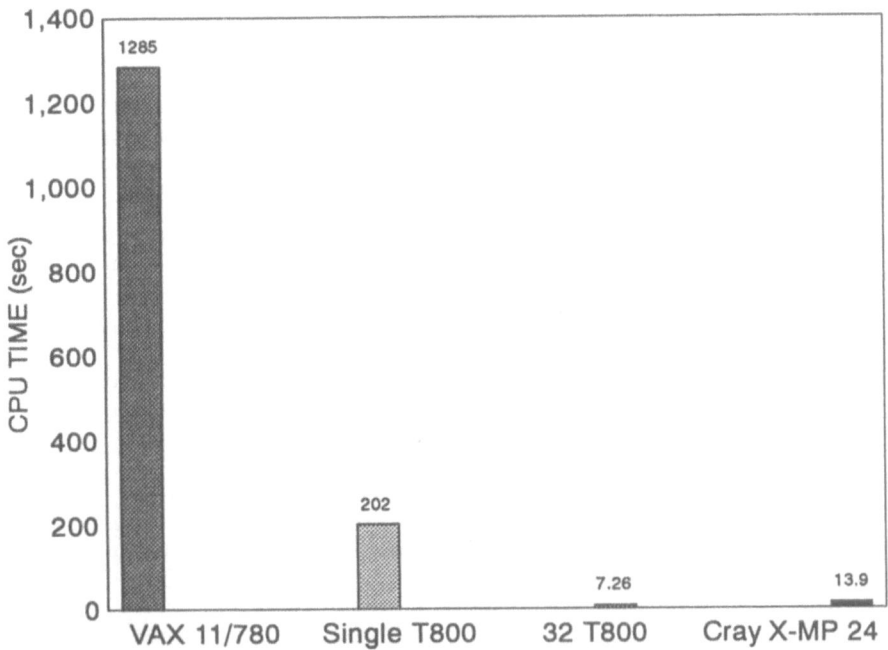

Figure 8. *Level of performance for mainframes and parallel networks after Favenesi et al [33]*

- The number of elements within each sub-domain (computational load).

- The length of communication path in case of non-adjacent sub-domain.

- The length of the pipe-line.

5.2 Parallel Equation Solvers

Farhat and Wilson [31] have reported the development of a parallel active column equation solver which is based on a Doolittle reduction in which each column of the stiffness matrix is assigned to a particular processor. The method distributes the storage of the stiffness matrix such as to ensure computational load balancing.

An alternative procedure for solving the stiffness equations was reported by Favenesi *et al* [33]. Their procedure was to assemble the global stiffness matrix node by node over a distributed set of transputers. Solution was undertaken using the preconditioned conjugate gradient method over the distributed processors. The level of performance of this type of solver implemented on a transputer based system and a conventional solver on various computers is shown in Figure 8.

6 Structural Re-Analysis

The structural re-analysis sub-problem was not shown in the flowchart of Figure 1. The objective of a re-analysis algorithm is to find the response of the structure after modifications using the original response of the structure such that the computational time of re-analysis is less than the complete analysis time. If the domain of the changes made during an optimization cycle is small then it may be possible to utilize a re-analysis algorithm. Sequential structural re-analysis methods have been reviewed in reference [1]. Little work except for that described in references [47, 70] has been undertaken on parallel re-analysis.

Recent work by Sziveri [70] on transputer re-analysis schemes was concerned with the theorems of structural and geometric variation where influence coefficients for changes in the structure are stored and manipulated on separate processors.

The efficiency of any re-analysis procedure has always been in doubt if the changes in the structure are large. For parallel schemes it is important that the computational load created by a fresh analysis is large since only then will parallel re-analysis schemes become generally efficient.

7 Sensitivity Analysis

Little work appears to have been done on parallel sensitivity analysis. It is apparent that this type of analysis is suited to processor farming. An illustrative example of the type of schemes that may be implemented was given by Moncrieff and Topping [53]. In their example the objective was to minimise the errors in a cutting pattern for stressed surface membrane structures. If the errors are not minimised it may lead to unsightly wrinkles in these architectural structures. The objective function was measured as the summation of all absolute principal prestress deviations. These deviations are the differences between actual predicted membrane prestress and the desired levels. Recursive quadratic programming was used for the solution of the non-linear programming problem. Since the behaviour of the membrane is highly non-linear a complete non-linear structural analysis must be performed for each design variable in order to calculate finite-difference derived gradients. A sequential scheme for this type of optimization problem is shown in Figure 9.

Each of the non-linear analyses for the gradient calculations is, however, entirely independent. This makes it possible for all sensitivity structural analyses to be performed concurrently on the distributed processors of a parallel computer. Figure 10 shows the resulting parallel scheme. This parallel optimization scheme was implemented on a MeiKo transputer array using processor farming.

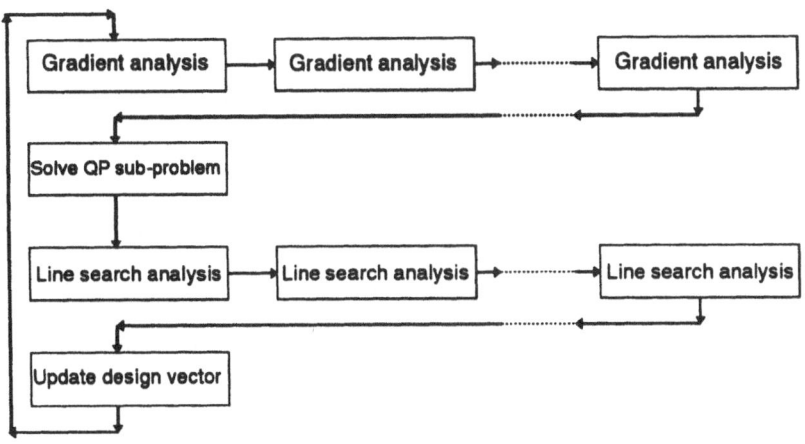

Figure 9. *Sequential Optimization Scheme*

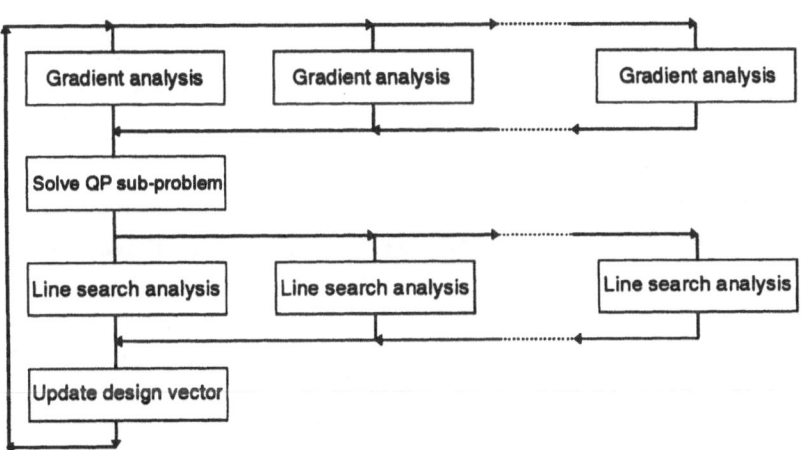

Figure 10. *Parallel Optimization Scheme*

8 Optimization

Research in parallelisation of optimization methods is still in its infancy. Linear programming stands as an established technique for structural optimization. Work has been undertaken for the parallelisation and vectorisation of the simplex method by Baddourah & Nguyen [7] and Beasley [8] on Cray super computers.

Baddourah and Nguyen [7] have applied a parallel searching algorithm on a CRAY-2 (4 processors) and Sequent(15 processors) supercomputers. Efficiencies of 90%, 87% and 75% were found respectively for 1500 by 1500, 1000 by 1000 and 500 by 500 equation systems on the four processors on the CRAY-2. The efficiency of the Sequent on a 500 by 500 system with fifteen processors was 96.6%. The Sequent does not have a vector capability and is much slower. Beasley [8] reports the implementation of linear programming on the Cray-1S and Cray X-MP/48 supercomputers. Greater computational efficiency is reported by the use of vector processing, especially in the case of the dual simplex problem. The effect on solution time of the choice of primal or dual simplex methods has been studied.

There is considerable scope for parallelisation of many non-linear programming methods. Research on the development of parallel algorithms for the following methods has been reported in reference [67]:

- Modified Newton method for unconstrained optimization;

- Nonlinear conjugate gradient methods;

- Global optimization methods (such as genetic algorithms); and

- Methods for nonlinear least squares.

9 Adaptive Idealization by Unstructured Mesh Generation

During the optimization of a structure idealised using the finite element method the mesh may become distorted due to the relocation of boundary nodes. Thus the structure may have to be re-meshed. It is most appropriate to use an automated re-meshing procedure such an adaptive unstructured re-meshing algorithm based on the advancing front technique. The aim of adaptive re-meshing is to determine an efficient mesh exhibiting a low and uniformly distributed domain error [81].

The adaptive process is repeated iteratively until the domain error becomes less than the specified value. Adaptive re-meshing is found to be the most time consuming task in adaptive analysis, after the finite element analysis. The parallelisation of the adaptive un-structured mesh generator was undertaken in order to render the adaptive algorithm computationally feasible for solving large problems.

The parallel adaptive mesh generator [41] follows the advancing front method for generation of two dimensional triangular element meshes. The advancing front algorithm requires searching over the background mesh for the local mesh parameter value [41] and the geometrical intersection of sides of the generated elements with the sides of the advancing front.

The parallel mesh generator implements concurrent element generation in different regions of the domain by dividing the domain into subdomains and mapping each subdomain to a different processor.

The requirement to carry out a search over the background mesh for interpolating the local mesh parameter values, is completely eliminated by specifying the background mesh elements as the sub-domains for the parallel scheme. The quantum of geometric searching for geometric intersections between the sides of the advancing front and the generated element is reduced owing to the formation of smaller advancing fronts within each subdomain.

Load balancing for each processor is difficult to achieve as the exact number of the generated elements is not known before the completion of the mesh generation hence the background mesh is passed through two pre-processing filters to weed out the triangular subdomains which will not undergo any element generation and to divide the subdomains exhibiting the prospect of large scale element generation into three subdomains by placing a node at the centroid of these elements. The non-generative nature of the subdomain is determined if the minimum nodal mesh parameter value for each side is greater than 51% of its length. The prolific nature of the subdomain is determined from the magnitude of the nodal mesh parameter values and geometrical properties of the subdomain [41].

The parallel mesh generator is based upon the processor farming technique since mesh generation may be carried out independently within each subdomain. No interprocessor communication is required among the slave processors, during the mesh generation phase. As soon the coordinates of a generated element become available the same are transmitted to the master or the ROOT processor.

The parallel scheme comprises a Master and Worker tasks as shown in Figures 11 and 12 respectively. The Master task distributes the work packets comprising the information for each subdomain. Prior to this distribution it applies the mesh filters described above. The Master task receives the generated elements, sent by the worker tasks and knits them together into an overall mesh. The Master then carries out the post processing of the compiled mesh.

Each slave processor receives a copy of the Worker task and on receipt of the information regarding the nodal coordinates and the nodal mesh parameters of a subdomain, proceeds with the adaptive mesh generation.

The Worker tasks at the end of each re-meshing of a sub-domain send a signal to the Master task. These signals are counted by the Master task. Until the total number of the signals recieved is equal to the total number of subdomains sent for re-meshing the Master task is descheduled except for its two concurrent processes communicating with the slave processors. This de-scheduling of the Master task enables more processor time for its communicating processes.

The processor farming environment is handled by a router facility and a flood-fill configurer provided within the *3L* Parallel C compiler [83] used for the purpose. The hardware comprised a TRAM mother board fitted into an expansion slot of an IBM compatible PC with an IMST800 transputer and 8MB DRAM and externally connected slave board having eleven IMST800 transputers each with 1MB DRAM.

As the parallel scheme does not require interprocessor communication among the slave processors, the "speed-ups" achieved using an increasing number of processors mainly depends on adequate load balancing of the computational load within each subdomain. The sequential and parallel remeshing of a square domain is shown in Figure 13.

10 Summary

This paper has:

- reviewed parallel processing systems;

- reviewed recent developments in transputer and associated hardware; and

- demonstrated that the sub-problems of the structural optimization design process have considerable scope for parallelisation.

Acknowledgment   The authors would like to acknowledge the useful discussions with other members of the Structural Engineering Computational Technology Group (SECT) at Heriot-Watt University. In particular the contribution of both Jørgen Stang and János Sziveri was greatly appreciated.

The support of Mike Cahill, Neil Clausen and Mark Wilson of Transtech Limited and Nakhee Nory and John Machar of *3L* Ltd for contributions to the SECT projects described in this paper is gratefully acknowledged.

References and Bibliography

[1] Abu Kassim, A.M., Topping, B.H.V., "Static Re-analysis: A Review", Proceedings of the American Society of Civil Engineers, Journal of Structural Engineering, vol. 113, no.5, 1029-1045, May, 1987.

[2] Adeli, H., (Editor), "Parallel and Distributed Processing in Structural Engineering", Proceedings of a session sponsored by the Structural Division of the American Society of Civil Engineers in conjunction with the ASCE National Convention, Nashville, Tennessee, May 11, 1988, American Society of Civil Engineers, New York, U.S.A., 1988.

[3] Adey, R.A., (Editor), "Parallel Processing in Engineering Applications", Computational Mechanics Publications, Southampton, England, 1990.

[4] Al-Nasra, M., Nguyen, D.T., "An Algorithm for Domain Decomposition in Finite Element Analysis", Computers and structures, v.39, no.3/4, 277-289, 1991.

[5] Alves-Filho, J.S.R., Kasper, K., Mitchell, G.P., Owen, D.R.J., "A Multi-Processor Implementation for Elastic-Plastic Finite Element Calculations", Proceedings of the Second International Conference in Computational Plasticity, COMPLAS II, Barcelona, September, 1989.

[6] Ameling, W., "Parallelism in Computer Architecture", in Ruschitzka M. et al. (Editors), in "Parallel and Large-Scale Computers: Performance, Architecure Applications", North Holland Publishing Co., 251-260, 1983.

[7] Baddourah, M.A., Nguyen, D.T., "Parallel-Vector processing for Linear Programming", Computers & Structures, v.38, 269-281, 1991.

[8] Beasley, J.E., " Linear Programming on Cray Supercomputers", J. Opl. Res. Spc., v.41, no.2, 133-139, 1990.

[9] Benantar, M., Flaherty, J.E., Shepard, M.S., "Parallel Computations for Automated Finite Element Modeling", in "Computer Utilization in Structural Engineering", Nelson J. K. Jr. (Editor), American Society of Civil Engineers, New York, NY, 197-200, 1989.

[10] Bettess, P., Downie, M.J., "Simple finite element codes in Occam on a transputer", Eng. Comput., v.6, 74-83, 1989.

[11] Bishop, P., "Fifth Generation Computers-Concepts implementations and uses", Ellis Horwood Limited, Chichester, England, 1986.

[12] Brebbia, C.A., Peters., A., (Editors), "Application of Supercomputers in Engineering: Fluid Flow and Stress Analysis Applications", Elsevier Science Publishers, Amsterdam, Netherlands, 1989.

[13] Brebbia, C.A., Peters, A., (Editors), "Application of Supercomputers in Engineering: Algorithms, Computer Systems and User Experince", Elsevier Science Publishers, Amsterdam, Netherlands, 1989.

[14] Carey, G.F., (Editor), "Parallel Supercomputing: Methods, Algorithms and Applications", John Wiley & Sons, Chichester, England, 1989.

[15] Carey, G.F., "Parallel Subdomain and Element-by-Element Techniques", Chapter 5, "Parallel Supercomputing: Methods, Algorithms and Applications", Carey, G.F., (Editor), John Wiley & Sons, Chichester, England, 1989.

[16] Carling, A., "Parallel Processing: The transputer and OCCAM", Sigma Press, Wilmslow, Cheshire, England, 1988.

[17] Chaing, K.N., Fulton, R.E., "Structural Dynamics Methods for Concurrent Processing Computers" Computers & Structures, v.36, no.6, 1031-1037.

[18] Chang, K.H., Santos, L.T., "Distributed Design Sensitivity Computations on a Network of Computers", Computers & Structures, v.37, no.3, 265-275, 1990.

[19] Coutinho, L.G.A., Alves, J.L.D., Ebeckn, N.F.F., Troina, L.M., "Conjugate Gradient Solution of Finite Element Equations on the IBM 3090 Vector Computer Utilizing Polynomial Preconditioning", Computer Methods in Applied Mechanics and Engineering, 84, 129-145, 1990.

[20] Deirlein, G.G., Abel, F., McGuire, W., "Some Interactive Graphics and Parallel Processing for Earthquake Engineering", in "Computer Utilization in Structural Engineering", Nelson J. K. Jr. (Editor), American Society of Civil Engineers, New York, NY, 438-477, 1989.

[21] Duff, I.S., "Architectures and Systems", Computer Physics Reports, v.11, 1-21, 1989.

[22] Duff, I.S., "Direct Solvers", Computer Physics Reports, v.11, 21-50, 1989.

[23] Evans, R.G., "The Joint Research Council's Supercomputer Unit", Bulletin of The Institution of Mathematics and its Applications, March, Vol.24, 49-51, 1988.

[24] Farhat, C., "A Simple and Efficient Automatic FEM Domain Decomposer", Computers and Structures v.28, no.5, 579-602, 1988.

[25] Farhat, C., "Which parallel finite element algorithm for which architecture and which

problem?" Engineering Computatations, v.7, 186-195, 1990.

[26] Farhat, C., "Computational Strategies for Finite Element Simulations on Supercomputers with 4 to 65,536 Processors", in "Computer Utilization in Structural Engineering", Nelson J. K. Jr. (Editor), American Society of Civil Engineers, New York, NY, 177-186, 1989.

[27] Farhat, C., Roux, F.-X., "A Method of Finite Element Tearing and Interconnecting and its Parallel Solution Algorithm", Report CU-CSSC-90-03, Center for Space Structures and Controls, College of Engineering, University of Colorado, U.S.A., February, 1990.

[28] Farhat, C., Sobh, N., "A Consistency Analysis of a Class of Concurrent Transient Implicit/Explicit Algorithms", Computer Methods in Applied Mechanics and Engineering, 84, 147-162, 1990.

[29] Farhat, C., Wilson, C., "Concurrent Iterative Solution of Large Finite Element Systems", Communications in Applied Numerical Methods, v.3, 319-326, 1987.

[30] Farhat, C., Wilson, C., "A New Finite Element Concurrent Computer Program Architecture", Int. J. for Numerical Methods in Engineering, v.24, 1771-1792, 1987.

[31] Farhat, C., Wilson, E., "A Parallel Active Column Equation Solver", Computers & Structures, v.28., n.2, 289-304, 1988.

[32] Farhat, C., Wilson, E., Powell, G., "Solution of Finite Element Systems on Concurrent Processing Computers", Engineering with Computers, v.2, 157-165, 1987.

[33] Favenesi, J., Daniel, A., Tombello, J., Watson, J., "Distributed Finite Element using a Transputer Network", Computing Systems in Engineering, vol. 1, Nos 2-4, 171-182, 1990

[34] Graham, I., King, T., (1990), "The Transputer Handbook", Prentice Hall, New York, U.S.A.

[35] Hinton E., Rao, N.V.R., Ozakca, M., " An Integrated Approach to Structural Shape Optimization of Linearly Elastic Structures. Part I: General Methodology.

[36] Hinton, E., Ozakca, M., Rao, N.V.R., " An Integrated Approach to Structural Shape Optimization of Linearly Elastic Structures. Part II: Shape Definition and Adaptivity.

[37] Hwang, K., Briggs, F.A., (1989), "Computer Architecture and Parallel Processing", McGraw Hill Book Co., New York, U.S.A.

[38] INMOS Limited, "Transputer Reference Manual", Prentice-Hall, Hertfordshire, U.K., 1988.

[39] INMOS Limited, "The Transputer Development and iq systems Databook", second edition, Inmos Ltd, 1991

[40] Kamat, M.P., (Editor), "Super and Parallel and Their Impact on Civil Engineering", Proceedings of a Session at Structures Congress'86, Amerciacn Society of Civil Engineers, New York, U.S.A., 1986.

[41] Khan, A.I., Topping, B.H.V., "Parallel Adaptive Mesh Generation", Computing Systems in Engineering, Vol 2, No. 1, 75-101, 1991.

[42] Kung, H.T., "Advances in Multicomputers", Computing Systems in Engineering, v.1, n.2-4, 153-162, 1990.

[43] Law, K.H., "A Parallel Finite Element Solution Method", Computers and Structures, v.23, n.6, 845-858, 1986.

[44] Law, K.H., Hammond, S. W., " Finite Element Calculations on a Linear Systolic Array", in "Computer Utilization in Structural Engineering", Nelson J. K. Jr. (Editor), American Society of Civil Engineers, New York, NY, 187-200, 1989.

[45] Lazou, C., "Supercomputers and their Use", revised edition, Oxford Science Publications, Oxford, England, 1988.

[46] Lions, P.L., "On the Schwarz alternating method Method III: A Varient of Non-overlapping subdomains", in Proceedings of the Third Int. Symp. on Domain Decomposition Methods for Partial Differential Equations, Houston, Texas, March, 1989, SIAM, Philadelphia, U.S.A., 202-223, 1990.

[47] Livesley, R.K., Modi, J.J., "The Re-Analysis of Linear Structures on Serial and Parallel Computers", Computing Systems in Engineering, v.2, n.4, 1991.

[48] Lou, J., Friedman, M.B., "A Parallel Computational Model for the Finite Element Method

on a Memory-Sharing Multiprocessor Computer", Computer Methods in Applied Mechanics and Engineering, 84, 193-209, 1990.

[49] Melli, P., Brebbia, C.A., (editors), "Supercomputing in Engineering Structures", Computational Mechanics, Southampton, 1989.

[50] Meurant, G., "Iterative Methods for Multiprocessor Vector Computers", Computer Physics Reports, v.11, 51-80, 1989.

[51] Modi, J.J., "Parallel Algorithms and Matrix Computation", Clarendon Press, Oxford, England, 1988.

[52] Moncrieff, E., "Stressed Membrane Structure Modelling", Edinburgh Concurrent Supercomputer Newsletter, n.6, Jan., 1989.

[53] Moncrieff, E., Topping, B.H.V., "The Optimization of Stressed Membrane Surface Structures using Parallel Computational Techniques" Eng. Opt., v.17, 205-218, 1991.

[54] Nguyen, D.T., Agarwal, T.K., "A Portable Parallel Equation Solver" in "Computer Utilization in Structural Engineering", Nelson J. K. Jr. (Editor), American Society of Civil Engineers, New York, NY, 158-167, 1989.

[55] Nguyen, D.T., Nui, K.T., "Structural Sensitivity Analysis on a Parallel Computer", in "Computer Utilization in Structural Engineering", Nelson J. K. Jr. (Editor), American Society of Civil Engineers, New York, NY, 98-112, 1987.

[56] Noor, A.K., (editor), "Parallel Computations and Their Impact on Mechanics", The American Society of Mechanical Engineers, New York, 1987.

[57] Noor, A.K., "New Computing Systems and Their Impact on Computational Mechanics", to be published in State-of-the-Art Surveys on Computational Mechanics, American Society of Mechanical Engineers, New York, 1989.

[58] Noor, A.K., "Parallel Processing in the Finite Element Structural Analysis", Engineeering with Computers, v.3, 225-241, 1988.

[59] Nour-Omid, B., Ortiz, M., "A Family of Concurrent Algorithms for Transient Finite Element solutions", in "Computer Utilization in Structural Engineering", Nelson J. K. Jr. (Editor), American Society of Civil Engineers, New York, NY, 143-151, 1989.

[60] Owen, D.R.J., "Parallel Finite Element Algorithms for Transputers", Proceedings of Mathematics of Finite elements and Applications VII, (MAFELAP' 90) Conference, April 1990.

[61] Owen, D.R.J., Alves-Filho, J.S.R., "Structural Shape Design by Transputer Based Finite Element Techniques", Materials and Engineering Design: The Next Decade, 143-150, The Institute of Metals, London, 1989.

[62] Owen, D.R.J., Alves-Filho, J.S.R., "Using Transputers in Finite Element Computations", NAFMS Benchmark, 39-41, October, 1989.

[63] Owen, D.R.J., Alves-Filho, J.S.R., Mitchell, G.P., Kasper, K., "The Implementation of Finite Element Computations in Transputer Based Systems", in Proceedings of the Symposium on Discretisation Methods in Structural Methods, IUTAM/IACM, Vienna, June, 1989.

[64] Parris, R.A., Gupta, S., "The Use of Supercomputers to solve large non-linear structural analysis problems" in Topping, B.H.V., (Editor), Proceedings of the Second International Conference on Civil and Structural Engineering Computing, Civil-Comp Press, Edinburgh, vol.2, 33-39, 1985.

[65] Przemieniecki, J.S., "Theory of Matrix Structural Analysis", McGraw Hill Book Co., New York, 1986.

[66] Schendel, U., "Introduction to Numerical Methods for Parallel Computers", Ellis Horwood Limited, Chichester, England, 1984.

[67] Schnabel, R.B., "Parallel Computing in Optimization", in Schittkowski, K., (Editor), "Computational Mathematical Programming", NATO ASI Series, v.F15, Springer Verlag, Berlin, Heidelberg, 1985.

[68] Stone, H.S., "High-Performance Computer Architecture", Second Edn., Addison-Wesley

Publishing Company, Reading Massachusetts, U.S.A., 1990.

[69] Storaasli, O., Poole, E., Ortega, J., Cleary, A., Vaughan, C., "Solution of Structural Analysis Problems on a Parallel Computer", AIAA/ASME/ASCE/AHS 29th Structures, Structural Dynamics and Materials Conference, Part I, 596-605, 1988.

[70] Sziveri, J., "Sequential and Parallel Formulations for the Theorems of Geometric Variation for the Re-analysis of Structures", Diploma thesis for Faculty of Civil Engineering, Technical University of Budapest, prepared at the Department of Civil Engineering, Heriot-Watt University, Edinburgh, 1991.

[71] TMB08 Installation and User Manual, Transtech Devices Ltd, Wye Industrial Estate, London Road, High Wycombe, Buckinghamshire.

[72] TMB12 Installation and User Manual, Transtech Devices Ltd, Wye Industrial Estate, London Road, High Wycombe, Buckinghamshire.

[73] Topping, B.H.V., Khan, A.I., "Parallel Dynamic Relaxation", submitted for publication, 1991.

[74] Vanderplaats, G.N., "Optimization and Supercomputing", in Melli, P., Brebbia, C.A., (editors), "Supercomputing in Engineering Structures", Computational Mechanics, Southampton, 1989.

[75] Wallace, D.J., "Supercomputing with Transputers", Computing Systems in Engineering, v.1, 131-141, 1990.

[76] White, D.W., Abel, J.F., "Bibliography on Finite Elements and Supercomputing", Communications in Applied Num. Methods, v.4, 279-294, 1988.

[77] Widlund, O.B., "Domain Decomposition Algorithms and the Bicentennial of the French Revolution", in Proceedings of the Third Int. Symp. on Domain Decomposition Methods for Partial Differential Equations, Houston, Texas, March, 1989, SIAM, Philaselphia, U.S.A., xv-xx, 1990.

[78] Willis, C.J., Wait., "Distributed Finite Element Calculations on Transputer Arrays and the DAP", Computing Systems in Engineering, v.2, n4, 421-424, 1991.

[79] Wilson, E.L., Farhat, E.L., "Linear and Nonlinear Finite Element Analysis on Multiprocessor Computer Systems", Communications in Applied Numerical Methods, v.4, 425-434, 1988.

[80] Yagawa, G., Soneda, N., Yoshimura, S., "A Large Scale Finite Element Analysis using Domain Decomposition Method on a Parallel Computer", Computers & Structures, v.38, n5/6, 615-625, 1991.

[81] Zienkiewicz, O.C., Zhu, J.Z., "A simple error estimator and adaptive procedure for practical engineering analysis", Int J for Num Methods in Eng., v. 24, 337-357, 1987.

[82] 3L Ltd., "Parallel Fortran User Guide", 3L Ltd., Livingston, Scotland, 1988.

[83] 3L Ltd., "Parallel C User Guide", 3L Ltd., Livingston, Scotland, 1988.

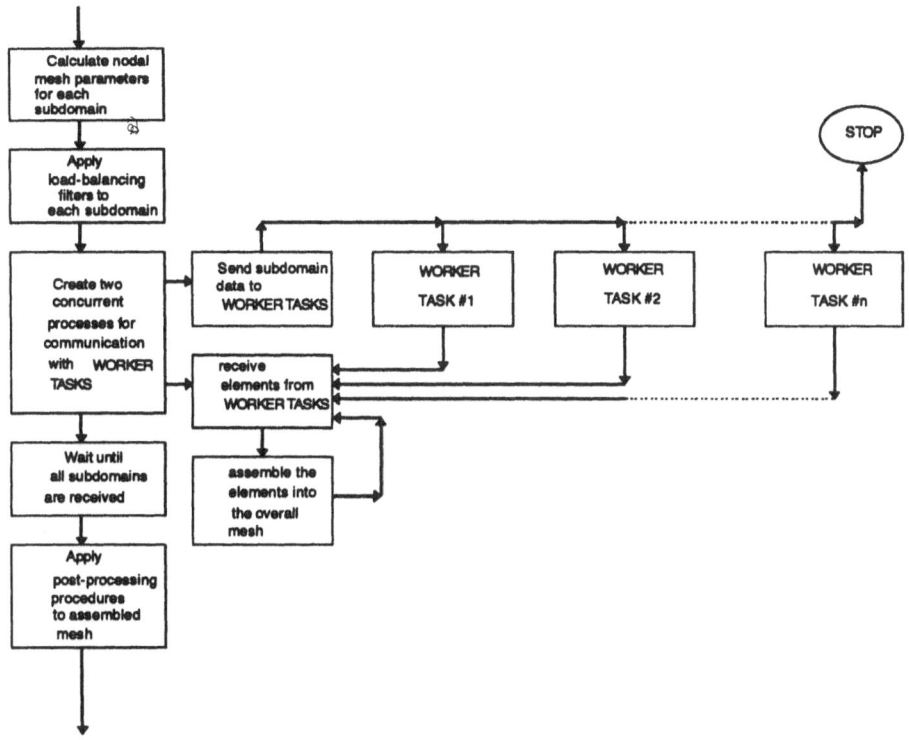

Figure 11. *Master task for Adaptive Remeshing*

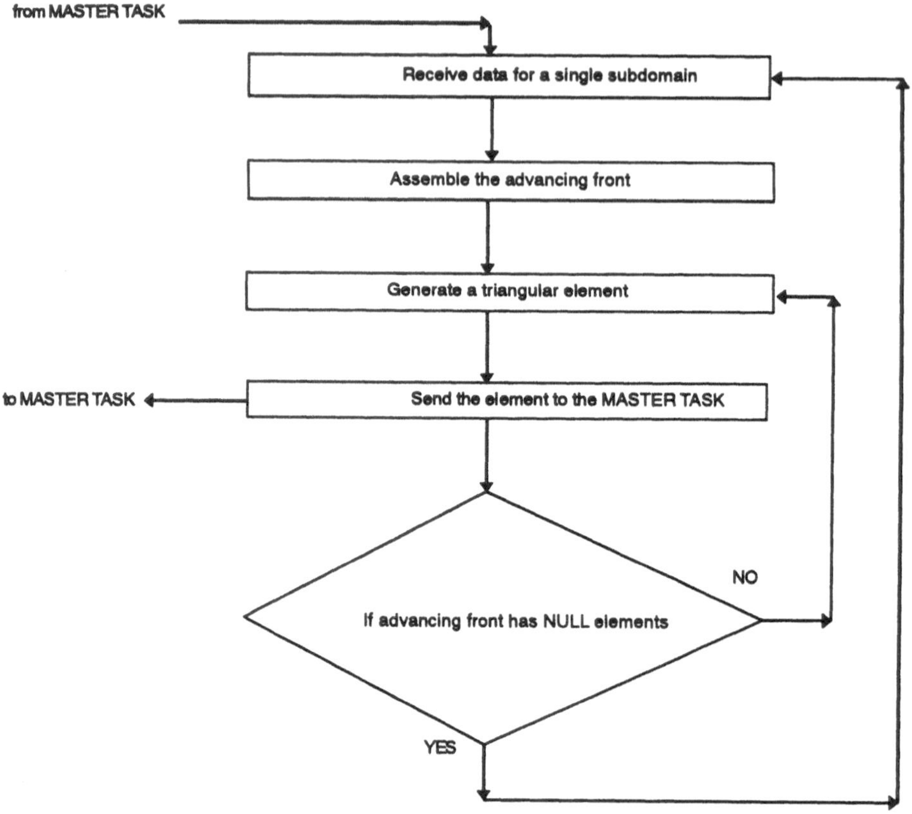

from MASTER TASK

Receive data for a single subdomain

Assemble the advancing front

Generate a triangular element

to MASTER TASK ← Send the element to the MASTER TASK

If advancing front has NULL elements

NO

YES

Figure 12. *Worker task for Adaptive Remeshing*

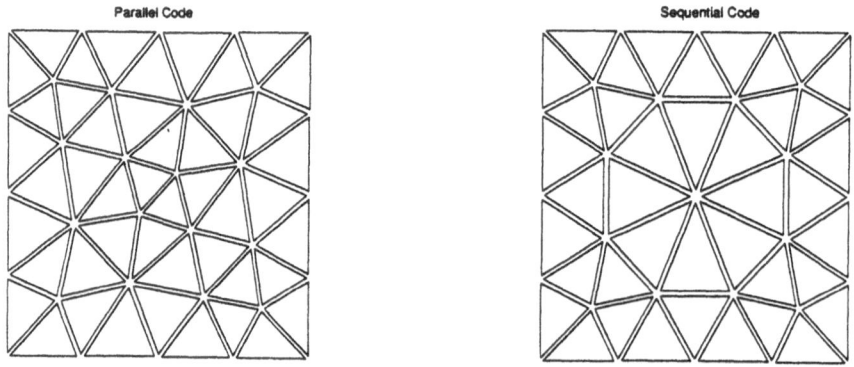

Parallel Code

Sequential Code

Figure 13. *Parallel and Sequential Remeshing of a rectangular domain*

MULTICRITERION STRUCTURAL OPTIMIZATION
State of the Art

JUHANI KOSKI
Tampere University of Technology
P.O. Box 527
33101 Tampere
Finland

ABSTRACT. In this article the present state of the multicriterion structural optimization is considered from the basis of about seventy publications which all are connected with the Pareto optimality concept. The completed works rather than open questions in the field are particularly emphasized. The basic concepts and the motivation of the multicriterion approach are briefly discussed. The classification of the multicriterion structural design process is proposed and it is used in describing the published applications.

1. INTRODUCTION

Usually several conflicting design requirements appear in designing a load supporting structure for practical purposes. Some of these requirements can be expressed in mathematical formulae while others are taken into consideration during the design process. The choice of the criteria and the constraints must be done in formulating the structural optimization problem. Sometimes only one dominating criterion, like the structural weight in air- and spacecraft design, may be a sufficient objective for minimization. This is true especially if the other requirements can be reliably presented by equality and inequality constraints. In some other branches of engineering, however, this kind of optimal design does not necessarily meet all the existing demands and it must be modified to some extent before subjecting it to the manufacturing process. A sensitivity analysis is commonly used in the modification of the optimal design but only a local information is obtained in this way. Accordingly, a single criterion or scalar optimization offers an elegant and sharp but relatively inflexible tool for the designer because the optimal solution is more or less fixed immediately after the problem formulation.

Multicriterion (multicriteria, multiobjective, vector, Pareto) optimization has recently been acknowledged as an advanced design technique also in structural optimization. One important reason for the success of this alternative approach is its natural property to allow the participation in the design process even after the

G. I. N. Rozvany (ed.), *Optimization of Large Structural Systems, Vol. II*, 793–809.
© 1993 *Kluwer Academic Publishers*.

formulation of the mathematical optimization model. It is generally considered that the multicriterion problem statement derives its origin from the end of the last century when Pareto (1848-1923) presented a qualitative definition of the optimality concept for economic problems with several noncomparable criteria [1]. Some other early contributors are discussed for example in ref. [2]. A wider interest in this subject concerning the fields of optimization theory, operations research and control theory was aroused at the end of the 1960s and since then the research work has been very intensive also in engineering design [3], [4]. Especially in mechanics applications have been published from the last half of the 1970s up to the present with an increasing speed [5], [6]. The purpose of this article is to discuss about the present state of the multicriterion structural optimization from the basis of about seventy papers which are available for the author.

2. MULTICRITERION PROBLEM AND PARETO OPTIMALITY

The main task in formulating an optimization problem is the choice of the design variables, the criteria and the constraints. The difference between the last two concepts is that the designer wants to improve the value of a criterion whereas this kind of desire is not associated with the constraints. Allowable values for the inequality constraints may be available in the design codes and other regulations but often they are difficult to choose. In addition, usually several competing and noncommensurable criteria appear in a typical real-life application and thus the designer is faced to a decision-making problem where the task is to find the best possible compromise solution between the conflicting requirements. As a natural result from the separation of the criteria f_i, $i = 1, 2, \ldots, m$, and the constraints the following multicriterion optimization problem is formulated

$$\min_{\mathbf{x} \in \Omega} [f_1(\mathbf{x}) \ f_2(\mathbf{x}) \ \ldots \ f_m(\mathbf{x})]^T \quad . \tag{1}$$

Here $\mathbf{x} = [x_1 \ x_2 \ \ldots \ x_n]^T$ represents a design variable vector and Ω is the feasible set defined by the equality and inequality constraints in the following way

$$\Omega = \{\mathbf{x} \in R^n \mid g(\mathbf{x}) \leq 0 \ , \ h(\mathbf{x}) = 0\} \quad . \tag{2}$$

Vector objective function $\mathbf{f}(\mathbf{x}) = [f_1(\mathbf{x}) \ f_2(\mathbf{x}) \ \ldots \ f_m(\mathbf{x})]^T$ contains all the m criteria as components which should be minimized simultaneously in $\Omega \subset R^n$. The image of the feasible set in the criterion space R^m is defined by

$$\Lambda = \{\mathbf{z} \in R^m \mid \mathbf{z} = \mathbf{f}(\mathbf{x}) \ , \ \mathbf{x} \in \Omega\} \tag{3}$$

and it is usually called the attainable (criteria) set. Usually there exists no unique point which would give an optimum for all m criteria simultaneously. Thus the common optimality concept used in scalar

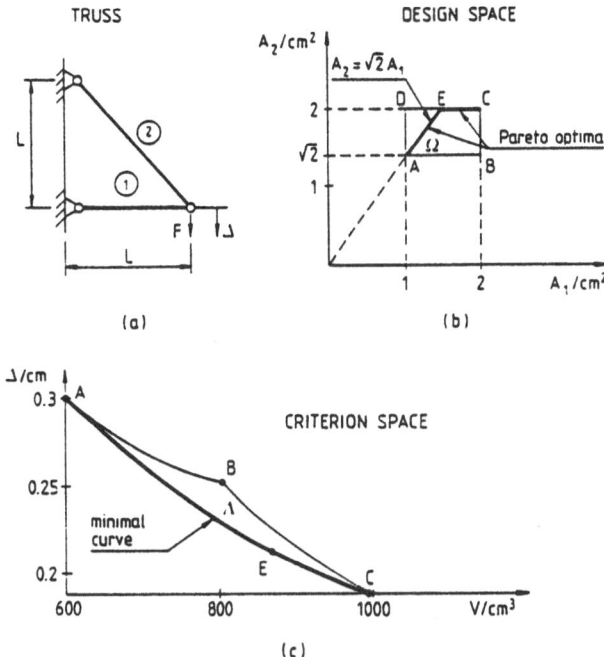

Figure 1. (a) Structure, loading and displacement criterion Δ; (b) feasible set Ω in design space and Pareto optimal polygonal line AEC; (c) attainable set Λ in criterion space and minimal curve AEC. Design data for the problem, given in kN and centimeters: F - 10 kN, ð - 10 kN/cm², Ā - 2 cm², L - 200 cm, $\underline{\sigma}$ - -10 kN/cm², $\underline{A}$ - 0.1 cm², E - 2 x 10⁴ kN/cm².

optimization must be replaced by a new one, especially adapted to multicriterion problem (1). Now that only a partial order exists in R^m it is natural to use the Pareto optimality concept in this context.

Definition 1   A vector $x^* \in \Omega$ is Pareto optimal for problem (1) if and only if there exists no $x \in \Omega$ such that $f_i(x) \leq f_i(x^*)$ for i - 1,2,...,m with $f_j(x) < f_j(x^*)$ for at least one j.

The basic concepts of the multicriterion optimization are illustrated in Fig. 1. where a bicriterion two-bar truss example has been considered. The material volume V and the vertical displacement Δ of the loaded node are the two criteria to be minimized subject to stress and member area constraints. Pareto optima locate both in the interior and on the boundary of the rectangular feasible set ABCD.

Verbally, definition 1 states that x^* is Pareto optimal if there exists no feasible vector x which would decrease some criterion without causing a simultaneous increase in at least one other criterion.

In the literature also some alternative expressions are used instead of the Pareto optimality. For example words like nondominated, non-inferior, efficient, functional efficient and EP-optimal solution have the same meaning. Here only the mathematical programming problem is shown but the corresponding control theory formulation can be found for example in refs. [3] and [4]. Vector $z^* = f(x^*)$, which represents the image of the Pareto optimum x^* in the criterion space, is called a minimal solution. In scalar optimization one optimal solution is usually characteristic of the problem, whereas there generally exists a set of Pareto optima as a solution to the multicriterion problem. Mathematically, problem (1) is regarded as solved immediately after the Pareto optimal set has been determined. In practical applications, however, it is necessary to order this set further because only one final solution is wanted by the designer. In this case some additional information is needed from the designer, who must take the decision-maker's role, to introduce preferences among Pareto optima.

3. MOTIVATION AND CLASSIFICATION

In the majority of the publications dealing with structural optimization the criterion, which should be minimized, has been the material volume or the structural weight. In some applications like air- and spacecraft structures it can be accepted as a simple and good measure for the economy of the design. In many civil and mechanical engineering applications, however, the manufacturing cost may be conflicting with the weight which does not reflect the total cost of the structure. Accordingly, from the purely economic viewpoint the weight represents only the material cost and additional criteria are needed.

When formulating the mathematical optimization problem the designer must consider which quantities are suitable for measuring the economy and the performance of the structure. Any quantity which has a tendency to improve or deteriorate is actually a criterion in nature. Those quantities which should only satisfy some imposed requirements are not criteria but they can be treated as constraints. In a strict sense, almost all design quantities have a criterion nature rather than a constraint nature because in the designer's mind they usually have better or worse values. As an example of a strict constraint the structural analysis equations can be mentioned. They represent equality constraints whereas norms, regulations, space limitations, available materials and manufacturing techniques impose inequality constraints. Thus the designer is usually faced with several conflicting and non-commensurable criteria in formulating the optimization problem. In order to reduce the computational and decision-making effort it is desirable to decrease the number of the criteria as far as possible by removing them into constraints. The choice of the constraint limits may be a difficult task in a practical design problem. These allowable values can be rather fuzzy even for such common quantities as displacements, natural frequencies and stresses. If the limit cannot be determined, it seems reasonable to treat the quantity in question as a criterion.

As a result, the multicriterion problem (1) is formulated and the best compromise solution should be found.

The previous motivation has been acknowledged in those about seventy publications which were available for the author in preparing this study. Next a classification of the methods used in these multicriterion applications is proposed. The lay-out of the multicriterion design process especially adapted for structural optimization, is shown in Fig. 2. This classification corresponds to the author's subjective view and other classifications may appear in the literature. The design process has been separated into three successive main phases which have further been decomposed into different methods. The first phase, which was discussed in the motivation section of this chapter, consists of the problem formulation. Here the criteria, the constraints and the design variables are chosen. After the original formulation it is sometimes useful to decrease the number of the criteria further by using some mathematical or physical arguments because the solution effort is proportional to the problem dimension m. The resulting problem should include at least two conflicting criteria.

The second phase consists of generating Pareto optimal solutions to the multicriterion problem. Here several methods are available and five different classes are shown in Fig. 2. Usually these methods include parameters which are varied in generating Pareto optima. Typically, each fixed parameter combination corresponds to one Pareto optimal solution. In the constraint methods one criterion is chosen as a scalar objective function and the others are removed into equality or inequality constraints which are parametrized. The distance methods (norm methods, l_p-methods, global criterion methods) are based on the minimization of the distance between the attainable set and some chosen reference point in the criterion space. The distance function

$$d_p = \left[\sum_{i=1}^{m} w_i \left(f_i(x) - \hat{z}_i \right)^p \right]^{1/p} \qquad (4)$$

is frequently used in structural optimization. Here $\hat{z}_i$ represent the chosen reference values and w_i as well as p are parameters. In most structural design applications $\hat{z}_i$ and p are fixed and w_i are the only parameters but also other possibilities exist [43]. The extreme case $p = \infty$ corresponds to the weighted minimax problem and the traditional linear weighting method is obtained by choosing $p = 1$ and $\hat{z}_i = 0$ for $i = 1, 2, \ldots, m$. In the latter case the scalar function

$$f(x) = \sum_{i=1}^{m} w_i \, f_i(x) \qquad (5)$$

is minimized in Ω by varying parameters w_i. Only in convex cases it is guaranteed that this method generates the whole Pareto optimal set. Also the direct application of the necessary conditions of the Pareto optimality is one possible strategy in generating Pareto optima. The last class in phase 2 consists of the utility function methods where some or all of the parameters are fixed in advance. Many different utility functions appear in the literature and usually only one Pareto

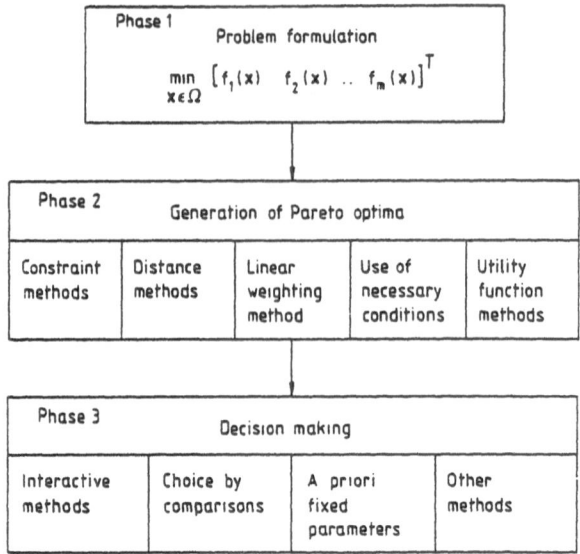

Figure 2. Classification of multicriterion structural design process. The best compromise solution is obtained as the result of this process.

optimum or some limited subset of the Pareto optima is obtained as a solution of the replacing problem. Often the utility function is given directly without formulating problem (1) first. All the four other classes offer more flexibility for the decision-maker than the utility function methods which may often be regarded as scalar optimization rather than multicriterion approaches.

Phase three represents the decision-making procedure where the best compromise solution or the preferred Pareto optimum is chosen. The most advanced methods here are different interactive approaches where the designer proceeds from each Pareto optimum to a better one until the satisfactory solution is achieved. These methods are especially flexible because the decision-maker can frequently participate in the design process. Another possibility is to compute a large collection of Pareto optima and to choose the best one by direct comparisons. Also here the designer's personal preferences, based on his experience and some other factors not included in the mathematical optimization problem (1), may be brought into the design process. The main drawback is the computation cost associated with the large number of Pareto optima which may be very high compared with the interactive approaches. The last two classes of the methods do not utilize the full capacity of the multicriterion formulation because some crucial parameters are fixed in advance. The third class, so-called a priori methods correspond to the utility function methods discussed in phase 2. If all the parameters are fixed i.e. all decisions have already been made, then a scalar optimization problem is obtained. Other methods, like the goal programming and the game theory approach for example, have been applied also in structural design. In this study they are only slightly discussed because their connection to the Pareto optimality concept is in many cases obscure.

4. APPLICATIONS IN STRUCTURAL OPTIMIZATION

In this chapter a brief review of almost seventy articles dealing with the multicriterion structural optimization is given. The purpose is to describe and evaluate different applications in this special field by using the classification shown in Fig. 2. Only articles written in English are included and the study has been restricted to those structural design applications where the Pareto optimality concept is utilized. A special attention has been devoted to the choice of the criteria, the methods for generating Pareto optima, the decision-making strategy and the presentation of the results. Also the problem dimension m is noticed but the number of the design variables and constraints as well as the numerical computation procedures are not discussed. First the few early articles published in the 1970s are considered and after that the summary of the production of each author is presented by using the alphabetic order.

The first papers dealing with multicriterion structural optimization were published in the late 1970s, most of them independently by a few authors far from each other. Stadler applied the control theory approach to different bicriterion problems with weight and stored energy as criteria and obtained analytical solutions for some structures, calling the results "natural structural shapes" [7], [9]. Also Leitmann used the control theory in solving some bicriterion problems in viscoelasticity [8]. Gerasimov and Repko used three criteria; weight, flexibility and easiness for construction, the applications consisting of a truss and a framed plate [10]. Koski formulated a vector optimization problem where the material volume and several chosen nodal displacements of a truss were chosen as the criteria [11]. In references [10] and [11] the mathematical programming approach was used for the computation of Pareto optima.

In the 1980s the number of publications increased considerably consisting now more than sixty papers which are available for the author of this study. The choice of the criteria varies from an article to another but usually physical quantities rather than economic are used. The most commonly used criterion is the weight or the material volume of a structure and the second in number is the flexibility which is usually measured by displacements or some energy expression. These two criteria are usually strongly conflicting and they offer an excellent possibility to study the multicriterion structural design problem with the lowest dimension m = 2. Among several different criteria the value of some natural frequency and the maximum stress are relatively popular. Most of the articles deal with a bicriterion problem which allows the graphic presentation of the minimal curve in the criterion space. In some articles the minimal surface has been depicted for a problem with three criteria (m = 3) but usually all the criteria have been combined into one scalar function if m > 2 and only one Pareto optimum has been computed. For certain specific problems the Pareto optimal set has been determined analytically but commonly some numerical optimization procedure has been applied. Interactive methods especially adapted to structural optimization are rare for the present but some general purpose approaches have been applied to structural design. Stochastic problems have been treated and plenty of industrial applications can be found. At least one comprehensive optimization package has been developed for the multicriterion design of large complex structural systems. Many different structures, such as beams, cables, trusses, frames, plates, shells and composites for example, under static and dynamic loading have been considered. In addition to the design problem the multicriterion approach has also been utilized in the plastic analysis of frames. Next a review of the contributions published in the 1980s and early 1990s is given by using a reference list where the main works of each author are included.

The weighting method has been applied to the computation of Pareto optima by Adali [12-14]. He has considered different problems including the shape optimization of a dynamically loaded beam and the design of an antisymmetrically laminated plate. The fundamental natural frequency, maximum stress, maximum displacement, buckling load and buckling temperature are used as criteria. Bicriterion problems have been solved and the results have been presented graphically in the criterion space. Baier [15,16] discusses about the optimal design of the spacecraft components and acknowledges the multicriterion nature of these problems. Several applications are treated by using such criteria as weight, thermal expansion, stiffness, natural frequency and the work done by the loads. Bicriterion examples have been considered and the weighting method as well as an interactive approach have been applied to solve the problems. He discussed also about the difficulties arising in the use of the weighting method in nonconvex laminate design problems. Bendsøe, Olhoff and Taylor [17,52] formulate some continuous and discrete parametrized minimax structural design problems and convert these into a min-form by introducing an additional parameter. From the resulting scalar problem called the variational formulation it is seen that this approach can be interpreted as one scalarization of a multicriterion problem. The bicriterion examples have been solved analytically by using the necessary conditions. Natural frequencies and buckling loads associated with different modes as well as the flexibility of a structure are used as the criteria. Bühlmeier [18] computes Pareto optima for laminated composites under time dependent loads and material behaviour by using some modifications of the weighting method. Several structural and thermal quantities appear as criteria in cases where m = 2 and 3. Carmichael [19,20] solves Pareto optima for a truss and frame example in bicriterion cases where weight and displacement are the criteria by using the constraint method. Diaz [21,22] concentrates on the numerical computation of the sensitivities of the criteria in the context of the distance methods. These figures are used in the interactive solution of some stuctural problems including such criteria as natural frequency, spring length, weight, relative displacement, safety factor against yielding and against fatigue.

Eschenauer and his co-workers present a large collection of practical applications in multicriterion structural optimization [23-31]. Plenty of different structures or components have been considered and generally the finite element method is used in the structural analysis. Many of these real life applications represent large scale optimization problems and several multicriterion methods as well as numerical procedures have been used. A comprehensive optimization program package SAPOP has been developed for solving industrial applications [30]. This system is multicriterion oriented allowing the optimization of many different criteria simultaneously and possessing a considerable flexibility in the choice of constraints and design variables. Both the computation of Pareto optima and some interactive approaches are included in the programs. Many criteria appear in the applications, among them surface accuracy, weight, displacement and stress. The number of the criteria varies from two to eight and in the bicriterion problems the minimal curve has been depicted in most cases. Such applications as the parabolic antenna of the 30 m radio telescope, novel solar energy collector and satellite tanks could be especially mentioned. Fu and Frangopol [32] consider a reliability based plastic design problems for a truss and a frame. Three and four criteria problems have been solved by the constraint method and as a nice result the minimal surfaces have been shown in the criterion spaces. The material volume and different probabilities of failure are chosen as criteria. Hajela and Shih [33] formulate a bicriterion problem for ordinary and composite beams by using the weight and one displacement as the criteria and they solve the problem by a distance method. Henrich and Schiffner [34] use the residual deformations after roller levelling as criteria and apply the d_2-distance method. Also Jendo [35,36] uses the same method for computing Pareto optima for certain civil engineering structures like trusses, beams and cables. The weight, strain energy and displacement are the criteria here. Kneppe [37] applies the constraint method to bicriterion air- and spacecraft problems by using weight and aeroelastic efficiency as well as heat flux and the fundamental frequency as criteria.

Koski [38-46] formulated a multicriterion problem where the material volume and several chosen nodal displacements are chosen as criteria. Most of the applications are trusses but also frame and plate are included, the latter one with the lowest natural frequency as one criterion. Usually the Pareto optimal set has been determined by using the linear weighting, the distance or the constraint method and the graphical representation of the results has been favoured. The isostatic truss problem where weight and one displacement are minimized subject to stress and Euler buckling constraints has been solved analytically in general terms (Koski and Silvennoinen (1982)) by using the necessary and sufficient conditions of Pareto optimality and a semi-analytic computation scheme has been given for the m-criterion case. An interactive design method for trusses has been proposed and the reduction of the dimension m of the multicriterion problem has been discussed. The failure of the weighting method in some nonconvex problems has been demonstrated by extremely simple truss examples. The design of a ceramic piston crown by using the material volume and the probability of failure as the criteria represents an industrial application. Müller-Slany [47] optimized the reflector by choosing different dynamic quantities as criteria.

Nafday, Corotis and Cohon [48,49] consider a limit analysis problem of plastic frames by generalizing the traditional static approach linear programming problem into a multicriterion form. The load factors are the criteria and the weakly minimal surface represents the limit state. Osyczka [50,51] mainly concentrates on mechanical design applications but he also formulated an early beam problem with weight and displacement as criteria. Pfeiffer [53] studies a dynamic mechanical design problem where the kinetic energy and two displacements are the criteria. Pareto optima are solved by the d_2-method and the minimal surface has been shown. Post [54] minimizes the weight and the maximum stress in time dependent composite problems. Rao and his co-workers [55-60] treat several different problems mainly applying the methods where the parameters are a priori fixed or using the goal programming and the game theory approach. Rozvany [61] combines the multicriterion approach with the optimality criteria method developed by Prager and himself.

Stadler [62-64] continues his studies on the natural structural shapes and gives a different application in a die design problem by solving some bicriterion cases analytically. Such new criteria as the total contraction and the total work of the plastic deformation are introduced here. Stenvers [65] chooses eleven criteria, mainly weights of components, temperatures and displacements, and combines them into one applying the d_2-method. Tseng and Lu [66] apply the trade-off, goal and compromise programming methods to four criteria truss problems by choosing weight, maximum stress, displacement and the fundamental natural frequency as criteria. Yoshimura and others [67,68] formulate certain machine tool design problems by choosing the weight and the flexibility as the main criteria and solve Pareto optima numerically by the linear weighting method and the necessary conditions. Also the manufacturing cost and the frequency response are considered as criteria. The minimal curves have been shown for the bicriterion examples.

5. CONCLUSIONS

The age of the multicriterion structural optimization is only about fifteen years but already now it has attracted the attention of many researchers. These early applications and results are encouraging because they clearly show that conflicting design objectives naturally exist also in structural design. The Pareto optimality concept has been widely accepted as a sound basis for the treatment of these problems. The applications cover a variety of structures from trusses and beams to composite shells and different branches of structural engineering are represented. Usually the articles concentrate on the analytical or numerical solution of some Pareto optimal structures and the computation of the relevant trade-off information. The number of industrial applications is increasing and computer programs which include the multicriterion module are emerging. Actually, the integration of any optimization procedure to a CAD work station could be done directly from the multicriterion viewpoint. In large scale problems, where both the finite element and sensitivity analysis are applied several times in calculating just one Pareto optimum, the computational cost may become high in generating the Pareto optimal set.

Multicriterion optimization may be viewed as a systematic sensitivity analysis of all those value judgements which are especially important for the designer. Consequently, it covers the problem formulation, the computation of Pareto optima and trade-offs, representation of the results and especially the solution strategy (interactive method, comparison, scalarization, etc.) for making good decisions. In structural optimization some characteristic features like accurate analysis equations may guide the research to a certain direction. For example the numerical computation of Pareto optima becomes more important in structural design than in some economic and social decision problems where the mathematical models are not so reliable. In the near future, however, also the multicriterion structural optimization may be expected to receive such dimensions which are only springing up in the present applications.

REFERENCES

1. Pareto, V. (1896-97) Cours D'Economie Politique, Volumes 1 and 2, F. Rouge, Lausanne.

2. Stadler, W. (1987) 'Initiators of multicriteria optimization - Recent advances and historical development of vector optimization', in J. Jahn and W. Krabs (eds.), Proceedings of an International Conference on Vector Optimization, Darmstadt, August 1986, Lecture Notes in Economics and Mathematical Systems 294, Springer-Verlag, Berlin.

3. Stadler, W. (ed.) (1988) Multicriteria Optimization in Engineering and in the Sciences, Mathematical Concepts and Methods in Science and Engineering 37, Plenum Press, New York.

4. Eschenauer, H., Koski, J. and Osyczka, A. (eds.), (1990), Multicriteria Design Optimization - Procedures and Applications, Springer-Verlag, Berlin.

5. Stadler, W. (1984) 'Multicriteria optimization in mechanics (A Survey)', Applied Mechanics Reviews 37, 277-286.

6. Stadler, W. (1986) 'Update to multicriteria optimization in mechanics', Applied Mechanics update, ASME, 417-420.

7. Stadler, W. (1977) 'Natural structural shapes of shallow arches', Journal of Applied Mechanics, Trans. ASME 44, 291-298.

8. Leitmann, G. (1977) 'Some problems of scalar and vector-valued optimization in linear viscoelasticity' Journal of Optimization Theory and Applications 23, 93-99.

9. Stadler, W. (1978) 'Natural structural shapes (the static case)' Quarterly Journal of Mechanics and Applied Mathematics 31, 169-217.

10. Gerasimov, E.N. and Repko, V.N.(1978) 'Multicriterial optimization' Soviet Applied Mechanics 14, 1179-1184.

11. Koski, J. (1979) Truss Optimization with Vector Criterion, Tampere University of Technology, Publications No. 6.

12. Adali, S. (1983) 'Pareto optimal design of beams subjected to support motions', Computers & Structures 16, 297-303.

13. Adali, S. (1983) 'Multiobjective design of an antisymmetric angle-ply laminate by nonlinear programming', Journal of Mechanisms, Transmissions, and Automation in Design, Trans. ASME 105, 214-219.

14. Adali, S. and Duffy, K.J. (1990) 'Multicriteria thermoelastic design of antisymmetric angle-ply laminates', in H. Eschenauer, J. Koski and A. Osyzka (eds.), Multicriteria Design Optimization - Procedures and Applications, Springer-Verlag, Berlin, pp. 417-428.

15. Baier, H. (1983) 'Structural optimization in industrial environ-ment', in H. Eschenauer and N. Olhoff, Optimization Methods in Structural Design, Proceedings of the Euromech-Colloquium 164, University of Siegen, FR Germany, Bibliographisches Institut A6, Zürich, pp. 140-145.

16. Baier, H. (1990) 'Multicriteria design of spacecraft structures with special emphasis on mass and stiffness', in H. Eschenauer, J. Koski and A. Osyczka (eds.), Multicriteria Design Optimization - Procedures and Applications, Springer-Verlag, Berlin, pp. 244-259.

17. Bendsøe, M.P., Olhoff, N. and Taylor, J.E. (1983-84) 'A variational formulation for multicriteria structural optimization', Journal of Structural Mechanics 11, 523-544.

18. Bühlmeier, J. (1990) 'Design of laminated composites under time dependent loads and material behaviour', in H.A. Eschenauer, C. Mattheck and N. Olhoff (eds.), Engineering Optimization in Design Processes, Proceedings of the International Conference, Karlsruhe Nuclear Research Center, Germany, Lecture Notes in Engineering 63, Springer-Verlag, Berlin, pp. 107-125.

19. Carmichael, D.G. (1980) 'Computation of Pareto optima in structural design (short communication) 'International Journal for Numerical Methods in Engineering 15, pp. 925-929.

20. Carmichael, D.G. (1981) Structural Modelling and Optimization, Ellis Horwood Limited, Chichester.

21. Diaz, A. (1987)'Interactive solution to multiobjective optimization problems', International Journal for Numerical Methods in Engineering 24, 1865-1877.

22. Diaz, A. (1987) 'Sensitivity information in multiobjective optimization', Engineering Optimization 12, 281-297.

23. Eschenauer, H. (1983) 'Vector optimization in structural design and its application on antenna structures', in H. Eschenauer and N. Olhoff (eds.), Optimization Methods in Structural Design, Proceedings of the Euromech-Colloquium 164, University of Siegen, FR Germany, Bibliogaphisches Institut A6, Zürich, pp. 146-155.

24. Eschenauer, H., Kneppe, G. and Stenvers, K.H. (1986) 'Deterministic and stochastic multiobjective optimization of beam and shell structures', Journal of Mechanisms, Transmissions, and Automation in Design, Trans. ASME 108, 31-37.

25. Eschenauer, H.A. (1988) 'Multicriteria optimization techniques for highly accurate focusing systems' in W. Stadler (ed.), Multicriteria Optimization in Engineering and in the Sciences, Plenum Press, New York, pp. 309-354.

26. Eschenauer, H.A., Schäfer, E. and Bernau, H. (1988) 'Application of interactive vector optimization methods with regard to problems in structural mechanics' in H.A. Eschenauer and G. Thierauf (eds.), Discretization Methods and Structural Optimization - Procedures and Applications, Proceedings of a GAMM-Seminar, Siegen, FRG, Lecture Notes in Engineering 42, Springer-Verlag, Berlin, pp. 110-117.

27. Bremicker, M., Eschenauer, H. and Wodtke, H.-W. (1990) 'Optimal layouts of chilled cast-iron high quality rollers', in H. Eschenauer, J. Koski and A. Osyczka (eds.), Multicriteria Design Optimization - Procedures and Applications, Springer-Verlag, Berlin, pp. 319-338.

28. Eschenauer, H., Fuchs, W. and Post, P.U. (1990) 'Fibre-reinforced plate and shell structures under various loads - Proposals for their optimization', in H. Eschenauer, J. Koski and A. Osyczka (eds.), Multicriteria Design Optimization - Procedures and Applications, Springer-Verlag, Berlin, pp. 397-416.

29. Eschenauer, H.A. and Vietor, T. (1990) 'Some aspects on structural optimization of ceramic structures', in H.A. Eschenauer, C. Mattheck, and N. Olhoff (eds.), Engineering Optimization in Design Processes, Proceedings of the International Conference, Karlsruhe Nuclear Research Center, Germany, Lecture Notes in Engineering 63, Springer-Verlag, Berlin, pp. 145-154.

30. Bremicker, M., Eschenauer, H.A. and Post, P.U. (1990) 'Optimization procedure SAPOP - a general tool for multicriteria structural designs', in H. Eschenauer, J. Koski and A. Osyczka (eds.), Multicriteria Design Optimization - Procedures and Applications, Springer-Verlag, Berlin, pp. 35-69.

31. Fuchs, W.J., Karandikar, H.M., Mistree, F. and Eschenauer, H.A. (1988) 'Compromise: an effective approach for designing composite conical shell structures', in S.S. Rao (ed.), Advances in Design Automation, ASME Des. tech. conf. 14, pp. 279-286.

32. Fu, G. and Frangopol, D.M. (1990) 'Reliability-based vector optimization of structural systems', ASCE Journal of Structural Engineering 116, 2143-2161.

33. Hajela, P. and Shih, C.J. (1990) 'Multiobjective optimum design in mixed integer and discrete design variable problems', AIAA Journal 28, 670-675.

34. Henrich, L. and Schiffner, K. (1990) 'A mechanical model for the optimization and simulation of the metal forming process in roller levelling of sheets', in H. Eschenauer, J. Koski and A. Osyczka (eds.), Multicriteria Design Optimization - Procedures and Applications, Springer-Verlag, Berlin, pp. 339-353.

35. Jendo, S., Marks, W. and Thierauf, G. (1985) 'Multicriteria optimization in optimum structural design', Large Scale Systems 9, 141-150.

36. Jendo, S. (1990) 'Multicriteria optimization of concrete beams, trusses and cable structures', in H. Eschenauer, J. Koski and A. Osyczka (eds.), Multicriteria Design Optimization - Procedures and Applications, Springer-Verlag, Berlin, pp. 356-375.

37. Kneppe, G. (1990) 'Multicriteria optimal layouts of aircraft and spacecraft structures', in H. Eschenauer, J. Koski and A. Osyczka (eds.), Multicriteria Design Optimization - Procedures and Applications, Springer-Verlag, Berlin, pp. 229-243.

38. Koski, J. (1981) 'Multicriterion optimization in structural design', Proceedings of the 11th ONR Naval Structural Mechanics Symposium on Optimum Structural Design, University of Arizona, Tucson, Arizona.

39. Koski, J. and Silvennoinen, R. (1982) 'Pareto optima of isostatic trusses', Computer Methods in Applied Mechanics and Engineering 31, 265-279.

40. Koski, J. (1984) 'Multicriterion optimization in structural design', in E. Atrek, R.H. Gallagher, K.M. Ragsdell and O.C. Zienkiewicz (eds.), New Directions in Optimum Structural Design, John Wiley & Sons, Ltd., New York, pp. 483-503.

41. Koski, J. (1984) Bicriterion optimum design method for elastic trusses, Acta Polytechnica Scandinavica, Mechanical Engineering Series No. 86, Dissertation, Helsinki.

42. Koski, J. (1985) 'Defectiveness of weighting method in multi-criterion optimization of structures', Communications in Applied Numerical Methods 1, 333-336.

43. Koski, J. and Silvennoinen R. (1987) 'Norm methods and partial weighting in multicriterion optimization of structures', International Journal for Numerical Methods in Engineering 24, 1101-1121.

44. Koski, J. (1988) 'Multicriteria truss optimization', in W. Stadler (ed.), Applications of Multicriteria Optimization in Engineering and in the Sciences, Mathematical Concepts and Methods in Science and Engineering 37, Plenum Press, New York, pp. 263-307.

45. Koski, J., Silvennoinen, R. and Lawo, M. (1988) 'Multicriterion plate optimization' in G.I.N. Rozvany and B.L. Karihaloo (eds.), Structural optimization, Proceedings of the IUTAM Symposium on Structural optimization, Melbourne, Australia, Kluwer Academic Publishers, Dordrecht.

46. Koski, J. and Silvennoinen R. (1990) 'Multicriteria design of ceramic piston crown', Engineering Costs and Production Economics 20, 175-189.

47. Müller-Slany, H.H. (1990) 'Modelling of multibody systems by means of optimization procedures', in H. Eschenauer, A. Osyczka and J. Koski (eds.), Multicriteria Design Optimization - Procedures and Applications, Springer-Verlag, Berlin, pp. 193-204.

48. Nafday, A.M., Corotis, R.B. and Cohon, J.L. (1988) 'Multiparametric limit analysis of frames: Part I - Model', ASCE Journal of Engineering Mechanics 114, 377-386.

49. Nafday, A.M., Corotis, R.B. and Cohon, J.L. (1988) 'Multiparametric limit analysis of frames: Part II - Computations', ASCE Journal of Engineering Mechanics 114, 387-403.

50. Osyczka, A. (1981) 'An approach to multi-criterion optimization for structural design' Proceedings of the 11th ONR Naval Structural Mechanics Symposium on Optimum Structural Design, Tucson, Arizona.

51. Osyczka, A. (1984) Multicriterion Optimization in Engineering, Ellis Horwood Limited, Chichester.

52. Olhoff, N. (1989) 'Multicriterion structural optimization via bound formulation and mathematical programming' Structural Optimization 1, 11-17.

53. Pfeiffer, F.(1990) 'On the optimal synthesis of a automotive drive train', in H. Eschenauer, J. Koski and A. Osyczka (eds.), Multi-criteria Design Optimization - Procedures and Applications, Springer-Verlag, Berlin, pp. 184-192.

54. Post, P.U. (1990) 'Optimization of the long-term behaviour of composite structures under hygrothermal loads', in H.A. Eschenauer, C. Mattheck and N. Olhoff (eds.), Engineering Optimization in Design Processes, Proceedings of the International Conference, Karlsruhe Nuclear Research Center, Germany, Lecture Notes in Engineering 63, Springer-Verlag, Berlin, pp. 99-106.

55. Rao, S.S. (1984) 'Multiobjective optimization in structural design with uncertain parameters and stochastic processes', AIAA Journal 22, 1670-1678.

56. Rao, S.S. and Hati, S.K. (1986) 'Pareto optimal solutions in two criteria beam design problems', Engineering Optimization 10, 41-50.

57. Rao, S.S. (1987) 'Game theory approach for multiobjective structural optimization', Computers & Structures 25, 119-127.

58. Rao, S.S., Venkayya, V.B. and Khot, N.S. (1988) 'Optimization of actively controlled structures using goal programming techniques', International Journal for Numerical Methods in Engineering 26, 183- 197.

59. Rao, S.S., Venkayya, V.B. and Khot, N.S. (1988) 'Game theory approach for the integrated design of structures and controls', AIAA Journal 26, 463-469.

60. Dhingra, A.K., Rao, S.S. and Miura, H. (1990) 'Multiobjective decision making in a fuzzy environment with applications to helicopter design', AIAA Journal 28, 703-710.

61. Rozvany, G.I.N. (1989), Structural Design via Optimality Criteria, Kluwer Academic Publishers, Dordrecht.

62. Stadler, W. (1988) 'Natural structural shapes (A unified optimal design philosophy)' in W. Stadler (ed.), Multicriteria Optimization in Engineering and in the Sciences, Plenum Press, New York, pp. 355-390.

63. Stadler, W. and Krishnan, V. (1989) 'Natural structural shapes for shells of revolution in the membrane theory of shells' Structural Optimization 1, 19-27.

64. Stadler, W. (1990) 'Optimal die design for symmetric strip drawing', in H. Eschenauer, J. Koski and A. Osyczka (eds.), Multicriteria Design Optimization - Procedures and Applications, Springer-Verlag, Berlin, pp. 303-318.

65. Stenvers, K.-H. (1990) 'Multicriteria optimization and advanced materials in telescope design', in H. Eschenauer, J. Koski and A. Osyczka (eds.), Multicriteria Design Optimization - Procedures and Applications, Springer-Verlag, Berlin, pp. 429-446.

66. Tseng, C.H. and Lu, T.W. (1990) 'Minimax multiobjective optimization in structural design', International Journal for Numerical Methods in Engineering 30, 1213-1228.

67. Yoshimura, M., Hamada, T., Yura, K. and Hitomi, K. (1984) 'Multi-objective design optimization of machine-tool spindles' Journal of Mechanisms, Transmissions, and Automation in Design 106, 46-53.

68. Yoshimura, M. (1990) 'Application of multicriteria optimization methods to machine tool structural design', in H. Eschenauer, J. Koski and A. Osyczka (eds.), Multicriteria Design Optimization - Procedures and Applications, Springer-Verlag, Berlin, pp. 261-281.

CONTROL AUGMENTED STRUCTURAL SYNTHESIS

H.L. THOMAS and G.N. VANDERPLAATS
VMA Engineering
5960 Mandarin Ave., Suite F
Goleta, CA 93117

ABSTRACT. Methods for the synthesis of structures augmented with noncollocated direct output feedback control systems are presented. Noncollocated control systems generate nonsymmetric system stiffness and viscous damping matrices and require that the real part of the system complex eigenvalues be constrained to be negative. In this work the approximation concepts approach to system synthesis is used. High quality approximations for system complex eigenvalues and steady state harmonic displacements and control forces are presented. Examples of structures composed of an assemblage of frame members, concentrated masses, and noncollocated control elements are used to demonstrate the method and the quality of the approximations.

1. Introduction

The traditional approach to control augmented structural system design is to first design the structure and then design a control system for the structure. This approach can lead to a structure that is too flexible, requiring large control forces in order to keep responses at an acceptable level. In this case both the structure and its control system must be redesigned. Another drawback to this iterative approach is that it does not account for the interaction of structural and control system dynamics. This is especially likely to occur in large space structures where the structural frequencies can be quite low.

This work presents methods for the simultaneous synthesis of both the structure and its control system. In this manner the synergistic effect between the two can be used to enhance the design performance, rather than degrade it. A noncollocated direct output position and velocity feedback type of control system is used. The structural member cross sectional dimensions and properties and the control system position and velocity gains are treated simultaneously as independent design variables. The system mass or total control effort is minimized while constraints are place on both the real and imaginary parts of the system complex eigenvalues and on the steady state harmonic displacements and control system forces.

The approximation concepts approach to system synthesis is used to efficiently improve the system. The concepts of intermediate design variables and intermediate response quantities are used to generate high quality approximations. These approximations allow for the use of larger move limits from design cycle to design cycle and lead to faster design convergence. Another benefit derived from using these new approximation concepts is that an approximation can be developed that captures the effect of resonance as system frequencies approach loading frequencies.

811

G. I. N. Rozvany (ed.), Optimization of Large Structural Systems, Vol. II, 811–828.
© *1993 Kluwer Academic Publishers.*

2. Background

There have been various works published on simultaneous structure-control system synthesis (see [1-12]). In [1-7] the control gains are treated as dependent design variables and are determined by use of the Riccati Equation. In [8-12] the control system gains are treated as independent design variables. The approximation concepts approach was first applied to structure-control system synthesis problems in [12] and [13].

In the approximation concepts approach an explicit approximate optimization problem is formulated and solved during each design cycle. In the mid 1970's approximate representations for constraints and objective functions were generated using first order Taylor series expansions in terms of direct or reciprocal sizing design variables (see [14-15]). More accurate approximations can be constructed using approximations of intermediate response quantities, which were introduced in [15]. In this approach simple approximations (e.g. linear, reciprocal, or hybrid (see [16])) of intermediate response quantities (such as forces in the case of stress constraints and modal energies in the case of frequency constraints) in terms of the design variables can be used while retaining the explicit nonlinear dependence of the constraints on the intermediate response quantities. The intermediate response quantity idea was applied to stress constraints in [17] and frequency constraints in [18]. The idea was also applied to steady state dynamic displacements calculated by the modal solution method in [19] and steady state dynamic displacements and control forces calculated by the direct method in [20] and [21]. The use of intermediate response quantities in approximations for complex eigenvalues was also presented in [21]. The intermediate response quantity concept can be stated mathematically as:

$$\bar{R}_I = f(X) \tag{1}$$

where the approximate intermediate responses ($\bar{R}_I$) are a simple function of the design variables (X). The approximate value of the constrained response ($\bar{R}_C$) is then calculated as:

$$\bar{R}_C = g(\bar{R}_I, X) \tag{2}$$

In other words, given the design vector X, the approximate values of the intermediate responses ($\bar{R}_I$) are calculated first using Eq. 1. Then the constrained response $\bar{R}_C$ corresponding to the $\bar{R}_I$ and X is evaluated using Eq. 2.

Approximations made with respect intermediate design variables (as introduced in [22] for static displacement constraints and [23] for natural frequency constraints) can also be used to improve the quality of approximations in control augmented structures as shown in [21].

This work is a summary of [19] and [21] in which the approximation concepts approach is used to solve the control augmented structural synthesis problem.

3. Problem Statement

Two practical approaches to control augmented structural synthesis are discussed here. The first is to minimize the structural mass with constraints on the total control force (sum of the control force amplitudes) and on the dynamic displacements of nodes at critical locations. The problem is then stated as:

Minimize
$$M(Y) \tag{3}$$

subject to $g_q(Y) \leq 0$; $q \in Q$

$$\sum_j F_j(Y) \leq F_T^U \quad \text{with bounds} \quad Y_n^L \leq Y_n \leq Y_n^U \quad ; \quad n \in N \tag{4}$$

where F_T^U is the upper bound on the total control force and separate constraints on the dynamic displacements at critical locations are included in the set Q.

The other practical approach is to minimize the total control force with constraints on the structural mass and dynamic displacements. This problem is stated as:

Minimize
$$\sum_j F_j(Y) \tag{5}$$

subject to $g_q(Y) \leq 0$; $q \in Q$

$$M(Y) \leq M^U \quad \text{with bounds} \quad Y_n^L \leq Y_n \leq Y_n^U \quad ; \quad n \in N \tag{6}$$

where M^U is the upper bound on the structural mass and separate constraints on the dynamic displacement at critical locations are included in Q.

The problem statement defined by Eqs. 3 and 4 leans towards the structural designers point of view but still takes into account constraints on the control system. The problem statement defined by Eqs. 5 and 6 is more like the approach taken by the control system designer, but it also includes constraints on the structural mass and the structural response.

An efficient approach to design synthesis is to generate and then solve a sequence of approximate optimization problems. These problems are nonlinear but explicit and therefore they are inexpensive to solve compared to the solution of the original problem. A description of the approximate problem is presented in the next section. Each approximate optimization problem is solved using CONMIN (see [24]) to generate a new design point, which is then analyzed at the beginning of the next cycle. This process is continued until the objective function changes less than some prescribed value for two consecutive cycles. The entire synthesis process is summarized in Fig. 1.

4. Approximate Problem Formulation

It is often the case that the behavior constraints g_q and the objective function used to describe a design optimization problem can be written as explicit nonlinear functions of intermediate response quantities (R), intermediate design variables (X), and actual design variables (Y). In general the foregoing statement can be expressed as follows:

$$g_q(\mathbf{R}, \mathbf{X}, \mathbf{Y}) \le 0 \tag{7}$$

In the dynamic response analysis the sine and cosine components of the actuator forces (F_s and F_c) as well as the real and imaginary components of the dynamic displacements (c_R and c_I) are taken as intermediate response quantities in (R). The explicit approximations of these intermediate response quantities (F_s, F_c, c_R, c_I) in terms of intermediate design variables X, can be linear, reciprocal, or hybrid at the users option. The amplitude of the actuator force is approximated by

$$F \approx \bar{F} = \sqrt{\bar{F}_s^2 + \bar{F}_c^2} \tag{8}$$

and the amplitude of the dynamic displacement can be approximated by

$$u_D \approx \bar{u}_D = \sqrt{\bar{c}_R^2 + \bar{c}_I^2} \tag{9}$$

It should be emphasized that the use of appropriate intermediate response quantities captures the explicit nonlinearity inherent to the definition of force and displacement amplitude.

In the case of natural frequencies the modal strain energy (U_n) and modal kinetic energy (T_n) are taken as the intermediate response quantities in (R). The square of the natural frequency for mode n is approximated as

$$\omega^2 \approx \bar{\omega}_n^2 = \frac{\bar{U}_n}{\bar{T}_n} = \frac{\{\phi_n\}^T [\bar{K}] \{\phi_n\}}{\{\phi_n\}^T [\bar{M}] \{\phi_n\}} \tag{10}$$

The equations of motion governing the dynamic behavior of the control augmented structures considered here can be written as follows:

$$[M]\{\ddot{u}\} + [C_A]\{\dot{u}\} + [K_A + i\gamma K]\{u\} = \{P\} \tag{11}$$

where $[C_A] = [C] + [H_v]$ denotes the control augmented damping matrix, $[K_A] = [K] + [H_p]$ denotes the control augmented stiffness matrix and $[i\gamma K]$ represents preassigned structural damping. Since noncollocated actuator-sensor pairs are considered the matrices $[C_A]$ and $[K_A]$ are not symmetric.

An approximation of complex eigenvalues is generated by looking at the second order eigenproblem

$$\lambda^2[M]\{\phi_p\}_R + \lambda[C_A]\{\phi_p\}_R + ([K_A] + i\gamma[K])\{\phi_p\}_R = 0$$

(12)

Premultiplying by $\{\phi_p\}_L^T$ and solving for λ gives

$$\lambda = \sigma + i\omega_d = \frac{-S^\wedge \pm [(S^\wedge)^2 - 4U'T']^{1/2}}{2T^\wedge}$$

(13)

where

$$U^\wedge = \{\phi_p\}_L^T ([K_A] + i\gamma[K]) \{\phi_p\}_R = U_r^\wedge + i U_i^\wedge$$

$$T^\wedge = \{\phi_p\}_L^T [M] \{\phi_p\}_R = T_r^\wedge + i T_i^\wedge$$

(14)

$$S^\wedge = \{\phi_p\}_L^T [C_A] \{\phi_p\}_R = S_r^\wedge + i S_i^\wedge$$

Selecting the + sign option in Eq. 13, so that $\omega_d \to \omega > 0$ in the undamped case, substituting Eqs. 14 into Eq. 13, and multiplying the numerator and denominator by $(T_r^\wedge - i T_i^\wedge)$ leads to

$$\sigma = \frac{T_r^\wedge(-S_r^\wedge + \alpha) + T_i^\wedge(-S_i^\wedge + \beta)}{2[(T_r^\wedge)^2 + (T_i^\wedge)^2]} \quad \text{and} \quad \omega_d = \frac{T_r^\wedge(-S_i^\wedge + \beta) - T_i^\wedge(-S_r^\wedge + \alpha)}{2[(T_r^\wedge)^2 + (T_i^\wedge)^2]} \tag{15a, 15b}$$

where α and β are real and can be expressed explicitly in terms of $U_r^\wedge$, $U_i^\wedge$, $T_r^\wedge$, $T_i^\wedge$, $S_r^\wedge$ and $S_i^\wedge$. High quality approximations for $\bar{\sigma}$ and $\bar{\omega}_d$ are now formed by constructing explicit approximations for the intermediate response quantities $(U_r^\wedge, U_i^\wedge, S_r^\wedge, S_i^\wedge, T_r^\wedge, T_i^\wedge)$ and substituting them into Eqs. 15a and 15b. Furthermore, the damping ratio approximation can be formed using $\bar{\sigma}$ and $\bar{\omega}_d$, as

$$\xi = \frac{-\bar{\sigma}}{\sqrt{\bar{\sigma}^2 + \bar{\omega}_d^2}}$$

(16)

Finally, it is noted that in the limiting case where all damping vanishes, Eqs. 15a and 15b reduce to $\sigma = 0$ and $\omega_d = \sqrt{U_r/T_r}$.

5. Approximations for Dynamic Displacements (Modal Solution Method)

The matrix equation of motion for an undamped structure is:

$$[M]\{\ddot{u}\} + [K]\{u\} = \{P\}$$

(17)

Assuming a sinusoidal loading and response at frequency Ω:

$$\{P\} = \{p\} \sin\Omega t \tag{18}$$

$$\{u\} = \{a\} \sin\Omega t \tag{19}$$

Equation 17 can be transformed into the frequency domain:

$$(-\Omega^2 [M] + [K]) \{a\} = \{p\} \tag{20}$$

In the usual approach to approximating the dynamic displacements Eq. 20 is solved directly for $\{a\}$ and the derivatives of the a_i are found by implicitly differentiating Eq. 20:

$$\frac{\partial}{\partial x_j} \{a\} = [-\Omega^2 M + K]^{-1} \left(\frac{\partial}{\partial x_i} \{p\} - \left[-\Omega^2 \frac{\partial M}{\partial x_i} + \frac{\partial K}{\partial x_i} \right] \{a\} \right) \tag{21}$$

The dynamic displacements are then approximated as direct, reciprocal, or hybrid functions of the design variables so that

$$\bar{u}_i = \bar{a}_i \sin\Omega t = f(X) \sin\Omega t \tag{22}$$

The hybrid approximation uses either a direct or reciprocal expansion in each of the design variables, selecting term by term the alternative that is most conservative (see [16]). It is commonly used for static and dynamic displacement constraints and it will be employed here to construct the full order solution for comparison purposes.

The approximation shown in Eq. 22 has two drawbacks. The first is that it is based on the direct solution of Eqs. 20 and 21 which is very expensive for large problems. The second drawback is that it is a poor approximation for the response whenever the loading frequency (Ω) is near a natural frequency of the structure, because of the strong nonlinear effects of resonance.

In order to determine the dynamic response of large structures, modal analysis is often used to reduce the order of Eq. 20. In modal analysis the response of the structure is approximated as a linear sum of an orthogonal set of basis vector $\{\phi_n\}$, that is

$$\{a\} = \sum_{n=1}^{N} \{\phi_n\} z_n = [\Phi] \{z\} \tag{23}$$

where the z_n are called the modal participation coefficients. The first N natural vibration modes of the structure (eigenvectors) are usually chosen as the basis vectors $\{\phi_n\}$ and that approach is used in this work. Substituting Eq. 23 into Eq. 17 and pre-multiplying by $[\Phi]^T$ gives

$$[\Phi]^T (-\Omega^2[M] + [K]) [\Phi] \{z\} = \{f\} \quad \text{where} \quad \{f\} = [\Phi]^T \{p\} \tag{24}$$

Using the modal potential and kinetic energies defined in Eq. 10 makes it possible to rewrite Eq. 24 in the following form

$$(-\Omega^2 [T] + [U]) \{z\} = \{f\} \tag{25}$$

Note that the order of Eq. 25 is N which is much smaller than the order of the full system. The value of N is chosen to be the number of modes needed to accurately represent the structural response.

When the basis vectors $\{\phi_n\}$ are the natural mode shapes of the structure, they are orthogonal to both $[K]$ and $[M]$ so that $[U]$ and $[T]$ are diagonal matrices. Equation 25 is then decoupled and the individual z_n are calculated using the following expression

$$z_n = \frac{f_n}{U_n - \Omega^2 T_n} \tag{26}$$

The approximation introduced in this work is constructed as follows. Evaluate the derivatives of U_n and T_n with respect to the intermediate design variables (X) assuming the eigenvectors are invariant, so that:

$$\frac{\partial U_n}{\partial x_j} = \{\phi_n\}^T \frac{\partial [K]}{\partial x_j} \{\phi_n\} \quad \text{and} \quad \frac{\partial T_n}{\partial x_j} = \{\phi_n\}^T \frac{\partial [M]}{\partial x_j} \{\phi_n\} \tag{27}$$

Approximate U_n and T_n as linear functions of the intermediate design variables:

$$\tilde{U}_n = U_{n_0} + \sum_{j=1}^{NDVI} \frac{\partial U_n}{\partial x_j} \left(x_j - x_{j_0}\right) \quad \text{and} \quad \tilde{T}_n = T_{n_0} + \sum_{j=1}^{NDVI} \frac{\partial T_n}{\partial x_j} \left(x_j - x_{j_0}\right) \tag{28}$$

where NDVI is the number of intermediate design variables. Calculate the approximate modal participation coefficients as

$$\tilde{z}_n = \frac{f_n}{\tilde{U}_n - \Omega^2 \tilde{T}_n} \tag{29}$$

and finally the approximate amplitudes of displacement as

$$\{\tilde{a}\} = [\Phi] \{\tilde{z}\} \tag{30}$$

The error associated with this approximation is small and comes from two assumptions. The first is that the displacements can be represented by a truncated set of modes (N). This is the error associated with the analysis and it can be controlled by choosing a satisfactory value for N. The other assumption is that the mode shapes are invariant with changes in the design variables. This error can be controlled by putting move limits on the design variables at each design stage. In the example section of this paper it is shown that 60% move limits are not unreasonable.

The reason for the accuracy of the approximation near resonance is now examined. If the numerator and denominator of the right hand side of Eq. 29 are divided by T_n the result is

$$\bar{z}_n = \frac{f_n/\bar{T}_n}{\bar{U}_n/\bar{T}_n - \Omega^2} \tag{31}$$

Note that $\bar{U}_n/\bar{T}_n$ is the Rayleigh Quotient Approximation (see [18]) for the structural eigenvalue ($\bar{\lambda}_n$) corresponding to n^{th} natural mode. Therefore, Eq. 31 can be rewritten as

$$\bar{z}_n = \frac{f_n/\bar{T}_n}{\bar{\lambda}_n - \Omega^2} \tag{32}$$

Note that as the structural eigenvalue for mode n ($\bar{\lambda}_n$) approaches the loading frequency (Ω), the modal participation coefficient for the nth mode ($\bar{z}_n$) becomes very large, which is exactly what happens at resonance. The full order approximation in Eq. 22 cannot capture this effect.

It should be recognized that an approximation based on the direct solution of Eq. 20 which does capture the effect of resonance was presented in [25]. However, this approximation cannot be employed when modal analysis is used to solve for the dynamic response of the structure.

6. Examples

The effect of using intermediate design variables and intermediate response quantities on the quality of the approximate problem is shown in the following examples.

6.1 PROBLEM 1 - CANTILEVER BEAM: MASS MINIMIZATION

The first problem is that of finding the minimum mass design of the 10 meter cantilevered beam shown in Fig. 2 and described in [21]. The design variables for this problem are the web and flange thicknesses (t_A, t_b) of the beam and the position and velocity gains of a single collocated control element located at the tip of the beam (see Fig. 2). The beam is loaded by a vertical harmonic load of 4000 N at 3.9 Hz applied to the tip.

Three runs are made for this example problem, all starting from the same initial structural design. In the first run no control elements are used. In the second run an axial controller is located at the tip of the beam in order to control the vertical displacement at the tip. The initial values of the position and velocity gain design variables in the second and third runs are h_p = 20.0 N/cm and h_v = 2.0 N-sec/cm. While there are no constraints placed on the control gains, the magnitude of the control force is constrained to be less than 1000 N. The third run is the same as the second except that the real part of the complex eigenvalue (stability measure) of the first mode is constrained to be less than -1.0 rad/sec.

The final designs and final design response ratios for all three runs are given in Tables 1 and 2. The iteration history plots are shown in Fig. 3. In all three cases the tip displacement constraint is critical. This is to be expected because the minimization of the mass of the structure results in lower structural stiffness. In the second and third runs, where the controller is used, the control force constraint is also critical. This is because the control force increases as the structural stiffness decreases. Note that the addition of critical actuator force constraints does not slow down the convergence of the synthesis process.

The iteration history of the first run (Uncontrolled from [12]) is also shown in Fig. 3. Note the faster convergence rate of the new result reported here. This is due to the increased accuracy of the dynamic displacement constraint which is achieved through the use of approximations of intermediate response quantities.

In the final design of the second run the value of the real part of the complex eigenvalue of the first mode is -0.35 rad/sec. The stability constraint in the third run requires almost three times as much damping in this mode [i.e. $\sigma_1 \leq -1.0$ rad/sec]. Note that the addition of critical dynamic stability constraints does not slow down the convergence of the synthesis process. This can be attributed to the quality of the improved approximations for complex eigenvalue constraints presented in this work.

6.2 PROBLEM 2 - DRAPER/RPL STRUCTURE: CONTROL FORCE MINIMIZATION

The Draper/RPL structure consists of a massive central hub surrounded by four flexible appendages that have non-structural mass attached to their free ends (see Fig. 4 and [10] and [21]). The entire structure is free to rotate about the central axis of the hub. There are four sensors on each arm. These sensors measure displacement and velocity in the circumferential direction. There is an additional rotational position and velocity sensor located at the axis of the hub. There are three actuators that apply torques to the structure. Actuator 1 is located at the axis of the hub and actuator 2 and 3 are located on arms 1 and 2 respectively. Twenty-one controllers are used to control the structure. This actuator-sensor configuration is the same as that used in [21].

In this example the total control force is minimized subject to constraints on the complex eigenvalues of the first nine modes and a tip displacement constraint. The tip displacement of Arm 2 is constrained to be less than 0.1 in. The iteration history plot is shown in Fig. 5. At the final design five of the nine damping ratio constraints are critical. The tip displacement constraint is also critical.

In this problem, because the objective function is total control force, the force and hence the gains, of the individual control elements are driven towards zero. The control force is a very nonlinear function of the gains when they are near zero. Therefore the high quality nonlinear approximation, based on the concept of approximate intermediate response quantities, must be used in order to achieve rapid convergence. The effect of this approximation on the convergence rate is shown graphically in Fig. 5. In this plot the iteration history is shown, for the same problem, prior to the introduction of the refined approximations (see [13]). The results reported in [13] were generated without the use of approximations based on intermediate response quantities. Note the oscillation near the optimum. This is caused by the poor quality of the approximations used in [13]. The introduction of refined approximations not only reduces the number of analyses needed for convergence by a factor of two but it also leads to an objective function value that is 13.3% lower (16.32 in-lb compared with 18.82 in-lb).

6.3 PROBLEM 3 - ANTENNA: MASS MINIMIZATION

This problem is the mass minimization of the antenna structure shown in Fig. 6 and described in [19]. The structure is modeled with 10 beam type finite elements has 24 degrees of freedoms, and five modes are used for the modal analysis. The elements are linked to produce a symmetric structure with five design elements (see Fig. 6). Each design element cross section (see Fig. 2) has 2 design variables (t_k and t_b) for a total of 10 design variables. The response of node 7 in the y-direction due to a $500\,N$ load applied in the y-direction at node 9 with various forcing frequencies is shown by the solid line in Fig. 7. The off center load will excite both bending and torsion in the structure. Note the response peaks near the first natural frequency of 0.43 Hz (first bending mode) and the second natural frequency of 1.04 Hz (first torsional mode).

In this example problem the structure is loaded at 0.7 Hz, which is away from the resonance peaks. The displacement amplitudes of nodes 5 and 7 in the y-direction are constrained to be less than 1 cm. There are no other constraints on the structure. The design histories for both the modal and direct approximations for 30% and 60% move limits (M.L.) are plotted in Fig. 8. Note that with 30% move limits both approximations perform reasonably well and with 60% move limits the modal approximation gives much faster design convergence. The response of the final design at various loading frequencies is shown by the solid line in Fig. 7. In the final design the constraint on node 5 is active.

6.4 PROBLEM 4 - ANTENNA: MASS MINIMIZATION

The fourth example problem is the same as the third except the loading frequency is now 0.5 Hz, which is near resonance, and displacement amplitudes of nodes 5 and 7 in the y-direction are constrained to be less than 10 cm. The design histories are plotted in Fig. 9. Note that with 30% move limits the modal approximation converges rapidly while the direct approximation overshoots the optimum and oscillates with designs that have about 7% infeasibility with respect to the displacement amplitude constraints. Even when 60% move limits are used, the modal approximation is quite accurate and a near final design is achieved after only 4 iterations. When 60% move limits are used with the direct approximation the optimization process fails. At the final design the constraint on node 5 is active.

7. Conclusions

It has been shown that the introduction of high quality explicit approximations for behavior constraints on dynamic harmonic displacements and control forces, structural frequencies, closed loop eigenvalues, and modal damping ratios significantly improves the convergence characteristics of the approximation concepts method when it is extended for application to control augmented structural synthesis problems. Noncollocated control configurations and dynamic stability constraints have been considered.

The modal approximation for dynamic displacement response is quite accurate even when the design is near resonance and large move limits are used. The modal approximation presented in this work can be extended to damped response problems when modal analysis is used. In these problems the quantities u_i, a_i, z_n, $\{\phi_n\}$, U_n and T_n are all complex but the development of the approximation is quite similar. Transient response of damped structures can also be approximated in the manner presented in this work.

In constructing improved explicit approximations for various behavior constraint functions, the flexibility of intermediate response quantity and intermediate design variable concepts has been exploited. It should be recognized that in general each type of behavior constraint can be approximated using different intermediate response quantities and different intermediate design variables, chosen to enhance the quality of approximation. As long as the actual constraints are explicit functions of the intermediate response quantities and the intermediate design variables are explicit functions of the actual design variables, all the final constraint approximations will be explicit functions of the actual design variables spanning the space where the sequence of approximate design optimization problems is solved.

8. Acknowledgements

The authors would like to thank Dr. A.E. Sepulveda and Prof. L.A. Schmit of UCLA for their contributions to this work.

9. References

1. Khot, N.S., Venkayya, V.B. and Eastep, F.E. (1986) 'Optimal structural modifications to enhance the active vibration control of flexible structures', *AIAA Journal*, Vol. 24, pp. 1368-1374.

2. Hale, A.L., Lisowski, R.J. and Dahl, W.E. (1985) 'Optimal simultaneous structural and control design of maneuvering flexible spacecraft', *Journal of Guidance, Control and Dynamics*, Vol. 8, pp. 86-93.

3. Miller, D.F. and Shim, J. (1987) 'Gradient based combined structural and control optimization', *Journal of Guidance, Control and Dynamics*, Vol. 10, pp. 291-298.

4. Onoda, J. and Haftka, R.T. (1987) 'An approach to structure/control simultaneous optimization for large flexible spacecraft', *AIAA Journal*, Vol. 25, pp. 1133-1138.

5. Rao, S.S., Venkayya, V.B. and Khot, N.S. (1988) 'Game theory approach for the integrated design of structures and controls', *AIAA Journal*, Vol. 26, pp. 463-469.

6. Belvin, W.K. and Park, K.C. (1988) 'Structural tailoring and feedback control synthesis: an interdisciplinary approach', *Proceedings of the AIAA/ASME/ASCE/AHS 29th Structures, Structural Dynamics and Materials Conference*, AIAA, Washington, D.C., pp. 1-8.

7. Messac, A. and Turner, J. (1984) 'Dual structural-control optimization of large space structures', *Proceedings of the NASA Symposium on Recent Advances in Multidisciplinary Analysis and Optimization*, Hampton, VA, NASA CP 2327, Part 2, pp. 775-802.

8. Haftka, R.T., Martinovic, Z.N., Hallauer, W.L. Jr. (1985) 'Enhanced vibration controllability by minor structural modification', *AIAA Journal*, Vol. 23, pp. 1260-1266.

9. Khot, N.S., Öz, H., Grandhi, R.V., Eastep, F.E. and Venkayya, V.B. (1988) 'Optimal structural design with control gain norm constraint', *AIAA Journal*, Vol. 26, pp. 604-611.

10. Bodden, D.S. and Junkins, J.L. (1985) 'Eigenvalue optimization algorithms for structure/controller design iterations', *Journal of Guidance and Control*, Vol. 8, pp. 697-706.

11. Onoda, V. and Watanabe, N. (1989) 'Integrated direct optimization of structural/regulator/observer for large flexible spacecraft', *Proceedings of the AIAA/ASME/ASCE/AHS/ASE 30th Structures, Structural Dynamics and Materials Conference*, AIAA, Washington, D.C., pp. 1336-1344.

12. Lust, R.V. and Schmit, L.A. (1988) 'Control-augmented structural synthesis', *AIAA Journal*, Vol. 26, pp. 86-94.

13. Thomas, H.L. and Schmit, L.A. (1991) 'Control augmented structural synthesis with dynamic stability constraints', *AIAA Journal*, Vol. 29, pp. 619-626.

14. Schmit, L.A. and Farshi, B. (1974) 'Some approximation concepts for efficient structural synthesis', *AIAA Journal*, Vol. 12, pp. 692-699.

15. Schmit, L.A. and Miura, H. (1976) 'Approximation concepts for efficient structural synthesis', NASA CR 2552.

16. Starnes, J.R. Jr. and Haftka, R.T. (1979) 'Preliminary design of composite wings for buckling, stress and displacement constraints', *Journal of Aircraft*, Vol. 16, pp. 564-570.

17. Vanderplaats, G.N. and Salajegheh, E. (1989) 'A new approximation method for stress constraints in structural synthesis', *AIAA Journal*, Vol. 27, pp. 352-358.

18. Canfield, R.A. (1990) 'High-quality approximation of eigenvalues in structural optimization', *AIAA Journal*, Vol. 28, pp. 1116-1122.

19. Thomas, H.L., Sepulveda, A.E., and Schmit, L.A. (1990) 'Improved approximations for dynamic displacements using intermediate response quantities', *Proceedings of the Third Air Force/NASA Symposium on Recent Advances in Multidisicplinary Analysis and Optimization*, San Francisco, CA, pp. 95-104.

20. Manning, R.A., Lust, R.V., and Schmit, L.A. (1986) 'Behavior sensitivities for control augmented structures', Sensitivity Analysis in Engineering, NASA CP 2457, *Proceedings of a Conference Held at Langley Research Center*, Hampton, VA, pp. 33-57.

21. Thomas, H.L. and Schmit, L.A. (1990) 'Improved approximations for control augmented structural synthesis', *Proceedings of the AIAA/ASME/ASCE/AHS/ASC 31st Structures, Structural dynamics and Materials Conf.*, AIAA, New York, pp. 227-294, accepted for publication in *AIAA Journal*.

22. Yoshida, N. and Vanderplaats, G.N. (1988) 'Structural optimization using beam elements', *AIAA Journal*, Vol. 26, pp. 454-462.

23. Vanderplaats, G.N. and Salajegheh, E. (1988) 'An efficient approximation technique for frequency constraints in frame optimization', *International Journal for Numerical Methods*, Vol. 26, pp. 1057-1069.

24. Vanderplaats, G.N. (1973) 'CONMIN - A fortran program for constrained function minimization', NASA TM X-62,682.

25. Miura, H. and Chargin, M. (1986) 'Automated tuning of airframe vibration by structural optimization', *Proceedings of the 42nd Annual Forum and Display American Helicopter Society*, Washington, D.C.

Table 1. Final Designs for Problem 1
Cantilevered Beam: Mass Minimization

Element Type	Element Number	Design Variables	Final Design (cm, N/cm, N-sec/cm)		
			Uncontrolled	Controlled	Controlled Stability Constraint
Frame	1-10	t_b	1.995	1.588	1.617
		t_h	0.500[a]	0.500[a]	0.500[a]
Control	1	h_p		100.324	86.48
		h_v		0.217	2.067
Mass (kg)			541.4	453.5	459.8

[a] denotes lower bound value

Table 2. Final Design Response Ratios for Problem 1
Cantilevered Beam: Mass Minimization

Constraint	Response Ratio		
	Uncontrolled	Controlled	Controlled Stability Constraints
Tip Displacement	1.000[a]	0.996[a]	0.998[a]
Frequency	0.653	0.705	0.700
Control Force		1.001[a]	1.000[a]
Stability			0.999[a]

[a] indicates critical constraint

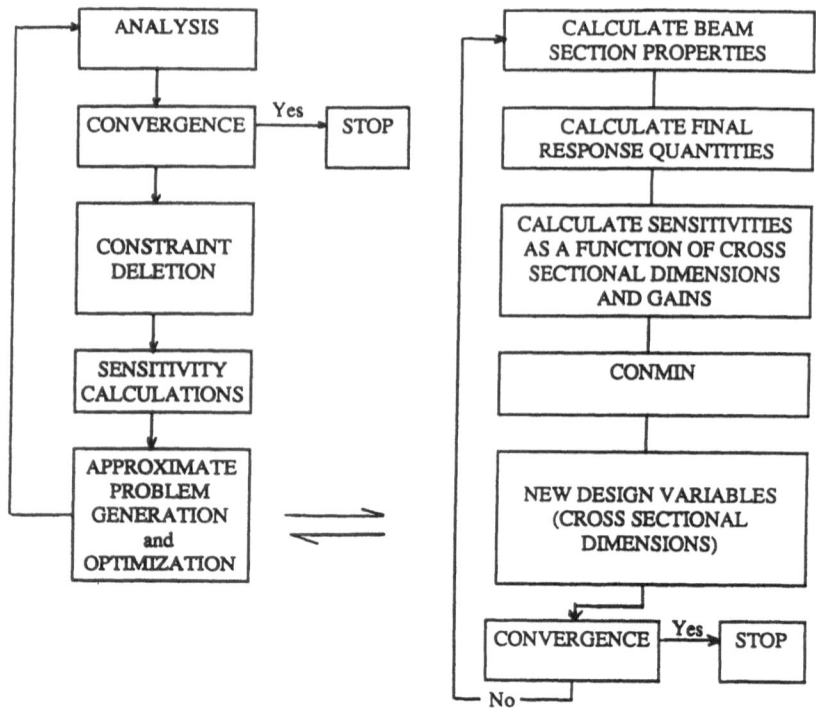

Figure 1. Optimization Procedure

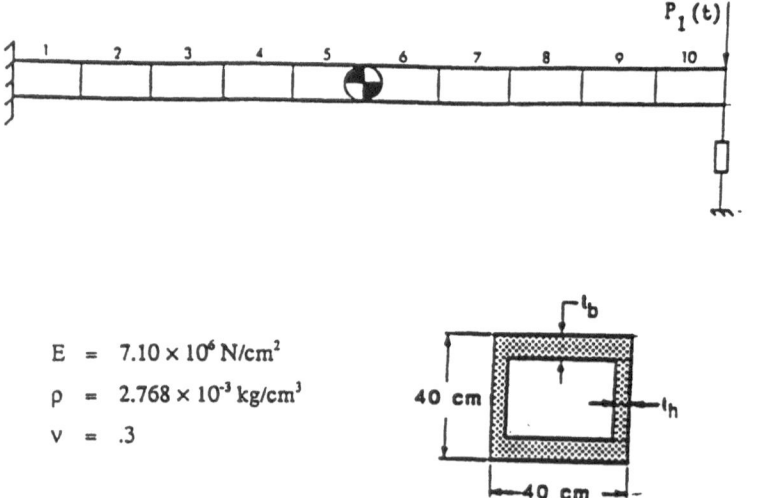

$E = 7.10 \times 10^6 \, \text{N/cm}^2$

$\rho = 2.768 \times 10^{-3} \, \text{kg/cm}^3$

$v = .3$

Figure 2. Cantilevered Beam, Problem 1

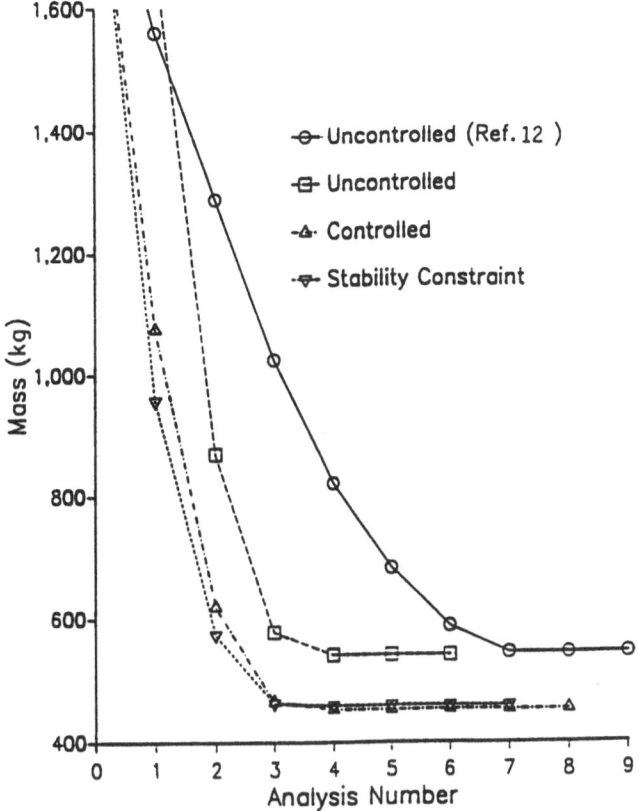

Figure 3.   Iteration Histories for Problem 1
Cantilevered Beam: Mass Minimization

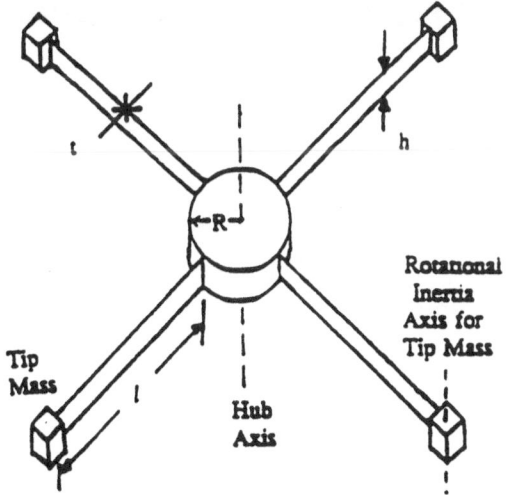

Figure 4. Draper/RPL Structure and Analysis Model

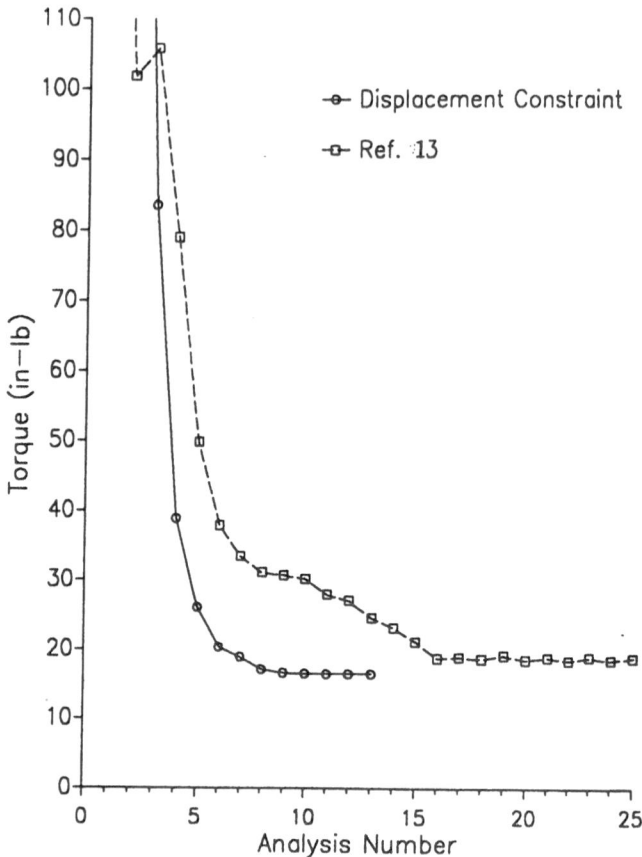

Figure 5. Iteration History Comparison for Problem 2
Draper/RPL Structure: Control Force Minimization

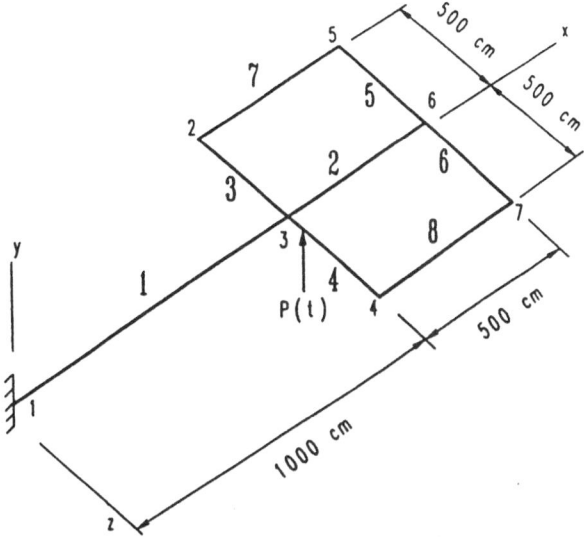

Figure 6. Antenna Structure

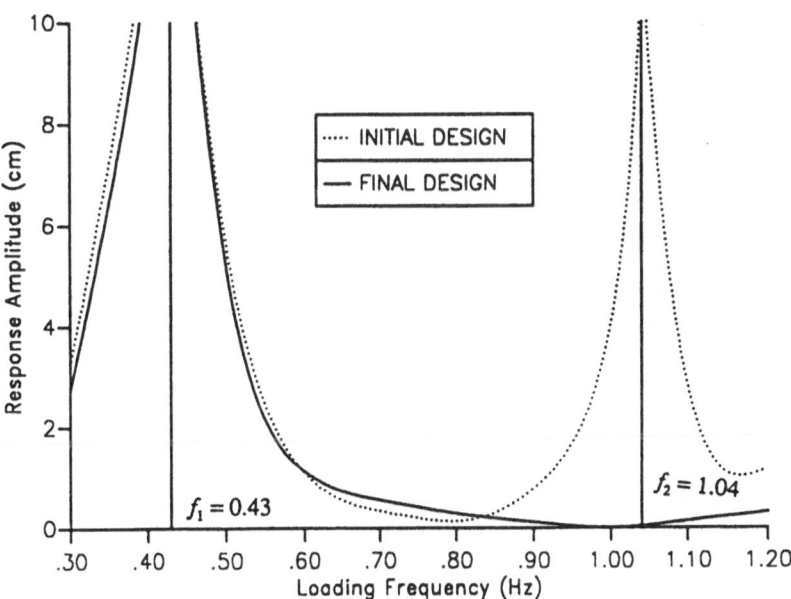

Figure 7. Structural Response at Various Frequencies

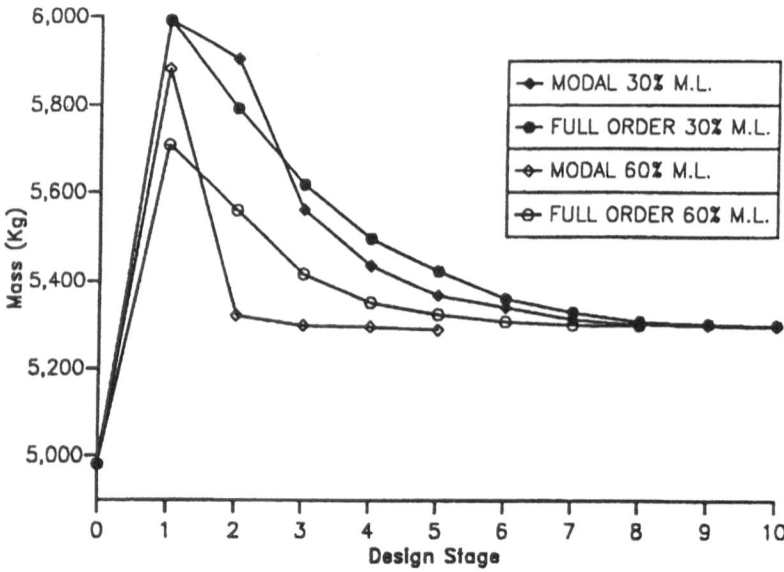

Figure 8. Mass Iteration Histories for Problem 3

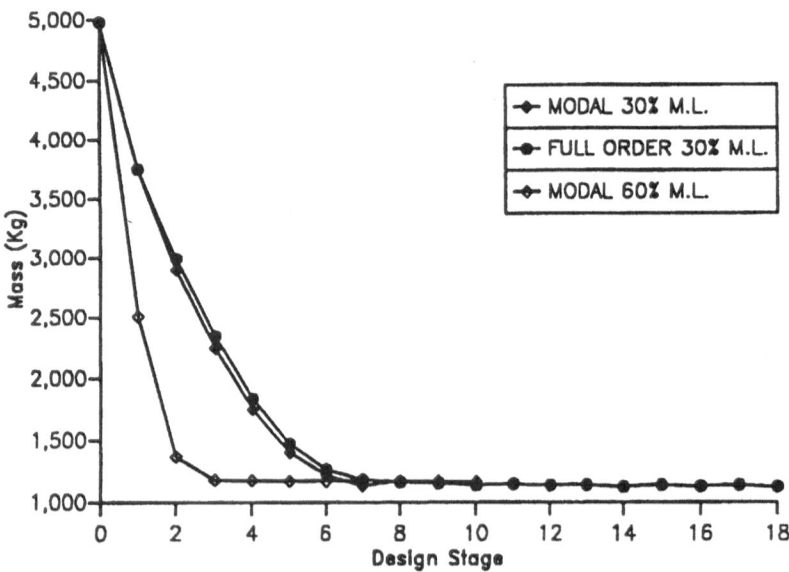

Figure 9. Mass Iteration Histories for Problem 4

STRUCTURAL AND CONTROL OPTIMIZATION

N. S. KHOT
Structures Division, Flight Dynamics Directorate (WL/FIBRA)
Wright-Patterson Air Force Base, Ohio 45433-6553

ABSTRACT. In this paper we have examined the integrated approach to the simultaneous optimization of the structure and control system. The approach is illustrated for a linear quadratic regulator control applied to a flexible structure. Analytical sensitivity expressions are used. An illustrative example is solved with constraints on the closed-loop eigenvalues and the spectral radius which defines the robustness characteristics of the control design.

1. Introduction

The conventional approach to the design of a controlled structural system is to design the structure first by satisfying its requirements and then to design the control system for this structure. The structure is designed with constraints on weight, allowable stresses in the elements, displacements at node points, buckling of local elements, general instability, frequency distribution, etc. When the selection of the geometry, cross-sectional areas of the members and material is completed in a structure then its structural frequencies and vibration modes become input to the control design. The control system is designed with constraints on closed-loop eigenvalues, available control energy, location of actuators and sensors, robustness characteristics, etc. This sequential approach would require many iterations the of structural design and control design and might not lead to the optimum system. Since the structural design affects the control design, the natural process would be to integrate these two design processes in order to obtain an efficient structure and control system. This can be achieved by posing it as an optimization problem with constraints on the objective functions and response quantities from structural as well as control design requirements. In the last decade a number of investigators have written papers discussing integrated design approaches. A summary of the work in this field may be found in [1].

The integrated design approach can be formulated as a mathematical optimization problem by defining the objective function and constraints and solving by using an optimization program. In the present formulation, it will be assumed that the geometry of the structure is fixed and the design variables are the cross-sectional areas of the members. The weight of the structure will be the primary objective function with constraints on the closed-loop frequencies which determine the performance of the controler. In addition, a constraint will be specified on the robustness parameter which determines the stability of the control system under the presence of structured uncertainties in the plant matrix. These are generally due to the variations between the real and analytical frequencies and vibration modes

829

of the structure. For simplicity, the controler will be designed based on the full order system using the linear quadratic regulator with constant feedback. The design variables for the controller will be the elements of the weighting matrices in the performance index. The design problem can be mathematically stated as: Find the vector of the specified design variables such that the objective function W is minimized subject to

$$g_j \leq 0 \qquad j = 1, 2, \ldots, n_g \tag{1}$$

$$h_j = 0 \qquad j = 1, 2, \ldots, n_h \tag{2}$$

where g_j and h_j are functions of the design variables. The numerical results presented in this chapter are obtained by using NEWSUMT-A [2] which is based on an extended interior penalty function method with Newton's method of unconstrained minimization.

2. Dynamic Analysis

The equations of motion for a flexible structure with no external disturbance can be written as

$$\mathbf{M}\ddot{u} + \mathbf{E}\dot{u} + \mathbf{K}u = \mathbf{D}f \tag{3}$$

where $\mathbf{M}$ is the mass matrix, $\mathbf{E}$ is the damping matrix, and $\mathbf{K}$ is the total stiffness matrix. These matrices are $n \times n$ where n is the number of degrees of freedom of the structure. In Eq. 3, $\mathbf{D}$ is the applied load distribution matrix relating the control input vector f to the coordinate system. The number of elements in the f vector is equal to the number of actuators, p. The vector u in Eq. 3 defines the structural response.

Introducing the coordinate transformation

$$u = [\phi]\,\eta \tag{4}$$

where η is the modal coordinate system and $[\phi]$ is the $n \times n$ modal matrix, Eq. 3 can be transformed into n uncoupled equation. The uncoupled equation can be written as

$$\overline{\mathbf{M}}\ddot{\eta} + \overline{\mathbf{E}}\dot{\eta} + \overline{\mathbf{K}}\eta = [\phi]^T \mathbf{D}f \tag{5}$$

where

$$\overline{\mathbf{M}} = \mathbf{I} = [\phi]^T \mathbf{M}[\phi] \tag{6}$$

$$\overline{\mathbf{E}} = [2\varsigma\omega] = [\phi]^T \mathbf{E}[\phi] \tag{7}$$

$$\overline{\mathbf{K}} = [\omega^2] = [\phi]^T \mathbf{K}[\phi] \tag{8}$$

The matrices $\overline{\mathbf{M}}$, $\overline{\mathbf{E}}$ and $\overline{\mathbf{K}}$ are diagonal square matrixes. ω is the vector of structural frequencies and ς is the vector of modal damping. The modal matrix $[\phi]$ is normalized with respect to the mass matrix.

3. Control Design

The control analysis is performed by reducing the second order uncoupled equation (Eq. 5) to a first order equation. This can be achieved by using the transformation.

$$x_{2n} = \begin{bmatrix} \eta \\ \dot{\eta} \end{bmatrix}_{2n} \tag{9}$$

where $\underset{\sim}{x}$ is the state variable vector of size $2n$. This gives

$$\dot{\underset{\sim}{x}} = \mathbf{A}\underset{\sim}{x} + \mathbf{B}\underset{\sim}{f} \tag{10}$$

where $\mathbf{A}$ is a $2n \times 2n$ plant matrix and $\mathbf{B}$ is a $2n \times p$ input matrix. The plant and the input matrices are given by

$$\mathbf{A} = \left[\begin{array}{c|c} 0 & \mathbf{I} \\ \hline -\omega^2 & -2\varsigma\omega \end{array} \right] \tag{11}$$

$$\mathbf{B} = \begin{bmatrix} 0 \\ \phi^T \mathbf{D} \end{bmatrix} \tag{12}$$

The state output equation is given by

$$\underset{\sim}{y} = \mathbf{C}\underset{\sim}{x} \tag{13}$$

where $\underset{\sim}{y}$ is a $q \times 1$ output vector and $\mathbf{C}$ is $q \times 2n$ output matrix, q is equal to the number of sensors. If the number of sensors and actuators is equal and if they are colocated, then $q = p$ and

$$\mathbf{C} = \mathbf{B}^T \tag{14}$$

In order to design a controller using a linear quadratic regulator, a performance index J is defined as

$$J = \int\limits_0^\infty \left(\underset{\sim}{x}^T \overline{\mathbf{Q}} \underset{\sim}{x} + \underset{\sim}{f}^T \overline{\mathbf{R}} \underset{\sim}{f} \right) \, dt \tag{15}$$

where $\overline{\mathbf{Q}}$ and $\overline{\mathbf{R}}$ are state and control weighting matrices. The matrix $\overline{\mathbf{Q}}$ has to be positive semidefinite ($\underset{\sim}{x}^T \overline{\mathbf{Q}} \underset{\sim}{x} \geq 0$) and $\mathbf{R}$ has to be positive definite ($\underset{\sim}{f}^T \overline{\mathbf{R}} \underset{\sim}{f} > 0$). The dimensions of $\overline{\mathbf{Q}}$ and $\overline{\mathbf{R}}$ depend on the size of vectors $\underset{\sim}{x}$ and $\underset{\sim}{f}$ respectively. The selection of elements of $\overline{\mathbf{Q}}$ and $\overline{\mathbf{R}}$ determines the amount of closed-loop damping and the time required to control the disturbances. Instead of assuming that all the elements of $\overline{\mathbf{Q}}$ and $\overline{\mathbf{R}}$ are design variables, let us for simplicity assume that these matrices are comprised of a design variable multiplied by a specified constant matrix which does not get modified during optimization. Thus, weighting matrices $\overline{\mathbf{Q}}$ and $\overline{\mathbf{R}}$ can be written as

$$\overline{\mathbf{Q}} = \delta \mathbf{Q} \tag{16}$$

and

$$\overline{\mathbf{R}} = \frac{1}{\gamma}\mathbf{R}^{-1} \tag{17}$$

where δ and γ are the design variables and $\mathbf{Q}$ and $\mathbf{R}^{-1}$ are the constant matrices.

The performance index represents a compromise between minimum error and minimum energy criteria. The result of minimizing the performance index and satisfying Eq. 10 gives the state feedback control law

$$\underset{\sim}{f} = -\mathbf{G}\underset{\sim}{x} \tag{18}$$

where $\mathbf{G}$ is the optimum gain matrix given by

$$\mathbf{G} = \gamma\mathbf{R}\mathbf{B}^T\mathbf{P} \tag{19}$$

where $\mathbf{P}$ is a positive definite matrix called the Riccati matrix and is obtained by the solution of the algebraic Riccati equation

$$0 = \delta\mathbf{Q} + \mathbf{P}\mathbf{A} + \mathbf{A}^T\mathbf{P} - \gamma\mathbf{P}\mathbf{B}\mathbf{R}\mathbf{B}^T\mathbf{P} \tag{20}$$

The open-loop system is given by

$$\dot{\underset{\sim}{x}} = \mathbf{A}\underset{\sim}{x} \tag{21}$$

and the closed-loop system is given by

$$\dot{\underset{\sim}{x}} = \overline{\mathbf{A}}\underset{\sim}{x} \tag{22}$$

where

$$\overline{\mathbf{A}} = \mathbf{A} - \mathbf{B}\mathbf{G} \tag{23}$$

Eq. 22 can be obtained by substituting Eq. 18 in Eq. 10. The eigenvalues of the closed-loop matrix $\overline{\mathbf{A}}$ are a set of complex conjugate pairs written as

$$\lambda_i = \tilde{\sigma}_i \pm j\tilde{\omega}_i \qquad i = 1,\ldots,n \tag{24}$$

where $j = \sqrt{-1}$. The sign of $\tilde{\sigma}_i$ must be negative for all i in order to make the system asymptotically stable. The closed-loop damping factors are given by

$$\xi_i = -\frac{\tilde{\sigma}_i}{\left(\tilde{\sigma}_i^2 + \tilde{\omega}_i^2\right)^{\frac{1}{2}}} \tag{25}$$

The magnitudes of ξ_i associated with each mode determine the extent to which those modes are damped.

For the initial condition $\underset{\sim}{x}(o)$, solutions to Eqs. 21 and 22 are given by

$$\underset{\sim}{x}(t) = \exp\left(\mathbf{A}t\right)\underset{\sim}{x}(0) \tag{26}$$

and

$$\underset{\sim}{x}(t) = \exp\left(\overline{\mathbf{A}}t\right)\underset{\sim}{x}(0) \tag{27}$$

where

$$\exp\left(\mathbf{A}t\right) = 1 + \frac{\mathbf{A}t}{1!} + \frac{\left(\mathbf{A}t\right)^2}{2!} + \cdots \tag{28}$$

and

$$\exp\left(\overline{\mathbf{A}}t\right) = 1 + \frac{\overline{\mathbf{A}}t}{1!} + \frac{\left(\overline{\mathbf{A}}t\right)^2}{2!} + \cdots \tag{29}$$

Eqs. 26 and 27 can be used to determine the transient response of the open-loop and closed-loop systems.

4. Sensitivities of Closed-Loop Eigenvalues

The sensitivity of the closed-loop eigenvalues λ_i with respect to the structural design variables A_l is given by

$$\lambda_{i,l} = \beta_{\sim i}^T \overline{\mathbf{A}}_{,l} \alpha_i \tag{30}$$

The closed-loop matrix defined in Eq. 23 can be rewritten as

$$\overline{\mathbf{A}} = \mathbf{A} - \mathbf{X}\mathbf{P} \tag{31}$$

where

$$\mathbf{X} = \gamma \mathbf{B}\mathbf{R}\mathbf{B}^T \tag{32}$$

In Eq. 30, $\beta_{\sim i}^T$ and α_i are the normalized left-hand and right-hand eigenvectors of $\overline{\mathbf{A}}$.

The partial derivatives of $\overline{\mathbf{A}}$ with respect to the structural design variables $\mathbf{A}_l$ can be obtained by differentiating Eq. 31 with respect to $\mathbf{A}_l$. This gives

$$\overline{\mathbf{A}}_{,l} = \mathbf{A}_{,l} - \mathbf{X}_{,l}\mathbf{P} - \mathbf{X}\mathbf{P}_{,l} \tag{33}$$

The matrices $\mathbf{A}$ and $\mathbf{X}$ in this equation are functions of the structural frequencies ω_j and the modal matrix ϕ. Eq. 33 needs the sensitivities of the Riccati matrix $\mathbf{P}$. Differentiating Eq. 20 with respect to the design variables A_l and using Eq. 32 gives

$$\overline{\mathbf{A}}^T \mathbf{P}_{,l} + \mathbf{P}_{,l}\overline{\mathbf{A}} = \tilde{\mathbf{B}} \tag{34}$$

where

$$\tilde{\mathbf{B}} = -\mathbf{A}_{,l}^T \mathbf{P} - \mathbf{P}\mathbf{A}_{,l} + \mathbf{P}\mathbf{X}_{,l}\mathbf{P} \tag{35}$$

The sensitivity of the Riccati matrix $\mathbf{P}_{,l}$ is given by the solution to the Lyapunov equation, Eq. 34.

The sensitivities of the closed-loop eigenvalues with respect to the control design variables δ and γ are given by

$$\lambda_{i,\delta} = \beta_i^T \overline{\mathbf{A}}_{,\delta} \alpha_i \tag{36}$$

and

$$\lambda_{i,\gamma} = \beta_i^T \overline{\mathbf{A}}_{,\gamma} \alpha_i \tag{37}$$

The sensitivities of the closed-loop matrix $\overline{\mathbf{A}}$ with respect to δ and γ can be obtained by partial differentiation of Eq. 31. This gives

$$\overline{\mathbf{A}}_{,\delta} = -\mathbf{X}\mathbf{P}_{,\delta} \tag{38}$$

and

$$\overline{\mathbf{A}}_{,\gamma} = -\mathbf{X}_{,\gamma}\mathbf{P} - \mathbf{X}\mathbf{P}_{,\gamma} \tag{39}$$

Differentiating Eq. 20 with respect to δ and γ gives

$$\overline{\mathbf{A}}^T\mathbf{P}_{,\delta} + \mathbf{P}_{,\delta}\overline{\mathbf{A}} = -\mathbf{Q} \tag{40}$$

and

$$\overline{\mathbf{A}}^T\mathbf{P}_{,\gamma} + \mathbf{P}_{,\gamma}\overline{\mathbf{A}} = -\mathbf{P}\mathbf{B}\mathbf{R}\mathbf{B}^T\mathbf{P} \tag{41}$$

The solutions to the two Lyapunov equations, Eqs. 40, and 41, give the sensitivities of the Riccati matrix $\mathbf{P}$ with respect to δ and γ, respectively.

5. Robustness Parameter

Robust control design is one which performs in an acceptable manner even in the presence of modelling errors. There are different measures of robustness depending on the structured and unstructured uncertainties treated within the time doman and state space framework. The perturbed closed-loop state space model can be written as

$$\dot{\underline{x}} = \left(\overline{\mathbf{A}} + \varepsilon\right)\underline{x} \tag{42}$$

where $\overline{\mathbf{A}}$ is a stable matrix and ε is a perturbation matrix. The perturbed system would be stable if the elements ε_{ij} of matrix ε satisfy the relation

$$\varepsilon_{ij} < \frac{1}{\sup\limits_{p \geq 0} \rho\left[\left|\left(jp\mathbf{I} - \overline{\mathbf{A}}\right)^{-1}\right|\mathbf{U}_e\right]} \cdot \mathbf{U}_{e_{ij}} = \frac{1}{\rho_s} \cdot \mathbf{U}_{e_{ij}} \tag{43}$$

according to the robustness measure proposed by Juang [4] et al. In Eq. 43 $|(\cdot)|$ denotes an absolute matrix and $\rho[\cdot]$ denotes the spectral radius of the matrix $[\cdot]$. $\mathbf{U}_{e_{ij}}$ are the elements of the perturbation identification matrix $\mathbf{U}_e \cdot SUP$ represents the supremium of the matrix over a range of p. The elements of $\mathbf{U}_e$ have assigned values depending upon the relative perturbations allowed for those elements of $\overline{\mathbf{A}}$. The spectral radius is equal to the maximum modulus of the complex eigenvalues for a specified operating frequency p. The maximum spectral radius value amongst all possible values of p gives the critical value of p_s. For structural problems, it has been shown in [4] that peaks in the spectral radius plot occur only at the p equal to the modulus of the closed-loop frequencies of the matrix $\overline{\mathbf{A}}$. This assumption can be used to determine the critical frequency and associated spectral radius.

The elements of the perturbation matrix ε are proportional to the elements of the perturbation identification matrix $\mathbf{U}_e$ and inversely proportional to the spectral radius ρ_s. If any element of $\overline{\mathbf{A}}$ is constant or zero, then the associated element of $\mathbf{U}_e$ would be zero allowing no perturbation in that element. Setting all the elements of $\mathbf{U}_e$ equal to unity would allow the same amount of perturbation in all the elements of $\overline{\mathbf{A}}$. This would be too conservative for some elements and could be avoided by setting $\mathbf{U}_e$ equal to the normalized $\overline{\mathbf{A}}$. This gives

$$U_{e_{ij}} = \frac{|\overline{\mathbf{A}}_{ij}|}{|\overline{\mathbf{A}}_{pq}|} \qquad i > n \qquad (44)$$

where $|\overline{\mathbf{A}}_{pq}|$ is the absolute value of a specific element of $\overline{\mathbf{A}}$. The ratio of ε_{ij} and the elements of $|\overline{\mathbf{A}}_{ij}|$ defined as ε_r for all the elements is given by

$$\varepsilon_r = \frac{1}{\rho_s |\overline{\mathbf{A}}_{pq}|} \qquad (45)$$

The percentage allowable deviation in the elements of $\overline{\mathbf{A}}$ is $100\varepsilon_r$. It may be noted that for the elements of $\mathbf{U}_e$ as defined in Eq. 44, even though ε_{ij} is different for different elements of $\overline{\mathbf{A}}$, the percentage allowable deviation is the same.

In the derivation of the sensitivities of the spectral radius and for the elements of the perturbation matrix, it was assumed that these quantities are continuous functions of the design variables in the vicinity of the current design. It will also be assumed that the sensitivities are calculated for a specified critical operating frequency p_s and it was invariant for small changes in the design variables. A detailed derivation of the sensitivities of the robustness parameters may be found in [5] and [6].

6. Optimization Procedure

The major steps involved in the integrated structural control design optimization problem discussed in this chapter are as follows:

1. For the specified values of the cross-sectional areas of the members, the structural frequencies ω_j and the vibration modes ϕ_j and their sensitivities with respect to the cross-sectional area of the members are calculated.
2. The plant matrix $\mathbf{A}$ and the input matrices $\mathbf{B}$, in the state space input equation and the output matrix $\mathbf{C}$ in the state output equation are assembled.
3. The linear optimum regulator control problem is solved and the closed-loop eigenvalues, eigenvectors and damping factors are determined.
4. The spectral radii for the operating frequencies p, which are equal to the modulus of the eigenvalues of the closed-loop matrix $\overline{\mathbf{A}}$, are evaluated. The maximum value amongst all these values is equal to ρ_s and the associated operating frequency is equal to p_s.
5. The sensitivities of the closed-loop eigenvalues, damping parameters and ρ_s are calculated as needed.
6. The design variables are modified by using a suitable optimization algorithm.
7. With the new values of the design variables, steps 1 through 6 are repeated until the optimum solution satisfying all the specified constraints is obtained. Repetition of the above steps is required because of the nonlinear nature of the problem.

7. Numerical Examples

In order to illustrate the application of an integrated approach to the optimization of a structure-control system, the ACOSS-FOUR structure shown in Fig. 1 was optimized with constraints on the closed-loop frequencies and the spectral radius. This structure is frequently used to study the performance of different control algorithms. The optimization problem is defined to minimize the weight, improve the robustness characteristics of the design and maintain a specified distribution of closed-loop frequencies.

The ACOSS-FOUR model represents a flexible structure pointing system attached to a rigid base. The structure has twelve translational and no rotational degrees of freedom since the members are assumed to carry only axial loads. The coordinates of the node points are given in Table 1. The dimensions of the structure are specified in consistent nondimensional units. Young's modulus is equal to unity and the density of the material is assumed to be 0.001. A nonstructural mass of 2 units is located at node points 1 through 4. In the assembly of the mass matrix M in Eq. 3, it is assumed that the contribution of the structural mass is zero. Thus the mass matrix is a diagonal matrix. The structural modal damping matrix ς in Eq. 15 is also assumed to be zero for all frequencies. The six actuators and sensors are colocated in the six bipods. For simplicity, all the degrees of freedom, of the structural modal are included for the control design. Since the structure has twelve degrees of freedom, the matrices A and B would be 24×24 and 24×6 respectively. The weighting matrices $\overline{Q}$ and $\overline{R}$ would be 24×24 and 6×6 respectively. The constant part of the weighting matrices Q and R^{-1} are assumed to be identity matrices.

The cross-sectional areas of the members for the initial design are those of Design A. The ACOSS-FOUR structure was designed with constraints on $\tilde{\omega}_1$ and $\tilde{\omega}_2$ and ρ_s with weight as the objective function. $\tilde{\omega}_1$ and $\tilde{\omega}_2$ are the imaginary parts of the lowest two frequencies of the closed-loop matrix $\overline{A}$. ρ_s is the spectral radius which determines the robustness of the control design. The two constraints on frequencies are

$$\tilde{\omega}_1 - 1.34 \geq 0 \tag{46}$$

$$\tilde{\omega}_2 - 1.6 \geq 0 \tag{47}$$

The intention of enforcing the constraints on these frequencies is to prevent the decrease of the lowest frequency and not to allow the frequencies to coalesce. In addition to these two constraints, a constraint on the spectral radius ρ_s is imposed to improve the robustness of the structure-control design.

7.1. DESIGN A

The details of the initial design used for optimization are given in Table 2. The controller is designed with δ and γ both equal to unity in the weighting matrices. The spectral radius ρ_s is calculated for two normalizing elements $|\overline{A}_{24,12}|$ and $|\overline{A}_{13,1}|$ for defining the perturbation identification matrix U_e in Eq. 44. These two elements of the closed-loop system are associated with the maximum and minimum structural frequencies. In these two perturbation identification matrices, the elements $U_{e24,12}$ and $U_{e13,1}$ are equal to unity respectively. The spectral radius ρ_s for the two identification matrices is 0.4771 and 44.35 respectively. Thus, the allowable perturbation in the elements $\overline{A}_{24,12}$ and $\overline{A}_{13,1}$ for the two cases would be $\frac{1}{0.4771} = 2.045$ and $\frac{1}{44.35} = 0.0225$ respectively. Even though the spectral radius ρ_s for the two normalizing approaches are different, the ratio ε_r defined in Eq. 45 is equal to 0.126 for both cases. Both approaches indicate that this design permits 1.26% deviation in the elements of the closed-loop matrix. The initial design weighs 43.69 units.

7.2. DESIGN B

In this design, the cross-sectional areas of the elements as well as the parameters δ and γ in the weighting matrices are the design variables. In addition to the two constraints on the imaginary parts of the closed-loop frequencies $\tilde{\omega}_1$ and $\tilde{\omega}_2$ as specified in Eqs. 46 and 47, a constraint is specified on the spectral radius as

$$0.04771 - \rho_s \geq 0 \tag{48}$$

This constraint makes the spectral radius for the optimum design one tenth of that of the initial design. The elements of the perturbation identification matrix are obtained by normalizing $|\mathbf{A}|$ with respect to $|\bar{\mathbf{A}}_{24,12}|$ which is the largest element. This would make element $\mathbf{U}_{e_{24,12}}$ equal to unity with all other elements of $\mathbf{U}_e$ smaller than one. Thus the constraint on the spectral radius requires $\varepsilon_{(24,12)}$ to be greater or equal to $\frac{1}{0.04771} = 20.95$. The details of the optimum design are given in Table 3. This structure weighs 19.45 units. The optimization algorithm particularly satisfies the constraint on the spectral radius by increasing the cross-sectional area of element number seven to 2024.9 which increases ω_{12} to 19.07. The control design variables δ and γ for the optimum design are both equal to 5.4103. It is interesting to see that both design variables have the same value. The maximum allowable perturbation in element $\bar{\mathbf{A}}_{24,12}$ is equal to $\frac{1}{0.04767} = 20.97$. The ratio ε_r for this design is equal to .0576 permitting 5.76% deviation in all the elements of the closed-loop matrix.

7.3. DESIGN C

This problem was formulated in order to achieve a 10% permissible deviation in all the elements of the closed-loop matrix. The perturbation identification matrix was calculated by normalizing the absolute values of the elements of the closed-loop matrix with respect to the absolute value of the element $\bar{\mathbf{A}}_{13,1}$. Thus, for the optimum design $\varepsilon_r = 0.1$. Using Eq. 45, the design value of spectral radius $\bar{\rho}_s$ can be written as

$$\rho_s = \frac{1}{0.1|\bar{\mathbf{A}}_{13,1}|} \tag{76}$$

Now let us assume that $|\bar{\mathbf{A}}_{13,1}|$ is nearly equal to $\tilde{\omega}_1^2$ which is constrained by Eq. 46. This is a true assumption if the contribution from the elements of the product $\mathbf{BG}$ is small compared to the elements of $\mathbf{A}$. If this assumption is not valid, an iterative process based on the initial results would be needed. Thus, in addition to the two constraints on the imaginary parts of the closed-loop frequencies given in Eqs. 40 and 47, the constraint on the spectral radius is specified as

$$\frac{1}{0.1 * 1.341^2} - \rho_s = 5.560 - \rho_s \geq 0 \tag{77}$$

The details of the optimum design are given in Table 4. This design weighes 16.01 units. The control design variables δ and γ are equal to 3.8906. The cross-sectional area of element seven is of the same order of magnitude as that of the other elements. The ratio ε_r for the optimum design is 0.09999 which is sufficiently close to the desired value of 0.1. This small variation is due to the difference between $|\bar{\mathbf{A}}_{13,1}|$ and $\tilde{\omega}_1^2$. This example illustrates the advantage of defining the perturbation matrix by normalizing it with respect to the element which is related to the frequency constraint.

The iteration history for the two designs using the NEWSUMT-A mathematical optimization program is given in Table 5.

References

1. Khot, N.S. (1991), "Optimization of Controlled Structures," "Advances in Design Optimization" edited by Prof H. Adeli, Chapman & Hall Ltd, London.
2. Thareja, R. and Haftka, R.T., (1985), "NEWSUMT-A. A Modified Version of NEWSUMT for Inequality and Equality Constraints," VPI Report 148, Aerospace Engineering Department, Virginia Tech., Blacksburg VA.
3. Juang, Y., Kuo, T. and Hsu, C., (1987), "New Approach to Time-Domain Analysis for Stability Robustness of Dynamic Systems," Intl. J. of Systems Science, Vol. 8, No. 7, pp. 1363-1376.
4. Grandhi, R.V., Haq, I. and Khot, N.S., "Enhanced Robustness in Integrated Structural/Control Systems Design," proceedings of 31st AIAA Structures, Structural Dynamics and Materials Conference, pp. 247-257, Long Beach CA, April 2-4, 1990. To be published in AIAA Journal.
5. Khot, N.S. and Veley, D.E. (1990), "Robustness Characteristics of Optimum Structural/Control Design," To be published in the Journal of Guidance and Controls.
6. Khot, N.S. and Veley, D.E., (1990), "Use of Robusntess Constraints in the Optimum Design of Space Structure," First Joint U.S./Japan Conference on Adaptive Structures, Maui, Hawaii, Nov. 1990. To be published in the Journal of Intelligent Material Systems and Structures.

Table 1

Node Point Coordinates for ACOSS-Four

Node	X	Y	Z
1	0.0	0.0	10.165
2	-5.0	-2.887	2.0
3	5.0	-2.887	2.0
4	0.0	5.7735	0.0
5	-6.0	-1.1547	0.0
6	-4.0	-4.6188	0.0
7	4.0	-4.6188	0.0
8	6.0	-1.1547	0.0
9	-2.0	5.7735	0.0
10	2.0	5.7735	0.0

Table 2

DESIGN A

Cross-Sectional Areas of Members

Element	1	2	3	4	5	6
Area	1000.0	1000.0	100.0	100.0	1000.0	1000.0
Element	7	8	9	10	11	12
Area	100.0	100.0	100.0	100.0	100.0	100.0

Weight = 43.69
Control Design Variables
$\delta = 1.0 \quad \gamma = 1.0$

Structural Frequencies ω_j

Mode	1	2	3	4	5	6
ω_j	1.342	1.664	2.890	2.950	3.398	4.204
Mode	7	8	9	10	11	12
ω_j	4.662	4.755	8.539	9.250	10.284	12.905

Closed-Loop Damping

Mode	1	2	3	4	5	6
ξ_{ji}	0.0546	0.0653	0.0737	0.0801	0.0839	0.0864
Mode	7	8	9	10	11	12
ξ_{ji}	0.0760	0.0723	0.03417	0.0298	0.0207	0.0064

Robustness Parameters

$\rho_s = 0.4771$

$\epsilon_r = 0.0126$

$\bar{A}_{pq} = \bar{A}_{24,12}$

$\rho_s = 44.35$

$\epsilon_r = 0.0126$

$\bar{A}_{pq} = \bar{A}_{13,1}$

Table 3
DESIGN B
Structural Design Variables

Element	1	2	3	4	5	6
Area	199.53	140.89	215.07	157.67	261.22	204.06
Element	7	8	9	10	11	12
Area	2024.9	105.25	107.22	168.95	157.86	131.06

Weight = 19.45
Control Design Variables
δ=5.4103 γ=5.4103

Structural Frequencies ω_j

Mode	1	2	3	4	5	6
ω_j	1.351	1.609	2.842	2.937	3.626	4.856
Mode	7	8	9	10	11	12
ω_j	4.876	5.494	6.099	6.123	6.617	19.07

Closed-Loop Damping

Mode	1	2	3	4	5	6
ξ_{5i}	0.2498	0.1873	0.1955	+0.3034	0.1909	0.3779
Mode	7	8	9	10	11	12
ξ_i	0.3029	0.3236	0.2774	0.2977	0.1145	0.1003

Robustness Parameters

$\rho_s = 0.04767$

$\varepsilon_r = 0.05768$

$\bar{A}_{pq} = \bar{A}_{24,12}$

Table 4

DESIGN C

Structural Design Variables

Element	1	2	3	4	5	6
Area	216.4	138.5	165.5	168.0	154.5	468.7
Element	7	8	9	10	11	12
Area	174.6	144.8	155.6	235.4	143.2	170.5

Weight = 16.01
Control Design Variables
δ=3.8906 γ=3.8906

Structural Frequencies ω_j

Mode	1	2	3	4	5	6
ω_j	1.355	1.607	2.865	3.003	3.654	5.175
Mode	7	8	9	10	11	12
ω_j	5.390	5.718	6.147	6.383	6.922	7.884

Closed-Loop Damping

Mode	1	2	3	4	5	6
ξ_i	0.1411	0.1364	0.1262	0.1517	0.1618	0.2544
Mode	7	8	9	10	11	12
ξ_i	0.2009	0.2365	0.2165	0.1974	0.1713	0.1131

Robustness Parameters

ρ_s = 5.560

ε_r = 0.09999

$\bar{A}_{pq} = \bar{A}_{13,1}$

Table 5

Iteration History

Iteration No.	DESIGN B	DESIGN C
1	43.69	43.69
2	1492.6	32.01
3	608.49	32.01
4	63.97	25.91
5	26.97	22.12
6	22.08	19.14
7	20.27	16.58
8	19.94	16.11
9	19.48	16.02
10	19.45	16.01
11	19.40	16.01

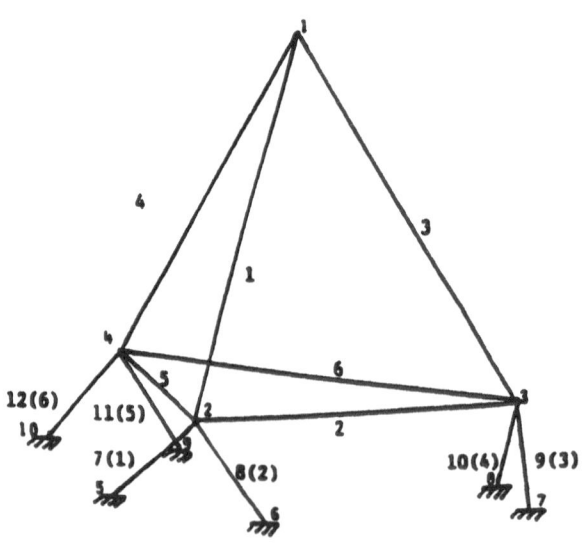

Figure 1.   Finite Element Model of ACOSS-FOUR

THEORY AND PRACTICE OF STRUCTURAL OPTIMIZATION

M. Z. COHN
Department of Civil Engineering
University of Waterloo
Waterloo, Canada   N2L 3G1

ABSTRACT

Despite its progress over the last three decades, there is an obvious imbalance between the extraordinary growth of structural optimization theory and its negligible application to professional practice. What are the causes of this imbalance ? How can it be corrected ?

This paper attempts to answer these questions by surveying authoritative opinions on the subject. A reappraisal of the optimization nature, theory and practice suggests that the disappointing penetration of structural optimization in professional engineering is largely due to the confusion between theoretical and engineering models. It is suggested that an important step forward could be taken by focusing on *engineering* rather than *mathematical optimization*, i. e. by starting from real problems in search of their optimal solutions, instead of starting from optimization algorithms in search of practical applications, as currently is often the case.

1. Introduction

During the last three decades the optimization literature has known a spectacular development, as illustrated by specialty periodicals (*Jl.Optimization Theory and Applications, Engng Optimization, Structural Optimization*) and numerous topical conferences, symposia and workshops. We note among the monographs on structural optimization those by Prager (1971), Carmichael (1981), Kirsch (1981), Siddall (1983), Farkas (1984), Haftka et al. (1985, 1990, 1992), Vanderplaats (1984), Save and Prager (1985), Arora (1989), Rozvany (1989), Borkowski and Jendo (1990); and among the edited compilations those by Cohn (1969), Gallagher and Zienkiewicz (1973), Sawczuk and Mroz (1975), Lev (1981), Morris (1982), Atrek et al. (1984), Brandt (1984 &1989) and Mota Soares (1987). Optimization from a systems engineering viewpoint finds its place in numerous texts such as those by Rubinstein (1976), Blockley (1980), Carmichael (1981), DeNeufville (1990) and others. Significant optimization work is connected to structural safety, decision theory and reliability-based optimization, as in the books by Turkstra (1971), Lind (1972), Madsden et al. (1986), Melchers (1987), Thoft-Christensen (1987,1989) and Frangopol (1988).

The field has been frequently and widely surveyed from different angles. Among the more insightful and comprehensive surveys we note (somewhat subjectively) those by Schmit (1960, 1969, 1981), Lev (1981), Ashley (1982), Vanderplaats (1982) and Olhoff and Taylor (1983). Some major surveys on special topics are due to to Rosenblueth and

G. I. N. Rozvany (ed.), Optimization of Large Structural Systems, Vol. II, 843–862.

Esteva (1972), Venkayya et al. (1973), Gellatly and Dupree (1976), Moses (1976), Fleury (1981), Templeman (1983, 1988), Hörnlein (1987) and Kirsch (1989). References listed by surveys and exhaustive bibliographies by Brandt (1984), Rozvany (1989) and Thoft-Christensen (1990) place a very conservative estimate of the optimization literature at some 50 surveys, 150 books, over 500 papers on reliability-based and code optimization and over 2000 papers on deterministic optimization.

Can potential users cope with this veritable information explosion ? Do they benefit from the promise of optimization theory ? It is fair to say that from the viewpoint of practicing structural engineers at large, and particularly those engaged in civil engineering work, the answers to both questions are negative. Convincing testimonies on the virtual absence of structural optimization use by civil and aeronautical designers emerge from direct questionnaires to the profession presented by Lev (1981) and Ashley (1982). The latter points-out that :

..." the number of optimization users is shockingly small ..."   and " thus the evidence accumulates regarding the curious gulf between what might be and what is ..." *(Ashley, 1982, p.9)* .

..."Despite the demonstrated success of structural synthesis it remains difficult to identify industrial organizations that utilize formal optimization techniques to a significant and continuing degree. It must be a source of frustration to developers of this technology that millions of dollars are currently invested in computer graphics and computer-aided design, while the opportunity to fully automate major portions of the design process is being virtually ignored " *(Vanderplaats, 1982, p.996)*.

The imbalance between the extraordinary advances of optimization theory and its modest application has not passed unnoticed: on the occasion of a NATO - ASI (Mota Soares, ed. 1986), an ample panel discussion debated the acceptance of optimization technology, with contributions from leading authorities in the field (Sobieski et al., 1986). Berke's concise but comprehensive discussion addresses the essential questions : for whom, why/ why not, where and how ?   He identifies the potential clients (major aero industries, civil engineering consultants and small companies, benefits (feasible, optimal and/ or automated designs), difficulties (knowledge, training expense, mistrust of CAD), doubts (effective benefits, reliability, methodologies), fields of application (conceptual studies, preliminary or final designs, special problems), and means for expanded use (involvement of governments, universities and researchers, use of artificial intelligence and knowledge-based systems and development of "user-seductive" codes). Roles of artificial intelligence, user-friendly codes, interactive CAD, computer graphics and of algorithmic features were among the additional topics raised by other panel participants.

Developments over the last three decades suggest that the widening gap is due to :
(i) the nature of optimization, (ii) the problems of its theory, and (iii) the special problems of its practical applications. To these three categories pertain some questions to which clear answers are expected by structural designers :
•what are the advantges of structural optimization ? what are its shortcomings and limitations ?
• what has optimization to offer the engineering practice ?
• what are the classes of problems it can solve and who are its potential users ?
• how good and how relevant are optimal design solutions ?
• how does a structural designer start using optimization techniques when faced with a literature of over 150 books and 2500 papers ?
•how can a designer decide on the theoretical and operational puzzles unsolved by the experts?
• more specifically, how does one approach a specific structural problem, when

845

confronted with a bewildering range of concepts, formulations and methods ?
• how does one select the right approach and software for any given structural design problem ?
• what is the role of structural standards in optimization ?
• what are the practical lessons from the last 30 years of research ?
• what are the prospects of optimization applications in the short and long terms ?

The object of the survey is to document some possible answers to the above questions. This is attempted by reexaminimg the nature, theory and practice of structural optimization as reflected by the published opinion of recognized authorities on the subject.

2. The Nature of Structural Optimization

2.1 ENGINEERING VS. MATHEMATICAL OPTIMIZATION

Optimization is a quest for excellence in design and optimization methods are the paths to achieve it. Optimization methods are applicable to all problems for which alternative solutions are possible. Common to all optimization problems is that, being invested with a specific goal , their solutions are directed toward relative " perfection". Although "perfection" is not a realistic possiblity, it represents a guiding ideal and a yardstick against which feasible but non-optimal alternatives can be measured.

The nature of structural optimization is readily apparent from the optimization process scheme in fig.1(a). The "best" structural design may be part of a more complex engineering system. The actual structural problem is idealized into a mathematical model which operates with deterministic, probabilistic or fuzzy sets of variables. The model links the geometric, loading and material variables with some function(s) of merit and constraints.

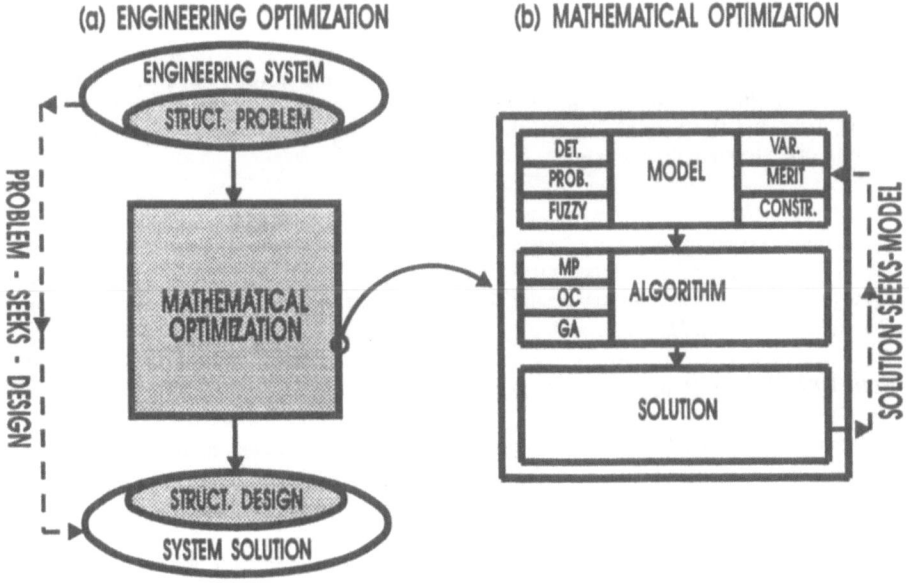

Fig. 1. *Engineering vs. mathematical optimization*

The mathematical optimization problem consists of finding the variable values that yield extreme figures of the merit and do not violate the constraints. This is achieved by mathematical programming, MP (Schmit 1969, Borkowski and Jendo 1990), optimality criteria, OC, (Venkayya et al., 1973, Save and Prager, 1985) or genetic algorithms, GA, (Goldberg, 1988). Fig.1(b) suggests that mathematical optimization is a processor of the model data that yields a mathematical solution by using an appropriate algorithm. Finally, the optimal solution of the mathematical model is evaluated and adapted to the actual structural problem.

It is important to note that books on structural optimization (e.g. Carmichael, 1981, Siddall, 1983, Vanderplaats, 1984, Arora, 1989) are organized around and mainly deal with the central part of the schematic process in fig.1(a), i.e. the mathematical optimization. Although the literature covers various aspects of this central part, its major preoccupation is the development of successful algorithms. Practical demonstration of their success starts from the existence of a mathematical solution in search of an engineering problem that fits the mathematical model. This may be called a " solution-seeks-model" approach (fig.1b).

To the professional designer mathematical optimization is only a vehicle on the road that leads from a structural problem to its optimal design. To him engineering optimization starts with a *real*, physical problem and ends with a *real* design, in what may be referred to as a " problem-seeks-design" approach (fig.1a). In this approach mathematical optimization is an operator that searches the desired optimal design for the given structural problem, i.e. it is only a link between the real problem and its real solution.

The most serious roadblocks in the application of optimization theory to structural design practice appear to be the complexity of mathematical optimization (an indispensable tool), and the identification problem, stemming from the difference between engineering and mathematical optimization approaches ("solution-seeks-problem" vs. "problem -seeks-solution"). In a broader sense these difficulties reflect the essential differences between the worlds of science and engineering, or the theory and practice (Blockley, 1980, Duddeck,1987, Addis,1990).

2.2 OPTIMIZATION ROLES

Figure 1 has identified optimization as a process whereby structural input data (geometry, loadings, materials) are synthesized into optimal output via a mathematical operator. It is this functional feature that opens the objective approaches of system and control theories to structural design.

...." In other engineering branches optimal control theory has elevated the "art" of design to a status approaching a systematic and exact "science". This has occured despite the everpresent yet necessary "engineering judgement" which recognizes the existence in all designs of certain intangible quantities that defy precise mathematical statements. This theory offers some advantages to structural design..." *(Carmichael, 1981, p.91).*

Some authors caution against over-confidence in automated optimization.

..."It cannot be over-emphasized that no general method of optimization, no matter how robust and powerful, can be used as a substitute for good engineering intuition. The best strategy is one that emphasizes both as complements of each other " *(Haftka and Kamat, 1991, p.13).*

More generally, what can structural optimization do for the design practice ? Its roles, or functions are related to all major aspects (i. e. (i) problem formulation (modelling), (ii) processing and (iii) solution), and may be summarized as follows :

(i) systematization of design approach ; goal-orientation ; goodness standard setting ;
(ii) problem solving ; tedium reduction / elimination ;
(iii) decision-making help ; trend finding ; preliminary design guidance.

How good and valid are optimal solutions if the process functions properly ?

2.3 VALUE OF OPTIMIZATION

The value of optimization follows from its goal (best) and object (design). Unlike the scientific search for understanding (what is ?), design search for best (what ought to be ?) is more concerned with goodness than truth. How true a design is ? – is a meaningless question. However, how valid or good a design is ? – are pertinent questions, worthy of discussion in their relative context. We focus on the latter question first and briefly examine the former in the next section.

The main considerations on optimal solutions value are : (i) problems relevance, (ii) measure of "goodness" and (iii) solution relativity. Frequent references to these aspects are made in the literature.

...." it must be emphasized that the "best" is selected from the alternative models conceived. For most problems of reasonable complexity there is no assurance that the best alternative has been included in the list of feasible alternatives, namely the list is not exhaustive..." *(Rubinstein, 1975, p.222).*

..."Seldom if ever the precise , absolute best design will be obtained using numerical optimization. Expectations of achieving the absolute "best" design will invariably lead to maximum disappointement".... *(Vanderplaats, 1984, p. 21).*

Clearly the crucial ingredient that determines the value of a solution is the problem relevance , as identified by the variable hierarchy and its modelling. Solving the "right" problem appears to be a trivial , self-evident, but often forgotten requirement of any design. Or, as graphically but aptly a wise old engineer asked :

...." what good it does to optimize the hell out of a bad idea ? "........ *(Ashley, 1982, p.22).*

The suitability of any design may be established on a "goodness scale" with regard to the optimal solution obtained under identical conditions. As measure of goodness the differences or ratios between the costs, weights, performance or other merit indices may be considered.

We note that just as any design may be evaluated relative to an optimum design, so the intrinsic value of the optimum design itself is relative to the specific variables, constraints and merit functions considered. The relativity of an optimal solution restricts its value to the confines of the physical and mathematical validity of its models, or simply put an optimal design can only be as good as each and all of its ingredients.

Is then an absolute optimum conceivable ? Is it possible ?

2.4 VALIDITY OF OPTIMIZATION

An *absolute optimum* design could only be conceived if , in addition to a *perfectly correct algorithm* , all *components* of the process in fig.1 *are accounted for, are relevant, realistic and accurate* . For example : optimization of a component representing only 10% of a system may be irrelevant ; omitting some critical constraints may result in a meaningless, albeit mathematically correct optimum ; over-simplified material models may be easier to handle, but correponding optimum may not be representative of the real structure ; similarly, optimal solutions of dubious safety or general validity may result from adoption of improper target constants for materials, behavior states, etc.. or of error-prone optimization algorithms.

Thus, it is clear that, although conceivable, an absolute optimum design that would satisfy all the above requirements is a virtual impossibility. Again, any optimal structure is a relative optimum, i.e. it is the best only with regard to its specified merit and well defined constraints.

The relative value of optima and the uncertainties related to their search have raised some questions of principle about the validity and accuracy of optimization.

..." Essentially the optimization procedure is demanding high precision about a complex system. Zadeh, in 1973, had something to say in this matter in his *principle of incompatibility* :

As the complexity of the system increases, our ability to make precise and yet significant statements about its behavior diminishes until a threshold is reached beyond which precision and significance (or relevance) become almost mutually exclusive characteristics "....

....." In structural optimization procedures, precision or uncertainty, member location and connectivity, material parameters, loading and boundary conditions are essential ; the analytical tools are an example in precision ; the best of all worlds that the solution provides is a precise answer of doubtful significance." *(Brown, 1990, p. 160).*

The literature offers ample reference on the irrelevance of excessive rigor in optimization and the essential difference between scientific and engineering endeavour :

...." It has never been the aim of engineering sciences to portray nature. We want to find a model that works, that covers the entire complexity of reality for a class of structures and whose accuracy is consistent with the overall behavior scatter of a great number of similar structures "....
(Duddeck, 1977).

..." The difference between the work of the engineering designer and that of the scientific researcher is ... due to differing consequences of error in the predictions they make. Structural designers are interested in cautious, safe theories ; scientists are interested in accurate theories ; both are interested in solving problems. Structural engineering scientists tend to be dominated by their interest of accuracy and, as a result, frown upon " rules of thumb" as intelectually inferior. Structural designs rely upon " rules of thumb", when organized science lets them down, as it frequently does".... *(Blockley, 1980, p.76).*

..."Perfectionists....deplore using the word " optimal" to describe the results of searches wherein some mathematical formalities are disregarded, and for.....convenient approximations....are employed to speed convergence or reduce the volume of computation....The operative facts for engineering utility are that designs produced in such " sub-optimal " ways are often superior to those produced by standard methods, notably in cases where rigorous optimization might be impractical....." *(Ashley, 1981, p.8)*

We may conclude that the validity of structural optimization is dependent on the balance between the mathematical models and algorithms it uses, and the physical reality of the structural problem and its practical solutions. Excess of rigor is no virtue when adequate reliability, function and economy are the governing design criteria.

Finally, given the nature of optimization, its roles, value and validity and that conventional design is a process of optimization by trial-and-error, it would be of interest to consider the "credits and debits " of structural optimization, its advantages and disadvantages (Templeman (1976), Blockley (1980), Ashley (1981), Vanderplaats (1984), and others).

3. Optimization Theory

3.1 VARIABLE UNCERTAINTY

Consideration of uncertainty has introduced probabilistic aspects in structural design and developed into what is known today as Reliability-Based-Optimization (Moses, 1969, Madsden et al., 1986, Thoft-Christensen, 1987, 1989). If a set of variables X of known statistical properties are to be found so that some merit function Z and probabilities of survival P_s should satisfy their target merit Z_o and reliability R, respectively, a typical minimum-cost, reliability-based optimization for a single limit-state may be formulated as :

$$\min \{ Z(X) \mid P_s(X) \geq R \}$$

It can be shown that a large variety of constrained and unconstrained optimization types result if more than one limit state is considered and some intricate problems arise in the probabilistic analyses of limit state constraints, their correlation, the selection of target reliabilities and the related solution algorithms (Cohn and Parimi, 1973, Parimi and Cohn, 1975).

A sample of the alternative optimization problems is shown in Table 1 for plane frames of known member layout, probable loading and variations of member strength means M_p, such that adequate reliabilities R_1 and R_2 be ensured against structural collapse and/or unserviceability and cost function not to exceed a target value Z_o. Optimal performance corresponds to maximum (+)probability of survival (+ P_{s1} or +P_{s2}) and optimal economy corresponds to minimum (-) cost (-Z) (Cohn and Parimi, 1973).

TABLE 1 *Alternative Formulations of Probabilistic Structural Design (two limit states)*

Prob. (No.)	Specified Parameters	Objective Function	Constraints	Solutions
1	R_1, R_2	$-Z$	$P_{s1} \geq R_1$) $P_{s2} \geq R_2$)	$\overline{M}_p, Z$
2	R_2, Z_0	$+P_{s1}$	$\{ P_{s2} \geq R_2 \}$ $\{ Z \leq Z_0 \}$	$\overline{M}_p, P_{s1}$
3	R_1, Z_0	$+P_{s2}$	$\{ P_{s1} \geq R_1 \}$ $\{ Z \leq Z_0 \}$	$\overline{M}_p, P_{s2}$
4	R_1	$+P_{s2}$	$P_{s1} \leq R_1$	$\overline{M}_p, P_{s2}$
5	R_1	$-Z$	$P_{s1} \geq R_1$	$\overline{M}_p, Z$
6	R_2	$-Z$	$P_{s2} \geq R_2$	$\overline{M}_p, Z$
7	C_1, C_2	$-[Z-C_1 P_{s1}-C_2 P_{s2}]$	—	$\overline{M}_p, Z, P_{s1}, P_{s2}$
8	C_2/C_1	$+[P_{s1}+(C_2/C_1) P_{s2}]$	$Z \leq Z_0$	$\overline{M}_p, P_{s1}, P_{s2}$
9	C_1, C_2	$-[Z-C_1 P_{s1}]$	$P_{s2} \geq R_2$	$\overline{M}_p, Z, P_{s1}$
10	C_2, R_1	$-[Z-C_2 P_{s2}]$	$P_{s1} \geq R_1$	$\overline{M}_p, Z, P_{s2}$
11	C_1	$-[Z-C_1 P_{s1}]$	—	$\overline{M}_p, Z, P_{s1}$
12	C_2	$-[Z-C_2 P_{s2}]$	—	$\overline{M}_p, Z, P_{s2}$

Important progress in reliability-based optimization is recorded in the bibliography compiled by Thoft-Christensen (1990) and some texts and conference proceedings on the subject (Thoft-Christensen 1982, 1987, 1989). Along with the research optimism expressed by these publications, some reservations are raised on the philosophical, theoretical and practical implications of reliability-based design and optimization (Templeman, 1988). These refer to the meaning of the probability of failure concept, the difficulties of system (as opposed to component) reliability analysis, the modelling uncertainties, relevance of the approach, etc...

...." Truly safe structures can only be assumed by design and fabrication procedures which thoroughly examine and accomodate the real causes of structural failures. The probability of failure approach forms only a small part of these."

..." There now seems to be an overwhelming momentum driving these methods towards becoming the basis of new structural design codes and a recommended method used in practical design.....A note of caution : though momentum might succeed in introducing a new design approach to the structural engineering profession, momentum alone does not guarantee validity"..... *(Templeman, 1988, p.274).*

We must also note attempts to adapt to structural design and optimization the concept of "fuzziness", in order to deal with problems the outcome of which is neither certain, nor uncertain, but imprecise. The fuzzy models of these problems range between those of deterministic and probabilistic optimization and, at this time appear to be mostly of academic interest (Blockley, 1980, Munro, 1984).

3.2 SINGLE OR MULTI-OBJECTIVE OPTIMIZATION

Some structural optimization problems require the satisfaction of more than one objective and are referred to as " multi-objective", " multi-criteria", or " vector optimization" (Carmichael, 1981, Duckstein, 1984, Eschenauer, 1990).

In some cases when desired objectives are conflicting (e. g. minimum cost and maximum safety) a real optimum solution may not exist and an " acceptable" solution or Pareto optimum is sought instead. This corresponds to a set of "non-inferior" or efficient solutions which cannot improve any objective function without worsening some other objective function. Multi-objective optimization enables the exploration of broader ranges of alternatives than single objective models, explicit trade-offs between conflicting objectives and inclusions of new design criteria (Duckstein, 1984).

3.3 SATISFICING AND OPTIMIZING

The value of strict optimization is questioned on practical and philosophical grounds by Vanderplaats (1982) and Brown (1990), respectively. They find sufficient reasons for recommending " near-optimum" or "trade-off" or "satisficing" instead of strict optimal solutions.

..." Although an inequality $g(x) \leq 0$ is mathematically active only when $g(x)=0$ precisely, in an engineering sense the value $g=-0.05$ may be considered critical ; this is because material properties, failure criteria and loading environments are virtually never known with greater precision. Consequently, from an engineering point of view, a mathematically precise optimum is seldom meaningful "...*(Vanderplaats, 1982, p. 996).*

....." At the moment optimizing is an act of functional and limited rationality ; satisficing has the

appearance of global rationality and respects the Principle of Bounded Rationality. Certainly an optimization scheme ignores the principle of incompatibility ; satisficing may respect the principle ".
...."Future research should attempt to introduce realism into the optimization scheme, but, in the process should anticipate a reduced precision in objective, constraints and solution. The measures for such studies should be the extent that realism co-exists with rigorous analysis. Cooperation betweeen practitioners and researchers seems inevitable if realism is to be tied to what goes on in practice. Indeed, studies that probe the extent the present-day design is a satisficing procedure with convergent and robust methodologies should be examined." (Brown, 1990, p.162).

The satisficing concept introduced by Simon (1957) is also discussed by Rosenblueth (1976) and Blockley (1980), but has yet to develop into practical operational procedures.

3.4   MATHEMATICAL PROGRAMMING, OPTIMALITY CRITERIA , GENETIC ALGORITHMS

Explicit formulations of design objectives and constraints have resulted in the *direct methods* of optimization, described by a vast number of *mathematical programming* studies. Convincingly advocated by Schmit (1963,1969, 1981) these have been amply covered and are surveyed in many existing monographs (e.g. Kirsch, 1981, Vanderplaats, 1984, Borkowsky and Jendo, 1990, etc..). Thought to be the most general and powerful optimization approach, mathematical programming has severe limitations, mostly related to convergence and computational efficiency, for large-scale problems.
The indirect (or optimality criteria) methods are problem-dependent, focus on the known or assumed features of the optimum, and search for a solution in its vicinity.

..."The basic concept behind optimality criteria is the rejection of the generality of mathematical programming and the utilization of the physical characteristics of the structural optimization problem to generate an approach of somewhat limited applicability, but of the greatest computational efficiency." (Gellatly and Dupree, 1976, p.79).

Seminal contributions to this approach by Prager (1971), Venkayya et al. (1973), Save and Prager (1985), Rozvany (1989), etc.. have demonstrated its remarkable potential for practical applications. The optimization literature tends to present side-by-side mathematical programming and optimality criteria (Hörnlein, 1987), generally as conflicting methods (Morris, 1982, Brandt, 1984, 1989, Arora, 1989, etc).
Attempts at explaining the deep connection between mathematical programming and optimality criteria approaches are found in a definitive study by Fleury (1982) and in Carmichael's book (1981). Although the theoretical equivalence (duality) of the two methods is well established, the practical implications of their "reconciliation" still remain to be explored.

..." It appears that a combination of approximate (MP) concepts and dual (OC) methods has provided the "best of both worlds" for a large class of design problems"... (Vanderplaats 1982, p.995).

It is surprising that a method simulating the "brute force" approach to optimization shows extraordinary promise in both conceptual simplicity and computational efficiency : the *genetic algorithm* approach (Goldberg, 1989) systematically modifies tentative solutions of a design problem, so that each trial is of improved fitness, with regard to the preceding ones. (Note that stopping the algorithm short of reaching a real optimum ensures a "satisficing" process, with a possible near-optimum solution).

...." Four differences -- direct use of costing, search from a population, blindness to auxiliary information and randomized operators -- contribute to a genetic algorithm's robustness and resulting advantages over other commonly used techniques"... (*Goldberg*, 1989, p.10).

Optimization is critically dependent on the existence and capabilities of digital computers. Which software should be preferred ?

" Comparison of optimization algorithms and software is a multi-criteria decision problem which has qualitative as well as quantitative aspects" (*Ragsdell* , 1984).

Surveys of comparative results from tests on optimization codes are presented by Schittkowski (1980), Ragsdell (1984), Hörnlein (1987) and others. The measures of merit are : (i) generality, (ii) robustness, (iii) efficiency (computer time), (iv) user-friendliness, (v) reliability, (vi) global convergence, (vii) sensitivity to initial design, (viii) accuracy of final design. Additional technical parameters that may be considered are : (i) number of function and constraint evaluations, (ii) number of iterations, (iii) basic operations count and (iv) storage requirements.

It is noted that, however measured, the relative performance of optimization codes depends on the computer and compiler used for testing. Using a carefully selected set of tests and percentage weights to nine merit criteria, Schittkowski (1980) has ranked 26 types of nonlinear programming. Ragsdell (1984) surveys the results, gives indications on code availability and hopes that, " although all of these codes are doomed to obsolescence, much more powerful algorithms and codes will be available in the future". A listing of 30 codes and their main features is appended in a study by Hörnlein (1987).

3.6 VARIABLE HIERARCHY, OPTIMIZATION LEVELS

It has been long recognized that the variables to be found through optimization determine the type of problem to be solved. In civil engineering structures loadings and sometimes materials are preassigned parameters. In descending generality, geometry (and possible material) variables include : (i) structural systems ; (ii) structure layout (configuration) ; (iii) member sizes and (iv) section detailing.

The higher the variable level in this hierarchy, the broader the scope of optimization problem and the better its solution. On the other hand its formulation will be more complex and the solving process more difficult. This is why much of the published optimization work refers to the lower levels in the hierarchy and very little is available on the optimization at the higher levels.

The role of variable hierarchy in optimization was recognized by Schmit (1963, 1969) and thoroughly discussed among others by Templeman (1976), Gellatly and Dupree (1976) and Svanberg (Morris, 1982, chapter 14).

3.7 STRUCTURAL CODE OPTIMIZATION

Existing structural codes of practice set minimum standards for civil engineering design and compliance with their provisions is usually supposed to achieve the best a professional designer can do in his current practice. They also play a role in specifying constraints and safety targets in structural optimization. Commenting on the interaction between research, codes and practice, Siess (1960) had this to say :

...." Although the interposition of a code between the sources of knowledge and the people who must use that knowledge in practice is undoubtedly restrictive, it is not necessarily a universal evil. Although it is most desirable that knowledge gained from any source be utilized as fully as possible and for the benefit of all, it is equally important that this knowledge be used correctly and its misuse not be permitted. For example, all research is not equally good ; all research results are not equally valid, and much research cannot be applied directly to practice except by an engineer having considerable experience and a sense of caution based on that experience.

Similarly, the knowledge gained from experience is frequently limited in scope, and there is always the possibility that it may be extrapolated to applications for which it is not valid. In many instances, the role of research is to determine the range of applicability of knowledge gained from experience, or to extend that range, or to define its limits so that it can be used safely by all. Thus, we see that meeting of research and practice in the code is potentially capable of deriving the best from each and presenting to the profession a synthesis of contemporary knowledge and current practices"... (*Siess, 1960, 1108-1109*).

However admirable the idea of codes as a synthesis of engineering research and practice, the quest for such an ideal does not justify the almost perpetual changes of national codes, their diverging provisions on the same subject and conflicting principles of structural design for different materials. At the present state-of-the-art in material sciences and structural mechanics, how can identical structures, under identical conditions, result in vastly different designs when subjected to different national standards ? What are the benefits derived from the all too frequent code changes ?

The most recent structural codes are based on the concept of "limit states", (serviceability impairment or structural failure) and specified minimum reliabilities against them. Most of the existing codes require that for any part of a structure, under the worst conceivable loading conditions, their effects should not exceed the corresponding strengths or deformation capacities. These criteria are satisfied at the price of some gross simplifications (e. g. linear combination of service effects assumed to predict ultimate limit state response of structures compared to plastic capacities of members or sections, etc.).

Basic design criteria are cast in a deterministic format, with some specified (partial load and resistance) factors based on probabilistic considerations, aiming at achieving some desired target reliabilities. There is an established "reliability hierarchy" of structural codes (Melchers, 1987) of which the above is the first level. A second level introduces "second-moment area methods" , based on normal distributions and resulting in nominal failure probabilities. The third level, " exact methods" makes use of more accurate probabilistic analysis and yields "true" failure probabilities. Level 4, "decision method", combines advanced probabilistic techniques with economic and decision analysis to result in minimum cost or maximum benefit.

There is considerable knowledge on the structural reliability and its standard code implications (e.g. Turkstra, 1970, Madsden et al., 1986, Melchers, 1987) which deserves a separate, in-depth examination. Here we note two basic flaws in the current code developments :

(i) All existing codes, supposed to ensure reliability of complete structural systems, only require that reliability of their " weakest-link" be no less than a minimum target value ;

(ii) Target reliabilities are determined so that designs by a new code be equivalent to those of the preceding code, through a process referred to as "calibration".

It thus appears that an improved code deals only with a part, rather than the whole structure, and that the expected improvements are mostly academic, because potential practical benefits are negated by the calibration process. More recently the question of probability "validation" (i. e. how reliable is the probabilistic approach ?) has been raised :

..." Actual factors of safety and safety margins in the deterministic code are random variables. Probabilistic studies attempt to bridge the gap between the code and reality at the price of adding another layer of complexity to traditional methods".

..." The..... study can be applied.......to arrive finally at statements that code provisions do or do not appear consistent. Nothing is said about the actual reliability of such probability statements except that *they appear reasonable".... (Benjamin, 1989, p.313).*

"Calibration" and "validation" reflect the need to test results based on new codes against the pragmatic results of old codes and against the theoretical truth. There appears to be considerable uncertainty both at the operational and principial levels of reliability application to structural design and codes. Simply put, reliability-based design is neither as rational, nor as rigorous as it appears.

Despite the conceptual difficulties of existing codes (Blockley, 1980), efforts are continuing for the development of codes at levels 3 and 4 in the above mentioned hierarchy. These are intended as fundamental model-codes, are addressed to reliability experts only and do not aim at promoting the general understanding of reliability concepts (Ditlevsen and Madsden, 1989).

It would appear that there is obvious merit in systematically improving future codes of practice by providing them with a sounder scientific basis and methodology. Some theoretical foundations of structural code optimization are given in the pioneering work by Lind (1972), Rosenblueth (1976), Rosenblueth and Esteva (1972). Further work towards an "ideal" code type via optimizing or satisficing techniques should bring new relevance to the structural codes of the future.

We close this section by noting that code optimization, not only is the proper connection between optimization research and practice (hence also a proper transition to the next section), but an extension of the variable hierarchy at § 3.6 to include *classes of structures* in the realm of optimization variables. A further and realistic extension of this hierarchy would add *complex engineering systems* to the most meaningful problems of engineering optimization. An example of such a problem would be the global optimization of a building, including its electrical, mechanical and environmental systems, as well as its structural system (ASCE, 1978).

4. Optimization Practice

4.1 STATE-OF-THE PRACTICE IN STRUCTURAL OPTIMIZATION

The great strides in the early development stages of structural optimization theory were accompanied by great hopes on their engineering application. Vanderplaats (1982) pertinent assessment of the situation at the time seems to retain its essential validity to this day :

..." By late 1960's it was becoming apparent that structural synthesis was not being embraced by the professional community, as many people, including this new convert, expected it would be. Some plausible explanations can be offered.

First, design is far more complex than analysis , and at that time the finite element method was just becoming generally accepted after approximately 15 years of development. A new design methodology takes longer to gain general acceptance.

Second, structural synthesis represents an integration of engineering and operations research disciplines. Because MP were unknown to the vast majority of engineering researchers, educators and practitioners, it was unreasonable to expect immediate and widespread acceptance, particularly since NLP itself was not a mature discipline. Each explanation was reasonable and required only time and patience to overcome. It was also becoming recognized, however, that there might be a fundamental limit of this new technology. The simplest problem often needed to be analyzed hundred of times during optimization. If this analysis were time-consuming, as is often the case for large FEM, the cost of optimization quickly become prohibitive.

Even if the efficiency of the optimization algorithms could be improved to an acceptable level, there were good theoretical arguments to indicate that the cost of optimization was not constant and did not even increase linearly as the number of variables increased, but instead increased at a quadratic, or in some cases exponential rate."

" Although the generality of MP made this a most attractive design tool, practicality dictated that it was limited to problems with only a few design variables." *(Vanderplaats, 1982, p.993)*

At about the same time , Templeman (1983) notes :

..." The major reason why so little of the vast research output in structural optimization has filtered through to design practice is that very little of it satisfied the specific needs of its potential users".... *(Templeman, 1983, p.2421).*

More recently research in structural optimization has expanded considerably, while highlighting some new directions.However, with the exception of proprietary designs in specialized industries (aeronautical, auto, off-shore, nuclear, etc.)there still are very few documented engineering applications ; among these we note the optimization software developed for tall building steel structures (Grierson and Cameron, 1987) and reinforced concrete structures (Spires and Arora, 1989).

Published information on practical applications of optimization shows much progress on structural members (Farkas, 1984), but less on structural systems (Lev, 1981), limited to mostly academic examples that illustrate algorithmic use. A catalogue of all published optimization applications, to appear elsewhere, would reveal that, despite the fine prospects of structural optimization foreseen by many early promoters, its engineering applications had continued to remain insignificant.

Even if the practical value of optimization were positively demonstrated, the structural designer ready for it would be faced with its vast literature and, before proceeding, would have to make some puzzling choices :

• deterministic, probabilistic or fuzzy optimization ?
• single or multiple objectives ?
• optimizing or satisficing ?
• mathematical programming, optimality criteria or genetic algorithms ?
• automated or interactive, general-purpose or problem-specific software ?
• which is the optimal package for the problem at hand ?
• member, structure, structural configuration or complex systems optimization ?

Deciding how to solve a problem would seem a hopeless task unless some simple recommendations could guide the designer in facing his major difficulties, i.e. : (i) problem identification ; (ii) mathematical modelling and (iii)   solution implementation.

4.2 STRUCTURAL PROBLEM IDENTIFICATION

We mentioned that most of the published optimization literature reflects the mathematical core of the design problem (fig.1) and, consequently, optimization examples are only algorithmic illustrations. Structural engineers willing to optimize their designs have no alternative but to learn many optimization procedures and then decide which of these fit their real problems.

Clearly, this "solution-seeks-model" approach is unrealistic and probably is one of the main reasons why practice cannot take advantage of optimization theory. The design process must start with the identification of a problem by its physical aspects, continue by matching the appropriate algorithm and stop when an optimum structural design is found.

It is felt that presentation of optimization applications following the "problem-seeks-design" approach could expand the understanding and use of optimization by focusing designers' attention to both physical and mathematical aspects of their problems, with due priority to the former. This suggests that a structural optimization problem may be identified by its physical features in Table 2 and then solved by an appropriate algorithmic match.

TABLE 2   *Factors in Problem Identification*

Uncertainty Level	Variable and Parameters			Formulation			Algorithm Code
	Geometry	Loading	Material	Objective	Limit States	Constraints	
Deterministic	Section	*Static*	Steel	*Single*	Single	SLS • Stress • Deflection • Cracking • Fatigue • Buckling •*Local damage*	*M P*
	Member		RC /PC /PPC				
			Composite	Multiple	*Multiple*		
	Structure						O C
Probabilistic	Struct.Layout	Dynamic	Elastic	Performance Economy • Weight • *Cost*	SLS	ULS	
			Elastic-Plastic			• *Collapse*	
	ComplexSyst.		Str.- Hard.	CostParameter •Relative costs •Expected life	ULS	• Ductility	G A
			Str.- Soft.		Other	• Instability	

A particular structural optimization problem is defined when appropriate choices in each column of Table 2 are made. For example the italicized entries in Table 2 identify a *probabilistic optimization* , of a *given structure,* under *static loading* made of *reinforced concrete (elastic-plastic model*) ; for a *single objective (minimum cost), multiple limit states (serviceability and ultimate limit states)* with *local damage* as serviceability constraints and total collapse as ultimate limit state constraint, and the solution obtained by an *MP algorithm* (Cohn and Parimi, 1973).

4.3   MODELLING AND SOLVING PRACTICE

Problem identification, modelling and solving are closely related inpractice.Oftenapproximate models are substituted for the design of real structures because some efficient solvers are available, or simpler algorithms are adopted to cope with computational inefficiency of complex structural systems. In examining the trade-off between basic algorithmic features, Arora notes that while approximate methods are efficient, but unreliable, inaccurate and lacking generality, globally convergent (robust) methods are reliable, accurate and general, but need more computational effort. He recommends relaxation of the efficiency requirement for the optimization of practical systems (Sobieski et al. 1987)

When these related aspects are clear to the structural engineer, optimization can deliver on the promise of a logical approach to automated design, unbiased by undue reliance on intuition. When heuristic considerations affect the design outcome, or features of intermediate feasible solutions are of interest, the interactive process becomes very useful. Haug points out that "human control of the optimization process can (i) aid in formulation refinement, (ii) identify onset of mesh distortion in shape optimization and (iii) evaluate progress and / or difficulties and permit the termination of the iterative optimization process" (Sobieski et al., 1987).

Some hopeful discussion centers around the possible impact of artificial intelligence and expert systems on optimization practice (Brotchie et al., 1990, Sobieski, Fleury, Morris, 1987, etc.). Morris notes that expert systems concepts could aid in modelling the structure, ordering the algorithms and controlling the utility routines.

Although much of the expectations raised by expert systems are yet to materialize, there is little doubt that they are the best and most natural broker between the needs of the engineering practice and the accumulated knowledge in optimization theory. In this regard, a singularly important role of expert systems may be that of providing the ideal medium for proper problem-algorithm matching.

Opinions are divided on the practical merits of general vs. problem-specific codes (e.g. Arora, 1987, Hörnlein, 1987, etc). It is probably fair to say that there are uses for both types, but specific codes should be more appealing to structural designers and more conducive to expanded applications of optimization.

In brief, optimization practice has yet to clarify its preferred options on approximate or exact methods, automated or interactive operation, possible uses of expert systems, optimal code selection, general or specialized programs, etc.

4.4 NEEDS, NEW DIRECTIONS

Numerous recommendations offered for the application of optimization link their optimistic expectations to the features of software distribution and development (robustness, efficiency, user-friendliness, documentation, practicality) and optimization benefits (design sytematization, usefulness, speeding-up design process, assistance in decision-making) (Schmit, 1981, Vanderplaats, 1982, Templeman, 1983, Olhoff and Taylor, 1983, etc). However, the importance of the human factor is properly emphasized :

..." Unless researchers are prepared to step back from the research frontiers of structural optimization and become involved in providing practical design software and unless practicing design organizations are prepared to step forward and sponsor the writing of such software, both sides of the engineering profession are likely to miss-out on an exciting development within the profession, i.e. the incorporation of the computer into the routine work of the design engineer" (Templeman, 1983, p.2421).

In the same vein MP experts are challenged to be more involved in promoting the application of their methods and cooperate with their structural designers colleagues :

...." For a long time, attempts have been made to use the problem inherent properties in constructing the design model. Hence, the use of reciprocal variables, suitable constraint approximations, the linking of variables, even the application of "move limits" are more or less desperate attempts to make the "real life" problems more attractive to the established MP methods. Why not the other way around ?
Thanks to many years of experience in using MP methods, the experts in structural optimization now have such an intimate knowledge of MP theory that they ought to meet the MP experts in order to develop problem-specific procedures.... The school of MP ought to accept this challenge and signal its readiness for interdisciplinary collaboration".... (Hörnlein, 1987, p.912).

Perhaps the conclusions on the role of analysis-design coupling have an expanded, deeper meaning when read in the parallel context of theory-practice, or mathematical-engineering optimization :

...." The manner in which the optimization and analysis programs are coupled together to create the design program will determine the ultimate utility of these techniques, and the engineer's ability to match the optimization algorithm to the design problem usually has the most impact on the overall result "... *(Vanderplaats, 1982, p.996).*

In brief, it seems that future progress in application of optimization requires- first and foremost- an essential "rapprochement" between the specialists in engineering optimization and those in mathematical optimization, as well as a radical improvement of their interaction. A meeting of these two sides, with researchers making the major effort in trying to help professional designers, is indispensable. In this endeavour further applications of interactive optimization, expert systems (problem identification, modelling, software matching, etc.), genetic algorithms and other new research directions may prove their usefulness in time.

Until much more confidence and knowledge building is achieved and simpler, "user-seductive", problem-specific codes are available to better educated and motivated structural designers, optimization will continue to remain an exclusive tool reserved for exceptional problems where it is uniquely beneficial .

5. Summary and Conclusions

5.1 SUMMARY

In this survey, we have deliberately chosen the engineering path along the tortuous alleys of optimization, insisting on the priority and finality of practice while recognizing the indispensability of theory. For indeed, in the words of the wise, *practice without theory is like a body without spirit, i.e. a corpse ; and theory without practice is like a spirit without body, i.e. a ghost.*

1. Optimization is nothing short of a revolutionary design concept that replaces the conventional trial-and-error approach by a systematic, goal -oriented design process. Its essential roles - in research, education, engineering- are as (i) *yardstick of "perfection"* , (ii) *conceptual framework* (problem identification, modelling, solving and evaluating), (iii) *operational device* (automatic or inter-active process) and (iv) *decision-making tool.*

2. There is a widening gap between the progress of optimization theory and optimization practice in structural engineering.

3. Application of optimization in structural design is generally desirable and almost axiomatically accepted as necessary and possible : such frequent views are not proven and need be challenged.

4. Large-scale, practical use of reliability-based optimization is unlikely and quite unjustified, at least in the foreseeable future.

5. The difficulties of transferring the structural optimization theory to practice are due to both objective and subjective causes, i.e. to its nature and the way it is presented to the public. These difficulties are readily traced to the different perspectives of optimization procedures (researchers) and its intended users (engineers).

6. Professional designers cannot cope with the broad spectrum of disciplines involved in optimization, and the complexities of the process and question the value of the efforts to find optimal solutions, given the uncertainties present at all levels.

7. Professional designers are perplexed by the vast optimization literature, its mostly mathematical coverage, the irrelevance of many optimization applications and the conflicting opinions of experts on various approaches to problem solving.

8. Like most consumer goods on the free market, structural optimization will find its natural application when potential users will discover it to be either *highly beneficial* or *indispensable.*

5.2 CONCLUSIONS

Some time ago an optimistic forecast on the application of optimization to aero-space structures was encapsulated by an authority in the field as a *cosmic opportunity* (Ashley, 1982). More realistically, with regard to earth-bound civil engineering structures, we may expect that a sound *terrestrial opportunity* will be (progressively) enhanced by :

In the short term

1. Use of optimization in areas of considerable benefits (off-shore, nuclear and complex structural systems), or where it can produce preliminary designs not otherwise possible by intuition or approximate techniques.
2. Defining trends and solution ranges for classes of optimized structures.
3. Development of optimization software for specific sub-systems and structural components (as opposed to general-purpose packages).
4. Coordinated efforts by specialty journals to compile yearly up-dates of topical bibliographies, optimization software listings and success stories of practical optimization applications.

In the long term

5. A broad, early education of structural engineers in optimization concepts and techniques.
6. Development of "*user-seductive*" codes by expanding both inter-active and automated aspects of the design process, possibly via expert systems for problem identification, formulation, modelling, algorithm selection and solution evaluation.
7. Increasing motivation for and and accessibility of optimization methods by placing special emphasis on realistic engineering applications, large-scale structures and "problem-seeks-solution" (rather than the opposite) types of methods.
8. Researcher-developed, but practice-oriented, user-friendly optimization codes for application to specific classes of structural systems.
9. Development of "near-optimization" or "satisficing" principles and techniques as viable alternatives to formal optimization.
10. Development of "optimal standard codes" theory and of structural standards conforming to it.

Acknowledgments

The financial support of the Natural Sciences and Research Council of Canada under Grant A-4789 is gratefully acknowledged. The author is indebted to D. E. Grierson for the reference on genetic algorithms.

6. References

Addis, W. (1990), Structural Engineering. The Nature of Theory and Design, Ellis Horwood , Chichester, England.
Arora, J. S. (1989), Introduction to Optimum Design, McGraw-Hill, New York.
ASCE (1978), ' Optimization of tall concrete buildings', in ASCE-IABSE monograph on 'Planning and Design of Tall Buildings', vol. CB, 111-144.
Ashley, H. (1982), ' On making things the best- aeronautical uses of optimization', *J. Aircraft*, 19, 5-28.

Atrek, E., Gallagher, R .H.,Ragsdell, K. M. and Zienckiewicz, O. C. (Eds. 1984), New Directions in Optimum Structural Design, John Wiley & Sons, New York.

Benjamin, J. R. (1989), ' The validation problem- historical developments', in A. H. S. Ang (Ed.) " Structural design, analysis and testing', Proc. ASCE Structures Congress, ASCE, N. Y.

Blockley, D. J. (1980) , The Nature of Structural Design and Safety, Ellis Horwood, Chichester, England.

Borkowski, A. and Jendo, S. (1990), Structural Optimization , vol. 2 , Mathematical Programming, Plenum, New York.

Brandt, A. (Ed. 1984), Criteria and Methods of Structural Optimization, Nijhoff Publishers, Dordrecht, The Netherlands.

Brandt, A. (Ed. 1989), Foundations of Optimum Design in Civil Engineering, Nijhoff Publishers, Dordrecht, The Netherlands.

Brotchie, A., Ehmke and Sharpe (1990), 'Intelligent systems for tall buildings', in L. S. Beedle and D. B. Rice (Eds.),' Tall buildings : 2000 and Beyond', Council on Tall Buildings and Urban Habitat, Lehigh University, Bethlehem, PA.

Brown, C. B. (1990), 'Optimizing and satisficing', Structural Safety, 7, 155-163.

Carmichael, D. J. (1981), Structural Modelling and Optimization, Ellis Horwood, Chichester, England.

Cohn, M. Z. (Ed. 1969), An Introduction to Structural Optimization, SM Study No. 1, University of Waterloo, Waterloo, Ont.

Cohn, M. Z. and Parimi, S. R. (1973), ' Multi-criteria probabilistic structural design', Jl. Structural Mechanics, 1, 479-496.

De Neufville, R. (1990), Applied System Analysis, McGraw-Hill, New York.

Ditlevsen, O. and Madsden, H. O. (1989), 'Proposal for a code for the direct use of reliability methods in structural design', IABSE Periodica, Zürich, Switzerland, Nov., 1-20.

Duckstein, L. (1984), ' Multiobjective optimization in structural design : the model choice problem', in Atrek, E. et al. (Eds.), New Directions in Optimum Structural Design, John Wiley & Sons, New York, 459-481.

Duddeck, H. (1977), 'The role of research models and technical models in engineering sciences', in ICOSSAR 2 Proc., München, Germany, 115-118.

Eschenauer, H ., Koski, J. and Osyczka, A. (Eds. 1990), Multicriteria Design Optimization, Springer Verlag, Berlin, Germany.

Farkas, J. (1984), Optimum Design of Metal Structures, Ellis Horwood, Chichester, England.

Fleury, C. (1982), 'Reconciliation of mathematical programming and optimality criteria', in Morris, A. J. (Ed.), Foundations of Structural Optimization, J. Wiley & Sons, New York.

Frangopol, D. M. (Ed. 1988), New Directions in Structural System Reliability', Proc. of Workshop, University of Colorado, Boulder, Colo.

Gallagher, R. H. and Zienkiewicz, O. C. (Eds. 1973), Optimum Structural Design, John Wiley & Sons, New York.

Gellatly, R. A. and Dupree, D. M. (1976), 'Examples of computer-aided optimal design of structures', in Introductory Report, 10th IABSE Congress, Tokyo, Japan, Sept. 1976, IABSE, Zürich, Switzerland, 77-105.

Goldberg, D. E. (1989), Genetic Algorithms in Search Optimization and Machine Learning, Addison-Wesley, Reading, Mass.

Grierson, D. E. and Cameron, G. (1989), 'SODA- Structural Optimal Design and Analysis', User's Manual, WES, New York-Waterloo, Sept.

Haftka, R. T., Gurdal, Z. and Kamat, M. (1990), Elements of Structural Optimization, 2nd Ed., Kluwer Academic Publishers, Dordrecht, The Netherlands.

Hörnlein, H. (1987), 'Take-off in optimum structural design', in Mota Soares, C.A. (Ed.), Computer-Aided Optimal Design, Springer Verlag, Berlin, Germany, 901-927.

Kirsch, U. (1981), Optimum Structural Design, McGraw-Hill, New York.

Kirsch, U. (1989), 'Optimal topologies of structures', *Applied Mechanics Reviews*, 42, 223-239.

Lev, O. (Ed., 1981), Structural Optimization, ASCE Publications, New York.

Lind, N. C. (1972), Theory of Codified Structural Design, SMD, University of Waterloo, Waterloo, Ont.

Madsden, H. O., Krenk, S.and Lind, N. C. (1986), Methods of Structural Safety, Prentice-Hall,Englewood Cliffs, New Jersey.

Melchers, R. E. (1987), Structural Reliability, Analysis and Prediction, Ellis Horwood, Chichester, England.

Morris, A. J. (Ed. 1982), Foundations of Structural Optimization, A Unified Approach, J. Wiley & Sons, New York.

Moses, F. (1968), ' Optimum design for structural safety', Prelim.Publ. 8th IABSE Congress, N. Y. , IABSE Zürich, Switzerland, 163-175.

Moses, F. (1976) , 'System and geometrical optimization for linear and nonlinear structural behavior', in Introductory Report, 10th IABSE Congress, Tokyo, Japan, Sept. 1976, IABSE, Zürich, Switzerland, 61-75.

Mota Soares, C. A. (Ed. 1987), Computer-Aided Optimal Design, Springer Verlag, Berlin, Germany.

Munro, J. (1984), 'Systems engineering', in G. A. O. Davies (Ed.), Mathematical Methods in Engineering, J. Wiley & Sons, New York.

Olhoff, N. and Taylor, J. E. (1983), 'On structural optimization', *ASME J. Applied Mechanics*, 50, 1139-1151.

Parimi, S. R. and Cohn, M. Z. (1975), 'Optimal criteria in probabilistic structural design', in Sawczuk, A. and Mroz, Z. (Eds.), Optimization in Structural Design, Springer Verlag, Berlin, Germany, 278-293.

Prager, W. (1971), 'Introduction to Structural Optimization', Int'l Centre for Mechanical Sciences, CISM, Courses and Lectures No.212. Udine, Italy, Springer Verlag, Berlin, Germany.

Ragsdell, K. M. (1984), ' The utility of nonlinear programming methods for engineering design', in Atrek, E. et al. (Eds) New Directions in Optimum Structural Design, John Wiley & Sons, New York, 385-412.

Rozvany, G. I. N. (1989) Structural Design Via Optimality Criteria. The Prager Approach to Structural Optimization , Kluwer Academic Publishers, The Hague, The Netherlands.

Rosenblueth, E. (1976), ' Towards optimum design through building codes', *ASCE J'l. Struct. Div.*, 102, 591-607.

Rubinstein, M. (1976) , Patterns of Problem Solving, Prentice-Hall, Englewood Cliffs, N.J.

Save, M. and Prager, W. (1985), Structural Optimization, vol.1, Optimality Criteria, Plenum, New York.

Sawczuk, A. and Mroz, Z. (Eds.,1975), Optimization in Structural Design, Springer Verlag, Berlin, Germany.

Schittkowski, K. (1980), ' Nonlinear Programming Codes : Information, Tests, Performance', Lecture Notes in Economics and Mathematical Systems, vol.183, Springer Verlag, Berlin, Germany.

Schmit, L. A. and Mallett, R. H. (1963), ' Structural synthesis and design parameter hierarchy', Proc. Second Electronic Computation Conference, *ASCE, Jl. Struct. Div.*, 89, ST4, 269-299

Schmit, L. A. (1969), 'Problem formulation, methods and solutions in the optimum design of structures', in Cohn, M. Z. (Ed.), An Introduction to Structural Optimization, SM Study No. 1, University of Waterloo, Waterloo, Ont., 19-46.

Schmit, L. A. (1981), 'Structural synthesis-its genesis and development', *AIAA Jl.*, 19, 1249-1263.

Siddall,J. N. (1983), Optimal Engineering Design, Principles and Applications, Marcel Dekker Inc., New York.

Siess, C. P. (1960), 'Research , building codes and engineering practice', *ACI Jl.*, 31, 1105-1122.

Simon, H. A. (1957), Models of Man, John Wiley & Sons, New York.

Sobieski, J., Berke, L., Fleury, C., Haug, E. J., Hörnlein, H., Lecina, G., Morris, A. and Taylor, J. E.(1987), ' Panel discussion : trends in computer-aided optimal design', in Mota Soares, C. A. (Ed. 1987) , Computer-Aided Optimal Design, Springer Verlag, Berlin, Germany, 1018-1029.

Spires, D. B. and Arora, J. S. (1989), ' Optimal design of tall reinforced concrete tube buildings', in Nelson, J. K. (Ed.) , " Computer optimization in structural engineering', Proc. ASCE Struct. Congress, 1989, ASCE, N. Y., 479-489.

Templeman, A. B. (1976), 'Optimization concepts and techniques in structural design', in Introductory Report, 10th IABSE Congress, Tokyo, Japan, Sept. IABSE, Zürich, Switzerland, 41-60.

Templeman, A. B. (1983), 'Optimization methods in structural design practice', *ASCE Jl. Struct. Eng.*, 109, 2420-2433.

Templeman, A. B. (1988), ' Some reservations about reliability-based design', in Frangopol, D. M. (Ed.) ' New Directions in Structural System Reliability', Proc. of Workshop, University of Colorado, Boulder, Colo, 265-276.

Thoft-Christensen, P. (Ed. 1987), Reliability and Optimization of Structural Systems, Proc. First IFIP WG 7.5 Conf., Aalborg, Denmark, May 1987 , Lecture Notes in Engineering, v. 33, Springer Verlag, Berlin, Germany.

Thoft-Christensen, P. (Ed. 1989), Reliability and Optimization of Structural Systems, Proc. Second IFIP WG 7.5 Conf., London, U. K., Sept. 1988, Lecture Notes in Engineering, v. 48, Springer Verlag, Berlin, Germany.

Thoft-Christensen, P. (1990), On Reliability-Based Structural Optimization, Univ. of Aalborg, Structural Reliability Series, Paper No. 70, July.

Thoft-Christensen, P. and Baker, M. J. (1982), Structural Reliability Theory and its Applications, Springer Verlag, Berlin, Germany.

Turkstra, C. J. (1970), Theory of Structural Design Decisions, SM Study No. 2, Solid Mechanics Division, Univ. of Waterloo Press, Waterloo, Ont.

Vanderplaats, G. N. (1982), 'Structural optimization : past, present, future', *AIAA Jl.*, 20, 992-1000.

Vanderplaats, G. N. (1984), Numerical Optimization Techniques, Mc Graw-Hill, New York.

Venkayya, V. B., Khot, N. S. and Berke, L. (1973), ' Application of optimality criteria to automated design of large practical systems', Second AGARD Symposium, Milan, Italy.

DESIGN OPTIMIZATION OF TALL STEEL BUILDING FRAMEWORKS

D.E. GRIERSON and C.-M. CHAN
Department of Civil Engineering
University of Waterloo, Waterloo,
Ontario, Canada N2L 3G1

ABSTRACT. This lecture presents an efficient computer-based method for the optimum design of tall steel building frameworks. Specifically, an Optimality Criteria method is applied to minimize the weight of a lateral load-resisting structural system of fixed topology subject to constraints on interstorey drift. A range of steel framework examples is presented to illustrate the features of the design optimization method.

1. Introduction

Once the topology and support conditions have been established for a framework, the main design effort involves sizing the individual beam, column, and bracing members. For tall slender buildings, the design of structural members is generally governed by lateral drift criteria rather than by stress criteria. The allowable value for overall building drift or interstorey drift is usually specified as some ratio of the building or storey height (e.g., 1/400). It is generally a complex task to size the members so that the drift criteria is satisfied.

This lecture presents a computer-based Optimality Criteria (OC) method for the least-weight design of tall slender lateral load-resisting steel frameworks of fixed topology subject to constraints on interstorey drift, [1,2]. The OC method employs a pseudo-discrete optimization strategy to directly size members using commercial-standard steel sections. The explicit design problem is first formulated, and then the details of the OC method are developed. Finally, a range of steel framework examples is presented to illustrate the features of the design method.

2. The Explicit Design Problem

Consider a slender steel framework having i=1,2,...,n members (or member fabrication groups) and j=1,2,...,m storeys. The problem to find the minimum weight design of the framework under lateral loads can be generally stated as the following discrete optimization problem

G. I. N. Rozvany (ed.), Optimization of Large Structural Systems, Vol. II, 863–872.
© 1993 *Kluwer Academic Publishers.*

$$Minimize: \qquad W(a_i) = \sum_{i=1}^{n} w_i \, a_i \qquad\qquad (1a)$$

$$subject \ to: \qquad (\, \delta_j - \delta_{j-1} \,) \, / \, h_j \ \leq d_j^{\,U} \qquad (j=1,2,...,m) \qquad (1b)$$

$$a_i \in A_i \qquad (i=1,2,...,n) \qquad (1c)$$

Equation (1a) defines the weight for the structure, where a_i is the axial cross-section area for member i and w_i is the corresponding weight coefficient (= material density X member length); Eqs. (1b) define the interstorey drift constraints for the structure, where δ_j and δ_{j-1} are the lateral deflections at two adjacent storey levels, h_j is the corresponding storey height and $d_j^{\,U}$ is the allowable interstorey drift; Eqs. (1c) require each cross-sectional area a_i to belong to the set of areas $A_i = \{ \, a_1, \, a_2,... \, \}_i$ prevailing for the standard steel section profile specified for member i (e.g., W-shape).

The interstorey drift constraints Eqs. (1b) are but implicit functions of the member sizing variables a_i. It remains to express these constraints as explicit functions of the a_i in order to facilitate computer solution of the design optimization problem. To that end, by the principle of virtual work, the displacement at any point of interest of a general three-dimensional building framework having $i=1,2,...,n$ members can be expressed as

$$\delta_j = \sum_{i=1}^{n} \int_0^{L_i} (\, \frac{F_x f_x}{Ea} + \frac{F_y f_y}{GA_Y} + \frac{F_z f_z}{GA_Z} + \frac{M_x m_x}{GI_X} + \frac{M_Y m_Y}{EI_Y} + \frac{M_z m_z}{EI_Z} \,) \ dx \qquad (2)$$

where: E, G = axial and shear elastic material moduli; a, A_Y, A_Z = axial and shear areas for the cross-section; I_X, I_Y, I_Z = torsional and flexural moments of inertia for the cross-section; F_X, F_X, F_Z = axial and shear forces, f_X, f_Y, f_Z = virtual axial and shear forces; M_X, M_Y, M_Z = torsional and flexural moments; m_X, m_Y, m_Z = virtual torsional and flexural moments.

Now, for commercial standard steel sections, the cross-section properties A_Y, A_Z, I_X, I_Y and I_Z may all be expressed in terms of the axial area a as follows, [1,2]

$$1 \, / \, A_Y = C_{AY} \, (\, 1 \, / \, a \,) + C_{AY}' \qquad\qquad (3a)$$

$$1 \, / \, A_Z = C_{AZ} \, (\, 1 \, / \, a \,) + C_{AZ}' \qquad\qquad (3b)$$

$$1 \, / \, I_X = C_{IX} \, (\, 1 \, / \, a \,) + C_{IX}' \qquad\qquad (3c)$$

$$1 \, / \, I_Y = C_{IY} \, (\, 1 \, / \, a \,) + C_{IY}' \qquad\qquad (3d)$$

$$1 \, / \, I_Z = C_{IZ} \, (\, 1 \, / \, a \,) + C_{IZ}' \qquad\qquad (3e)$$

where the coefficients C and C' are determined by linear regression analysis and have different values depending on the type and size of the section (e.g. see Reference 1 for the C and C'

values for several typical standard steel sections used for tall building design [3]). Having the relations Eqs. (3), the displacement δ_j from Eq. (2) can be concisely expressed solely in terms of the member cross- section areas a_i as

$$\delta_j = \sum_{i=1}^{n} \left(\frac{C_{ij}}{a_i} + C'_{ij} \right) \tag{4}$$

where, from Eqs. (2) and (3), the coefficients C_{ij} and C'_{ij} are given by

$$C_{ij} = \int_0^{L_i} \left(\frac{F_x f_x + M_Y m_Y C_{IY} + M_Z m_Z C_{IZ}}{E} + \frac{F_Y f_Y C_{AY} + F_Z f_Z C_{AZ} + M_X m_X C_{IX}}{G} \right)_i dx \tag{5a}$$

$$C'_{ij} = \int_0^{L_i} \left(\frac{M_Y m_Y C'_{IY} + M_Z m_Z C'_{IZ}}{E} + \frac{F_Y f_Y C'_{AY} + F_Z f_Z C'_{AZ} + M_X m_X C'_{IX}}{G} \right)_i dx \tag{5b}$$

From Eq. (4), the left-hand sides of the interstorey drift constraints Eqs. (1b) may be formulated as explicit functions of the sizing variables a_i to obtain

$$d_j(a_i) = (\delta_j - \delta_{j-1}) / h_j = \sum_{i=1}^{n} \left(\frac{c_{ij}}{a_i} + c'_{ij} \right) \qquad (j=1,2,...,m) \tag{6}$$

where the coefficients

$$c_{ij} = (C_{ij} - C_{ij-1}) / h_j \tag{7a}$$

$$c'_{ij} = (C'_{ij} - C'_{ij-1}) / h_j \tag{7b}$$

Therefore, from Eqs. (1) and (6), the explicit form of the design optimization problem in terms of member sizing variables a_i is

Minimize:
$$W(a_i) = \sum_{i=1}^{n} w_i \, a_i \tag{8a}$$

subject to:
$$\sum_{i=1}^{n} \left(\frac{c_{ij}}{a_i} + c'_{ij} \right) \le d_j^U \qquad (j=1,2,...,m) \tag{8b}$$

$$a_i \in A_i \qquad (i=1,2,...,n) \tag{8c}$$

3. Optimality Criteria Algorithm

Temporarily ignoring the discrete sizing constraints Eqs. (8c), the design optimization problem can be reformulated as the minimization of the Lagrangian function

$$L(a_i, \lambda_j) = \sum_{i=1}^{n} w_i a_i + \sum_{j=1}^{m} \lambda_j \left[\sum_{i=1}^{n} \left(\frac{c_{ij}}{a_i} + c_{ij}' \right) - d_j^U \right]$$ (9)

where the Lagrange multipliers are such that $\lambda_j > 0$ if constraint j is active, or $\lambda_j = 0$ if constraint j is inactive. Temporarily assume that all drift constraints are active.

Differentiate Eq. (9) with respect to the sizing variables a_i, and rearrange terms to obtain the stationary conditions [4,5]

$$\sum_{j=1}^{m} \lambda_j \frac{c_{ij}}{w_i a_i^2} = 1 \qquad (i=1,2,...,n)$$ (10)

Equations (10) are optimality criteria having the physical meaning that the weighted sum of the, so called, virtual energy densities for each member i is equal to unity for the optimum structure.

A recursive algorithm is applied to find the sizing variables a_i satisfying Eqs. (10) at the optimum. To that end, multiply both sides of Eqs. (10) by a_i^{η} and take the ηth root and, then, apply a first-order binomial expansion to obtain the linear recursive relation

$$a_i^{v+1} = a_i^v \left[1 + \frac{1}{\eta} \left(\sum_{j=1}^{m} \frac{\lambda_j c_{ij}}{w_i a_i^2} - 1 \right)_v \right] \qquad (i=1,2,...,n)$$ (11)

where η is a step-size parameter that controls convergence of the recursive process [4,5], while $v+1$ and v indicate successive iterations ($v = 0$ corresponds to the set of cross-section areas a_i^0 prevailing at the beginning of the recursive process).

Consider the n discrete sizing constraints Eqs. (8c) and assume, without any loss in generality, that discrete standard section sizes a_i^d have been assigned to the last ξ members after v iterations of the recursive process ($\xi = 0$ initially), i.e., they become fixed section sizes that are 'passive' to the recursive process, such that for these members

$$a_i^{v+1} = a_i^d \qquad (i=n-\xi+1,...,n)$$ (12)

for the (v+1)th iteration and all succeeding iterations (the technique used to assign discrete section sizes to the members is described later in the lecture).

In order to apply Eqs. (11) to find the new sizing variables a_i^{v+1}, the current values of the Lagrange multipliers λ_j^v must first be determined. To that end, for the (v+1)th iteration of the recursive process, consider the change $(d_k^{v+1} - d_k^v)$ in the kth drift constraint due to changes $(a_i^{v+1} - a_i^v)$ in the n-ξ active sizing variables, i.e.,

$$(d_k^{v+1} - d_k^v) = \sum_{i=1}^{n-\xi} (\frac{\partial d_j}{\partial a_i})_v (a_i^{v+1} - a_i^v) \tag{13}$$

where, from Eqs. (6) for drift constraint k,

$$(\frac{\partial d_k}{\partial a_i})_v = - (\frac{c_{ik}}{a_i^2})_v \qquad (i=1,2,...,n-\xi) \tag{14}$$

and, from Eqs. (11) for the active sizing variables,

$$a_i^{v+1} - a_i^v = \frac{a_i^v}{\eta} (\sum_{j=1}^{m} \frac{\lambda_j c_{ij}}{w_i a_i^2} - 1)_v \qquad (i=1,2,...,n-\xi) \tag{15}$$

Suppose that drift constraint k becomes active after the (v+1)th iteration so that $d_k^{v+1} = d_k^U$, and then substitute from Eqs. (14) and (15) into Eq. (13) and rearrange terms to obtain, for all m drift constraints,

$$\sum_{j=1}^{m} \lambda_j \sum_{i=1}^{n-\xi} (\frac{c_{ik} c_{ij}}{w_i a_i^3})_v = \sum_{i=1}^{n-\xi} \frac{c_{ik}}{a_i^v} - \eta \, (d_k^U - d_k^v) \qquad (k=1,2,...,m) \tag{16}$$

Equations (16) are a system of m simultaneous linear equations in terms of m Lagrange variables λ_j.

Equations (15) and (16) together form the basis for an Optimality Criteria method to solve the discrete optimization problem posed by Eqs. (8). The details of the OC method, including the means to establish active drift constraints and how to assign discrete section sizes to members, are as follows, [1,2,6,7]:

1. Set $\xi = 0$ and $v = 0$ and adopt an initial set of sizing variables a_i^v (e.g., for the steel section shapes specified for the members, select the largest cross-section areas allowed by the discrete sizing constraints Eqs. (8c)).
2. For the current a_i^v, analyze the structure and establish the coefficients c_{ij} and c_{ij}.
3. For the current active a_i^v, use the Gauss-Seidel iterative technique to solve Eqs. (16) for the corresponding set of Lagrange multipliers λ_j^v; i.e., iteratively apply the recursive relation

$$\lambda_k^{\tau+1} = \frac{1}{e_{kk}} (b_k - \sum_{j=1}^{k-1} e_{kj} \lambda_j^{\tau+1} - \sum_{j=k+1}^{m} e_{kj} \lambda_j^{\tau}) \qquad (k=1,2,...,m) \tag{17a}$$

where

$$e_{kk} = \sum_{i=1}^{n-\xi} (\frac{c_{ik}^2}{w_i a_i^3})_v \qquad ; \qquad e_{kj} = \sum_{i=1}^{n-\xi} (\frac{c_{ik} c_{ij}}{w_i a_i^3})_v \tag{17b,c}$$

$$b_k = \sum_{i=1}^{n-\xi} \frac{c_{ik}}{a_i^\nu} - \eta \ (\ d_k^U - d_k^\nu \ ) \tag{17d}$$

Any $\lambda_k^{\tau+1}$ value found to be negative at any stage of applying Eqs. (17a) is temporarily set to zero to reflect the fact that the corresponding drift constraint is presently not active. Upon establishing all $\lambda_k^{\tau+1} = \lambda_k^\tau$ through Eqs. (17a), the Gauss-Seidel technique has converged to the solution $\lambda_j^\nu = \lambda_j^{\tau+1}$ (j=1,2,...,m) of Eqs. (16).

4. For the current active a_i^ν and current λ_j^ν, find the new set of active sizing variables $a_i^{\nu+1}$ from Eqs. (15).

5. Check convergence of the OC recursive process: if all $a_i^{\nu+1} = a_i^\nu$ and $\lambda_j^{\nu+1} = \lambda_j^\nu$, go to step 6; otherwise, update Eqs. (16) for the current $a_i^{\nu+1}$ values, set $\nu = \nu+1$ and return to step 3.

6 For the current active $a_i^{\nu+1}$ values from step 4, identify the active member i of length L_i and material density ρ_i that results in the smallest increment in structure weight when it is assigned the next available discrete section size a_i^d satisfying Eqs. (8c), i.e., identify the active member i for which

$$(\ a_i^d - a_i^{\nu+1} \ )\rho_i L_i = \min \ ! \qquad (i=1,2,...,n-\xi) \tag{18}$$

and assign that member the discrete section size a_i^d (the member size then becomes a passive variable for the remainder of the current design stage).

7. Set $\xi = \xi+1$ and check convergence of the pseudo-discrete optimization process: if $\xi = n$, thereby indicating that all n members have been assigned a discrete section size, go to step 8; otherwise, set $\nu = \nu+1$ and return to step 3.

8. Check convergence of the overall design process: if the structure weight from Eq. (8a) is the same for two successive design cycles, terminate with the minimum weight structure; otherwise, set $\xi = 0$ and $\nu = 0$ and return to step 2.

4. Design Examples

Consider the range of eight planar steel frameworks of 30 to 60 storeys shown in Figure 1, [1,2]. Five of the frameworks are rigid frames for which lateral stiffness is supplied solely through the racking action of the beams and columns. Three of the frameworks are rigid frames with shear trusses, where additional storey shear capacity is supplied by the axially-loaded diagonal truss members. One of the frameworks is a rigid frame with both shear and outrigger trusses, where the latter trusses provide additional lateral stiffness by further mobilizing the exterior columns to resist overturning moments. The frameworks involve from 330 members for the 30-storey rigid frame to 1128 members for the 60-storey frame with shear and outrigger trusses. For a bay width of 20 ft. and a storey height of 12 ft., the frameworks have height-to-width aspect ratios that vary from 3.6 to 5.4.

The frameworks are to be designed such that interstorey deflection does not exceed h/400 (0.36 in.) under the lateral wind loading shown in Figure 1 (i.e., 20 psf over 0-100 ft; 25 psf over 100-300 ft; 30 psf over 300-600 ft; and 35 psf over 600 ft).

Members are to be designed using AISC standard sections, as follows: columns are to have

Members are to be designed using AISC standard sections, as follows: columns are to have W14 shapes selected from among 36 discrete sections over the range W14X22 to W14X730; girders are to have W24 shapes selected from among 24 discrete sections over the range W24X55 to W24X492, [3]. To satisfy practical fabrication requirements, girders are grouped together as having a common section for each storey, while exterior columns are grouped together over two adjacent storeys, as are symmetrical pairs of interior columns (for the 45-storey frameworks, the top storey exterior columns and pairs of interior columns form separate fabrication groups). Steel material properties are given in Figure 1.

The previously described OC pseudo-discrete optimization method is applied to determine least-weight designs of the frameworks while ensuring acceptable lateral stiffness. First-order behaviour is taken as the response basis for the design of all frameworks (while second-order behaviour is readily accounted for, and is certainly required for practical design, first-order behaviour is an acceptable basis for a comparative study of the results for the various frameworks). For all frameworks, the largest available section sizes for the columns (W14X730) and girders (W24X492) were selected for the initial trial design to commence the optimization process.

The total weight of steel in kips for the least-weight design of each framework found by the OC method is given in Figure 2. Also given, assuming a square building, is the approximate unit weight of steel in pounds per square foot of floor area, which is a useful index for comparing the relative economy of the different structural configurations. For example, for the two 5-bay rigid frames the steel weight increases from 11.35 to 20.05 psf as the number of stories increases from 30 to 45. On the other hand, incorporating shear trusses into the design of the 5-bay, 45-storey framework reduces the steel weight from 20.05 to 15.02 psf. The positive influence of the shear trusses on structure weight is also evident in Figure 2 for the 7-bay frameworks having 45 and 60 storeys. Finally, for the 7-bay, 60-storey frameworks with truss stiffening systems, incorporating outrigger trusses into the design in addition to shear trusses further reduces the steel weight from 16.90 to 15.86 psf.

On average for all frameworks, the OC method converged to the least-weight structure in 2 to 3 iterations of the design process. In fact, even though the largest available section sizes are taken for the initial design, the structure weight after the first design cycle is generally within five percent of the final weight. All designs were conducted on a SUN Sparc 1 station. The 30-storey rigid frame design required 20 minutes of execution time, while the design of the 60-storey frame with shear and outrigger trusses required 2 hours.

5. Concluding Remarks

The Optimality Criteria (OC) pseudo-discrete design optimization technique will generally converge quite rapidly for building frameworks because the member force distributions for such structures are somewhat insensitive to changes in member sizes. The efficiency of the OC technique is mainly influenced by the number of constraints and is weakly dependent on the number of variables, and as such, the method provides an effective strategy for the optimal design of tall building structures involving many sizing variables and comparatively few drift constraints.

This work was sponsored by the Natural Sciences and Engineering Research Council of Canada, and is based upon research conducted by the second author under the supervision of the first author for the degree of Doctor of Philosophy in Civil Engineering at the University of Waterloo, Canada.

References

1. Chan, C.-M. and Grierson, D.E., "An Efficient Resizing Technique for the Design of Tall Steel Buildings Subject to Multiple Drift Constraints", Journal of Structural Engineering, ASCE, (submitted 1991).
2. Chan, C.-M., "An Automatic Resizing Technique for the Design of Tall Steel Building Frameworks", PhD Thesis, University of Waterloo, Canada (in progress 1991).
3. Manual of Steel Construction - Load and Resistance Factor Design, American Institute of Steel Construction, Inc., Chicago, Illinois, USA, First Edition, 1986.
4. Berke, L. and Knot, N.S., "Use of Optimality Criteria Methods for Large Scale Systems", AGARD-LS-70, pp. 1-1 to 1-29, Sept. 1974.
5. Knot, N.S., Berke, L. and Venkayya, V.B., "Comparison of Optimality Criteria Algorithms for Minimum Weight Design of Structures", AIAA/ASME 19th Structures, Struct. Dynamics and Material Conf., Bethesda, Md., USA, pp. 37-46, 1978.
6. Grierson, D.E. and Chan C.-M., " Design Optimization of Tall Steel Buildings", Proceedings of OPTI'91, International Conference on Structural Optimization, Boston, USA, June 25-27, 1991.
7. Cameron, G., Chan, C.-M., Xu L. and Grierson, D.E., "Alternative Methods for the Optimal Design of Slender Steel Frameworks", Computational Structures Technology Conference, Heriot-Watt Univ., Edinburgh, UK, August 20-22, 1991.

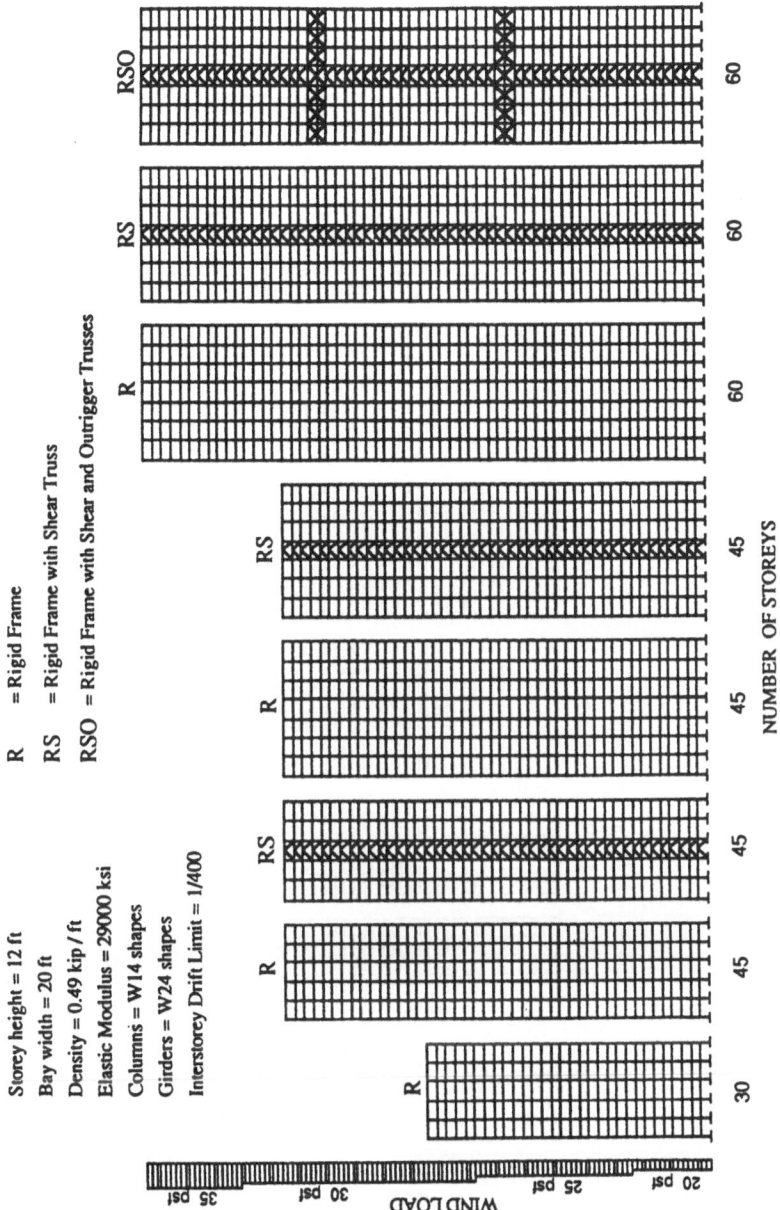

Figure 1: Example Multistorey Frameworks

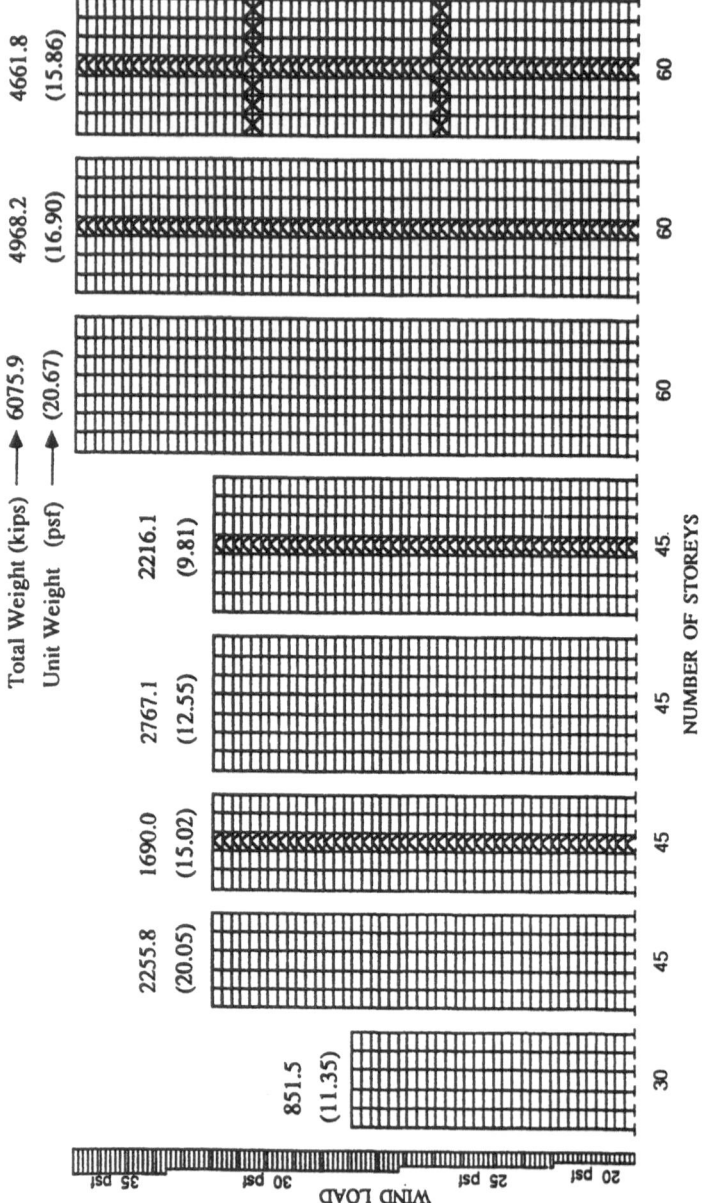

Figure 2: Least-Weight Design Results

DESIGN OPTIMIZATION OF STEEL FRAMEWORKS ACCOUNTING FOR SEMI-RIGID CONNECTIONS

D.E. GRIERSON and LEI XU
Department of Civil Engineering
University of Waterloo, Waterloo,
Ontario, Canada N2L 3G1

ABSTRACT. This lecture presents an efficient computer-based method for the optimum design of steel frameworks accounting for the behaviour of semi-rigid connections. The Dual method is applied to minimize the 'cost' of the connections and members for the structure subject to constraints on stresses and displacements under specified design loads. A steel framework example is presented to illustrate the features of the design optimization method.

1. Introduction

For conventional analysis and design of steel framed structures, the actual behaviour of beam-to-column connections is simplified to the two idealized extremes of either rigid-joint behaviour or pinned-joint behaviour. In fact, experimental investigations of actual joint behaviour have clearly demonstrated that a pinned-joint connection possesses a certain amount of rotational stiffness, while a rigid-joint connection possesses some degree of flexibility [1]. Thus, in reality, steel frame connections should be treated as 'semi-rigid' connections for the purposes of analysis and design.

In the past three decades, considerable research has been carried out to assess the actual behaviour of steel frameworks accounting for the effect of connection flexibility, e.g.[2,3,4,5]. Much of this research has been limited to the analysis problem with little attention being paid to the more important problem of design, e.g.[6,7]. This lecture presents a systematic method for the optimum design of steel frameworks accounting for the behaviour semi-rigid connections [8,9]. The design having the minimum combined 'cost' of members and connections is sought, while ensuring that stresses and displacements are within acceptable limits. Members are sized using discrete standard steel sections, while connections are selected on the basis of their continuous-valued moment-rotation stiffnesses. The continuous-discrete Dual method [10,11] is applied for the design optimization of an example steel framework.

2. Analysis Of Frameworks With Semi-Rigid Connections

For the design of steel frameworks with semi-rigid connections it is necessary to have the

G. I. N. Rozvany (ed.), Optimization of Large Structural Systems, Vol. II, 873–881.

capability to conduct corresponding analysis. To that end, the effects of connection flexibility are modelled by attaching rotational springs of moduli K_1 and K_2 to the ends 1 and 2 of a member, as shown in Figure 1. For first-order analysis, the stiffness matrix of a member i with semi-rigid restraint at the ends can be represented by the stiffness matrix for the member taken to have rigid connections modified by a correction matrix[2], i.e.,

$$K_i = S_i \cdot C_i \tag{1}$$

where K_i is the local-axis stiffness matrix of member i accounting for semi-rigid end-connections, S_i is the local-axis stiffness matrix of the member taken to have fixed ends, and C_i is the required correction matrix. For a planar member, the matrix C_i is conveniently expressed as a function of parameters r_1 and r_2 defined as 'fixity factors' at the two ends 1 and 2 of the member, and which are related to the corresponding rotational spring stiffnesses K_1 and K_2 as

$$r_1 = \frac{1}{1 + \dfrac{3EI}{K_1 L}} \quad ; \quad r_2 = \frac{1}{1 + \dfrac{3EI}{K_2 L}} \tag{2}$$

where E is Young's modulus, and L and I are the length and cross-section moment of inertia of the member, respectivly. The fixity factor r defines the stiffness of a connection relative to the attached member. For a unit end-moment, r is the ratio of the rotation α of the end of the member to the combined rotation ϕ of the member and the connection (i.e. $r = \alpha/\phi$, see Figure 2,[5]). For a pinned connection, the value of the fixity factor is zero ($r = 0$). For a fully-fixed connection, the fixity factor is unity ($r = 1$). A semi-rigid connection has a fixity factor between zero and unity ($0 < r < 1$).

Having the local-axis member stiffness matrices through Eq.(1), for specified values of fixity factors r reflecting connection stiffness, the analysis of frameworks with semi-rigid connections is directly carried out using the conventional Displacement method.

3. Optimization Of Frameworks With Semi-Rigid Connections

It is required to design a steel framework for minimum combined 'cost' of members and connections while accounting for semi-rigid behaviour of the connections. Herein, the cost of each member i is represented by its weight, while the cost of each connection j is taken to be directly related to its rotational stiffness. Therefore, the total cost of a member i with two end-connections j=1,2 may be expressed as

$$Z_i = w_i a_i + \sum_{j=1}^{2} \left(\beta_{ij} K_{ij} + \beta_{ij}^{\circ} \right) \tag{3}$$

where a_i and w_i are the member cross-section area and weight coefficient (w_i = material density x member length), K_{ij} and β_{ij} are the connection stiffness and cost coefficient, and β_{ij}° is the cost of a pinned connection having zero rotational stiffness.

Each member i is to be sized using a commercial standard steel section and, as such, its cross-section area a_i is a 'discrete' variable to the design. On the other hand, each connection j may be fabricated anywhere in the range from being fully-pinned to fully-fixed and, as such, its rotational stiffness K_{ij} is a 'continuous' variable to the design. That is, the design variables a_i and K_{ij} are restricted by the design as

$$a_i \in A_i \qquad ; \qquad K_{ij}^L \le K_{ij} \le K_{ij}^U \qquad (4a,b)$$

where Eq.(4a) requires the cross-section area a_i of member i to belong to the discrete set of areas $A_i \equiv \{a_1, a_2, ...\}$ prevailing for the standard section shape specified for the member (e.g., W-shape), while Eq.(4b) imposes specified lower and upper bounds K_{ij}^L and K_{ij}^U on the connection rotational stiffness K_{ij}.

The optimal, member plus connection, design of a framework of i=1,2,...,n members having semi-rigid connections, subject to stress and displacement constraints, may be generally stated as[8,9],

$$Minimize : \quad Z = \sum_{i=1}^{n} \left(Z_i - \sum_{j=1}^{2} \beta_{ij}^{\circ} \right) = \sum_{i=1}^{n} \left(w_i a_i + \sum_{j=1}^{2} \beta_{ij} K_{ij} \right) \qquad (5a)$$

Subject to :

$$\sigma_k^L \le \sigma_k \le \sigma_k^U \qquad\qquad (k = 1, 2, ..., s) \qquad (5b)$$

$$\delta_l^L \le \delta_l \le \delta_l^U \qquad\qquad (l = 1, 2, ..., d) \qquad (5c)$$

$$K_{ij}^L \le K_{ij} \le K_{ij}^U \qquad (i = 1, 2, ..., n ; j = 1, 2) \qquad (5d)$$

$$a_i \in A_i \qquad\qquad (i = 1, 2, ..., n) \qquad (5e)$$

where, from Eq.(3), the objective function Eq.(5a) is a measure of the combined 'cost' of the members and connections, Eqs.(5b) and (5c) define constraints on stresses σ_k and displacements δ_l for the structure, and from Eqs.(4) the restrictions Eqs.(5d) and (5e) control the values of the continuous-valued connection stiffnesses K_{ij} and the discrete-valued member sizes a_i, respectively.

To enable computer solution of the design optimization problem posed by Eqs.(5), it is first necessary to formulate each stress σ_k and displacement δ_l in Eqs.(5b) and (5c) as an explicit function of the design variables. Since stresses and displacements vary inversely with member size a_i and connection stiffness K_{ij}, upon adopting the reciprocal variables

$$x_i = \frac{1}{a_i} \qquad ; \qquad f_{ij} = \frac{1}{K_{ij}} \qquad (6a,b)$$

where f_{ij} is connection 'flexibility', a good quality explicit approximation of each stress σ_k is provided by the first-order Taylors series

$$\sigma_k = \sigma_k^o + \sum_{i=1}^{n} \left[\left(\frac{\partial \sigma_k}{\partial x_i} \right)^o (x_i - x_i^o) + \sum_{j=1}^{2} \left(\frac{\partial \sigma_k}{\partial f_{ij}} \right)^o (f_{ij} - f_{ij}^o) \right] \qquad (7a)$$

while that for each displacement δ_l is

$$\delta_l = \delta_l^o + \sum_{i=1}^{n} \left[\left(\frac{\partial \delta_l}{\partial x_i} \right)^o (x_i - x_i^o) + \sum_{j=1}^{2} \left(\frac{\partial \delta_l}{\partial f_{ij}} \right)^o (f_{ij} - f_{ij}^o) \right] \qquad (7b)$$

where the superscript zero(^{o}) defines known quantities for the current structure (e.g., for the initial 'trial' design), while x_i and f_{ij} are the variables to the design. The various gradients in Eqs.(9) are found using virtual-load sensitivity analysis [8,9].

For the current structure, it is readily shown that

$$\sigma_k^o = \sum_{i=1}^{n} \left[\left(\frac{\partial \sigma_k}{\partial x_i} \right)^o x_i^o + \sum_{j=1}^{2} \left(\frac{\partial \sigma_k}{\partial f_{ij}} \right)^o f_{ij}^o \right] \qquad (8a)$$

$$\delta_l^o = \sum_{i=1}^{n} \left[\left(\frac{\partial \delta_l}{\partial x_i} \right)^o x_i^o + \sum_{j=1}^{2} \left(\frac{\partial \delta_l}{\partial f_{ij}} \right)^o f_{ij}^o \right] \qquad (8b)$$

Therefore, from Eqs.(6),(7) and (8), the design optimization problem Eqs.(5) expressed explicitly in terms of intermediate design variables x_i and f_{ij} is

$$\text{Minimize :} \quad Z = \sum_{i=1}^{n} \left(\frac{w_i}{x_i} + \sum_{j=1}^{2} \frac{\beta_{ij}}{f_{ij}} \right) \qquad (9a)$$

Subject to :

$$\sigma_k^L \leq \sum_{i=1}^{n} \left[\left(\frac{\partial \sigma_k}{\partial x_i} \right)^o x_i + \sum_{j=1}^{2} \left(\frac{\partial \sigma_k}{\partial f_{ij}} \right)^o f_{ij} \right] \leq \sigma_k^U \qquad (k = 1, 2, ..., s) \qquad (9b)$$

$$\delta_l^L \leq \sum_{i=1}^{n} \left[\left(\frac{\partial \delta_l}{\partial x_i} \right)^o x_i + \sum_{j=1}^{2} \left(\frac{\partial \delta_l}{\partial f_{ij}} \right)^o f_{ij} \right] \leq \delta_l^U \qquad (l = 1, 2, ..., d) \qquad (9c)$$

$$f_{ij}^L \leq f_{ij} \leq f_{ij}^U \qquad\qquad (\ i = 1, 2, ..., n \ ; j = 1, 2 \ ) \qquad (9d)$$

$$x_i \in X_i \qquad\qquad (\ i = 1, 2, ..., n \ ) \qquad (9e)$$

where, from Eqs.(6), the bounds $f_{ij}^L = 1/K_{ij}^U$ and $f_{ij}^U = 1/K_{ij}^L$, and the components of each discrete set $X_i = \{x_1, x_2, ...\} = \{1/a_1, 1/a_2, ...\}$.

The continuous-discrete optimization problem Eqs.(9) is solved at each design stage using the Dual method [10,11] to find new values of x_i and f_{ij} for which the combined member plus connection 'cost' of the structure is reduced relative to that for the previous design stage. The stress and displacement gradients are then updated for this new structure, and the design optimization Eqs.(9) is conducted again. This iterative process is repeated until cost convergence occurs for successive design stages, at which point the minimum 'cost' design has been found.

4. Example Application

The two-bay, three-storey steel framework loaded as shown in Figure 3 has semi-rigid connections at the ends of the beam members. Connection rotational stiffness is specified to be the same for all semi-rigid connections at each storey level, and is to be constrained to lie in the range from $K_{ij}^L = 0.0$ kips-in/rad to $K_{ij}^U = 5 \times 10^6$ kips-in/rad, [1]. Each member cross-section area is to be selected consistent with the requirement that column members at each storey level are to have the same American standard W-section, as are beam members at each storey level. The frame is to be designed in accordance with the strength/stability (stress) requirements of the AISC-ASD code [6] while ensuring that the top-storey lateral sway does not exceed 1.08in (i.e., h/400).

The material density $\rho = 0.283 \times 10^{-3}$ kip/in^3 and, as such, the weight coefficient in Eq.(9a) for each column member is $w_i = 0.283 \times 10^{-3} \times 144 = 40.75 \times 10^{-3}$ kip/in^2, while that for each beam member is $w_i = 0.283 \times 240 \times 10^{-3} = 67.92 \times 10^{-3}$ kip/in^2. The cost coefficient for each semi-rigid connection is taken to be $\beta_{ij} = 0.25 \times 10^{-6}$ rad/in., [8,9].

Initial member cross-section areas are adopted consistent with the initial design W-sections indicated in Figure 3. A fixity factor $r = 0.9$ is adopted to define the initial rotational stiffness of each of the semi-rigid connections for the framework. Upon applying the iterative design optimization procedure, the optimal design of the frame having minimum member plus connection cost is found after four design stages. The optimal member cross-section areas correspond to the final design W-sections shown in Figure 4. The optimal rotational stiffnesses of the semi-rigid connections are $K_1 = 0.609 \times 10^6$ kip-in/rad for the first storey, $K_2 = 0.543 \times 10^6$ kip-in/rad for the second storey, and $K_3 = 0.203 \times 10^6$ kip-in/rad for the third storey (these values are entirely consistent with those for actual connections in steel frameworks, [1]). Figure 4 illustrates the history of the design optimization process. Also shown in Figure 4 is the design history when all connections are assumed to be fully rigid. The fully-rigid connection design weighs more than the design found when the influence of semi-rigid connections is accounted for.

Acknowledgements

This work was sponsored by the Natural Sciences and Engineering Research Council of Canada, and is based upon research conducted by the second author under the supervision of the first author for the degree of Doctor of Philosophy in Civil Engineering at the University of Waterloo, Canada.

References

1. Gerstle, K.H.(1988), "Effects of Connections on Frames", in Steel Beam-to-Column Building Connections(edited by Chen, W.F.), Elsevier Applied Science.
2. Monforton, G.R. and Wu, T.S.(1963), "Matrix Analysis of Semi-Rigidly Connected Steel Frames", ASCE, Structural Division, Vol. 89, No. ST6, pp13-42.
3. Frye, M.J. and Morris, G.A.(1975), "Analysis of Flexible Connected Steel Frames", Canadian Journal of Civil Engineering, Vol. 2, No. 3, pp280-291.
4. Lui, E.M. and Chen, W.F.(1986), "Analysis and Behaviour of Flexibly-Jointed Frames", Engineering Structures, No. 8, pp107-118.
5. Cunningham, R.(1990), "Some Aspects of Semi-Rigid Connections in Structural Steelwork", Structural Engineering, Vol. 68, No. 5, pp85-92.
6. "Manual of Steel Construction" (1978), American Institute of Steel Construction, Inc., Chicago, Illinois, USA, Eighth Edition.
7. "Manual of Steel Construction-Load and Resistance Factor Design" (1986), American Institute of Steel Construction, Inc., Chicago, Illinois, USA, Eighth Edition.
8. Xu, L.(1991), "Optimal Design of Steel Frameworks with Semi-Rigid Connections", PhD. Thesis, University of Waterloo, Canada (in progress).
9. Xu, L. and Grierson, D.E.(1991), "Computer-Automated Design of Semi-Rigid Steel Frameworks", Journal of Structural Division, ASCE (submitted).
10. Fleury, C. and Sander, G.(1978), "Structural Optimization by Finite Element", LTAS Report SA-58, University of Liege, Belgium.
11. Schmit, L.A. and Fleury, C.(1979), "An Improved Analysis/Synthesis Capability based on Dual Methods-ACCESS 3", AIAA Paper No. 79-0721, Structures Volume,AIAA/ASME/ASCE/AHS 20th Structures, Structural Dynamics and Materials Conference, St. Louis, Missouri, USA.

Figure 1.   Planar Semi-Rigid Member

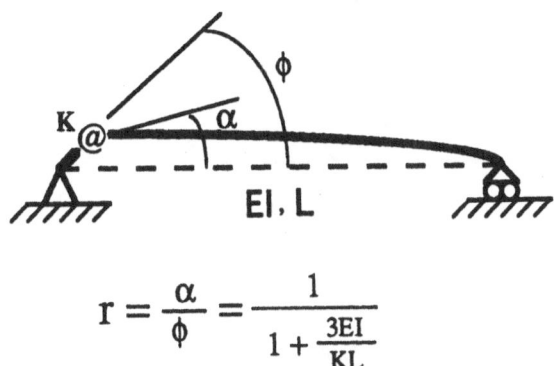

$$r = \frac{\alpha}{\phi} = \frac{1}{1 + \dfrac{3EI}{KL}}$$

Figure 2.   Fixity Factor

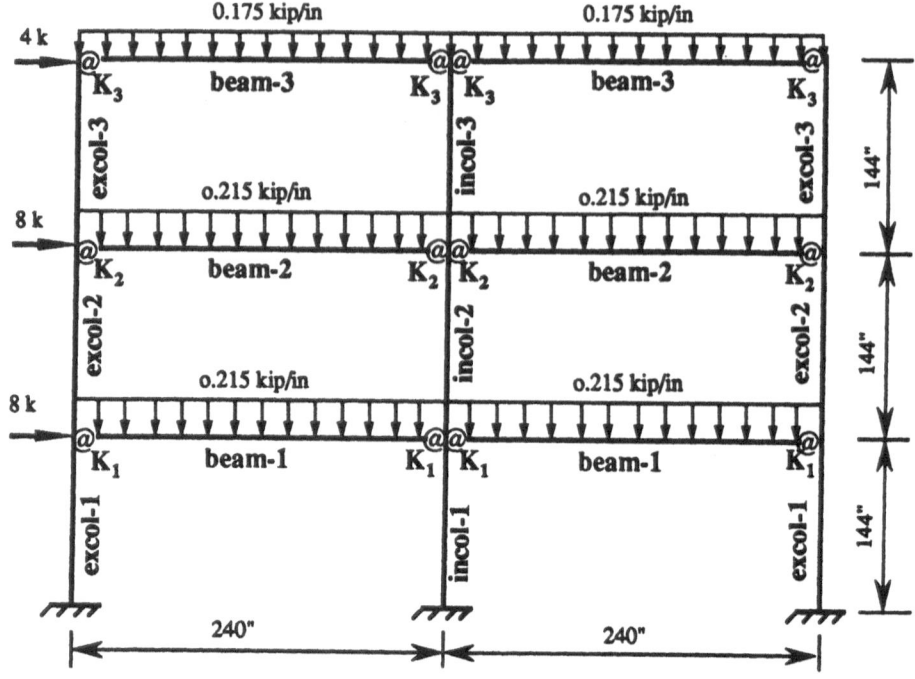

Figure 3.   Two-Bay Three-Storey Steel Frame

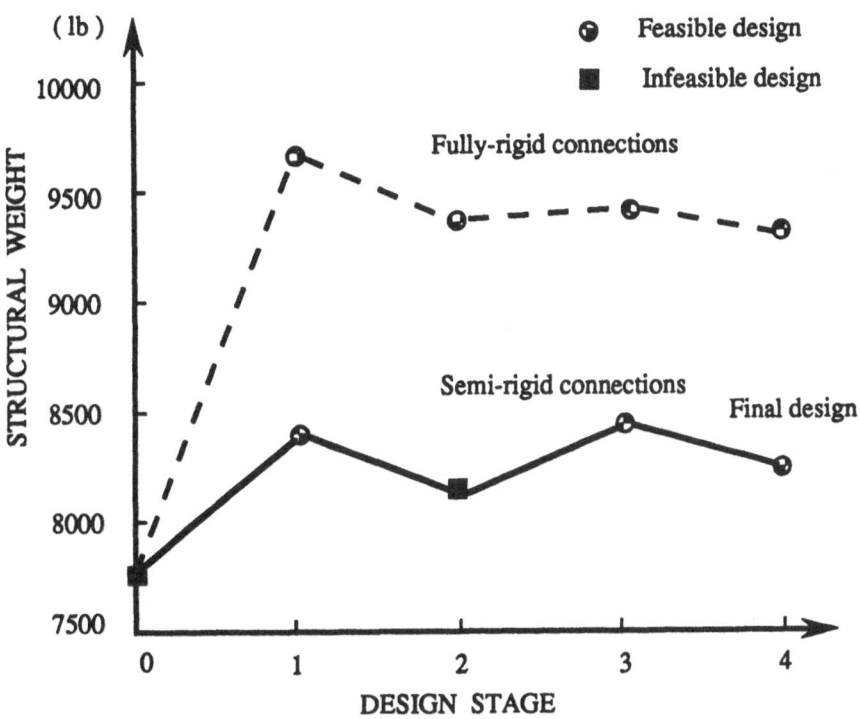

Figure 5 . Design History

DESIGN OPTIMIZATION OF REINFORCED CONCRETE BUILDING FRAMEWORKS

Donald E. Grierson and Hamid Moharrami
Department of Civil Engineering
University of Waterloo,
Waterloo Ontario, Canada, N2L 3G1

ABSTRACT

This lecture presents an efficient computer-based method for the optimal design of reinforced concrete building frameworks. An Optimality Criteria method is applied to minimize the cost of concrete, steel and formwork subject to constraints on strength and stiffness. A reinforced concrete framework example is presented to illustrate the features of the design optimization method.

1. Introduction

A number of studies have been concerned with the optimal determination of section dimensions and reinforcement within the context of an assembled reinforced concrete framework under gravity and lateral loads e.g.,[1,2,3,4]. Such a design problem involves numerous design variables and constraints, even for modest size structures. This poses no difficulties to the Optimality Criteria (OC) method, which is readily applied for the solution of large scale optimization problems involving many design variables and constraints,[5,6,7].

This lecture presents a computer-based OC method for the minimum cost design of reinforced concrete frameworks under gravity and lateral loading. The optimum width, depth, and longitudinal reinforcement of member sections are sought, while ensuring that stresses and displacements are within acceptable limits. A reinforced concrete framework example is presented to illustrate the features of the design method.

2. The Explicit Design Problem

Consider a reinforced concrete framework having i=1,2,...,n beam and column members. Shear walls and floor slabs are not explicitly considered, except that the flanges of T-beams are assumed to be formed by floor slabs of known thickness. All columns are taken to have rectangular cross-section of width b, depth h and steel reinforcement area a_s. The parameters

883

G. I. N. Rozvany (ed.), Optimization of Large Structural Systems, Vol. II, 883–896.
© 1993 Kluwer Academic Publishers.

b, h and a_s similarly apply for beams, which may have rectangular, T or L shape cross-section (for T and L sections, b and h define the width and depth of the web). The problem to find the minimum cost design of the framework under static gravity and lateral loads can be generally stated as the following optimization problem, [8,9]

Minimize :

$$Z = \sum_{i=1}^{n} C(b, h, a_s)_i \qquad (1a)$$

Subject to :

$$F_j - S_j \leq 0 \qquad (j=1,..., m) \qquad (1b)$$

$$D_k - D_k^{U} \leq 0 \qquad (k=1, ..., s) \qquad (1c)$$

$$b_l^{L} \leq b_l \leq b_l^{U}$$
$$h_l^{L} \leq h_l \leq h_l^{U} \qquad (l=1,..., n) \qquad (1d)$$
$$a_{s_l}^{L} \leq a_{s_l} \leq a_{s_l}^{U}$$

Equation (1a) defines the cost of the concrete, reinforcing steel and formwork for the structure (the cost term C_i has different forms depending on the type of member section, such as a rectangular column or beam, or a T-beam, [8,9]). Eqs.(1b) define the m strength constraints for the design, where F_j is an internal force acting on section j (i.e., axial force P, bending moment M, or shear force V) and S_j is the corresponding strength of the section ; Eqs.(1c) define the s stiffness constraints for the design, where D_k is beam deflection or lateral column sway and D_k^u is the corresponding allowable value; Eqs.(1d) define the lower and upper-bound sizing constraints on section width b, depth h and reinforcement area a_s for the n members of the structure.

The constraint Eqs.(1b) and (1c) are implicit nonlinear functions of the design variables, and in order to facilitate computer solution they must be written as explicit functions of the variables b, h, and a_s. For the purposes of this study, this is achieved by adopting the generic variable notation

$$x_{i1} = b \; ; \; x_{i2} = h \; ; \; x_{i3} = a_s \qquad (i=1, ..., n) \qquad (2)$$

and then expressing constraint Eqs.(1b) and (1c) as explicit linear functions of the variables x_{ik}(i=1, ..., n; k=1,2,3) using first-order Taylor series expansions, as follows

$$F_j^o - S_j^o + \sum_{i=1}^{n} \sum_{k=1}^{3} \left(\frac{\partial F_j}{\partial x_{ik}} - \frac{\partial S_j}{\partial x_{ik}} \right) (x_{ik} - x_{ik}^o) \leq 0 \qquad (j=1,...,m) \qquad (3a)$$

$$D_l^o + \sum_{i=1}^{n} \sum_{k=1}^{3} \left(\frac{\partial D_l}{\partial x_{ik}} \right) (x_{ik} - x_{ik}^o) - D_l^U \leq 0 \qquad (l=1,..., s) \qquad (3b)$$

where the superscript zero (o) denotes known or calculated quantities for the current design (e.g., the initial 'trial' design), and the x_{ik} are the variables to the design.

3. Design Sensitivity

To complete the explicit statement of the design optimization problem, it remains to identify the sensitivities $\partial D_l / \partial x_{ik}$, $\partial F_j / \partial x_{ik}$ and $\partial S_j / \partial x_{ik}$ in Eqs.(3) as explicit functions of the design variables x_{ik} [8,9]. In the following, the sensitivities of structure displacements D_l and member forces F_j are first given. Then, the sensitivities of member section strengths S_j are given.

3.1 DISPLACEMENT SENSITIVITY

The sensitivities of displacements Δ to changes in design variables x_{ik} are found as, [8,9]

$$\left\{ \frac{\partial \Delta}{\partial x_{ik}} \right\} = K^{-1} \left\{ \sum_{i=1}^{n} \sum_{k=1}^{3} \left[\frac{\partial q}{\partial A_i} \cdot \frac{\partial A_i}{\partial x_{ik}} - \left(\frac{\partial K}{\partial A_i} \cdot \frac{\partial A_i}{\partial x_{ik}} + \frac{\partial K}{\partial I_i} \cdot \frac{\partial I_i}{\partial x_{ik}} \right) \Delta \right] \right\} \qquad (4)$$

where Δ= nodal displacement vector in global coordinates; K= global-axis stiffness matrix for the structure; q= load vector in global coordinates (including selfweight loads); A_i=area and I_i= moment of inertia of each member i.The sensitivity of lateral sway D of the structure to change in any design variable, $\partial D/\partial x_{ik}$, is obtained by evaluating the corresponding right hand side of Eq.(4) for the current values of the design variables, accounting for gross values of the section parameters A_i and I_i. On the other hand, the sensitivity of beam deflection is obtained using the effective moment of inertia of the member section.

3.2 FORCE SENSITIVITY

The sensitivities of the internal forces f_{ij} at end-section j of member i to change in design variables x_{ik} are found as, [8,9]

$$\frac{\partial f_{ij}}{\partial x_{ik}} = \left(\frac{\partial k_{ii}}{\partial A_i}\frac{\partial A_i}{\partial x_{ik}} + \frac{\partial k_{ij}}{\partial I_i}\frac{\partial I_i}{\partial x_{ik}}\right)T_i\Delta +$$

$$k_{ij}T_iK^{-1}\left\{\sum_{i=1}^{n}\sum_{k=1}^{3}\left[\frac{\partial q}{\partial A_i}\frac{\partial A_i}{\partial x_{ik}} - \left(\frac{\partial K}{\partial A_i}\frac{\partial A_i}{\partial x_{ik}} + \frac{\partial K}{\partial I_i}\frac{\partial I_i}{\partial x_{ik}}\right)\Delta\right]\right\} \tag{5}$$

where T_i is a transformation matrix that relates the local-axis end displacements for member i to the global-axis nodal displacements Δ, and k_{ij} is a stiffness matrix that relates the local-axis forces f_{ij} at end section j of member i to the local-axis end displacements for the member. For a planar member, f_{ij} is a 3×1 vector of the bending moment M, shear force V and axial force P at end-section j of member i. The sensitivity of a particular member force F_j (i.e., F_j=M, V or P) to change in any design variable, $\partial F_j/\partial x_{ik}$, is obtained by evaluating the corresponding right hand side of Eq.(5) for the current design of the structure.

3.3 STRENGTH SENSITIVITY

The sensitivity of section strength S_j to change in any design variable, $\partial S_j/\partial x_{ik}$, is given in the following for different shape sections and different measures of section strength, [8,9].

For a rectangular singly reinforced section, the sensitivities of its nominal flexural strength M_n to changes in the cross section design variables b, h and a_s are found as

$$\frac{\partial M_n}{\partial b} = \frac{a_s^2 f_y^2}{1.7b^2 f_c'} \tag{6a}$$

$$\frac{\partial M_n}{\partial h} = a_s f_y \tag{6b}$$

$$\frac{\partial M_n}{\partial a_s} = (h-d')f_y - \frac{a_s f_y^2}{0.85b f_c'} \tag{6c}$$

where d' is the depth of concrete cover over the tensile reinforcement, and f_y and f_c are the yield stress for steel and the uniaxial crushing strength for concrete, respectively.

A doubly reinforced section is adopted by this study when a singly reinforced section with maximum allowable tensile reinforcement $0.75a_{sb}$ is not capable of tolerating the applied force or controlling the deflection for acceptable section dimensions b and h (where a_{sb} is the reinforcement for the balanced strain condition). In addition, it is assumed that the required additional reinforcement is added to the section in both the tension and compression zones such that the position of the neutral axis remains unchanged. The amount of compression steel a''_s can then be obtained from $a''_s = \psi a_{s,add}$ where $a_{s,add}$ identifies the reinforcement in excess of that allowed for a singly reinforced section, and the parameter ψ is a proportioning factor depending on the concrete cover over the compressive steel and the location of the neutral axis. The sensitivities of the nominal flexural strength of the member section to changes in the design variables b, h and a_s are found as

$$\frac{\partial M_n}{\partial b} = 0.75a_{s_s} \frac{(h-d')}{b} f_y - \frac{(0.75a_{s_s} f_y)^2}{1.7b f_c'} \tag{7a}$$

$$\frac{\partial M_n}{\partial h} = (1.5a_{s_s} + (1-\psi)a_{s_{ad}})f_y - \frac{(0.75a_{s_s} f_y)^2}{0.85b(h-d')} \tag{7b}$$

$$\frac{\partial M_n}{\partial a_s} = (1-\psi)(h-d'-d'')f_y \tag{7c}$$

where d" is the depth of concrete cover over the compression steel.

The sensitivities of the nominal shear capacity V_n of a beam section to changes in the design variables b and h are found as (evidently, $\partial V_n/\partial a_s = 0$)

$$\frac{\partial V_n}{\partial b} = 2\sqrt{f_c'} \ (h-d') + 50(h-d') \tag{8a}$$

$$\frac{\partial V_n}{\partial h} = 2\sqrt{f_c'} \ b + 50b \tag{8b}$$

The sensitivities of the shear capacity V_n of a column section to change in the design variables b and h are found as (again, $\partial V_n/\partial a_s = 0$)

$$\frac{\partial V_n}{\partial b} = (2\sqrt{f_c'} + 50)(h-d') \tag{9a}$$

$$\frac{\partial V_n}{\partial h} = \frac{P_u}{1000h^2}\sqrt{f_c'}\, d' + (2\sqrt{f_c'} + 50)\, b \tag{9b}$$

where P_u is the ultimate compressive axial force.

The axial force capacity of a rectangular column with equal tensile and compressive steel reinforcement subject to uniaxial bending is given by, [10]

$$P_n = 0.85\beta_1\, f_c'\, b\, y - \frac{1}{2}\, f_s\, a_s + \frac{1}{2}\, f_s'\, a_s \tag{10}$$

where β_1=a coefficient based on the ultimate concrete crushing stress, y=distance the neutral axis is from the extreme compressive fibres of the section, f_s=stress in tensile reinforcement, and f'_s=stress in compressive reinforcement. Column axial force sensitivities may be evaluated using the following finite-difference technique. Determine the column axial force capacity from Eq.(10) for the current design {b, h, a_s} and the six neighbouring designs {b+δb, h+δh, a_s+δa_s} and {b-δb,h-δh, a_s-δa_s}, where δb, δh and δa_s are small specified increments in the design variables. Good quality estimates of the sensitivities of column axial force capacity are then found as

$$\frac{\partial P_n}{\partial b} = \frac{P_n(b+\delta b) - P_n(b-\delta b)}{2\delta b} \tag{11a}$$

$$\frac{\partial P_n}{\partial h} = \frac{P_n(h+\delta h) - P_n(h-\delta h)}{2\delta h} \tag{11b}$$

$$\frac{\partial P_n}{\partial a_s} = \frac{P_n(a_s+\delta a_s) - P_n(a_s-\delta a_s)}{2\delta a_s} \tag{11c}$$

4. Optimality Criteria Method

Equations (3) to (9) and Eq.(11) are employed to formulate the design optimization problem Eqs.(1) explicitly in terms of the k=1,2,3 design variables $x_{ik} \equiv \{b, h, a_s\}_i$ for each of the i=1,2, ..., n beam and column members comprising the reinforced concrete framework. An Optimality Criteria (OC) method is then employed to solve the explicit design optimization problem. To facilitate the description of the OC solution algorithm, the explicit form of Eqs.(1) is concisely expressed as

Minimize :
$$Z = Z(x_{ik}) \tag{12a}$$

Subject to :
$$g_j(x_{ik}) \leq 0 \qquad\qquad (j=1,2,...,m) \tag{12b}$$

$$x_{ik}^L \leq x_{ik} \leq x_{ik}^U \qquad\qquad (i=1,2,...,n \; ; \; k=1,2,3) \tag{12c}$$

where Eq.(12a) represents the objective function Eq.(1a), Eqs.(12b) collect together all the strength and displacement constraints Eqs.(3a) and (3b), and Eqs.(12c) define the sizing constraints Eqs.(1d).

Temporarily ignoring the sizing constraints Eqs.(12c), the design optimization problem can be reformulated as the minimization of the Lagrangian function

$$L(x_{ik},\lambda_j) = Z(x_{ik}) + \sum_{j=1}^{m} \lambda_j \, g_j(x_{ik}) \tag{13}$$

where the Lagrange multipliers are such that $\lambda_j > 0$ if constraint j is active, or $\lambda_j = 0$ if constraint j is inactive. Temporarily assume that all constraints are active.

Differentiate Eq.(13) with respect to the design variables x_{ik}, and rearrange terms to obtain the stationary conditions

$$1 = - \sum_{j=1}^{m} \lambda_j \frac{\dfrac{\partial g_j}{\partial x_{ik}}}{\dfrac{\partial Z}{\partial x_{ik}}} \qquad\qquad (i=1,..., n \; ; \; k=1,2,3) \tag{14}$$

A recursive algorithm is applied to find the design variables x_{ik} satisfying Eqs.(14) at the optimum. To this end, multiply both sides of Eqs.(14) by x^η_{ik} and take the ηth root and, then, apply a first-order binomial expansion to obtain the linear recursive relation

$$x_{ik}^{v+1} = x_{ik}^{v} \left[1 - \frac{1}{\eta} \left(1 + \sum_{j=1}^{m} \lambda_j \frac{\dfrac{\partial g_j}{\partial x_{ik}}}{\dfrac{\partial Z}{\partial x_{ik}}} \right) \right]_v \qquad (i=1,2,...,n \ ; \ k=1,2,3) \qquad (15)$$

where η is a step-size parameter that controls convergence of the recursive process, while $v+1$ and v indicate successive iterations ($v=0$ corresponds to the initial design stage).

In order to apply Eqs.(15) to find the new design variables x_{ik}^{v+1}, the current values of the Lagrange multipliers λ_j^v must first be determined. To that end, for the $(v+1)$th iteration of the recursive process, consider the change Δg_l in the lth constraint due to changes Δx_{ik} in the design variables, i.e.,

$$\Delta g_l = g_l (x_{ik}^v + \Delta x_{ik}) - g_l(x_{ik}^v) = \sum_{i=1}^{n} \sum_{k=1}^{3} \frac{\partial g_l}{\partial x_{ik}} \Delta x_{ik} \qquad (16)$$

Assuming the constraint becomes active after the change such that $g_l(x_{ik}^v + \Delta x_{ik}) = 0$, and noting from Eqs(15) that

$$\Delta x_{ik} = x_{ik}^{v+1} - x_{ik}^v = -\frac{x_{ik}^v}{\eta} \left(1 + \sum_{j=1}^{m} \lambda_j \frac{\dfrac{\partial g_j}{\partial x_{ik}}}{\dfrac{\partial Z}{\partial x_{ik}}} \right)_v \qquad (i=1,2,...,n \ ; \ k=1,2,3) \qquad (17)$$

we have from Eqs.(16) and (17) that

$$g_l(x_{ik}) = \sum_{i=1}^{n} \sum_{k=1}^{3} \frac{\partial g_l}{\partial x_{ik}} \frac{x_{ik}^v}{\eta} \left(1 + \sum_{j=1}^{m} \lambda_j \frac{\dfrac{\partial g_j}{\partial x_{ik}}}{\dfrac{\partial Z}{\partial x_{ik}}} \right)_v \qquad (l=1,2,...,m) \qquad (18)$$

Rearranging Eqs.(18), we obtain

$$\sum_{j=1}^{m} \lambda_j \sum_{i=1}^{n} \sum_{k=1}^{3} x_{ik}^v \left(\frac{\dfrac{\partial g_l}{\partial x_{ik}} \dfrac{\partial g_j}{\partial x_{ik}}}{\dfrac{\partial Z}{\partial x_{ik}}} \right) = \eta g_l(x_{ik}^v) - \sum_{i=1}^{n} \sum_{k=1}^{3} x_{ik}^v \frac{\partial g_l}{\partial x_{ik}} \quad (l=1,2,...,m) \quad (19)$$

which is a system of simultaneous linear equations in terms of m lagrange variables λ_j.

The Optimality Criteria method to solve Eqs.(12) involves the sequential application of Eqs.(15) and (19) in an iterative fashion until the structure cost converges to a minimum value. The details of the design optimization procedure are as follows:

1. set $v=0$ and adopt an initial set of design variables x_{ik}^v (i=1,2,...,n; k=1,2,3).
2. for the current x_{ik}^v, establish the gradient vector $\partial Z / \partial x_{ik}$.
3. For the current x_{ik}^v, analyze the structure and establish the gradient vectors $\partial g_j/\partial x_{ik}$ (j=1,2,...,m) for the m constraints that are currently potentially active.
4. For the current active x_{ik}^v, use the Gauss-Seidel recursive technique to solve Eqs.(19) for the set of Lagrange multipliers λ_j^v (j=1,2,...,m); any $\lambda_j^{\tau+1}$ value found to be negative at any stage is temporarily set to zero to reflect the fact that the corresponding constraint is presently not active. When convergence of the Gauss-Seidel technique has occurred such that $\lambda_j^\tau = \lambda_j^{\tau+1}$, the solution of Eqs.(19) has been found as $\lambda_j^v = \lambda_j^{\tau+1}$
5. For the current active x_{ik}^v and current λ_j^v, find the new set of active design variables x_{ik}^{v+1} from Eqs.(15).
6. Check the convergence of the OC recursive process: if all $x_{ik}^{v+1} = x_{ik}^v$ and $\lambda_j^v = \lambda_j^{v-1}$, go to 7; otherwise, set $v=v+1$ and update Eqs.(19) for the current x_{ik}^v values and return to step 4.
7. Check convergence of the overall design process: if the structure cost from Eq(12a) is the same for two successive design cycles, terminate with the minimum cost structure; otherwise, set $v=0$ and return to step 2.

6. Design Example

Consider the simple two member reinforced concrete framework having the structural geometry, loading and material properties shown in Figure 1, [8]. The structure is subject to uniform gravity live loading of 2.4 kips/ft., plus distributed self-weight loading calculated on the basis of overall section area bh and concrete density =150 lb/ft^3 (note, then, that the self-weight loading is variable to the design). The depth d' of concrete cover on the reinforcing steel for the beam and column members is fixed at 2.5 and 2 inches, respectively. For ease of construction, it is required to have equal section width for the beam and column members. The design variables are width of beam and column section, b, depth of column section, h_1, depth of beam section, h_2, steel reinforcement area for the column, a_{s1}, and steel reinforcement areas at the left end, middle and right end of the beam, a_{s2}, a_{s3} and a_{s4}, respectively. The design optimization problem is to find the values of the design variables such as to minimize the cost of the structure, accounting for the costs of concrete, reinforcement and formwork, while satisfying all necessary performance constraints imposed by the ACI 318-89 code of practice, [11].

The ratio of the unit-volume cost of steel to that for concrete is taken as 60, while the ratio of the unit-area cost of formwork to the unit-volume cost of concrete is taken as 0.6. It is assumed that the column satisfies the short column definition in a braced system and that its moment magnification factor is equal to 1. The deflection of the beam is not accounted for in the performance conditions and, therefore, since sway does not occur for the structure, the design optimization problem involves strength and sizing constraints alone, as follows,

Minimize :

$$Z = [bh_1 + (60-1)a_{s_1} + 2\times0.6(b+h1)]L_{column} +$$

$$[bh_2 + (60-1)(0.2a_{s_2} + 0.6a_{s_3} + 0.2a_{s_4}) + 0.6(b+2h_2)]L_{beam} \qquad (20a)$$

Subject to :

$$P_{u_{column}} \leq \phi P_{n_{column}} \qquad ; \qquad V_{u_{column}} \leq \phi V_{n_{column}} \qquad (20b,c)$$

$$M_{u\ beam}^{L} \leq \phi M_{n\ beam}^{L} \qquad ; \qquad M_{u\ beam}^{M} \leq \phi M_{n\ beam}^{M} \qquad (20d,e)$$

$$M_{u\ beam}^{R} \leq \phi M_{n\ beam}^{R} \qquad ; \qquad V_{u_{beam}} \leq \phi V_{n_{beam}} \qquad (20f,g)$$

$$9 \leq b \leq 15 \quad ; \quad 12 \leq h_1 \leq 20 \quad ; \quad 14 \leq h_2 \leq 25 \quad (in) \qquad (20h,i,j)$$

$$1.2 \leq A_{s_1} \leq 15 \qquad ; \qquad 1 \leq A_{s_2} \leq 8 \qquad (in^2) \qquad (20k,l)$$

$$1 \leq A_{s_3} \leq 8 \qquad ; \qquad 1 \leq A_{s_4} \leq 10 \qquad (20m,n)$$

Equation (20a) is the objective function defining the total cost of the structure, assuming that the length of the steel reinforcement at the left end, middle and right end of the beam is 20, 60 and 20 percent of the beam length, respectively. Equations (20b) to (20g) are performance constraints on the axial strength capacity of the column, shear capacity of the column, moment capacity of the beam at the left end, middle and right end, and shear strength of the beam, respectively (for each such constraint, quantities with subscript u and n refer to the ultimate force and nominal capacity for the member section, respectively, and ϕ is the corresponding code-specified capacity reduction factor, [11]). Equations (20h) to (20n) are sizing constraints on section dimensions and reinforcement areas. The constraints on section width b and depth h reflect architectural and construction requirements. The upper and lower bounds on reinforcement areas are imposed in accordance with the ACI 318-89 design code [11] and, while they normally vary with the dimensions of the section, they are here taken as fixed constant values.

The previously described Optimality Criteria (OC) method is employed to solve the design problem posed by Eqs.(20). In order to evaluate the influence that the initial design point has on the OC solution technique, the design example is solved in two different ways: (1) starting with a feasible initial design point; (2) starting with an infeasible initial design point.

Figure 2(a) shows the feasible initial column and beam sections. Upon applying the OC technique for these initial sections, the final sections shown in Figure 2(b) were found to correspond to the least-cost design of the framework. The solution history for this design is shown in Figure 3, where it can be noted that the OC method converged smoothly to the optimum design in only four iterations. To gain an appreciation of the performance characteristics of the recursive OC technique for resizing the design variables, the variations of the design variables during the corresponding optimization process are shown in Figure 4, where it can be noted that the design variables also converged smoothly to their final values.

ACKNOWLEDGEMENTS

This work was sponsored by the Scholarship Department of the Ministry of Culture and Higher Education of Iran for the second author, and by the Natural Science and Engineering Research Council of Canada for the first author, and is based upon research conducted by the second author under the supervision of the first author for the degree of Doctor of Philosophy in Civil Engineering at the University of Waterloo, Canada.

REFERENCES

1. Shunmugavel,P., "Optimization of Two-Dimensional Reinforced Concrete Building Frames", Ph.D. Thesis, University of Illinois at Urbana, Champaign, 1974

2. Yang,M.F., "Optimization of Reinforced Concrete Structures", Ph.D. Thesis, University of Illinois at Urbana-Champaign, 1982

3. Booz,W. , Legewie,G. and Thierauf,G., "Optimization of Reinforced Concrete Structures According to German Design Regulations", in proceedings of the International Conference on Computer Aided Analysis and Design of Concrete Structures", 1984, Yugoslavia, pp. 761-773

4. Kanagasundaram,S. and Karihaloo,B.L. "Minimum Cost Design of Reinforced Concrete Structures", *Structural Optimization*, Vol.2, No. 3, 1990, pp. 173-184

5. Berke,L. and Khot,N.S., "Use of Optimality Criteria Methods For Large Scale

Systems", AGARD-LS-70, 1974.

6.   Khot,N.S., Berke,L. and Venkayya,V.B., "Comparison of Optimality Criteria Algorithms for Minimum Weight Design of Structures", AIAA/ASME/SAE 19th Structures, Structural Dynamics and Materials Conference, Bethesda, Md. 1978. pp. 37-46

7.   Venkayya,V.B., "Optimality Criteria: A Basic Multidisciplinary Design Optimization", *Computational Mechanics*, Vol. 5, 1989, pp. 1-21

8.   Moharrami, H. "Design Optimization of Reinforced Concrete Structures", Ph.D. Thesis, University of Waterloo, Canada (in progress 1991).

9.   Moharrami,H. and Grierson,D.E., "Computer Automated Design of Reinforced Concrete Frameworks", ASCE, *Journal of Structural Engineering* (submitted 1991).

10.   Wang, C.K. and Salmon, C.G. "Reinforced Concrete Design", Harper & Row, Publishers, inc., New York, 1979.

11.   Building Code Requirements for Reinforced Concrete, American Concrete Institute, (ACI 318-89) Detroit, 1989.

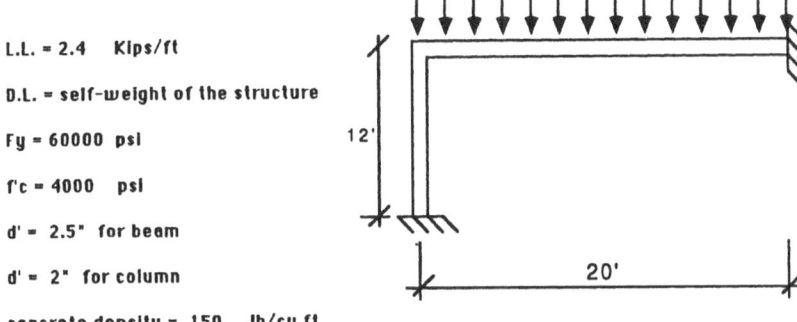

L.L. = 2.4   Kips/ft

D.L. = self-weight of the structure

Fy = 60000 psi

f'c = 4000   psi

d' = 2.5" for beam

d' = 2" for column

concrete density = 150   lb/cu ft

12'

20'

Figure 1: Structure Geometry and Loading

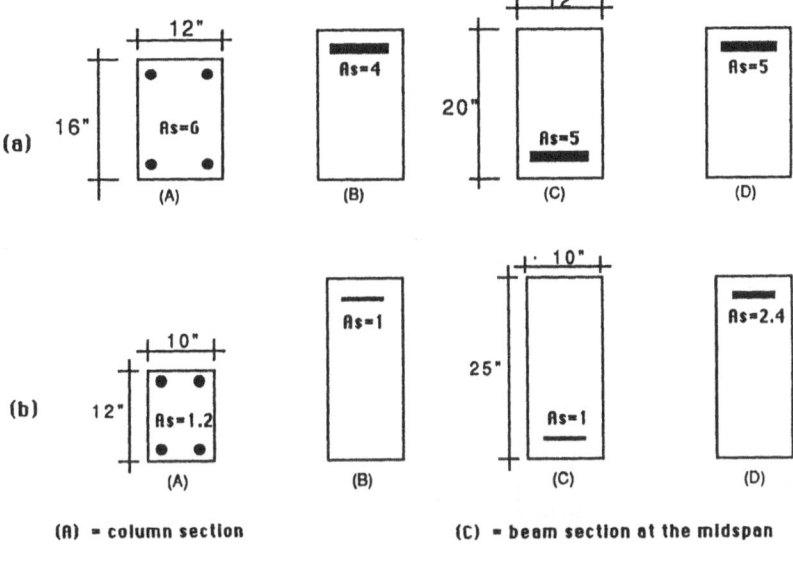

(a)

(b)

(A) = column section

(B) = beam section at the left side

(C) = beam section at the midspan

(D) = beam section at the right side

Figure 2 : (a) Feasible Initial Column and Beam Sections
(b) Final Design of Column and Beam Sections

896

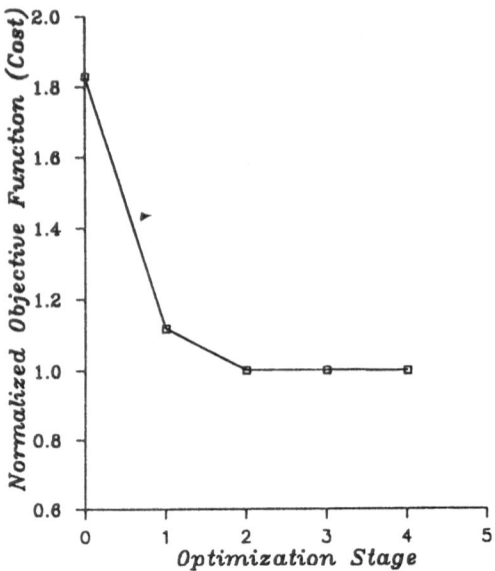

Figure 3 : Feasible start OC Solution History

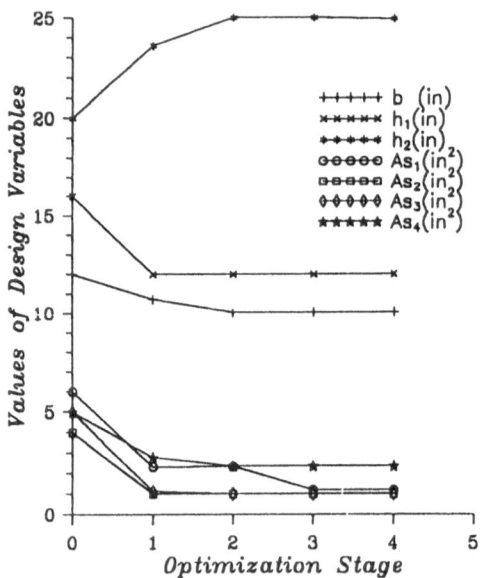

Figure 4 : Variation of Design Variables for
Feasible start OC Solution History

OPTIMUM DESIGN OF PLANE STRUCTURAL FRAMES BY NON-LINEAR PROGRAMMING

B.L. KARIHALOO and S. KANAGASUNDARAM
School of Civil & Mining Engineering
The University of Sydney
N.S.W. 2006
Australia

1. Introduction

This lecture will summarize the results of several recent publications by Kanagasundaram & Karihaloo (1985, 1988, 1990), and Karihaloo & Kanagasundaram (1986, 1987, 1988a,b,c, 1989) on the minimum weight design of plane structural frames under multiple load systems. Each load system consists of external concentrated and distributed loads, couples, *and* self-weight. It is required that the normal and shear stresses, the maximum transverse deflection and the critical load factor in buckling not violate prescribed limits under any load system. In order to obtain designs which are practically feasible, the variation of the mass (stiffness) along the frame members is restricted to splines of order zero or one. Within these restricted classes of variation, the optimization problem reduces to a non-linear programming problem whose solution is attempted by several methods. Throughout this work, it has been assumed that the material is linear and elastic and that the engineering beam-column theory (e.g. Timoshenko & Gere, 1961) is applicable.

The contents of this lecture are arranged so that the approach taken towards achieving the set goals unfolds itself gradually. With this in mind, the minimum weight optimization problem for a general beam-column member, isolated from a plane frame, is first formulated in Section 2 and then reduced to a non-linear programming problem (NLP). This reduction process is facilitated by the application of minimax variational approach to the differential stress constraints. In Section 3 this reduction process is illustrated on a typical plane frame member of different cross-sectional shapes. Section 4 presents an iterative scheme for the solution of the non-linear programming problem. A complete computer code in standard FORTRAN 77 for the scheme can be found in the paper by Kanagasundaram & Karihaloo (1990) and an up-to-date listing on a diskette is available on request.

In Section 5 the concepts introduced in Sections 2 to 4 are then used to obtain minimum-weight designs of statically indeterminate beams and plane structural

G. I. N. Rozvany (ed.), Optimization of Large Structural Systems, Vol. II, 897–926.
© 1993 Kluwer Academic Publishers.

frames. The frames are subjected to multiple loads specified in the design codes. The solution of the non-linear programming problems is attempted by several techniques. The Section ends with an elaborate discussion of the results obtained.

The concepts described in this lecture will be applied in the next lecture to minimum-cost reinforced concrete structures designed according to the limit-state philosophy.

2. Mathematical Formulation

We formulate now the minimum-weight design problem of a frame with constraints on strength, stiffness, stability and geometry. The strength and stiffness constraints are imposed on a typical plane frame member whereas the stability constraint is imposed on the frame as a whole. The strength constraints are simplified using the minimax approach. The cross-sectional area is represented by splines of suitable order, reflecting constant and linear variations. The original optimization problem is finally restated as a non-linear mathematical programming problem of minimizing a linear mass functional subject to non-linear inequality and equality constraints. Consider an isolated elastic member in flexure which forms a part of a plane frame (Figure 1). The member is subjected to reactive forces $P_i (i = 1, 2, \ldots, 6)$ from the joints of the parent frame, as well as external transverse forces $f(x)$ and self-weight $g(x)$. The external forces $f(x)$ consist of several individual force systems $\{f_\ell(x), \ell = 1, \ldots, NL\}$ each acting on the frame on separate occasions. The equilibrium state of the member of given stiffness under any load system is completely described by the corresponding bending moment $M_\ell(x)$ and shear force $Q_\ell(x)$, together with the appropriate statical boundary conditions. The force resultants are related as follows:

$$\frac{dM_\ell}{dx} = Q_\ell(x) \tag{1}$$

$$\frac{dQ_\ell}{dx} = -f_\ell(x) - g(x)\cos\theta \tag{2}$$

in which x refers to an arbitrary section along the member. It is also useful to describe the equilibrium of the member m in terms of its deformations since we will be interested in deflection and buckling constraints

$$EI_m(x)u_{xx} + M_\ell(x) = 0 \tag{3}$$

in which u_{xx} stands for d^2u/dx^2, and subscript x denotes differentiation with respect to the axial co-ordinate, $u(x)$ being the transverse deflection of the centreline and E the Young modulus. The geometric boundary conditions of the member must be supplied in order to obtain the unique solution of (3).

Throughout this paper it will be assumed that the mass of the member (which is proportional to the cross-sectional area $A_m(x)$) is related to its flexural stiffness (which is proportional to second moment of area $I_m(x)$) through

$$I_m(x) = cA_m^n(x) \tag{4}$$

in which c and n are constants determined by the cross-sectional shape. For instance, $n = 1$ represents a rectangular section of constant depth, $n = 2$ – a geometrically similar cross-section, say circular or square, and $n = 3$ – a rectangular section of constant width.

The strength (stress) constraints on each and every cross section of a member m $(m = 1, \ldots, NM)$ of given cross-sectional shape (n=1,2 or 3) under a load system $\ell(\ell = 1, \ldots, NL)$ may be written as differential inequalities

$$\sigma_i\{x, s, u_\ell, (u_x)_\ell, f_\ell, g, M_\ell, Q_\ell, A_\ell, (A_x)_\ell\} \le 0 \tag{5}$$

in which x and s are the co-ordinates of the cross-sectional fibre. The two stresses $(i = 1, 2)$ refer to normal and shear stresses. The normal stresses may be made up of two components due to axial force and bending. In the sequel, reference to the load system and member are omitted unless otherwise required.

The serviceability requirement limiting the magnitude of the maximum deflection u_{max} anywhere in the member under the load system ℓ to, say Δ_m is

$$|(u_{max})_\ell|_m \le \Delta_m \tag{6}$$

in which $u_{max} = \max u(x); 0 \le x \le L_m$, and L_m is the length of member m. The location of the maximum deflection need not be specified. The transverse deflection $u(x)$ is obtained by solving the differential equation (3) for a known stiffness distribution $I_m(x)$.

The strength and stiffness constraints are imposed on individual members, whilst the stability constraint is imposed on the frame as a whole. The latter constraint may be written as

$$(\lambda_{cr})_\ell/FS \ge 1; (\ell = 1, \ldots, NL) \tag{7}$$

where λ_{cr} is the critical load factor and FS is the factor of safety against buckling. In addition to the above constraints, the designs may be required to meet geometric constraints, for instance the value of the optimum sizing parameter may be required to be the same on either side of a joint other than a support. These constraints may be formally written as

$$G_j(A_m) = 0; (j = 1, \ldots, NC) \tag{8}$$

where NC is the total number of geometric constraints.

The optimization problem may now be stated as follows: Determine the design variables $A_m(x)(m = 1, \ldots, NM)$ and, thus $I_m(x)$ that meet the strength, stiffness, stability and geometric constraints (expressions 5–8) and minimize the total mass W of the frame

$$W = \sum_{m=1}^{NM} \gamma_m \int_0^{L_m} A_m(x)dx \qquad (9)$$

where γ_m is the mass density of the material used for member m.

2.1 THE MINIMAX APPROACH

In the minimax approach the differential game problem described by (5) and (9) is reduced to a minimax variational one. Assuming that the equilibrium equations, (1) and (2), or (3), with the appropriate boundary conditions, can be solved in a closed form, the expressions for the force resultants $M(x)$ and $Q(x)$ are known for each load system. It should be mentioned that $M(x)$ and $Q(x)$ depend not only on x, f and u but also on the control $A(x)$ itself through the compatibility conditions and self-weight. The control $A(x)$, in turn, depends on all the load systems. Substituting the (implicit) analytical expressions for M and Q, i.e. $M = M(x, u, f, g, A, A_x)$, $Q = Q(x, u_x, f, g, A, A_x)$, into the differential inequalities (5) gives, for each load system,

$$\sigma_i\{x, s, u, u_x, f, g, M(x, u, f, g, A, A_x), Q(x, u_x, f, g, A, A_x), A, A_x\} \leq 0 \qquad (10)$$

or, for brevity,

$$\Omega_i(x, s, u, u_x, f, A, A_x) \leq 0; (i = 1, 2) \qquad (11)$$

It should be mentioned that as $g(x) = \gamma A(x)$ it has not been repeated separately in (11). Next, the maxima of Ω_i with respect to s and f are determined for fixed values of x and $A(x)$. Let these be attained at $s = s_i^*$ and $f = f_i^*$, i.e.

$$\Omega_i(x, s_i^*, u, u_x, f_i^*, A, A_x) \equiv max_s max_f \Omega_i(x, s, u, u_x, f, A, A_x) \qquad (12)$$

and denote

$$\Omega_i^*(x, u, u_x, A, A_x) = \Omega_i\{x, s_i^*(x, A), u, u_x, f_i^*(x), A, A_x\} \qquad (13)$$

In practice, as the number of load systems (i.e. NL) is discrete, the maximum with respect to f is to be chosen among a finite set. Using the notation (13), the differential inequalities (5) reduce to

$$\Omega_i^*(x, u, u_x, A, A_x) \leq 0 \qquad (14)$$

Moreover, if it is assumed that the boundary value problem described by the equilibrium equation (3) and the associated boundary conditions has been solved so that the transverse deflection $u(x)$ and $u_x(x)$ are known for the particular distribution $A(x)$, the strength constraints (14) can be further simplified to read

$$\Omega_i^*(x, A, A_x) \leq 0 \qquad (15)$$

Next, each member is divided into a given number of segments $(NSEG_m)$ and the order of variation (NOR_m) of cross-sectional area $A_m(x)$ is prescribed using splines. For example, splines of order zero $(NOR_m = 0)$ are used to represent constant area within a segment and of order one and two – linear and parabolic variations (Figure 2). Irrespective of the order of spline chosen for each segment, $A_m(x)$ will be of the form

$$A_m(x) = A_m(x, \mu_k); (k = 1, \ldots, N_m) \tag{16}$$

where μ_k denote the unknown design variables at the knots and N_m is the total number of unknowns: $N_m = NSEG_m$ if $NOR_m = 0$ and $N_m = NSEG_m + 1$ if $NOR_m = 1 or 2$. The design variables may be the variable width $b_k(x)$ at the knots if $n = 1$ or the section size $r_k(x)$ if $n=2$ or the variable depth $h_k(x)$ if $n = 3$. Using the representation (16), the inequalities (15) become

$$\Omega_i^*(x, \mu_k) \leq 0 \tag{17}$$

Thus, the original optimization problem can now be restated as follows: Minimize

$$W = \sum_{m=1}^{NM} \gamma_m \int_0^{L_m} A_m(x, \mu_k) dx; (k = 1, \ldots, N_m)$$

subject to

$$\begin{aligned}
\Omega_i^*(x, \mu_k) &\leq 0; (i = 1, 2) \\
|(u_{max})_\ell|_m &\leq \Delta_m \\
(\lambda_{cr})_\ell / FS &\geq 1; (\ell = 1, \ldots, NL) \\
G_j(A_m) &= 0; (j = 1, \ldots, NC)
\end{aligned} \tag{18}$$

Application of the above procedure is illustrated below on an example.

3. Application to Plane Frame Members

To illustrate the procedure discussed in the preceding section, consider an arbitrary flexural member subjected to external loading, as well as reactive forces from the joints of the parent frame (Figure 1). Let the reactive forces due to each of the external load systems $f_\ell(x)(\ell = 1, \ldots, NL)$ be $P_{\ell i}(i = 1, \ldots, 6)$. It is assumed that a typical load system consists of a distributed load $q_\ell(x)$, concentrated loads $W_{\ell j}$ located at positions a_j along the member length and concentrated couples $C_{\ell j}$ acting at locations b_j, as shown in the figure. It should be emphasized that the treatment is equally applicable to a curved member, provided the distance x is measured along the arc joining the centres of the member sections.

902

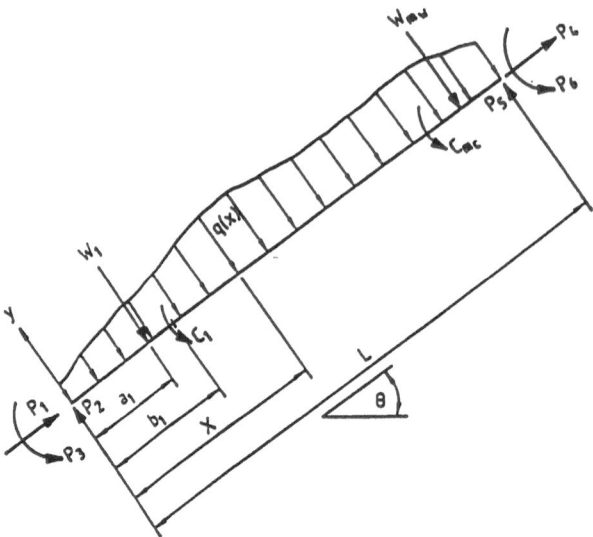

Figure 1: A typical plane frame member subjected to external load system $f_\ell(x)$ and reactive forces $P_i(i = 1, \ldots, 6)$ from the joints of the parent frame. Gravitational force due to self-weight is not shown

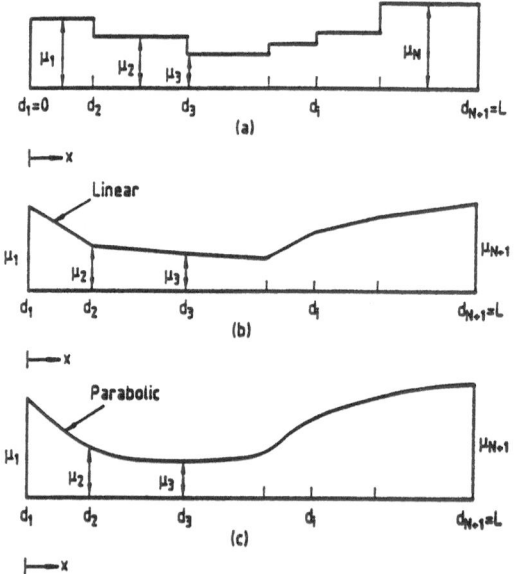

Figure 2: Representation of cross-sectional area $A_m(x, \mu_k)$ using splines

Assuming that the reactive forces $P_{\ell i}$ are known, the bending moment $M_\ell(x)$ and shear force $Q_\ell(x)$ at a section x are given by

$$
\begin{aligned}
M_\ell(x) \;=\;& P_{\ell 2}x + P_{\ell 1}u - \gamma_m \int_0^x \sin\theta[u(x) - u(\eta)]A(\eta)d\eta \\
& - \sum_{j=1}^{m_w} W_{\ell j} <x - a_j>^1 - P_{\ell 3} - \sum_{j=1}^{m_c} C_{\ell j} <x - b_j>^0 \\
& - \int_0^x [q_\ell(\eta) + \gamma_m A_m(\eta)\cos\theta](x - \eta)d\eta \qquad (19)
\end{aligned}
$$

$$
\begin{aligned}
Q_\ell(x) \;=\;& P_{\ell 2} + [P_{\ell 1} - \gamma_m \int_0^x A_m(\eta)\sin\theta d\eta]u_x - \sum_{j=1}^{m_w} W_{\ell j} <x - a_j>^0 \\
& - \int_0^x [q_\ell(\eta) + \gamma_m A_m(\eta)\cos\theta]d\eta \qquad (20)
\end{aligned}
$$

where m_w and m_c are respectively the total number of concentrated loads and couples acting on the member. The integral terms involving the section area $A_m(x)$ in each of the above equations express the contributions from the self-weight of the member.

The strength constraints limiting the magnitude of the normal (longitudinal) stress σ_{xx} and shear (transverse) stress τ_{xy} at the section x are

$$
\sigma_1 \equiv max_f \,|\sigma_{xx}| - \sigma_0 \leq 0 \qquad (21)
$$
$$
\sigma_2 \equiv max_f \,|\tau_{xy}| - \tau_0 \leq 0 \qquad (22)
$$

where σ_0 and τ_0 are known positive constants (allowable stresses). It is assumed that the magnitude of σ_0 is the same in tension and compression. Sometimes it is convenient to specify the following strength criterion

$$
max_f \,(\sigma_{xx}^2 + \alpha\tau_{xy}^2)^{1/2} - k \leq 0 \qquad (23)
$$

If $\alpha = 4$ it restricts the maximum shear stress and if $\alpha = 3$ – the value of the strain energy. Note that for $\alpha = 4$ and $\sigma_0 = 2\tau_0 = k$, (23) is identical to (21) and (22). From elastic beam-column theory it is known that

$$
|\sigma_{xx}| = \frac{|M_\ell(x)s(x)|}{I_m(x)} + \frac{|P_\ell(x)|}{A_m(x)} \qquad (24)
$$

$$
\tau_{xy} = K(x)\frac{\partial}{\partial x}\left\{\frac{M_\ell(x)a(x,s)}{I_m(x)}\right\} \qquad (25)
$$

where $P_\ell(x) = P_{\ell 1} - \gamma_m \int_0^x A_m(\eta)\sin\theta d\eta$ and $s(x)$ determines the position of the cross-sectional fibre. $K(x)$ and $a(x,s)$ are dependent on the cross-sectional shape. For instance, for a rectangular section of constant depth (h) and variable width $b(x)$, i.e. for $n = 1, K(x) = 1/b(x), a(x,s) = b(x)[(h^2/4) - s^2(x)]/2$ and $-h/2 \leq$

$s(x) \leq h/2$, whereas for a circular of radius $r(x)$ cross-section $(n = 2)$ $K(x) = 1/\{3[r^2(x) - s^2(x)]^{1/2}\}, a(x,s) = [r^2(x) - s^2(x)]^{3/2}$ and $-r(x) \leq s(x) \leq r(x)$; for a rectangular section of constant width (b) and variable depth $h(x)$, i.e. for $n = 3, K(x) = 1/b, a(x,s) = b[h^2(x)/4 - s^2(x)]/2$ and $-h(x)/2 \leq s(x) \leq h(x)/2$. For a square section of side $r(x)$, $K(x) = 1/h(x)$ and $a(x,s) = h(x)[h^2(x)/4 - s^2(x)]/2$. For other cross-sectional shapes, see the paper by Kanagasundaram & Karihaloo (1990). It should be noted that a more accurate expression for shear stress τ_{xy} than (25) can be written by considering the other shear stress component and satisfying compatibility of strains (see, e.g. Love, 1944). However, the expression for τ_{xy} in (25) based on the engineering design theory is adequate for most practical cases.

In view of the dependence of $K(x)$ and $a(x,s)$ on the shape parameter n, it is convenient to illustrate the procedure separately for each value of n.

3.1 RECTANGULAR SECTION OF CONSTANT DEPTH $(n = 1)$

In the case of a flexural member having rectangular section with constant depth h and variable width $b(x)$ the strength constraints (21,22) are

$$\sigma_1 \equiv \left\{ \frac{1}{b(x)h^3} \right\} max_s \, max_f \, [12|M_\ell(x) \, s(x)| + h^2|P_\ell(x)|] - \sigma_0 \leq 0 \qquad (26)$$

$$\sigma_2 \equiv \left\{ \frac{6}{b(x)h^3} \right\} max_s \left[\frac{h^2}{4} - s(x)^2 \right] max_f \, |Q_\ell(x)| - \tau_0 \leq 0 \qquad (27)$$

The maxima of σ_1 and σ_2 with respect to s are attained at $s_1^* = \pm h/2$ and $s_2^* = 0$ respectively, and so giving

$$\Omega_1^* \equiv \left\{ \frac{1}{b(x)h^2} \right\} max_f \, [6|M_\ell(x)| + h|P_\ell(x)|] - \sigma_0 \leq 0 \qquad (28)$$

$$\Omega_2^* \equiv \left\{ \frac{3}{2b(x)h} \right\} max_f \, |Q_\ell(x)| - \tau_0 \leq 0 \qquad (29)$$

It is interesting to note that the differential inequalities (15) degenerate into algebraic inequalities (28) and (29) for the cross-sectional shape under consideration. Therefore the lower bound on the optimum width of the member can be written by inspection

$$b(x) \geq max \left\{ \frac{\bar{M}^*(x)}{\sigma_0 h^2}, \frac{3\bar{Q}(x)}{2\tau_0 h} \right\} \qquad (30)$$

where

$$\begin{aligned} \bar{M}^*(x) &= max \, \{M_1^*(x), M_2^*(x), \ldots, M_{NL}^*(x)\} \\ \bar{Q}(x) &= max \, \{|Q_1(x)|, |Q_2(x)|, \ldots, |Q_{NL}(x)|\} \\ M_\ell^*(x) &= 6|M_\ell(x)| + h|P_\ell(x)| \end{aligned}$$

If it is assumed that the member consists of segments of linearly varying width $b(x)$, i.e. $NOR = 1$ (see Figure 2b) then for the ith segment $(d_i \leq x \leq d_{i+1})$ the stress constraints (30) give

$$\{1 - T(x)\} b_i + T(x) b_{i+1} \geq max \left\{ \frac{\bar{M}^*(x)}{\sigma_0 h^2}, \frac{3\bar{Q}(x)}{2\tau_0 h} \right\} \tag{31}$$

where $T(x) = (x - d_i)/(d_{i+1} - d_i)(i = 1, \ldots, NSEG_m)$ and b_i are the values of the optimum width at the knots. The mass of the member is

$$W_m = \left\{ \frac{\gamma_m h_m}{2} \right\} \sum_{i=1}^{NSEG_m} (b_i + b_{i+1})(d_{i+1} - d_i) \tag{32}$$

If the optimum member is desired to have segments of constant $(NOR_m = 0)$ width, the strength constraints and the mass of the member will become

$$b(x) = b_i \geq max\ (F_1, F_2); (i = 1, \ldots, NSEG_m) \tag{33}$$

$$W_m = (\gamma_m h_m) \sum_{i=1}^{NSEG_m} b_i(d_{i+1} - d_i) \tag{34}$$

where F_1 and F_2 are the expressions within the braces in (31).

3.2 GEOMETRICALLY SIMILAR SECTIONS $(n = 2)$

The procedure is illustrated on a member of circular cross-section of radius $r(x)$. The strength constraints (21,22) become (with $a(x) = 1/(\pi r^4(x))$)

$$\sigma_1 \equiv a(x)\ max_s\ max_f\ [4|M_\ell(x)\ s(x)| + r^2(x)|P_\ell(x)|] - \sigma_0 \leq 0 \tag{35}$$

$$\sigma_2 \equiv \frac{4\ a(x)}{3}\ max_s\ max_f\ \left| r(x)Q_\ell(x)[r^2(x) - s^2(x)] + M_\ell(x)\frac{dr}{dx}[4s^2(x) - r^2(x)] \right|$$
$$- \tau_0 \leq 0 \tag{36}$$

The maximum of σ_1 with respect to s is attained at $s_1^* = \pm r(x)$, whereas that of σ_2 is attained at either $s_2^{*2} = 0$ or $s_2^{*2} = r^2(x)$. The differential inequalities (12) therefore reduce to

$$\Omega_1^* \equiv \left\{ \frac{1}{\pi r^3(x)} \right\} max_f\ [4|M_\ell(x)| + r(x)|P_\ell(x)|] - \sigma_0 \leq 0 \tag{37}$$

$$\Omega_2^* \equiv max\ (\psi_1, \psi_2) - \tau_0 \leq 0 \tag{38}$$

where

$$\psi_1 = \left\{ \frac{4}{3\pi r^3(x)} \right\} max_f\ \left| Q_\ell(x)\ r(x) - M_\ell(x)\frac{dr}{dx} \right|;\ s_2^* = 0 \tag{39}$$

$$\psi_2 = \left\{ \frac{4}{\pi r^3(x)} \right\} max_f\ \left| M_\ell(x)\frac{dr}{dx} \right|;\ s_2^{*2} = r^2(x) \tag{40}$$

The algebraic inequality (37) simplifies to a cubic in $r(x)$, whereupon

$$r(x) \geq 2\,max_f \left\{ \sqrt{B_\ell(x)} \cos\left[\frac{C_\ell(x)}{3}\right] \right\}; \quad D_\ell^2(x) < 0 \tag{41}$$

$$\geq 2\,max_f \sqrt[3]{2M_\ell(x)/(\pi\sigma_0)}; \quad D_\ell^2(x) = 0$$

$$\geq \left\{\frac{2}{\pi\sigma_0}\right\}^{\frac{1}{3}} max_f \, \bar{M}_\ell^*(x); \quad D_\ell^2(x) > 0$$

where

$$B_\ell(x) = \frac{|P_\ell(x)|^3}{3\pi\sigma_0} \tag{42}$$

$$C_\ell(x) = \arccos\left\{6|M_\ell(x)|/\sqrt{B_\ell(x)}\right\} \tag{43}$$

$$D_\ell(x) = \sqrt{M_\ell^2(x) - \frac{|P_\ell(x)|^3}{108\pi\sigma_0}} \tag{44}$$

$$\bar{M}_\ell^*(x) = \sqrt[3]{|M_\ell(x)| + D_\ell(x)} + \sqrt[3]{|M_\ell(x)| - D_\ell(x)} \tag{45}$$

The differential inequalities (38) do not simplify as easily but may be treated as follows:

Consider the case when $\psi_1 > \psi_2$. Substituting for ψ_1 from (39) into the inequalities (38) gives

$$\left\{\frac{4}{3\pi r^3(x)}\right\} max_f \left| Q_\ell(x)r(x) - M_\ell(x)\frac{dr}{dx} \right| - \tau_0 \leq 0 \tag{46}$$

or

$$r(x) \geq max_f \sqrt[3]{\delta \left| Q_\ell(x)r(x) - M_\ell(x)\frac{dr}{dx} \right|}; \quad \delta = \frac{4}{3\pi\tau_0} \tag{47}$$

Similarly, if $\psi_2 > \psi_1$ we have

$$r(x) \geq max_f \sqrt[3]{3\delta \left| M_\ell(x)\frac{dr}{dx} \right|} \tag{48}$$

Assuming that the quantities within the modulus sign in equations (47) and (48) are known, we may combine the inequalities (41), (47) and (48) and write

$$r(x) \geq max \, (F_1, F_2, F_3) \tag{49}$$

where F_1, F_2 and F_3 are, respectively, the right hand sides of inequalities (41), (47) and (48). Depending on the relative magnitudes of F_1, F_2 and F_3, the governing inequality will be either linear or non-linear. However, if the stiffness is assumed to be constant within the segments (i.e. $NOR_m=0$) then $dr/dx = 0$ and the inequalities (49) are linearized

$$r(x) \geq max\ (F_1, F_4) \tag{50}$$

where

$$F_4 = \sqrt{\delta \bar{Q}(x)} \tag{51}$$

$$\bar{Q}(x) = max\ \{|Q_1(x)|, |Q_2(x)|, \ldots, |Q_{NL}(x)|\} \tag{52}$$

When the member consists of segments of constant radius (i.e. $NOR_m=0$), the stress constraints (35) and (36) for the ith segment ($d_i \leq x \leq d_{i+1}$) and mass of the member will take the form

$$r_i \geq max\ (F_1, F_4) \tag{53}$$

$$W_m = \pi \gamma_m \sum_{i=1}^{NSEG_m} r_i^2 (d_{i+1} - d_i) \tag{54}$$

where r_i are the values of the optimum radius at the knots.

On the other hand, if the member consists of segments with linearly varying (i.e. $NOR_m=1$) radius, then the stress constraints for the ith segment and mass of the member become

$$[1 - T(x)]\ r_i + T(x)\ r_{i+1} \geq max(F_1, F_2, F_3) \tag{55}$$

$$W_m = \left\{\frac{\pi \gamma_m}{3}\right\} \sum_{i=1}^{NSEG_m} (r_i^2 + r_i r_{i+1} + r_{i+1}^2)(d_{i+1} - d_i) \tag{56}$$

where $T(x) = (x - d_i)/(d_{i+1} - d_i)\ (i = 1, \ldots, NSEG_m)$.

One proceeds in a similar manner for a square section with side $r(x)$. Of course, the mathematical expressions are somewhat different from a circular section. The expressions corresponding to (35)–(56) will be found in the paper by Kanagasundaram & Karihaloo (1990).

3.3 RECTANGULAR SECTION OF CONSTANT WIDTH ($n = 3$)

In the case of a flexural member of rectangular section with constant width b and variable depth $h(x)$ the strength constraints (21,22) are

$$\sigma_1 \equiv \left\{\frac{1}{bh^3(x)}\right\} max_s max_f \left[12|M_\ell(x)\ s(x)| + h^2(x)|P_\ell(x)|\right] - \sigma_0 \leq 0 \tag{57}$$

$$\sigma_2 \equiv \left\{\frac{6}{bh^3(x)}\right\} max_s,\ max_f\ |Q_\ell(x)| \left[\frac{h^2(x)}{4} - s^2(x)\right]$$

$$+ M_\ell(x) \left[\frac{3s^2(x)}{h(x)} - \frac{h(x)}{4}\right] \frac{dh}{dx}| - \tau_0 \leq 0 \tag{58}$$

The maxima of σ_1 with respect to s are attained at $s_1^* = \pm h(x)/2$, whereas that of σ_2 are attained at either $s_2^{*2} = 0$ or $s_2^{*2} = h^2(x)/4$. The differential inequalities (5) therefore reduce to

$$\Omega_1^* \equiv \left\{\frac{1}{bh^2(x)}\right\} max_f \; [6|M_\ell(x)| + h(x)|P_\ell(x)|] - \sigma_0 \le 0 \tag{59}$$

$$\Omega_2^* \equiv max\,(\psi_1, \psi_2) - \tau_0 \le 0 \tag{60}$$

where

$$\psi_1 = \left\{\frac{3}{2bh^2(x)}\right\} max_f \; \left|Q_\ell(x)h(x) - M_\ell(x)\frac{dh}{dx}\right|; s_2^* = 0 \tag{61}$$

$$\psi_2 = \left\{\frac{3}{bh^2(x)}\right\} max_f \; \left|M_\ell(x)\frac{dh}{dx}\right|; s_2^{*2} = \frac{h^2(x)}{4} \tag{62}$$

The algebraic inequalities (59) may be re-written as

$$h(x) \ge \frac{max_f \; \left\{|P_\ell| + \sqrt{P_\ell^2 + 24b\sigma_0|M_\ell|}\right\}}{2b\sigma_0} \tag{63}$$

The differential inequalities (60) are rearranged such that if $\psi_1 > \psi_2$, we have

$$\left\{\frac{3}{2bh^2(x)}\right\} max_f \; \left|Q_\ell h - M_\ell\frac{dh}{dx}\right| - \tau_0 \le 0$$

or

$$h(x) \ge max_f \; \sqrt{\beta\left|Q_\ell h - M_\ell\frac{dh}{dx}\right|}; \beta = \frac{3}{2b\tau_0} \tag{64}$$

Similarly, if $\psi_2 > \psi_1$, we have

$$h(x) \ge max_f \; \sqrt{2\beta\left|M_\ell(x)\frac{dh}{dx}\right|} \tag{65}$$

Assuming that the quantities within the square root sign in inequalities (64) and (65) are known, we may combine the inequalities (63)–(65) and write

$$h(x) \ge max\,(F_1, F_2, F_3) \tag{66}$$

where F_1, F_2 and F_3 are, respectively, the right hand sides of inequalities (63)–(65). Depending on the relative magnitudes of F_1, F_2 and F_3, the governing inequality (66) will be either linear or non-linear. However, if the stiffness is assumed to be constant within the segments (i.e. $NOR = 0$) then $dh/dx = 0$ and the inequalities (66) are linearized

$$h(x) \ge max\,(F_1, F_4) \tag{67}$$

where

$$F_4 = \beta \, \bar{Q}(x)$$
$$\bar{Q}(x) = max \, \{|Q_1(x)|, |Q_2(x)|, \ldots, |Q_{NL}(x)|\}$$

When the member is assumed to consist of segments of constant depth (i.e. $NOR_m=0$) or linearly varying depth (i.e. $NOR_m=1$), the stress constraints and mass of the member are similar to those reported for $n=1$ in an earlier section. For instance if $NOR_m=0$, we have, for the ith segment ($d_i \leq x \leq d_{i+1}$)

$$h(x) = h_i \geq max(F_1, F_4) \tag{68}$$

whereas the mass of member is

$$W_m = \gamma_m b_m \sum_{i=1}^{NSEG_m} h_i(d_{i+1} - d_i) \tag{69}$$

If $NOR_m=1$, we have

$$[1 - T(x)]h_i + T(x)h_{i+1} \geq max \, (F_1, F_2, F_3) \tag{70}$$

$$W_m = \left\{ \frac{\gamma_m b_m}{2} \right\} \sum_{i=1}^{NSEG_m} (h_i + h_{i+1})(d_{i+1} - d_i) \tag{71}$$

where $T(x) = (x - d_i)/(d_{i+1} - d_i)(i = 1, \ldots, NSEG_m)$ and h_i are the values of the optimum depth at the knots.

3.4 STIFFNESS AND STABILITY CONSTRAINTS

The procedure used for imposing the deflection and buckling constraints is described below. In order to impose the constraint on the maximum deflection of a typical plane frame member, it is necessary to solve the differential equation (3) for a known distribution of stiffness $I_m(x)$ and specified system of loading $f_\ell(x)$ in an iterative way:

1. Assume $(u_{xx})_j = 0.01$ in the first iteration ($j = 1$).

2. Calculate $(u_x)_j = \int_0^x u_{zz}dz - v_3$ using the geometric boundary condition $u_x(0) = -v_3$ (Figure 3).

3. Calculate $u_j = \int_0^x u_z dz - v_2$ using the geometric boundary condition $u(0) = -v_2$ (Figure 3).

4. Calculate $M_\ell(x)$ from equation(19).

5. Update curvature using equation (3) $(u_{xx})_{j+1} = -M_\ell(x)/EI_m(x)$.

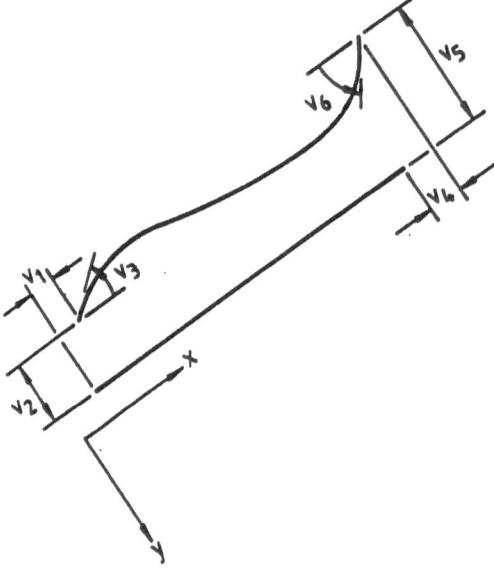

Figure 3: Deformations $v_i (i = 1, \ldots, 6)$ of a typical member

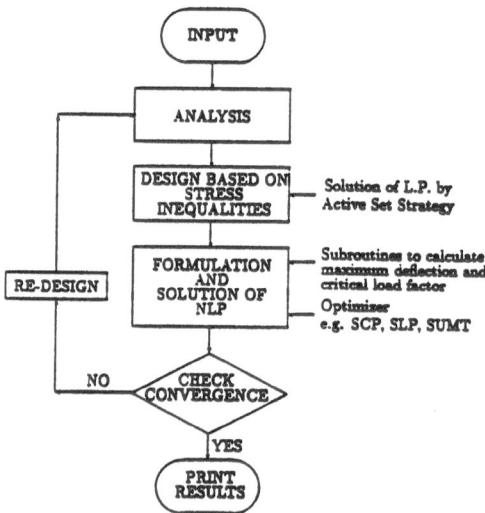

Figure 4: The iterative solution scheme for the optimization problem

6. Repeat steps 2–5 until $|(u_{xx})_{j+1} - (u_{xx})_j| \leq 10^{-6}$.

7. Locate the point of maximum deflection using a search procedure and calculate u_{max}.

8. Repeat for all loading cases $(\ell = 1, \ldots, NL)$.

The member length is divided into several intervals, and the integration is performed using Simpson's rule. The point of maximum deflection is assumed to coincide with one end of an interval and it is located by searching through the member length. It may be noted that the boundary values of slope and deflection $v_i(i = 1, \ldots, 6)$ are known beforehand, for example at exterior supports, or follow from the previous analysis stage.

The critical load factor $(\lambda_{cr})_\ell$ associated with the buckling of the frame under load system $f_\ell(x)$ is calculated by solving for the lowest eigenvalue of the characteristic equation (see, e.g. Howson, 1979)

$$\mathbf{K}(\lambda_\ell)\mathbf{D} = \mathbf{0} \tag{72}$$

where $\mathbf{K}(\lambda_\ell)$ is the overall stiffness matrix of the frame dependent on λ_ℓ and $\mathbf{D}$ is the displacement vector. The characteristic equation (72) is solved for all loading cases.

The geometric constraints are imposed on the design variables like the variable width ($n{=}1$), section size ($n{=}2$) or depth ($n{=}3$) of members. For instance, in the case of frames, as indeed of continuous beams and beam-columns it is customary to require that the width of the members meeting at an intermediate joint be the same, i.e.

$$G_j(b) = 0; (j = 1, \ldots, NC) \tag{73}$$

where NC is the number of intermediate joints. These constraints are linear.

The minimization of mass of the frame under strength, stiffness, stability and geometric constraints results in a non-linear programming problem and can be solved by an iterative scheme described below.

4. Solution of the Problem

The optimization problem of the frame under strength, deflection, buckling and geometric constraints is solved iteratively as follows:

1. Each member is divided into a given number of segments $(NSEG_m)$. The number of segments can vary between members, just as the latter can have different cross-sections, (i.e. different value of n).

2. For each member $A_m(x)$ is arbitrarily assumed in the first iteration $j = 1$. If $n=1$, it is sufficient to assume only $b_i(i = 1, \ldots, N_m)$, r_i if $n=2$ or h_i if $n=3$. A prismatic design is a reasonable first approximation.

3. The frame is analysed by a standard stiffness matrix routine (see, for example Kanagasundaram, 1984) for each of the load systems $f_\ell(x)$ to obtain $M_\ell(x)$ and $Q_\ell(x)$ for all the members using equations (19), (20).

4. Each segment is divided into p intervals and the following linear programming (LP) problem is solved for a typical member: Minimize W_m subject to only stress constraints (31), (49) or (66). Let the solution of the above problem be $\bar{A}_m$. Repeat this step for each of the members.

5. Using the values of $\bar{A}_m$ from step 4 as lower bounds, the non-linear programming problem (18), but without the stress constraints is solved for the entire frame. The procedure for the calculation of the maximum deflection u_{max} and critical load factor λ_{cr} required in this step was explained in the previous section. Let the solution of this non-linear programming problem be $(A_m)_{j+1}$.

6. Steps 3–5 are repeated until either $|(A_m)_{j+1} - (A_m)_j| \leq \epsilon \, (= 0.5\%)$ for each of the members or the total volume of the frame does not differ by more than ϵ in successive iterations.

The formulation of the optimization problem is so versatile that it can allow various members of a frame to have different cross-sectional shapes. A flow chart of the above solution scheme is given in Figure 4. The scheme has been programmed in standard FORTRAN 77; a listing of the program FRAME and relevant user instructions can be found in the paper by Kanagasundaram & Karihaloo (1990) and an up-to-date copy on a diskette is available on request. The program uses an active set strategy (Best & Ritter, 1985) to solve the LP, and a new computer package called ADS(Automated Design Synthesis) by Vanderplaats & Sugimoto (1986) to solve the NLP. Several examples have been solved and the results are reported below.

5. Examples and Discussion

The concepts introduced in the previous sections will be used to obtain minimum-weight designs of statically indetrminate beams and plane structural frames. Several examples will be presented for three types of member cross-section ($n=1$, 2 or 3). The solution of the non-linear programming problems corresponding to beams is attempted only by the Sequential Linear Programming method (SLP) with move-limits, whereas those of the problems corresponding to frames is attempted by several techniques and namely Augmented Lagrangian Multiplier method (ALM),

Sequential Convex Programming with move-limits(SCP), Sequential Linear Programming with move-limits(SLP), Sequential Quadratic Programming (SQP), and Sequential Unconstrained Minimization Techniques(SUMT). The efficiency of each of the methods will be discussed. All the results reported below have been obtained by executing the program FRAME, a listing of which, and the user instructions, can be found in the paper by Kanagasundaram & Karihaloo (1990) and an up-to-date copy on a diskette is available on request .

5.1 STATICALLY INDETERMINATE BEAMS

An understanding of the optimization of simple beams provides an appreciation of minimum-weight design of highly indeterminate plane frames reported in the next section. The results for several single-span and two-span beams are shown in Figures 5–8. The following material properties were assumed in the examples: $\sigma_0 = 5MPa$, $\tau_0 = 0.5MPa$, $E = 20GPa$, $\gamma = 2450kg/m^3$. These values are typical for concrete. Data relating to the external load systems acting on the beams are shown on the respective figures. Each member was designed to consist of one-metre long segments, and the stress constraints were calculated at sections spaced at 0.05m intervals ($p=20$). For the sake of illustration, the beams shown in Figures 5–6 were subjected to only a single load system and, no constraints on maximum deflection and geometry were imposed. In other words, it was sought to obtain minimum-weight/maximum strength beams in terms of splines of suitable order. Figure 5 shows the minimum-weight design of a propped beam of variable width($n=1$), whereas Figure 6 shows the design of a continuous beam of variable depth($n=3$). It was found that the stress constraints were active only at discrete sections, and in all the designs the maximum deflection to span ratio of members was less than 1/350. A comparison of the volumes of optimum beams with increasing number of segments is presented in Table 1. Also indicated is the volume of the corresponding fully-stressed design (FSD) (Kanagasundaram & Karihaloo, 1985). It is clear that as the number of segments increases the volume decreases, eventually approaching the fully stressed volume.

TABLE 1. Comparison of volumes with increasing number of segments

Figure	$NSEG$	Volume, m^3		
		$NOR = 0$	$NOR = 1$	FSD
$5(n = 1)$	1	2.7654	2.0644	
	2	2.0629	1.2425	
	3	1.7327	1.2925	
	4	1.5606	1.2156	1.134
	5	1.4711	1.2046	
	6	1.4068	1.1450	

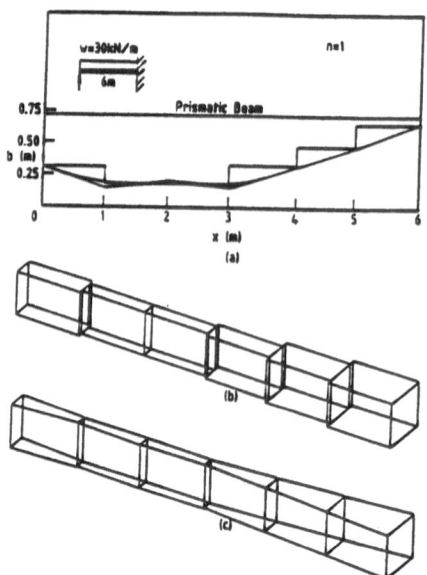

Figure 5: Minimum-weight design of pinned-clamped beam subjected to uniformly distributed load plus self-weight. The member has constant depth $h = 0.65\ m$, i.e. $n = 1$

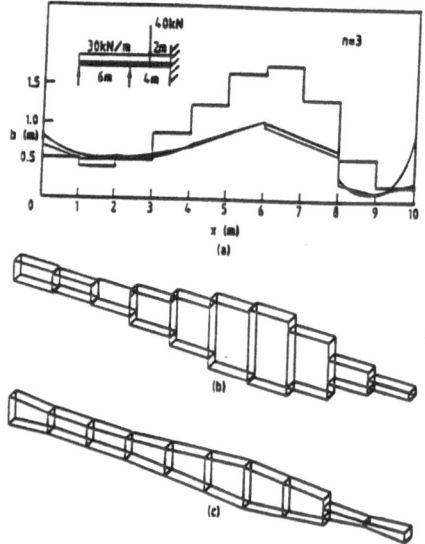

Figure 6: Minimum-weight design of a two-span continuous beam subjected to loading as shown, plus the self-weight. The members are of rectangular cross-section with constant width $b = 0.70\ m$, i.e. $n = 3$

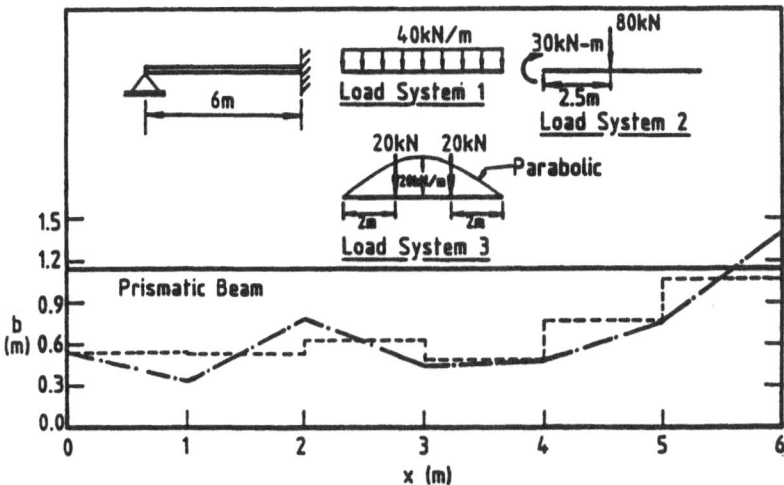

Figure 7: Minimum-weight design of a propped cantilever subjected to the load systems shown and self-weight. The member is rectangular in cross-section with constant depth $h = 0.55\ m$, i.e. $n = 1$

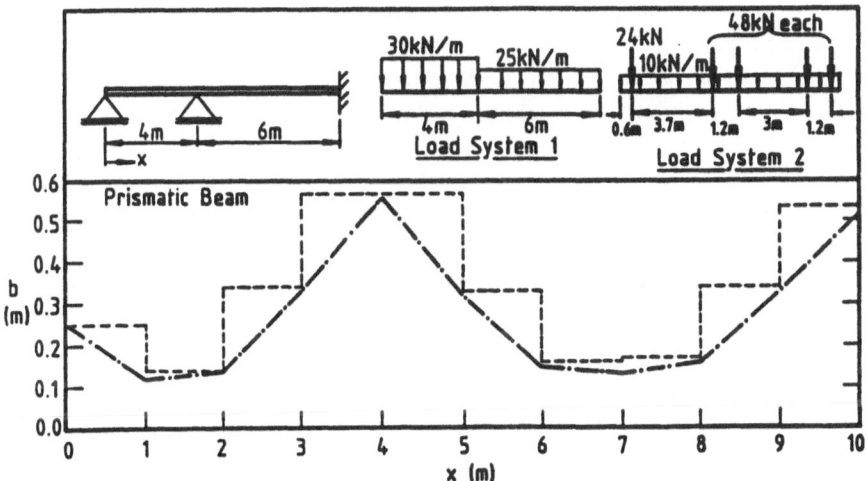

Figure 8: Minimum-weight design of a two-span beam subjected to the load systems as shown, plus the self-weight. Both members have $n = 1$ with $h = 0.50\ m$. The concentrated loads in system 2 represent a truck loading

It should be mentioned that the number of fully stressed sections in the minimum weight designs is equal to the number of design variables. For instance, there are seven such sections in the optimum designs with six segments obtained using splines of order one ($NOR = 1$). It would seem that with an increase in the number of segments, the number of fully stressed sections will increase, resulting eventually in a fully stressed design. The volumes of the minimum-weight designs are compared with those of fully-stressed and prismatic designs in Table 2.

TABLE 2. Comparison of volumes (m^3) of optimum designs

Figure	Optimum design		FSD	Uniform Design
	NOR=0	NOR=1		
5(n=1)	1.4068	1.1450	1.1340	2.7654
6(n=3)	2.6947	1.7679	1.6991	3.7571

The data file on the optimum designs was used to generate the isometric views shown in the Figures 5–6 using the ANVIL–4000 CAD-package. The data were fed into the CAD package using its menu system. In the existing menu system, the relevant points on the figure are defined and joined by splines of required order to obtain a plane figure, e.g. front elevation of a beam of constant width ($n = 3$). The plane figure is then projected together with the third dimension, e.g. the constant width ($n = 3$) along the span. Finally, the resulting three dimensional object is processed to provide the isometric view with surface finishes. The beams shown in Figures 7–8 were subjected to two loading cases, and the maximum deflection to span ratio of each member was limited to a stringent value of 1/3500. A comparison of volumes of these optimum beams having variable width (n=1) is presented in Table 3. The Table also shows volumes of the corresponding prismatic designs, and the location of the point of maximum deflection (x_0) and the corresponding load case (f_ℓ^*).

TABLE 3. Comparison of volumes (m^3) of optimum beams

Figure	NOR	Optimum Design	Uniform Design	$x_0(m)$	f_ℓ^*
7(n=1)	0	2.2176		2.40	1
	1	2.0671	3.7253	2.40	1
8(n=1)	0	2.9884		7.15	1
	1	2.7228	4.3060	7.15	1

It should be mentioned that in most of the designs the deflection constraint was active, but the stress constraints were active only at discrete sections. The total number of active constraints in each example was found to be less than or equal to the number of design variables. The geometric constraints on the beam shown in

Figure 8 are such that the optimum width($n=1$) has the same value on either side of the intermediate support. It should be noted that there was no need to include stability constraint in the formulation of the optimization problem of beams. Moreover, the deflection constraint was deliberately chosen to be very restrictive because it was found that the optimum designs for stress constraints alone could easily satisfy a less restrictive stiffness constraint in many cases. The non-linear programming problems were solved by Sequential Linear Programming method with move-limits (Pedersen, 1981; Vanderplaats & Sugimoto, 1986). The optimization of highly indeterminate plane frames is considered in the next section.

5.2 PLANE STRUCTURAL FRAMES

The results of an example of a portal frame are shown in Figures 9 and 10. The frame was subjected to multiple load cases as shown in the Figure 9, and its members were designed to consist of one-metre long segments. The maximum deflection to span ratio for each member was limited to a very stringent value of $1/1000$, and the factor of safety(FS) against the buckling of the frame was assumed to be 3. Moreover, The geometric constraints were imposed such that the beam and the column have the same optimum width($n=1$) or depth($n=3$) at each of the beam-column connections. The following material properties were assumed: $\sigma_0 = 5MPa, \tau_0 = 0.5MPa, E = 20GPa, \gamma = 2450kg/m^3$. The resulting minimum-weight designs are shown in Figures 9 ($n=1$) and 10 ($n=3$). It is seen that the designs are awkward and are unlikely to be accepted by practising engineers, for obvious reasons. Therefore, it was thought desirable to obtain more realistic optimum designs by treating each member as a segment, thus avoiding stiffness jumps. Such designs are reported in Figures 11–15. The multi-storeyed frames shown in Figures 11–15 were subjected to the following uniformly distributed loads on the beams in accordance with the recommendations of the Australian Standard (1981): Dead load(DL) = 27.6kN/m(21.6kN/m at roof level); Live load(LL) = 18kN/m(3kN/m at roof level). The wind loads are shown on the respective figures. The frames were designed to withstand the following load combinations: DL+LL, DL+LL+WL, DL+WL. The maximum deflection to span ratio for each member was limited to $1/350$ and the factor of safety(FS) against the buckling of the frame was assumed to be 3. Each member consisted of just one segment. The geometric constraints on the designs were also imposed. The designs with constant area segments (zeroth order spline) were to be such that the beams at any one level were of the same area; the columns at the same level were to be such that the outer columns had the same area, and the corresponding inner columns on either side of the line of symmetry at that level had the same area. The designs with linearly varying area (first order spline), on the other hand were to be such that all members meeting at any joint other than a support had the same value of the optimum

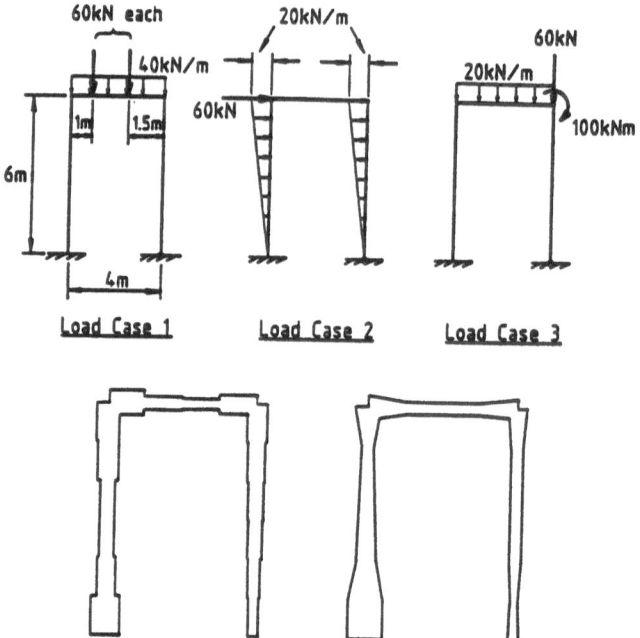

Figure 9: Minimum-weight design of a portal frame subjected to multiple load systems as shown plus self-weight. All members are rectangular in cross-section with constant depth $h = 0.8\ m$, i.e. $n = 1$

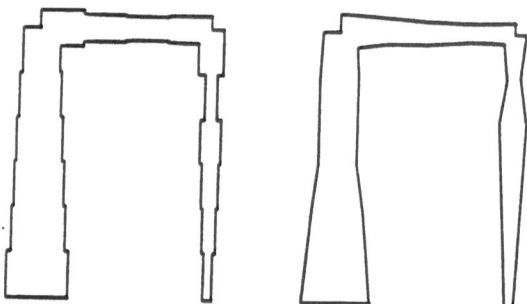

Figure 10: Minimum-weight design of the frame shown in Figure 9 but with members of constant width $b = 0.3\ m$, i.e. $n = 3$

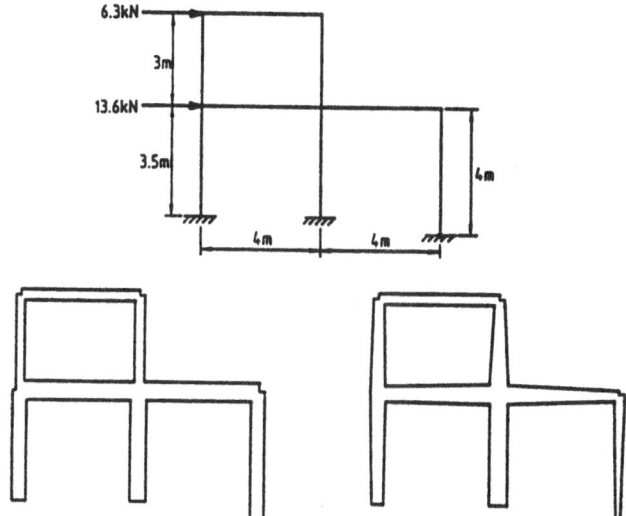

Figure 11: Minimum-weight design of a two-storey frame subjected to multiple load combinations indicated in the text, plus the self-weight. The wind loads are applied at floor levels as shown. All members have $n = 1$ with $h = 0.60\ m$

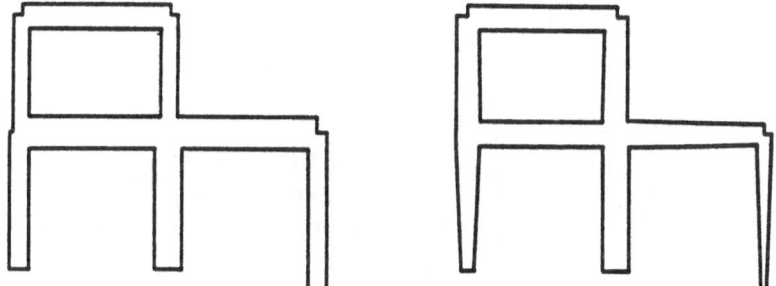

Figure 12: Minimum-weight design of the frame shown in Figure 11 but whose members have $n = 3$ with $b = 0.30\ m$

sizing parameter. The material properties were the same as those of the portal frame shown in Figure 9. To complete the presentation, the optimum designs of frames consisting of members of square cross-section ($n = 2$) were obtained. The loading for the frames shown in Figures 16–18 were the same as before, as were the material properties, but the maximum deflection to span ratio for each member was limited to 1/250.

The optimum designs (Figures 11–18) so obtained are realistic and are easier to fabricate. A comparison of the volumes of optimum frames is presented in Table 4 along with the volumes of the corresponding frames of uniform design.

TABLE 4. Comparison of Volumes (m^3) of Optimum Frames

					Optimum design				Uniform
Fig	n	NOR	N	M	ALM	SCP	SLP	SQP	Design
9	1	0	14(16)	12	*	5.9301	5.8894	*	9.6801
		1	17(19)	12	*	5.5276	5.5957	*	
10	3	0	14(16)	12	*	4.9906	5.0183	*	7.1719
		1	17(19)	12	*	4.6745	4.7338	*	
11	1	0	5(8)	27	7.7723	7.7723	7.7723	7.7723	10.6391
		1	8(16)	27	7.3342	7.3342	7.3342	7.3342	
12	3	0	5(8)	27	5.2581	5.2583	5.2581	5.2583	7.6328
		1	8(16)	27	6.3340	6.3341	6.3341	6.3992	
14	1	0	8(15)	48	20.5231	20.5276	20.5276	20.5276	63.1290
		1	13(30)	48	19.6648	19.6668	19.6668	19.6663	
15	3	0	8(15)	48	13.0310	13.0310	13.0310	13.0310	22.9505
		1	13(30)	48	14.6142	14.6149	14.6149	14.3693	
16	2	0	3(5)	18	3.4394	3.4358	3.4404	3.4546	5.6141
		1	6(10)	18	3.2669	3.3010	3.4834	3.3558	
17	2	0	4(6)	21	4.1392	4.1398	4.1394	4.1403	8.5397
		1	4(12)	21	4.9409	5.3134	5.3029	5.3199	
18	2	0	8(15)	48	12.4107	12.4107	12.4107	12.4107	19.8760
		1	13(30)	48	14.5182	14.5177	14.5177	14.5177	

* This technique was not attempted.

It is found that significant material saving is achieved by optimizing the frame designs. While material saving is achieved by adopting optimum designs, it is important to consider the cost of obtaining such designs. In this context, the importance of choosing the right numerical technique for the solution of optimization problems cannot be overemphasized. The right technique should be computationally effective and should use the least computer time. A comparison of the several techniques used in the solution of optimization problems of above frames is presented in the next section.

5.3 COMPARISON OF NLP TECHNIQUES

The solution of the non-linear programming problems arising in all of the frame examples (Figures 9–18) was attempted by several methods: (1) Augmented Lagrangian Multiplier method (ALM; Imai & Schmit, 1981); (2) Sequential Convex Programming (SCP) with move-limits (Fleury & Braibant, 1986); (3) Sequential Linear Programming with move limits (SLP) (Pedersen, 1981); (4) Sequential Quadratic Programming (SQP; Powell, 1978); and (5) Sequential Unconstrained Minimization Technique (SUMT) (Fiacco & McCormick, 1968). The algorithms used for all the methods are those available in a new general-purpose package called the ADS (Automated Design Synthesis) (Vanderplaats & Sugimoto, 1986).

TABLE 5. Comparison of Solution Techniques

Fig	N O R	Number of constraint function evaluations				CPU time, Seconds				Micro-VAX type
		ALM	SCP	SLP	SQP	ALM	SCP	SLP	SQP	
9	0	*	147	120	*	*	7110	5703	*	2000
	1	*	324	252	*	*	15604	12208	*	2000
10	0	*	133	225	*	*	6345	10530	*	2000
	1	*	504	666	*	*	23570	30817	*	2000
11	0	70	39	66	57	1412	830	1312	1214	4000
	1	147	86	144	164	2810	1702	2771	3115	4000
12	0	116	82	102	133	2278	1656	1998	2531	4000
	1	325	162	241	304	5977	3394	4706	5446	4000
14	0	160	78	108	100	6126	3164	4261	3927	4000
	1	268	165	224	235	9949	6239	8293	8655	4000
15	0	108	60	108	180	4001	2338	4007	6584	4000
	1	451	260	378	443	16784	9701	13830	18060	4000
16	0	270	121	97	151	10396	4929	3860	6027	2000
	1	140	133	119	164	2187	2141	1912	2574	4000
17	0	207	45	55	76	7923	2120	2484	3251	2000
	1	205	84	91	90	3149	1417	1504	1520	4000
18	0	317	173	243	400	30626	17485	23924	38241	2000
	1	401	204	294	399	38120	20051	28294	37740	2000

*   This technique was not attempted.

The solution of the non-linear programming problems by the SCP and SLP techniques provided quick convergence, and the computer times for solution were comparable (Table 5). The SUMT was the slowest of the methods in the only two examples attempted by this method (Figures 11,14); it was therefore decided not to

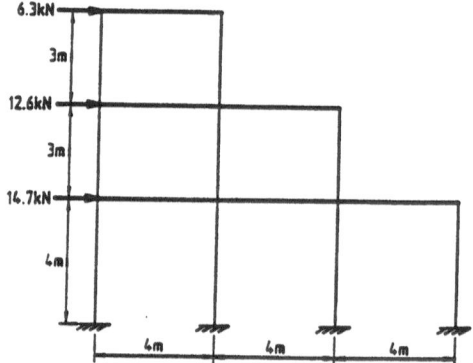

Figure 13: A multi-storey frame with wind loads applied at floor levels

Figure 14: Minimum-weight design of the frame shown in Figure 13 subjected to the load combinations indicated in the text. All members have $n = 1$ with $h = 0.60\ m$

Figure 15: Minimum-weight design of the frame shown in Figure 13 but with $n = 3$ and $b = 0.40\ m$

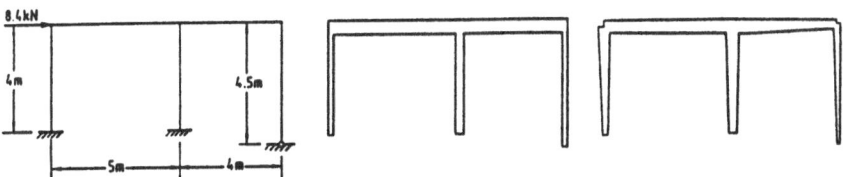

Figure 16: Minimum-weight design of a two-bay frame subjected to the load combinations indicated in the text. All members have a square section, i.e. $n = 2$

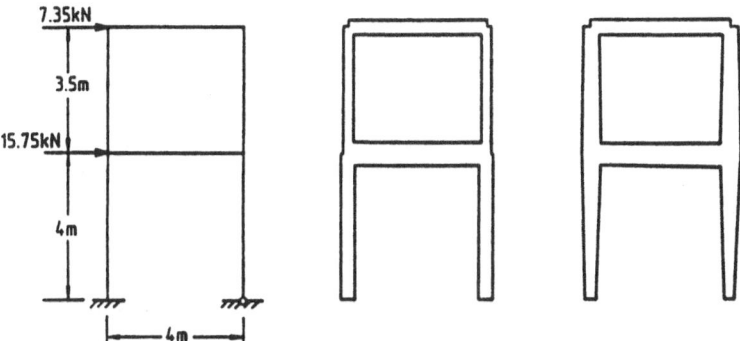

Figure 17: Minimum-weight design of a two-storey frame subjected to the load combinations indicated in the text. All members have a square section, i.e. $n = 2$

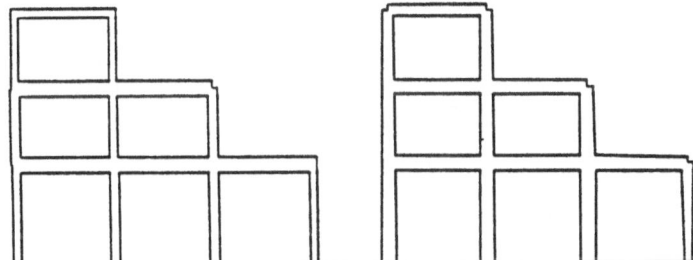

Figure 18: Minimum-weight design of the frame shown in Figure 13 but with members of square section, i.e. $n = 2$

persist with this method. For the frame of Figure 11 with $NOR = 0$, SUMT needed 2230 function evaluations and 83633 sec of CPU time on a microVAX 2000, whereas for the frame of Figure 14 with $NOR = 0$ it needed 4176 function evaluations and 305109 sec of CPU time.

Starting with the same uniform design, all the methods yielded practically identical optimum designs. Under the geometric constraints specified for the frames, the number of unknown design variables is reduced, because the simple equality constraints (8), can be omitted after adjusting the set of unknown variables. For instance, the total number of design variables (N) for the frame shown in Figure 9 reduces from 16 to 14 when $NOR = 0$, and from 19 to 17 when $NOR = 1$ (see Table 4, column 3). The number of non-linear constraints (M) in each of the examples was shown in Table 4. For the example frame shown in Figure 9, the number of design variables is greater than the number of constraints, but the converse applies to the remaining examples. It was found that in all the optimum designs, except those shown in Figures 9 and 10 the buckling and deflection constraints were inactive and the optimum designs were decided by the stress requirements alone. In the designs of Figures 9 and 10 the deflection constraints were active for some of the members, along with the stress constraints, although again the buckling constraints were inactive. In order to verify the correctness of the numerical computations, all the optimum frames were analysed, and the stresses and maximum deflection were found to be within the prescribed limits.

The CPU times (on microVAX 4000 or 2000, as indicated in the last column) for each of the methods are also shown in the Table. It should be noted that the CPU time is directly proportional to the number of constraint function evaluations. In most of the cases, SCP required the least number of evaluations and the solution was the fastest. At this point it is useful to mention that the gradients of the objective and constraint functions were calculated by finite differences and each set of evaluation of the constraint functions required analyses of the frame under all loading cases.

In conclusion, it may be said that of the several methods used in this study both the SCP and SLP were found to be the most suitable to optimize the frame designs under stress, deflection and buckling constraints, with the SCP slightly the better of the two. For this reason, only these two techniques were used to solve the NLPs corresponding to the optimization problems of the reinforced concrete members that will be discussed in the next lecture.

6. References

Australian Standard (1981) AS 1170–1981, Parts 1 and 2 'Minimum Design Loads on Structures'

Best, M J and Ritter, K. (1985) Linear Programming (Active Set Analysis and

Computer Programs), Prentice-Hall, Englewood Cliffs, NJ

Fiacco, A.V. and McCormick, G.P. (1968) Nonlinear Programming: Sequential Unconstrained Minimization Techniques, John Wiley, New York

Fleury, C. and Braibant, V. (1986) 'Structural optimization: A new dual method using mixed variables', Int. J. Num. Methods in Eng. 23, 409–428.

Howson, W.P. (1979) 'A compact method for computing the eigenvalues and eigenvectors of plane frames', in R. A. Adey (ed.) Engineering Software , Pentech Press, London. pp. 281–300.

Imai, K. and Schmit, L.A. (1981), 'Configuration optimization of trusses', ASCE J. Struct. Div. 107, 745–756.

Kanagasundaram, S. (1984) 'Optimum structural design (Application of differential game theory)', Ph.D Thesis, The University of Newcastle, Australia

Kanagasundaram, S. and Karihaloo, B.L. (1985), 'Maximum strength design of structural frames', ASCE J. Struct. Div. 111, 1267–1287.

Kanagasundaram, S. and Karihaloo, B.L. (1988), 'Optimum design of plane structural frames', in P.G. Lowe (ed.), Proc Eleventh Australasian Conference on the Mechanics of Structures and Materials, Auckland, New Zealand, pp. 109–113.

Kanagasundaram, S. and Karihaloo, B.L. (1990), 'Optimum design of frames under multiple loads', Comput. & Struct. 36, 443–489.

Karihaloo, B.L. and Kanagasundaram, S. (1986), 'Computer-aided minimum-weight design of statically indeterminate beams', Engng. Optimization 10, 139–156.

Karihaloo, B.L. and Kanagasundaram, S. (1987), 'Optimum design of statically indeterminate beams under multiple loads', Comput.& Struct. 26, 521–538.

Karihaloo, B.L. and Kanagasundaram, S. (1988a), 'Optimum structures under strength and stiffness constraints', Comput.& Struct. 28, 641–661.

Karihaloo, B.L. and Kanagasundaram, S. (1988b), 'Optimum design of statically indeterminate structures subject to strength and stiffness constraints and multiple loading', Comput.& Struct. 30, 563–572.

Karihaloo, B.L. and Kanagasundaram, S. (1988c), 'Comparative study of NLP techniques in optimal structural frame design', in G. I. N. Rozvany and B.L.Karihaloo (eds.), Structural Optimization, Kluwer Academic Publishers, Dordrecht pp. 143-150.

Karihaloo, B.L. and Kanagasundaram, S. (1989), 'Minimum-weight design of structural frames', Comput.& Struct. 31, 647–655.

Love, A. E. H. (1944) A Treatise on Theory of Elasticity, Dover Publications, New York.

Pedersen, P. (1981), 'The integrated approach to FEM-SLP for solving problems of optimal design', in E. J. Haug and J. Cea (eds.), Optimization of Distributed Parameter Systems, Vol I, Sijthoff & Noordhoff, The Netherlands, pp 739–756.

Powell, M.J.D. (1978), 'Algorithms for nonlinear constraints that use Lagrangian functions', Math. Programming, 14, 224–248.

Timoshenko, S.P. and Gere, J.M. (1961), Theory of Elastic Stability, McGraw-Hill Book Company Inc., New York

Vanderplaats, G.N. and Sugimoto, H. (1986), 'A general-purpose optimization program for engineering design', Comput.& Struct. 24, 13–21.

MINIMUM-COST DESIGN OF REINFORCED CONCRETE MEMBERS BY NON-LINEAR PROGRAMMING

B.L. KARIHALOO and S. KANAGASUNDARAM
School of Civil & Mining Engineering
The University of Sydney
N.S.W. 2006
Australia

1. Introduction

This lecture will summarize the results of several publications by Kanagasundaram & Karihaloo (1989, 1991) and Karihaloo & Kanagasundaram (1990) on the minimum-cost designs of reinforced concrete members that satisfy the requirements of the new Limit States Codes. The total cost of the structure includes the costs of concrete, reinforcing steel and formwork. The minimum cost structure is such that all its appropriate functional states(stability, strength, serviceability, durability and fire resistance) are within the allowable performance limits.

The strength and stability functional states have been considered under one set of design actions(loads), whereas serviceability and fire resistance states require consideration of different sets of design actions. Moreover, the strength functional state may involve not only flexure, but also shear and axial actions. Likewise the serviceability functional states may involve not only total deflections but also incremental ones, as well as effects of creep and shrinkage. On the other hand, fire resistance and durability functional states may be met simply by imposing geometrical constraints on the design, e.g. a minimum width of cross-section and a minimum cover on reinforcement.

The minimum mass designs of multi-storey frames subject to constraints on strength, serviceability and stability were considered in the previous lecture. The minimization problem was formulated as a non-linear mathematical programming problem and solved by several methods. The optimum designs involved only a single design parameter e.g. a linear cross-sectional dimension, although in practice the designers have more parameters at their disposal to control the outcome of the design process. The solution methods, however, are similar and are therefore adapted here to solve the minimum cost problem involving several design parameters. In the first instance, the design variables will be limited to the section sizes and steel ratio(s).

G. I. N. Rozvany (ed.), Optimization of Large Structural Systems, Vol. II, 927–949.

The minimum-cost problem is formulated as a non-linear programming problem whose solution is attempted by two techniques, namely sequential linear programming(SLP) and sequential convex programming(SCP). Both techniques give similar results. The user(designer) is not expected to have a knowledge of these solution techniques. Moreover, the designer is not required to estimate the self-weight of the structure in the calculation of dead load as this is performed by an iterative procedure. Several numerical examples of multi-span beams and of columns are given. A sensitivity analysis of the minimum-cost designs relative to the cost of formwork is also performed. The entire procedure of analysis, design and optimization attempts to imitate as far as possible the traditional structural design process, thereby adding a new dimension to the latter. It lends itself easily to programming. This has been done in standard FORTRAN 77. A graphics package in standard GKS(Graphics Kernal System) has also been included to produce structural drawings.

Finally, the influence of the quality of concrete, i.e. its crushing strength on the design is investigated by treating it also as a design variable. It is found that in this instance the minimum-cost designs of structural members use shallower sections and higher strength mixes but are still marginally cheaper than the designs based on a prescribed low strength mix. Such designs are therefore more durable and provide more headroom.

2. Simply Supported Beams

Consider a simply supported beam spanning a clear distance of L_n and resting on supports of width w at either end (Figure 1). The beam has a uniform cross-section of width b_w and total depth D, and is subjected to uniformly distributed dead load G and live load Q and any other appropriate design actions(the designer is not required to include the self-weight of the beam in the specification of G). For simply supported beams it is sufficient to base the design only on the sections at mid-span and near a support. The amount of longitudinal reinforcement governed by mid-span flexural considerations may be reduced towards the supports by curtailing some of the reinforcing bars as the bending moment envelope may allow. The spacing of stirrups, on the other hand, may be altered to the requirements of the shear force envelope. The detailing of reinforcement, however, will finally be subject to the requirements of the design code ,e.g. Australian Standard (1988). The strength design requirements on the sections at mid-span and near a support are such that the design ultimate strengths of the sections are not less than their respective design action effects. Mathematically, these requirements can be written as

$$\phi_f M_u \geq |M^*| \tag{1}$$

$$\phi_s V_u \geq |V^*| \tag{2}$$

in which M_u, ϕ_f and V_u, ϕ_s are the ultimate strength of a section and strength reduction factor in flexure and shear respectively. The bending moment $|M^*|$ at mid-span and the shear force $|V^*|$ near a support are chosen to be the maximum occurring among all the possible load combinations for strength design.

When the beam is cast as an integral unit of the slab it supports, a part of the slab can be considered to act together with the beam in resisting the bending moment. In such a situation the beam can be treated as having a T-section or an L-section (Warner et al., 1988), depending on whether the slab extends to both sides or only one side of the beam. The ultimate flexural strength of a T-section may be written as

$$M_u = p\bar{b}f_{sy}\left(1 - \frac{p\bar{b}f_{sy}}{1.7f_c'}\right)b_w d^2 \tag{3}$$

in which $p = A_{st}/b_w d$, $\bar{b} = b_w/b_{ef}$, $d = D - d_c$ and A_{st} is the area of the longitudinal reinforcing steel near the tension face of the section at mid-span, b_{ef} is the effective width of the beam as described in the Australian Standard (1988), d_c is the distance to the location of the centroid of the reinforcing steel from the tension(bottom) face, f_{sy} is the yield strength of reinforcing steel and f_c' is the crushing strength of concrete. On the other hand, if the beam is cast separately and is not an integral unit of the slab, the section at mid-span can be designed as a rectangular one. Its ultimate flexural strength is also given by(3), with b_{ef} replaced by b_w, i.e. $\bar{b} = 1$. The ultimate shear strength of a reinforced section limited by web crushing may be written as

$$V_u = 0.2f_c'b_w d \tag{4}$$

The serviceability requirement limiting the maximum deflection (Δ_{max}) includes the effects of the applied loads, as well as creep and shrinkage. In the calculation of Δ_{max} an effective second moment of area I_{ef} is to be used. This requirement may be written as

$$\Delta_{max} \leq \Delta \tag{5}$$

in which Δ is the allowable deflection. In a simply supported beam carrying a uniformly distributed load of intensity w per unit length, $\Delta_{max} = 5wL_{ef}^4/(384E_c I_{ef})$ in which L_{ef} is the effective length of the beam. Alternatively, if the intensity of live load Q is not greater than that of dead load G, the Australian code (1988) prescribes a maximum limit on the ratio of effective span L_{ef} to effective depth d, i.e.

$$\frac{L_{ef}}{d} \leq \left[\frac{k_1(\Delta/L_{ef})b_{ef}E_c}{k_2 F_{d.ef}}\right]^{\frac{1}{3}} \tag{6}$$

in which $F_{d.ef}$ is the effective design load per unit length, E_c is the modulus of elasticity of concrete, $k_1 = I_{ef}/b_w d^3$ and $k_2 = 5/384$. The deflection to be limited may either be the total deflection or the deflection that occurs after the addition or attachment of the partitions, or both.

The requirements of durability are met by using the grade of concrete appropriate to the exposure classification for the member surface and by ensuring that the cover on the reinforcing steel is not less than the minimum value c_d required for concrete placement and corrosion protection. The requirements of fire resistance, on the other hand, are met by proportioning the member in such a way as to provide the needed fire resistance period. For example, for a member with rectangular section whose upper surface is protected by a slab, the Standard prescribes a minimum width b_f and a corresponding value for cover $c_{f.min}$ for a given fire resistance period. However, the value of required cover may be reduced if the width of the member is greater than b_f. Therefore the requirements of durability and fire resistance for a member with rectangular section may be simply formulated as

$$b_w \geq b_f \tag{7}$$

provided that the cover c on the longitudinal reinforcement is at least equal to the greater of the two values $c_d + d_s$ and c_f, i.e.

$$c \geq max(c_d + d_s, \ c_f) \tag{8}$$

in which d_s is the diameter of stirrups and $c_f(\leq c_{f.min})$ is the cover required for a member of width b_w and a given fire resistance period.

In addition to the above constraints(1,2,5(or6),7,8) on strength, serviceability, durability and fire resistance, the designer may wish to prescribe constraints on the geometry of the beam section for architectural reasons, e.g. the depth to width ratio for a rectangular section may be limited to a maximum value K_1 or it may be set equal to a specific value K_2, or the width of the section may be prescribed. These constraints may be mathematically written as

$$\frac{D}{b_w} < K_1 \tag{9}$$

or,

$$\frac{D}{b_w} = K_2 \tag{10}$$

or,

$$b_w = b_0 \tag{11}$$

in which $b_0(\geq b_f)$ is the prescribed width. It is also possible to choose a combination of geometric constraints(10) and (11), such that $D/b_0 = K_2$. It is useful to ensure that the width of the section can accommodate all the reinforcing bars. This requirement for a section with a single layer of reinforcement may be written as

$$b_w \geq Nd_b + (N-1)d_n + 2c \tag{12}$$

in which d_n is the greater of d_b and d_{ag}, N and d_b are respectively the total number and nominal diameter of the reinforcing bars, d_{ag} is the nominal diameter of the aggregates, and c is the cover on the longitudinal reinforcement.

The Standard also prescribes minimum flexural strength at critical sections. This requirement is satisfied by providing minimum tensile reinforcement, i.e.

$$p \geq \frac{1.4}{f_{sy}} \tag{13}$$

For the purposes of ductility, it also suggests to avoid reinforced sections with the neutral axis parameter k_u greater than 0.4, i.e.

$$k_u \leq 0.4 \tag{14}$$

For a T-section, this requirement is met if $p \leq (0.34 \ \gamma \ f_c')/(\bar{b} \ f_{sy})$, in which γ is a constant which depends on the strength of concrete. This inequality is also valid for a rectangular section for which $\bar{b} = 1$.

The quantities of concrete V_c, and formwork A_f involved in the construction of a beam with a T-section may be written as

$$V_c = b_w(D-t)(L_n+2w) \tag{15}$$

$$A_f = 2(D-t)(L_n+2w) + b_w L_n \tag{16}$$

in which t is the thickness of the slab. For a beam with an L-section, V_c is the same as (15) but the quantity of formwork A_f is different $A_f = (2D-t)(L_n+2w)+b_w L_n$

For a rectangular section, V_c and A_f are given by

$$V_c = b_w D(L_n+2w) \tag{17}$$

$$A_f = 2D(L_n+2w) + b_w L_n \tag{18}$$

The quantity of reinforcement V_s in a beam includes the longitudinal reinforcement as well as the lateral reinforcement in the form of stirrups. We may, therefore, write

$$V_s = \sum_{i=1}^{NB} A_{st.i} \ l_i + A_{st.s} \ l_s \tag{19}$$

in which $A_{st.i}$ and $A_{st.s}$ are the cross-sectional area of a longitudinal bar $i(i = 1,\ldots,NB)$ and stirrups, respectively, and l_s is the equivalent length of all stirrups.

The total cost C of constructing a beam is therefore

$$C = C_c V_c + C_s V_s + C_f A_f \tag{20}$$

or

$$C = (V_c + C_{sc} V_s + C_{fc} A_f) C_c \tag{21}$$

in which C_c, V_c, C_s, V_s and C_f, A_f are respectively the cost(per m^3 or m^2 as the case may be) and quantity of concrete, reinforcement and formwork, and C_{sc} and C_{fc} the costs of reinforcement and formwork relative to the unit cost of concrete. The minimum-cost problem of a simply supported beam with a T-section may, now, be stated as follows: Determine the design variables b_w, D(or d) and p such that the cost(21) of the beam with V_c, A_f and V_s given respectively by (15,16,19) is minimized subject to the constraints on strength(1,2), serviceability(5 or 6), durability and fire resistance(7,8), geometry(9–12), minimum flexural strength(13) and ductility(14). The above statement is only slightly modified for a simply supported beam with a rectangular section or an L-section: for a simply supported beam with a rectangular section, $\bar{b}$ is set to 1 in (3) and V_c, A_f are replaced respectively by (17,18); and for a beam with an L-section, M_u in (1) is calculated taking into account skew bending and appropriate A_f.

3. Multi-span Beams

Consider the multi-span beam shown in Figure 2. Let NS be the number of members(spans) each of clear length $L_{nm}(m = 1, \ldots, NS)$ between supports and let $w_j(j = 1, \ldots, NS + 1)$ be the width of jth support. In the formulation of the optimization problem, the flexural strength constraints are included for sections over the supports and for sections in the span at which the span bending moment is maximum. For the sake of illustration, it will be assumed that the beam and slab are cast monolithically and that the slab extends to both sides of the beam. Hence, any section experiencing a positive bending moment can be treated as a T-section in the calculation of its flexural strength. The constraint on shear strength is imposed to avoid failure by web crushing. If p_i^+ and p_j^- represent the tensile reinforcing steel ratios at the critical section in the span i and over the support j respectively, the strength constraints may be written as

$$\phi_f \, p_i^+ \bar{b}_i f_{sy} \left(1 - \frac{p_i^+ \bar{b}_i f_{sy}}{1.7 f_c'}\right) b_w d^2 \;\geq\; |M_i^*|; \; (i = 1, \ldots, NS) \tag{22}$$

$$\phi_f \, p_j^- f_{sy} \left(1 - \frac{p_j^- f_{sy}}{1.7 f_c'}\right) b_w d^2 \;\geq\; |M_j^*|; \; (j = 1, \ldots, NS + 1) \tag{23}$$

$$\phi_s \, 0.2 f_c' b_w d \;\geq\; |V^*| \tag{24}$$

in which $p_i^+ = A_{st.i}^+/b_w d$, $p_j^- = A_{st.j}^-/b_w d$, $\bar{b}_i = b_w/b_{ef.i}$, $A_{st.i}^+$ and $A_{st.j}^-$ are the areas of longitudinal reinforcement at the sections of maximum positive bending moment $|M_i^*|$ and maximum negative bending moment $|M_j^*|$ respectively, and $|V^*|$ is the maximum shear force in the beam. Instead of adopting different positive steel ratios for each span, it may be convenient to keep the ratio constant(p^+) for all the spans. Similarly, the same negative steel ratio(p^-) may be adopted for all the support sections. With these simplifications, there will be only two flexural strength constraints for the entire multi-span beam.

The serviceability requirement on the maximum deflection in each span (Δ_m) once again includes the effects of creep and shrinkage and is based on an effective second moment of area I_{ef}:

$$\Delta_m \leq \Delta \ (m = 1, \ldots, NS) \tag{25}$$

Moreover, the maximum deflection is chosen to be the maximum occuring among all the possible load combinations for serviceability design. Alternatively, if the intensity of live load Q is not greater than that of dead load G, for each member a maximum limit on the ratio of effective span $L_{ef.m}$ to effective depth d may be prescribed as in equation(6)

$$\frac{L_{ef.m}}{d} \leq \left[\frac{k_1(\Delta/L_{ef.m})b_{ef.m}E_c}{k_2 F_{d.ef.m}} \right]^{\frac{1}{3}} \ (m = 1, \ldots, NS) \tag{26}$$

in which $F_{d.ef.m}$ is the effective design load per unit length of member m. For multi-span beams, where in adjacent spans the ratio of the longer to the shorter span does not exceed 1.2 and where no end span is longer than an interior span, k_2 may be taken as $1/185$ for an end span and $1/384$ for an interior span.

The requirements of durability and fire resistance (7,8) are the same as those for a simply supported beam except that the Australian code (1988) prescribes different values for c_f when $b_w > b_f$. In general, for a given fire resistance period and beam width b_w, the required cover c_f for a multi-span beam is less than that for a simply supported beam. In addition to the constraints on geometry(9–12), the minimum flexural strength of critical sections is prescribed as

$$p_i^+ \geq \frac{1.4}{f_{sy}} \ (i = 1, \ldots, NS) \tag{27}$$

$$p_j^- \geq \frac{1.4}{f_{sy}} \ (j = 1, \ldots, NS + 1) \tag{28}$$

Moreover, the ductility of the beam is ensured by limiting the depth of the neutral axis parameter at the critical sections, i.e.

$$k_{u.i}^+ \leq 0.4 \ (i = 1, \ldots, NS) \tag{29}$$

$$k_{u.j}^- \leq 0.4 \ (j = 1, \ldots, NS + 1) \tag{30}$$

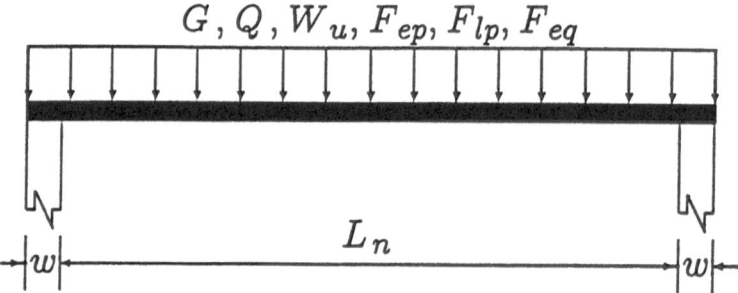

Figure 1. A simply supported beam subjected to several design actions.

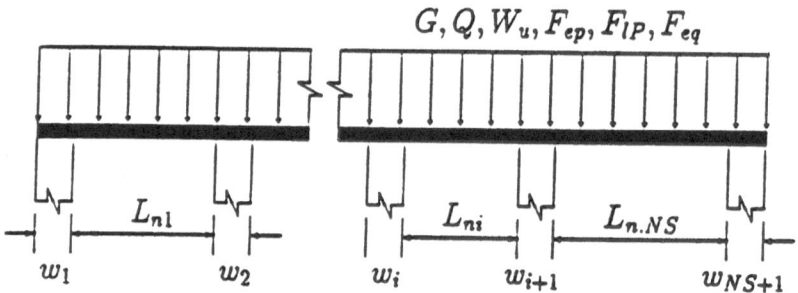

Figure 2. A multi-span beam subjected to several design actions.

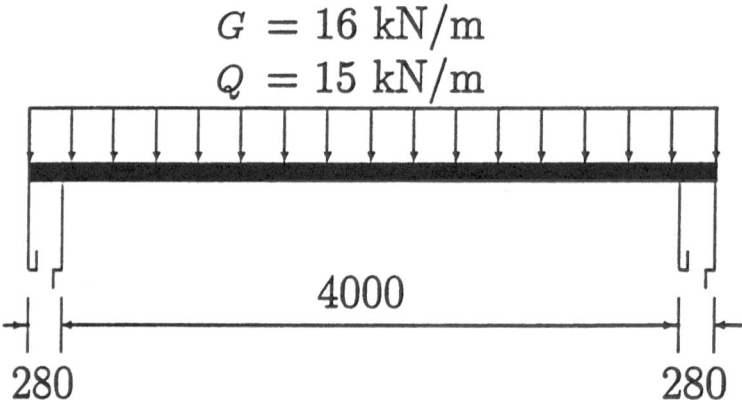

Figure 3. A simply supported beam subjected to uniformly distributed dead load G and live load Q. The value of dead load shown does not include self-weight.

in which $k_{u.i}^+$ and $k_{u.j}^-$ are the neutral axis parameters corresponding to the steel ratios p_i^+ and p_j^- respectively.

The quantities of concrete, steel and formwork in the beam cast monolithically with the slab, are

$$V_c = b_w(D - t)\left(\sum_{m=1}^{NS} L_{nm} + \sum_{j=1}^{NS+1} w_j\right) \tag{31}$$

$$V_s = \sum_{k=1}^{NB} A_{st.k}\, l_k + A_{st.s}\, l_s \tag{32}$$

$$A_f = 2(D - t)\left(\sum_{m=1}^{NS} L_{nm} + \sum_{j=1}^{NS+1} w_j\right) + b_w \sum_{m=1}^{NS} L_{nm} \tag{33}$$

The optimization problem for the multi-span beam may be stated as follows: Determine the width b_w, depth D(or d), and the steel ratios $p_i^+(i = 1, \ldots, NS)$ and $p_j^-(j = 1, \ldots, NS + 1)$ that minimize the cost of the beam (21) (in which V_c, V_s, A_f are given by (31–33)) and satisfiy the requirements on strength(22–24), serviceability(25) or (26), durability and fire resistance(7,8), geometry(9–12), minimum flexural strength(27,28), and ductility(29,30).

When the beam is cast separately and is not an integral unit of the slab it supports, the sections experiencing positive bending moments are designed as rectangular ones each of width b_w and depth D(excluding the thickness of slab t). The above statement is also valid for a multi-span beam designed as having a rectangular section or an L-section for positive bending moments with the following modifications: for a beam with a rectangular section, $\bar{b}$ is set equal to 1 in (22) and V_c(31), A_f(33) are replaced by

$$V_c = b_w D\left(\sum_{m=1}^{NS} L_{nm} + \sum_{j=1}^{NS+1} w_j\right) \tag{34}$$

$$A_f = 2D\left(\sum_{m=1}^{NS} L_{nm} + \sum_{j=1}^{NS+1} w_j\right) + b_w \sum_{m=1}^{NS} L_{nm} \tag{35}$$

and for a beam with an L-section, the ultimate flexural strength(left hand side of (22)) is calculated taking into account skew bending and A_f(33) is replaced by

$$A_f = (2D - t)\left(\sum_{m=1}^{NS} L_{nm} + \sum_{j=1}^{NS+1} w_j\right) + b_w \sum_{m=1}^{NS} L_{nm} \tag{36}$$

4. Columns

Columns are primarily compression members, and therefore reinforced columns are usually designed for strength limit state alone. The serviceability limit state based on cracking and deflection at service loads seldom controls the design. The treatment here is limited to columns subject to a single load system. The action effects at each end of a column will involve not only an axial compressive force but also a bending moment acting in one or two planes. The design criterion in the case of columns in combined axial compression and uniaxial bending is

$$\phi_a \, N_u \geq N^* \tag{37}$$

in which N_u is the ultimate strength of the column section, N^* the design compressive force and ϕ_a the strength reduction factor in combined bending and axial compression. The strength of a column of rectangular section subjected to an axial force and bending moments acting simultaneously about each principal axis should be such that

$$\left(\frac{M_x^*}{\phi_a \, M_{ux}}\right)^{\alpha_n} + \left(\frac{M_y^*}{\phi_a \, M_{uy}}\right)^{\alpha_n} \leq 1.0 \tag{38}$$

in which $\phi_a \, M_{ux}$ and $\phi_a \, M_{uy}$ are the design strengths in bending, calculated separately, about the major and minor axes under the design axial force N^*, M_x^* and M_y^* are bending moments about the major and minor axes, and $\alpha_n = 0.7 + [1.7N^*/(0.6N_{uo})]$ within the limits $1 \leq \alpha_n \leq 2$. Here, N_{u0} is the ultimate strength in compression without eccentricity. It should be mentioned that the design bending moment about each principal axis should be at least 0.05 times the product of the overall depth of the section in the plane of bending and the axial compressive force. Moreover, the ultimate strengths in compression(N_u, N_{uo}) and bending(M_{ux}, M_{uy}) of a reinforced column section will depend not only on the amount of reinforcement but also on its layout.

Minimum-cost columns can be designed as either short or long columns by limiting the slenderness ratio. For braced and unbraced short columns, the constraints on slenderness ratio may be written as

$$\frac{L_e}{r} \leq max\left\{25, \; 60\left(1 + \frac{M_1^*}{M_2^*}\right)\left(1.0 - \frac{N^*}{0.6N_{uo}}\right)\right\} \; (braced) \tag{39}$$

$$\frac{L_e}{r} \leq 22 \; (unbraced) \tag{40}$$

in which r is the radius of gyration($0.3D$ for a rectangular section), (M_1^*/M_2^*) is the ratio of the smaller to the larger of the design bending moments at the ends of the column, and L_e is the effective length. The ratio (M_1^*/M_2^*) is to be taken negative when the column is bent in single curvature and positive when it is bent in double curvature. Moreover, when the absolute value of M_2^* is equal to $0.05DN^*$ the ratio is to be taken as -1.0. In order to ensure that the column is durable and has adequate fire resistance, a good quality concrete has to be used and a suitable cover on the reinforcement has to be provided. The procedure is similar to that for beams(Section 2), and therefore we may write, for a rectangular section,

$$b \geq b_f \tag{41}$$

in which b is the smaller of the cross-sectional dimensions and b_f is the minimum cross-sectional dimension prescribed by the code for a given fire resistance period. Other geometric constraints similar to equations(8–12) may also be prescribed.

The reinforcement for column consists of longitudinal bars and lateral ties. The bounds on the longitudinal reinforcement ratio p may be written as

$$0.01 \leq p \leq 0.04 \tag{42}$$

in which $p = A_s/A_g$, A_s is the area of reinforcement and A_g is the gross cross-sectional area of column. It should be mentioned that the lower bound may be reduced under certain conditions and that the upper bound may be altered to effect proper placing and compaction of the concrete at splices and at junctions of the members (Australian Standard, 1988).

The quantities of concrete, steel and formwork for a column which is part of a frame may be written as

$$V_c = A_g L \tag{43}$$
$$V_s = A_s L + A_t L_t \tag{44}$$
$$A_f = SL \tag{45}$$

in which S is the perimeter of the cross-section and L is the centre to centre distance between the supporting beam members and A_t and L_t are respectively the cross-sectional area of the bar used to make ties and equivalent length of all the ties. Given the cross-sectional shape and layout of longitudinal steel of a uniform column, the minimum-cost design problem is to determine the cross-sectional dimensions(e.g. sides of a rectangular section, diameter of a circular section) and the amount of longitudinal reinforcement A_s such that the total cost (21) (with V_c, V_s, A_f given by (43–45)) is minimized subject to constraints on strength(37 or 38), slenderness ratio(39 or 40), durability and fire resistance(8,41), amount of longitudinal reinforcement(42) and geometry(9–12).

5. Solution Procedure, Examples and Discussion

The minimization problem formulated in each of the preceding three sections is a non-linear programming problem, and its solution can be sought by one of the two techniques that are known (Karihaloo & Kanagasundaram, 1988) to be efficient, namely sequential linear programming(SLP) (Pedersen, 1981) and sequential convex programming(SCP) (Fleury & Braibant, 1986). In both the techniques, the minimization starts at an arbitrary initial design point and proceeds until a convergence criterion is met. Moreover, the objective and constraint functions are replaced by their Taylor's series expansion using direct and/or reciprocal design variables. For instance, in SLP a constraint function $g(x)$ is replaced by its expansion in terms of direct design variables, at $x = x^0$, as

$$g(x) = g(x^0) + \sum_{i=1}^{NV} \left(\frac{\partial g}{\partial x_i}\right)_{x^0} dx_i \qquad (46)$$

in which x is the vector of design variables $x_i (i = 1, \ldots, NV)$. As an illustration, the shear strength constraint(24) will be replaced by its expansion at the design point $b_w = b^0$, $d = d^0$ as

$$\phi_s 0.2 f'_c \left(b^0 d^0 + d^0 \Delta b_w + b^0 \Delta d\right) \geq |V^*| \qquad (47)$$

in which $\Delta b_w = (b_w - b^0)$ and $\Delta d = (d - d^0)$. By this approach, the original non-linear optimization problem is reduced to a sequence of linear sub-problem in Δb_w, Δd and Δp. In SCP, on the other hand, the objective and constraint functions are replaced by their expansions in terms of direct and/or reciprocal design variables. The constraint function $g(x)$, at $x = x^0$, therefore, becomes

$$g(x) = g(x^0) + \sum_{+} \left(\frac{\partial g}{\partial x_i}\right)_{x^0} (x_i - x_i^0) + \sum_{-} \left(\frac{\partial g}{\partial x_i}\right)_{x^0} \frac{x_i^0}{x_i}(x_i - x_i^0) \qquad (48)$$

in which $\sum_+$ and $\sum_-$ stand for the summation of those terms for which the gradients $(\partial g/\partial x_i)$ are postive and negative, respectively. For example, the flexural strength constraint(23) will be replaced by

$$\phi_f p^0 f_{sy} \left(1 - \frac{p^0 f_{sy}}{1.7 f'_c}\right) [b^0 (d^0)^2 + (d^0)^2 \Delta b_w + 2b^0 d^0 \Delta d]$$

$$+ \phi_f f_{sy} \Delta p \, b^0 (d^0)^2 - \left(\frac{2\phi_f p^0 f_{sy}^2}{1.7 f'_c}\right) \frac{p^0}{p} \Delta p \, b^0 (d^0)^2 \geq |M_j^*| \qquad (49)$$

in which p refers to p_j^-, and p^0 is its current design value. By this approach, the original problem is reduced to a sequence of explicit, convex and separable sub-problems which may be solved by using dual methods of mathematical programming (Fleury & Braibant, 1986). In both the techniques, the solution of a sub-problem is restricted to a small region in the vicinity of the current design by specifying move-limits on design variables, i.e. on Δb_w, Δd and Δp. Both SLP and SCP solution algorithms are available as a part of the package ADS(Automated Design Synthesis, Vanderplaats & Sugimoto, 1986). The minimum-cost design of beams is obtained iteratively as follows:

1. Assume a starting design and calculate its self-weight G_i in iteration $i = 1$.

2. Analyse the beam and obtain bending moment and shear force envelopes.

3. Formulate and solve the minimization problem to obtain minimum-cost design.

4. Calculate the self-weight G_{i+1} of the minimum-cost design.

5. Repeat steps 2–4 until $100 \frac{|G_{i+1} - G_i|}{G_i} \leq \epsilon = 2\%$.

In case of columns subjected to a given axial force and bending moment at its ends step3 alone is executed. For the purposes of optimization, the area of cross-section of the reinforcing bars is assumed to be a continuous variable. However, it is recognized that the reinforcing bars are available only in discrete sizes, and therefore the sectional area of bars provided in the final design may be slightly greater than the calculated one.

The minimum-cost designs of simply supported and multi-span beams are shown in Figures 3–6. The details of geometry and loads are shown in the Figures 3 and 5. The value of dead load G shown does not include the self-weight, and the pattern of distribution of live load Q is fixed. Each beam supports a floor slab of thickness(t) 120mm in the interior of a domestic building. The fire resistance period of each beam is specified as 180 minutes and the maximum ratio(K_1) of depth to width of the cross-section as two (the width of the multi-span beam is given as 300mm). The following material properties and cost factors were used: f'_c=25 MPa, f_{sy}=400 MPa, C_{sc}=66, C_{fc}=0.42. The ratio of effective span to maximum deflection should not be less than 250 for total deflection, and 500 for incremental deflection. In all the Figures and Tables given below, linear dimensions are expressed in millimetres and area in square millimetres.

The solution of the non-linear programming problem of the beams was attempted using both SCP and SLP. Moreover, since there is no mathematical way of establishing the global nature of the minimum solution several starting designs were tried. A comparison of the minimum-costs of the simply supported beam(Figure 3) obtained by using both the techniques is given in Table 1. The starting designs have specified width and depth as shown, with 1% reinforcement, and some are feasible(F) while others are infeasible(NF). A design is considered feasible if it meets all the strength and serviceability requirements. It is found that both the techniques yield the same results, and several starting designs converge to the same minimum-cost design indicating the global nature of the design. To obtain the actual cost of design the above cost should be multiplied by the unit cost of concrete(per m^3), C_c. The efficiency of the techniques may be judged by the CPU time used on a microVAX/VMS as shown in Table 2.

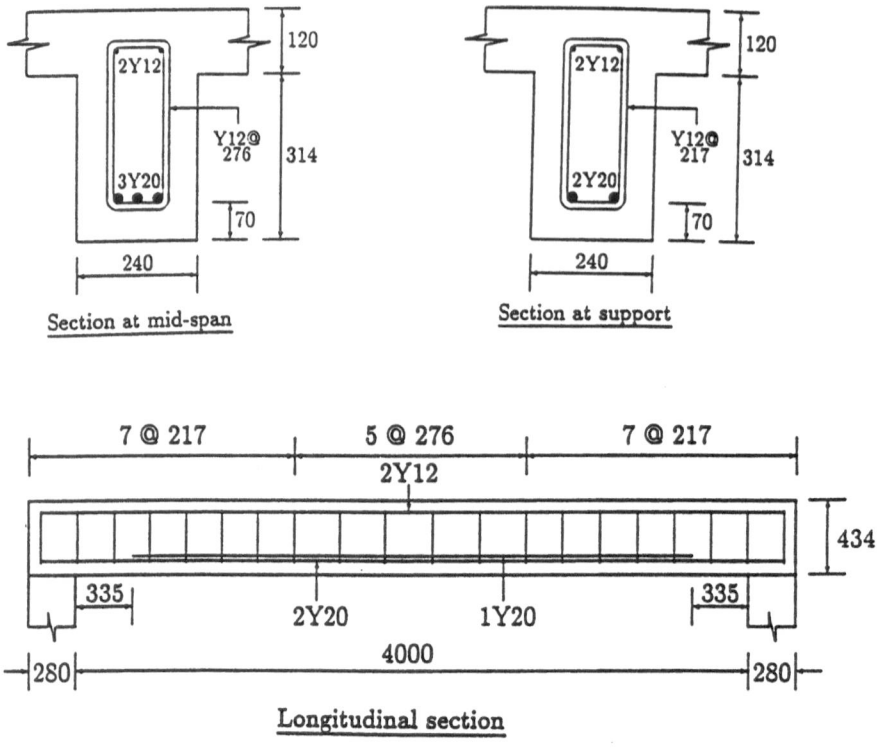

Figure 4. The minimum-cost design of the beam shown in Figure 3. The cross-sectional details do not include reinforcement in slab ($f'_c=25$ MPa).

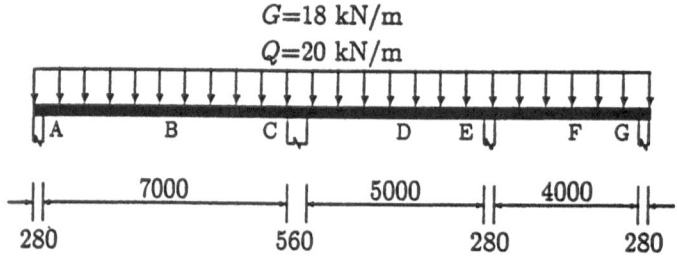

Figure 5. A multi-span beam subjected to uniformly distributed dead load G and live load Q. The value of dead load shown does not include self-weight.

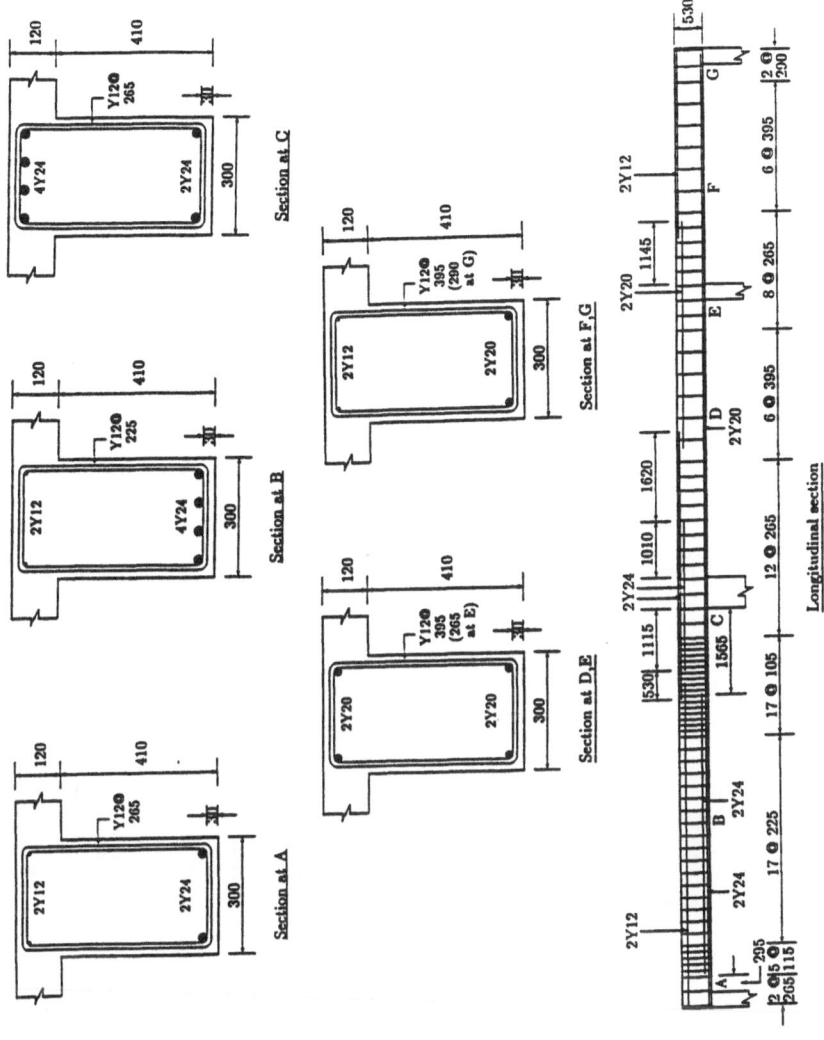

Figure 6. The minimum-cost design of the beam shown in Figure 5. The cross-sectional details do not include the reinforcement in slab. ($f'_c = 25$ MPa)

TABLE 1. Minimum-cost simply supported beam

Starting Design	Initial Cost	Number of Iterations	Minimum Cost	
			SCP	SLP
100x200 NF	1.0119	2	2.3629	2.3640
200x400 NF	2.1497	2	2.3635	2.3644
300x600 F	3.6432	2	2.3646	2.3656
350x500 F	3.2892	2	2.3652	2.3656
400x800 F	5.6419	3	2.3635	2.3648
500x1000 F	7.9457	3	2.3637	2.3648

TABLE 2. Convergence of NLP techniques (Simply supported beam)

Starting Design	Function Evaluations		CPU time, Seconds	
	SCP	SLP	SCP	SLP
100x200	51	56	4.92	5.42
200x400	32	28	2.17	1.79
300x600	36	48	3.46	3.47
350x500	43	36	3.20	2.64
400x800	55	60	4.17	2.49
500x1000	74	72	5.47	4.47

Table 2 also shows the number of evaluations of objective and constraint functions. In general, the fewer the number of function evaluations the better the technique. The properties of the minimum-cost design of the simply supported beam are: b_w=240mm, D=434mm, p=1.086x10^{-2}, c=70mm. The design was governed by the limits on the maximum total deflection, flexural strength at mid-span and minimum width required for fire resistance. Following the recommendations of the Standard, the flexural and shear reinforcement may be provided as shown in Figure 4. The relative contributions of the costs of concrete, reinforcing steel and formwork to the final cost are 0.3441, 0.3679, 1.6062 respectively, indicating the significant contribution of formwork cost to the total cost of construction. An evaluation of the minimum-cost design of the multi-span beam (Figure 5) is given in Table 3, whereas Table 4 compares the two solution techniques.

TABLE 3. Minimum-cost multi-span beam

Starting Design	Initial Cost	Number of Iterations	Minimum Cost SCP	Minimum Cost SLP
300x300 NF	7.8568	3	12.7928	12.7926
300x400 NF	10.1146	2	12.7900	12.7904
300x600 NF	14.7344	2	12.7946	12.7939
500x800 F	168.7860	3	12.7959	12.7924
1000x1000 F	619.3710	3	12.7942	12.7931
400x1500 NF	167.5470	3	12.7923	12.7926

TABLE 4. Convergence of NLP techniques (Multi-span beam)

Starting Design	Function Evaluations		CPU time, Seconds	
	SCP	SLP	SCP	SLP
300x300	95	104	7.92	8.64
300x400	55	64	4.95	5.38
300x600	56	64	4.84	4.63
500x800	80	120	8.95	8.65
1000x1000	132	128	10.74	8.61
400x1500	102	120	12.11	12.06

It is seen that both the techniques yield the same result irrespective of the starting design. The properties of the minimum cost design are: b_w=300mm, D=530mm, p_1^+= 1.143x10^{-2}, p_2^+=3.500x10^{-3}, p_3^+=3.867x10^{-3}, p_1^- = p_4^-=0, p_2^-=1.186x10^{-2}, p_3^-=3.500x10^{-3}, c=30mm. The design was governed by the limit on incremental deflection in the 7m span, flexural strength in all the three spans and at the interior supports, and by the specified width b_w. The beam may be reinforced as shown in Figure 6. As for the simple beam, the total cost of the beam was dominated by formwork(relative costs of concrete=2.1395, reinforcement=2.6438, formwork=8.0067, total cost=12.7900C_c). A sensitivity analysis of the minimum-cost designs of the beams was performed by varying the relative cost of formwork C_{fc} between 0.42 and 2. It was found that the minimum-cost designs remained unchanged.

TABLE 5. Minimum-cost column

Starting Design b_xxb_y	Initial Cost	Minimum cost design					
		SLP			SCP		
		b_y	$A_s/2$	C/C_c	b_y	$A_s/2$	C/C_c
300x300 NF	2.5950	522	783	3.4426	492	1056	3.4446*
300x400 NF	3.1487	522	783	3.4424	504	929	3.4378*
300x600 F	4.2562	522	784	3.4424	502	955	3.4389*
300x800 F	5.3637	522	784	3.4424	507	901	3.4368*
300x1000 F	6.4712	522	785	3.4428	521	782	3.4395
300x1500 F	9.2400	522	783	3.4424	521	782	3.4393
600x1000 F	9.8377	520	795	3.4415	520	788	3.4393

* the strength criterion(37) violated

TABLE 6. Comparison of NLP techniques (Column)

Starting Design	Function Evaluations		CPU time, Seconds	
	SLP	SCP	SLP	SCP
300x300	36	36	3.49	3.16
300x400	20	24	1.86	3.01
300x600	24	16	2.03	2.12
300x800	28	20	2.24	2.36
300x1000	32	19	2.45	1.64
300x1500	32	15	2.03	1.62
600x1000	40	24	3.86	2.46

Figure 7 shows a braced column subjected to an axial force and bending moment at its ends. It is to be designed as a short column and is to have a fire resistance period of 120 minutes. It is assumed that the column will have a rectangular cross-section with $b_x=300$mm and that longitudinal reinforcement will be placed only along the sides b_x. The relative cost factors are the same as those for the beams but $f'_c = 32$ MPa. The minimum-cost designs obtained using the two techniques from several different starting designs are compared in Table 5.

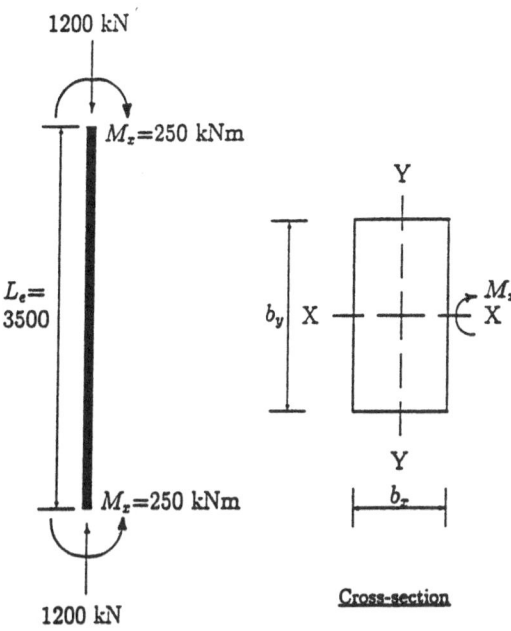

Figure 7. A column subjected to an axial force and a couple at its ends.

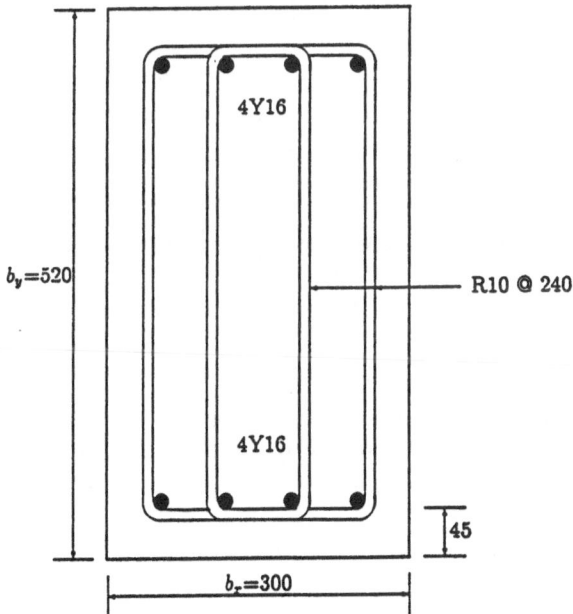

Figure 8. The minimum-cost design of the column shown in Figure 7 ($f_c'=32$ MPa).

The SLP converged to the same optimum design from several different starting designs. Although the SCP converged to a solution, it was found rather unreliable in the sense that for many starting designs the converged solution violated the strength constraint(37). It would seem, therefore, that the SLP is more suitable for the minimum-cost design of columns. The computational efficiency of the techniques on a microVAX/VMS may be judged by the various parameters listed in Table 6.

The minimum-cost column has the following properties: $b_x=300$mm, $b_y=520$mm, $c=45$mm, $A_s=1590$mm^2. It is interesting to note that the column has the minimum reinforcement ratio(p) of 0.01. The column design is governed by the strength criterion(37), the minimum steel ratio and the prescribed value for b_x. Moreover, the ultimate strength of the column section is achieved in a compression failure zone($N_u > N_{ub}$) of the interaction diagram for combined axial compression and uniaxial bending. The minimum-cost column may be reinforced as shown in Figure 8. The relative costs of concrete, reinforcement and formwork in final cost of the column are 0.5463, 0.5090 and 2.4116 respectively. An analysis of the sensitivity of the minimum-cost design of column to variations in the cost of formwork is shown in Table 7.

TABLE 7. Sensitivity analysis of minimum-cost column

C_{fc}	C/C_c	A_f (m^2)	Minimum-cost design	
			b_y	$A_s/2$
0.42	3.4415	5.7419	520	795
0.50	3.8920	5.6373	505	928
0.60	4.4448	5.4950	485	1142
≥ 0.65	4.7116	5.3675	467	1352

It is seen that as the cost of formwork C_{fc} increases the optimum design adjusts by reducing A_f. However, no further reduction is observed when C_{fc} reaches about 0.65. At this point it can be said that the procedure for obtaining the minimum-cost designs not only imitated mathematically the traditional design process, but more importantly resulted in practically feasible designs which are least expensive to construct. It can be incorporated into a knowledge-based system.

We shall now study the influence of the quality of concrete, i.e. its crushing strength f_c' on the minimum cost design.

6. Influence of f_c'

Besides the cross-sectional dimensions and steel ratio(s), f_c' will also be treated as a design variable. This will necessitate several changes to the formulation of the optimization problems, and these are listed below.

- The parameter γ appearing in the minimum tensile reinforcement requirement of a T-section (formula after eqn. (14)) is now a variable since it depends on the strength of concrete in the range $28 \leq f'_c \leq 56$ MPa.

- As the unit cost of ready-mix concrete, C_c, (eqn. 20) depends on its strength f'_c, it is no longer feasible to introduce unit costs of reinforcement and formwork relative to the unit cost of concrete, as in eqn (21). Instead, now the actual cost (eqn. 20) is minimized. In the Sydney market, $C_s = 8590$ $\$/m^3$ and $C_f = 55$ $\$/m^2$. The unit cost of concrete (in $\$/m^3$), not including the cost of transportation to site and placement, may be faithfully represented by the following regression equation in terms of f'_c in the range of $25 \leq f'_c \leq 60$ MPa with an error $<0.6\%$

$$C_c = 98.3 + 1.7120 f'_c - 0.0277 f'^2_c + 0.00041 f'^3_c. \tag{50}$$

- In the statement of each of the problems, f'_c is added to the list of design variables, and the number of design variables (if stated) increased by one.

- In the illustration of SLP following eqn (46), the shear strength constraint (47) will now be replaced by its expansion at the design point $b_w = b^0, d = d^0, f'_c = f^0$,

$$0.2\phi_s \left(b^0 d^0 f^0 + d^0 f^0 \Delta b_w + b^0 f^0 \Delta d + b^0 d^0 \Delta f'_c \right) \geq |V^*| \tag{51}$$

in which $\Delta b_w = (b_w - b^0)$, $\Delta d = (d - d^0)$, and $\Delta f'_c = (f'_c - f^0)$. The sequence of linear sub-problems now involves $\Delta b_w, \Delta d, \Delta f'_c$ and Δp.

- Likewise, in the illustration of SCP, the flexural strength constraint (49) will now read as follows

$$\phi_f p^0 f_{sy} \left(1 - \frac{p^0 f_{sy}}{1.7 f^0} \right) [b^0 (d^0)^2 + (d^0)^2 \Delta b_w + 2 b^0 d^0 \Delta d]$$
$$+ \phi_f f_{sy} \Delta p \ b^0 (d^0)^2 + \phi_f p^0 f_{sy} \left(1 + \frac{p^0 f_{sy}}{1.7 f^0} \right) \Delta f'_c \ b^0 (d^0)^2 -$$
$$\left(\frac{2 \phi_f p^0 f^2_{sy}}{1.7 f^0} \right) \frac{p^0}{p} \Delta p \ b^0 (d^0)^2 \geq |M^*_j| \tag{52}$$

The example problems of Figures 3, 5 and 7 were redesigned after treating f'_c as a design variable. In each example it was found that both SCP and SLP converged to the same solution from several starting designs using a 56 MPa mix.

The properties of the minimum-cost design of the simply supported beam (Figure 3) are: $b_w = 240$ mm, $D = 410$ mm, $p = 1.0 \times 10^{-2}$, $f'_c = 40.3$ MPa, $c =$

70 mm. The cost of construction is \$300, in which the cost of concrete has been evaluated from eqn (50) corresponding to the optimum value of $f_c' = 40.3\ MPa$. This should be compared with the minimum-cost design of Figure 4 for $f_c' = 25$ MPa which cost \$310 to construct.

The minimum-cost design of the multi-span beam (Figure 5) has the following properties: $b_w = 300$ mm, $D = 480$ mm, $f_c' = 47.5\ MPa$, $p_1^+ = 1.4 \times 10^{-2}$, $p_2^+ = 3.5 \times 10^{-3}$, $p_3^+ = 4.76 \times 10^{-3}$, $p_1^- = p_4^- = 0$, $p_2^- = 1.4 \times 10^{-2}$, $p_3^- = 4.14 \times 10^{-3}$, $c = 30$ mm. The cost of construction is \$1603, in which the cost of concrete has been calculated from eqn (50) corresponding to $f_c' = 47.5\ MPa$. This design should be compared with the minimum-cost design of Figure 6 for $f_c' = 25\ MPa$ which cost \$1663 to construct.

Finally, the minimum-cost column (Figure 7) has the following properties: $b_x = 300$ mm, $b_y = 470$ mm, $c = 45$ mm, $A_s = 1658$ mm^2, $f_c' = 48\ MPa$. The cost of construction is \$420, in which the cost of concrete has been calculated from eqn (50) corresponding to $f_c' = 48\ MPa$. This design should be compared with the minimum-cost design of Figure 8 for $f_c' = 32\ MPa$ which cost \$430 to construct.

The above study shows that when the crushing strength of concrete is treated as a variable together with the section sizes and steel ratios, the minimum cost designs use shallower sections and higher strength mixes, but are still marginally cheaper to construct. Therefore, these designs offer two distinct advantages over designs based on lower strength mixes. They will be more durable because they use higher strength mixes, and will provide more usable headroom in multi-storey construction because they use shallower sections.

7. References

Australian Standard (1988) AS3600: Concrete Structures , SAA, Sydney.

Fleury, C. and Braibant, V. (1986), 'Structural optimization: A new dual method using mixed variables', Int. J. Num. Methods in Eng. 23, 409–428.

Kanagasundaram, S. and Karihaloo, B.L. (1991) 'Minimum-cost design of reinforced concrete structures', Computers & Structures 41, 1357–1364.

Kanagasundaram, S. and Karihaloo, B.L. (1991), 'Minimum cost reinforced concrete beams and columns', Computers & Structures 41, 509–518.

Karihaloo, B.L. and Kanagasundaram, S. (1990), 'Minimum-cost design of reinforced concrete structures to AS 3600-1988', J. Structural Optimization 2, 83–94.

Karihaloo, B.L. and Kanagasundaram, S (1988), 'Comparative study of NLP techniques in optimum structural frame design', in G. I. N. Rozvany and B.

L. Karihaloo (eds.) Structural Optimization, Kluwer Academic Publishers, Dordrecht, pp. 143–150.

Pedersen, P (1981) 'The integrated approach to FEM–SLP for solving problems of optimal design', in E.J. Haug and J. Cea (eds.), Optimization of Distributed Parameter Systems, Vol I, Sijthoff & Noordhoff, The Netherlands, pp. 739–756.

Vanderplaats, G. N. and Sugimoto, H (1986) 'A general-purpose optimization program for engineering design', Computers & Structures 24, 13–21.

Warner, R.F., Rangan, B. V. and Hall, A.S. (1988), Reinforced Concrete, Third Edition, Longman Australia.

OPTIMIZATION OF SANDWICH STRUCTURES WITH RESPECT TO LOCAL INSTABILITIES WITH MBB-LAGRANGE

W. DOBLER, P. ERL
Messerschmitt-Bölkow-Blohm GmbH
Military Aircraft Division
P.O. Box 80 11 60
D-8000 München 80
Germany

H. RAPP
Messerschmitt-Bölkow-Blohm GmbH
Helicopter Division
P.O. Box 80 11 40
D-8000 München 80
Germany

ABSTRACT. MBB-LAGRANGE is a powerful optimization code which allows the optimization of structures. To achieve the optimal design of a structure all special requirements must be fulfilled, that means that constraints e.g. stress, strains or displacements have to be in a feasible range while the objective function e.g. the structural weight reaches a minimum value. The results are strongly dependent on the optimization constraints which must be fulfilled. In this context local and global stability constraints may be essential for the design process. In the continuing development of MBB-LAGRANGE stability constraints are one of the major topics the MBB-LAGRANGE team is busy developing. One topic is the realization of local stability constraints of sandwich structures which will be discussed in this paper. The theoretical background and the integration into the optimization code MBB-LAGRANGE will be shown. An example of a helicopter sandwich structure will demonstrate the influence of this constraint type in the design process while optimal results are found. These results satisfy all requirements: local stability, stress and dynamic constraints.

1. Introduction

Since 1984 MBB has been developing the structural optimization code MBB-LAGRANGE. For many design problems of space and aircraft structures it has already led to remarkable and significant improvements:

- increasing the product quality by finding optimal designs,

- relieving the engineer of time and cost consuming parameter studies to be able to perform more creative work,

- considering simultaneous requirements of statics, dynamics, aeroelastics and manufacturing.

To improve or modify a design with a computer-program like MBB-LAGRANGE different branches of applicable mathematics and structural mechanics are necessary. Figure 1 shows the main parts of MBB-LAGRANGE the optimization algorithms, the optimization model and the structural analysis with sensitivity calculation. With the help of these modules the optimization problem can be solved in an efficient way.

951

G. I. N. Rozvany (ed.), Optimization of Large Structural Systems, Vol. II, 951–972.
© 1993 *Kluwer Academic Publishers.*

952

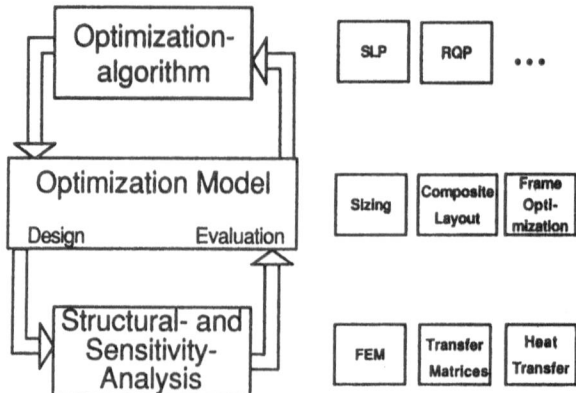

Figure 1: Main program modules of MBB-LAGRANGE

In an iteration loop the optimization procedure changes the chosen design variables in a way to achieve the best value for an objective while not violating defined constraints, which represent the boundaries of the design space. In the optimization process the *structural analysis* is started. Based on finite element methods (FEM), static, dynamic, aeroelastic, flutter and stability modules have been incorporated. These modules deliver the structural response e.g. displacements, stresses, strains, eigenfrequencies, eigenvalues, velocities, accelerations, aeroelastic efficiencies and stability values. It is possible to treat homogenous materials with isotropic, orthotropic and anisotropic behavior. The element library contains the types: truss, beam, membrane (3,4,8 nodes), shell (3,4 nodes) and volume elements. In addition shell structures can be analyzed with a special transfer matrix procedure. To perform the structural analysis a finite element structure model is a necessary requirement. The finite element structure must be described by INPUT-DATA, which describe the geometry, the properties and the materials. The INPUT-FORMAT is compatible as far as necessary to modern FEM-Packages and to pre- and postprocessor tools available on common software market.

To use the potential of modern mathematical optimization algorithms with rapid convergence gradient information is necessary. Therefore a further step in the optimization process is the *Sensitivity Analysis*. The gradient for the objective function f(x) and the constraint functions $g_j(x)$ with respect to the design variable x_i

$$dg_j(x)/dx_i \quad \text{and} \quad df(x)/dx_i$$

must be delivered. The sensitivity analysis can be performed by numerical difference formula or by analytical formulation. For large scale applications the analytical formulation is essential for the performance of the computer program. With the analysis modules incorporated in MBB-LAGRANGE it is possible to performe the sensitivity analysis in an efficient way (see chapter 3.2).

To provide the input for the optimization algorithm, the *Optimization Model* (figure 2), is necessary. Here the requirements for the structure are formulated as a general nonlinear programming problem.

The last step within the optimization loop is done by the *Optimization Algorithm*. It delivers new design variables which has to adjust to the optimization criteria. It is necessary

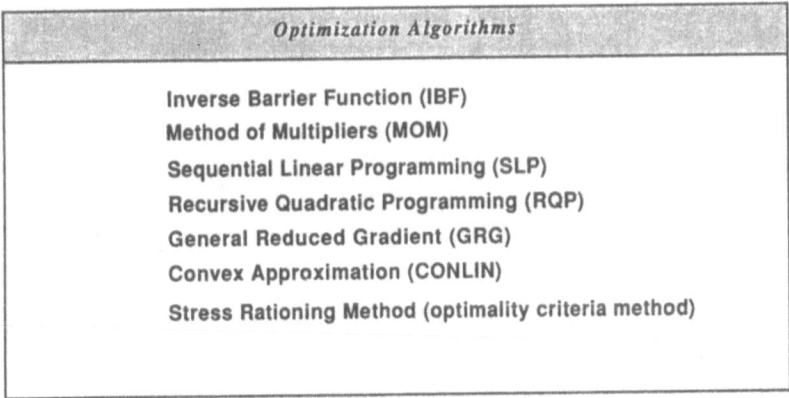

General Problem	Design Variables
min f(x) s.t. $g_j(x) \geq 0$ with $$x_i^L \leq x_i \leq x_i^u$$ I := number of design variables L := lower bound; u := upper bound	= > cross section areaes for trusses cross section areas for bars wall thickness for membrane elements wall thickness for shell elements wall thickness for composite elements layer angles for composite elements balance masses
Objective Function	**Optimization Constraints**
minimum structural weight general objective functions *	= > displacements, strains , stresses buckling * aeroelastic efficiencies flutter speed eigenfrequencies results of transient reponse result of frequency response manufacturing constraints bounds for design variables (gages) (* actual development)

Figure 2: Optimization model

to provide several different optimization algorithms because there is no known single algorithm which is suited to every type of problem. Figure 3 shows the available optimization algorithms in MBB-LAGRANGE.

Optimization Algorithms
Inverse Barrier Function (IBF) Method of Multipliers (MOM) Sequential Linear Programming (SLP) Recursive Quadratic Programming (RQP) General Reduced Gradient (GRG) Convex Approximation (CONLIN) Stress Rationing Method (optimality criteria method)

Figure 3: Optimization Algorithms in MBB-LAGRANGE

Today a variety of structural optimization problems of aerospace structures have already been investigated [5,6]. Discussions with the users of MBB-LAGRANGE have been shown that for further applications stability modules are indispensable constraints for lightweight structures. In figure 4 the chosen concept for the stability modules in MBB-LAGRANGE is shown. Two main parts have been picked out. One branch is to realize global and local buckling by the formulation of geometrical stiffness and solving the eigenvalue problem, the other branch is to integrate stability equations which take into account special local stability problems as discussed in the following chapters.

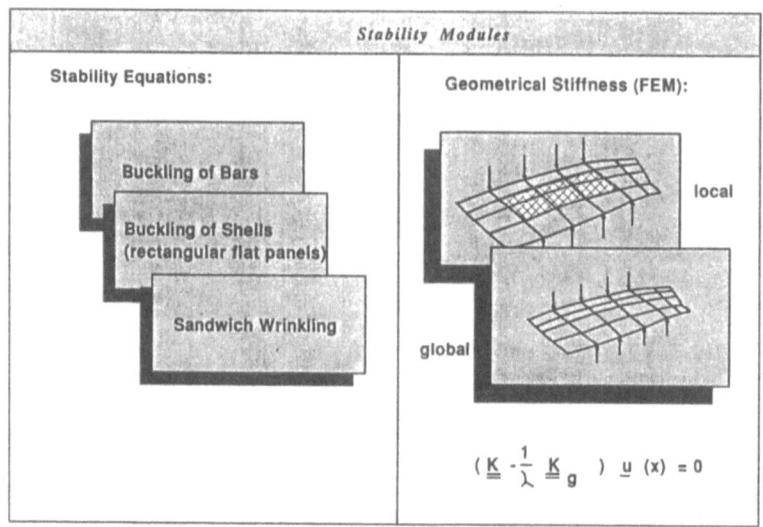

Figure 4: Stability modules in MBB-LAGRANGE

2. Local Stability of Sandwich Structures

Sandwich structures are very common in aerospace applications. Today sandwich plates are used for cowlings of engines, for panels in wings and stabilizers, for flaps and rudders and other aircraft components. In satellites nearly all structural components are made of sandwich panels: antennas, solar generators and the structure itself.

Sandwich plates mainly consist of three parts (figure 5):

1. the upper face sheet,
2. the sandwich core and
3. the lower face sheet.

The two face sheets are bonded to the sandwich core by adhesives. The sandwich core can either be a homogeneous material such as foam or an orthotropic material such as wood or honeycomb.

The basic concept of a sandwich is that the two facings carry all inplane and bending loads while the core carries the transverse shear load. This can be accomplished by the use of stiff and strong face sheets which are relatively thin and by a core material which is very light but which has a sufficient shear strength. So the sandwich concept produces very stiff and strong structures at a minimum weight.

When loading sandwiches, special failure mechanisms must be considered. Figure 6 shows the possible sandwich failure modes.

One failure mode of a sandwich panel is global buckling. But this instability is not sandwich specific. It has to be considered for every plate structure and can be handled in a usual

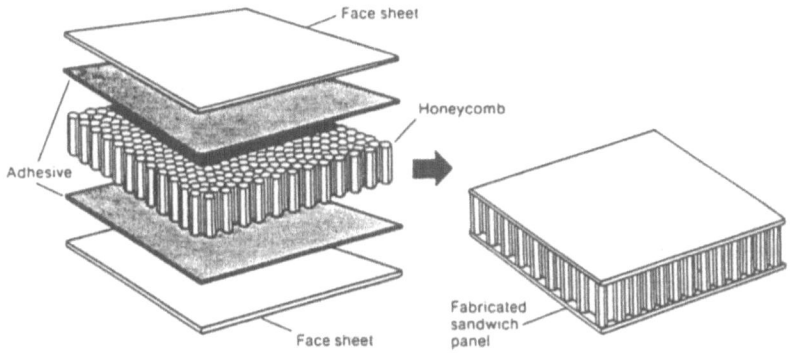

Figure 5: Typical sandwich configuration

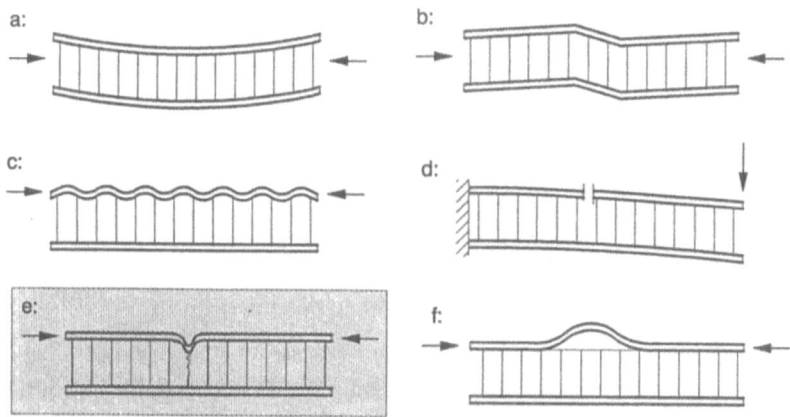

Figure 6: Sandwich failure modes

manner. The failure modes b (shear crimping), c (intercellular buckling or dimpling) and f (debonding of face sheets) are usually not the typical failure modes:

- Intercellular buckling (c) is not really a failure, only the stiffness of the sandwich decreases somewhat.
- Debonding (f) of face sheets indicates an incorrect fabrication.
- Shear crimping occurs, if the core material is very weak compared with the face sheets. With a correctly designed sandwich, shear crimping should never occur.

A failure in the face sheets themselves (d) is the typical failure mode for facings loaded by tension. It can be handled in a usual manner by application of failure criteria with the stress σ_f in the face sheets

$$\sigma_f = \frac{n_f}{t},$$

where n_f is the membrane force in the face sheet and t is the thickness of this face sheet.

For compression loads (or bending loads, where one facing is loaded by tension and one by compression) failure mode e in figure 6 is the essential failure mode. This mode is a local instability phenomenon and it is called face sheet wrinkling. In most applications the critical stress for face sheet wrinkling σ_{cr} is much lower than the ultimate stresses of the face sheets σ_{ultt} or σ_{ultc} (tension or compression), so that the strength of nearly all sandwich structures is determined by face sheet wrinkling.

In the following, the equations for determining the critical face sheet wrinkling stress σ_{cr} are derived. Some comments are given for the application of these equations for composite sandwiches.

2.1 THEORY OF SANDWICH FACE SHEET WRINKLING

When a sandwich is loaded by bending or compression at least one face sheet has to transfer a compression force. If this compression force exceeds a critical value the support of the face sheet by the the sandwich core is not strong enough and it fails locally. This failure always is a compression failure of the core either due to material strength (especially for isotropic foams) or due to stability of the very thin honeycomb core walls. This failure of the sandwich face sheet together with the sandwich core is called face sheet wrinkling.

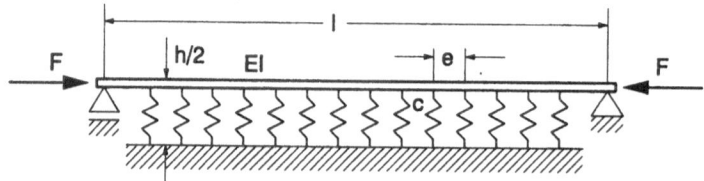

Figure 7: Bar on elastic foundation

The calculation of the critical face sheet wrinkling stress can be handled in analogy to the bar on an elastic foundation. For a bar supported by constant spaced springs and loaded by compression (figure 7) the differential equation for the deflection w reads

$$EIw'''' + Fw'' + \beta w = 0. \tag{1}$$

Herein EI denotes the bending stiffness of the bar and F the compressive force. β describes the elastic foundation and is defined as

$$\beta = \frac{c}{e}$$

with c as spring constant and e as distance between the springs.

With the assumption for the deflection w

$$w = C \sin m\pi\xi, \qquad \text{with} \quad \xi = \frac{x}{l}$$

follows for the critical load of the bar [1]:

$$F_{cr} = 2\sqrt{\beta EI}. \tag{2}$$

The real foundation of the face sheet in a sandwich is rather continuous than discrete as assumed in the above equations. To achieve this the spring constant c has to be replaced by the longitudinal stiffness of the core material perpendicular to the sandwich plane:

$$c_{cont} = \frac{2E_c be}{h}.$$

With this the elastic constant β reads

$$\beta = \frac{2E_c b}{h}, \tag{3}$$

where

E_c : Young's modulus of core material in the direction perpendicular to the sandwich plane,
b : width of bar and
h : height of sandwich core.

The bending stiffness of the bar can be written for a rectangular cross section as

$$EI = \frac{E_f bt^3}{12}, \tag{4}$$

where

E_f : Young's modulus of the bar (face sheet) material,
b : width of bar and
t : height of the bar (face sheet).

Introducing these equations into the equation for the critical load yields

$$F_{cr} = 2\sqrt{\frac{E_f E_c b^2 t^3}{6h}}. \tag{5}$$

From this follows, the critical stress in the bar is

$$\sigma_{cr} = 0.816\sqrt{\frac{E_f E_c t}{h}}. \tag{6}$$

With this formula the critical stress for a bar on an elastic foundation is known. It can be directly transferred to a sandwich plate with an orthotropic core (such as honeycomb core), eventually the Young's modulus of the face sheet E_f has to be exchanged by the membrane modulus $\overline{E_f} = E_f/(1 - \nu_{xy}\nu_{yx})$.

In practice due to imperfections of the sandwich face sheets the theoretical prefactor of 0.816 cannot be reached. So there is a recommendation to reduce this factor to a value of 0.33 [3]. This value is based on experiments and is valid for most materials and face sheet thicknesses. In special cases this factor has to be determined by tests.

For a sandwich with an isotropic core such as foam, the shear stiffness of the core has to be taken into account. This results in a different formula [4]:

$$\sigma_{cr} = 0.825\sqrt[3]{E_f E_c G_c}. \tag{7}$$

The above equations (6) and (7) are valid for a one-dimensional loading. A general two-dimensional theory for instabilities of sandwich plates is given in [2]. There the differential equation corresponding to (1) reads

$$D\nabla^4 w + N_x \frac{\partial^2 w}{\partial x^2} + N_y \frac{\partial^2 w}{\partial y^2} + N_{xy} \frac{\partial^2 w}{\partial xy} + \beta w = 0, \tag{8}$$

with

$$D \qquad : \text{ plate stiffnesses},$$
$$N_x, N_y, N_{xy} : \text{ normal and shear load of the plate}.$$

A general solution of this equation is not known. Therefore for the multiaxial loading of sandwich plates the following recommendation based on the one-dimensional solution is given [3]:

1. For a one-dimensional loading use equations (6) and (7) respectively.

2. In the case of a two-dimensional loading calculate the principal stresses σ_1 and σ_2. If only one stress is in compression then neglect the tension stress and handle the problem as one-dimensional.

3. If all two stresses are in compression then use the following interaction formula (9) to determine the strength. Failure does not occur, if the formula is valid. The critical stresses σ_{1cr} and σ_{2cr} are calculated in the directions of the principal stresses.

$$\frac{\sigma_1}{\sigma_{1cr}} + \left(\frac{\sigma_2}{\sigma_{2cr}}\right)^3 \leq 1, \qquad \text{with} \quad \sigma_2 < \sigma_1. \tag{9}$$

With these equations we can now determine the strength of any sandwich configuration.

2.2 APPLICATION FOR COMPOSITE SANDWICH PLATES

When applying above formulas to sandwich plates with face sheets made from composite materials some points need further explanation. If the face sheets are homogeneous in the thickness direction, the formulas (6) and (7) can be used directly. If the face sheets are not homogeneous, and in general this is the case with laminates, the formulas have to be transformed.

The lay-up of a general laminated sandwich is shown in figure 8.

As the face sheets are non-homogeneous, the local bending stiffness of the face sheet can not be calculated by

$$EI_f = E_f \frac{t^3}{12}.$$

Instead the effective stiffness has to be used. For the determination of the effective stiffness the classical laminate theory is used. It delivers a relation between the strains (ϵ_x, ϵ_y, γ_{xy}) and curvatures (κ_x, κ_y, κ_{xy}) and the corresponding forces (n_x, n_y, n_{xy}) and moments (m_x, m_y, m_{xy}):

$$\begin{pmatrix} n \\ m \end{pmatrix} = \begin{pmatrix} \mathbf{A} & \mathbf{B} \\ \mathbf{B} & \mathbf{D} \end{pmatrix} \left[\begin{pmatrix} \epsilon \\ \kappa \end{pmatrix} - \begin{pmatrix} \alpha_t \\ \kappa_t \end{pmatrix} \Delta T \right]. \tag{10}$$

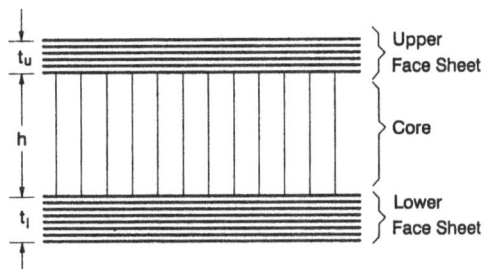

Figure 8: Lay-up of a general sandwich

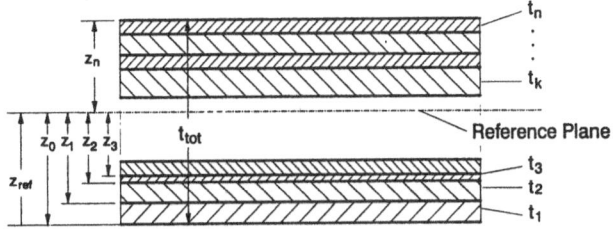

Figure 9: General laminate

The 3x3 submatrices **A**, **B** and **D** denote the stiffness coefficients of the laminate. **A** describes the longitudinal stiffness, **D** the bending stiffness and **B** the coupling between bending and stretching. Such coupling occurs, if the laminate lay-up is not symmetrical or if the reference plane is not the middle plane.

With the notations given in figure 9 the coefficients of these submatrices can be determined by the following relations ($[a_{ij}]_k$ denotes the stiffness coefficients of the k-th unidirectional layer in an i-j coordinate system):

$$
\begin{aligned}
A_{ij} &= \sum_{k=1}^{n}[a_{ij}]_k(z_k - z_{k-1}) = \sum_{k=1}^{n}[a_{ij}]_k t_k, \\
B_{ij} &= \frac{1}{2}\sum_{k=1}^{n}[a_{ij}]_k(z_k^2 - z_{k-1}^2) = \frac{1}{2}\sum_{k=1}^{n}[a_{ij}]_k t_k(z_k + z_{k-1}), \\
D_{ij} &= \frac{1}{3}\sum_{k=1}^{n}[a_{ij}]_k(z_k^3 - z_{k-1}^3) = \frac{1}{3}\sum_{k=1}^{n}[a_{ij}]_k t_k(z_k^2 + z_k z_{k-1} + z_{k-1}^2).
\end{aligned}
\tag{11}
$$

The local bending stiffness for calculating the wrinkling stress is given in the coefficients D_{11} and D_{22} of the bending submatrix **D**. But these coefficients represent the real stiffnesses only, when the corresponding coupling coefficients B_{11} and B_{22} are zero. Otherwise due to bending-stretching coupling the stiffnesses are too low. For a general laminate (figure 9) the coupling coefficients are not zero. The values of these coefficients depend on the position of the laminate reference plane when applying equations (11). Therefore the position of this reference plane has to be chosen in such a manner, that the coupling coefficient is zero. In general for the two directions 1 and 2 there will be a different reference plane. It can be determined for direction 1 by

$$z_{ref1} = \frac{B_{11}}{A_{11}} \tag{12}$$

and for direction 2 by

$$z_{ref2} = \frac{B_{22}}{A_{22}}. \tag{13}$$

In the following, only direction 1 will be considered. Then we can rewrite equation (2) for an orthotropic core (e.g. honeycomb core) with $EI = bD_{11}$

$$F_{cr} = 2\sqrt{\frac{2E_c b^2 D_{11}}{h}}. \tag{14}$$

Division of this critical force F_{cr} by the laminate thickness t_{tot} yields the critical face sheet wrinkling stress

$$\sigma_{cr} = \frac{2.82}{t_{tot}}\sqrt{\frac{E_c D_{11}}{h}}. \tag{15}$$

σ_{cr} is a principal stress and for the calculation of factors of safety it must be also compared with the principal stresses which act in that face sheet.

For sandwich plates with an isotropic core the corresponding equation for the wrinkling stress is

$$\sigma_{cr} = \frac{1.89}{t_{tot}}\sqrt[3]{D_{11}E_c G_c}. \tag{16}$$

The investigation of sandwich structures by means of a FEM code requires the following steps:

1. Calculate the forces in the sandwich elements of a structure by the use of a finite element package. Determine factors of safety for each face sheet laminate itself by using proper failure criteria.

2. Calculate the principal stresses in each of the two face sheets of the sandwich with the help of the classical laminate theory.

3. Determine the principal stresses and their directions.

4. Determine the critical face sheet wrinkling stresses in these directions by the use of equation (15) and (16) respectively under consideration of the correct position of the reference plane.

5. Calculate factors of safety against face sheet wrinkling by applying equation (9).

3. Integration of Stability Constraints in MBB-LAGRANGE

To take into account stability constraints special Input Cards have been defined. In figure 10 the essential modules are listed which will be executed during a design process.

The program code is divided in two main parts. In the Input Module the user of the program must supply the necessary model data. With the help of the Optimization Control Deck the use of the Buckling Module can be initiated. Within this Buckling Module different stability analysis modules (figure 4) can be selected, e.g. the sandwich wrinkling module. If the stability analysis is required special Optimization Data must be delivered. For local

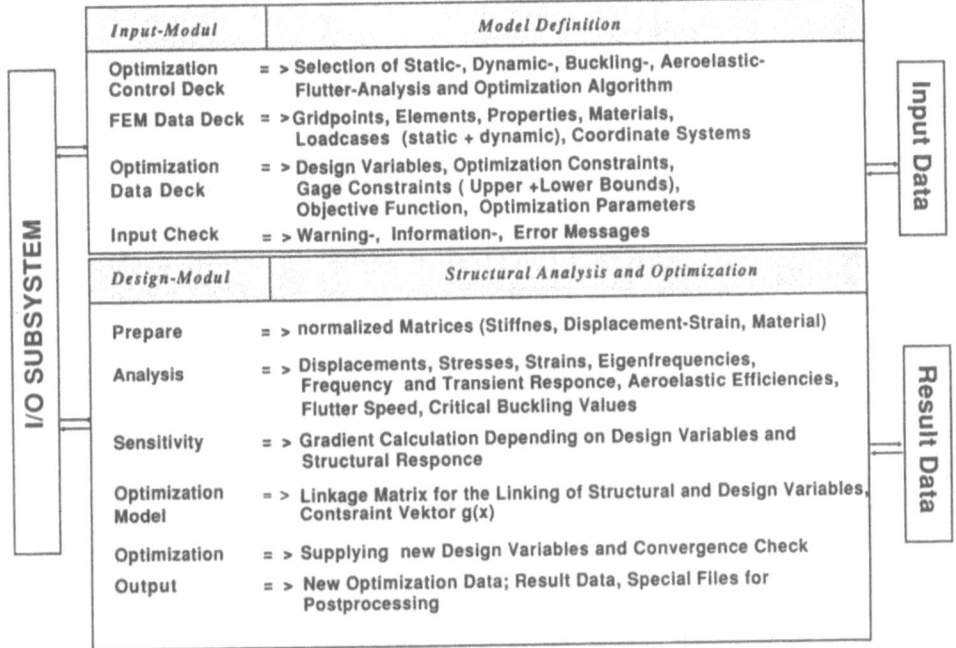

Input-Modul	Model Definition
Optimization Control Deck	= > Selection of Static-, Dynamic-, Buckling-, Aeroelastic-Flutter-Analysis and Optimization Algorithm
FEM Data Deck	= > Gridpoints, Elements, Properties, Materials, Loadcases (static + dynamic), Coordinate Systems
Optimization Data Deck	= > Design Variables, Optimization Constraints, Gage Constraints (Upper +Lower Bounds), Objective Function, Optimization Parameters
Input Check	= > Warning-, Information-, Error Messages

Design-Modul	Structural Analysis and Optimization
Prepare	= > normalized Matrices (Stiffnes, Displacement-Strain, Material)
Analysis	= > Displacements, Stresses, Strains, Eigenfrequencies, Frequency and Transient Responce, Aeroelastic Efficiencies, Flutter Speed, Critical Buckling Values
Sensitivity	= > Gradient Calculation Depending on Design Variables and Structural Responce
Optimization Model	= > Linkage Matrix for the Linking of Structural and Design Variables, Contsraint Vektor g(x)
Optimization	= > Supplying new Design Variables and Convergence Check
Output	= > New Optimization Data; Result Data, Special Files for Postprocessing

(I/O SUBSYSTEM) (Input Data) (Result Data)

Figure 10: Input and design modules of MBB LAGRANGE

stability analysis substructures must be selected defined by 3 or 4 nodes picked out of the FEM Model. In the following text we name these substructures buckling fields. Further information for a buckling field are the elements which lay in this substructure and the type of constraint (e.g. sandwich wrinkling) and some special parameters described in the user manual of MBB-LAGRANGE [7]. If the input check is passed without fatal error the design process can be started. The communication between these two programs is organized by the I/O Subsystem managing the data exchange. In the case of selected wrinkling analysis a special submodule of the analysis part will calculate the principal stresses σ_1, σ_2 and the critical stresses σ_{1cr}, σ_{2cr} for each element of the buckling field. By calculating these stress values equation (18) can be evaluated and part the constraint vector $g_j(x)$ where all constraints are summarized. A further step to integrate the wrinkling constraints has been done by the supplement of the Sensitivity Module. With the constraint vector $g_j(x)$ and the gradient vector dg_j/dx_i the Optimization Module is supplied with the necessary informations to solve the optimization problem.

3.1 INPUT DATA DEFINING THE OPTIMIZATION MODEL

To consider sandwich wrinkling constraints the required data must be supplied by input cards listed in figure 11. With this input data local stability constraints can be defined for substructures as illustrated in figure 12. To fulfill the constraint and the objective function the lay-up of a general laminated sandwich (figure 10, 9) is varied by changing the layer thicknesses or the layer angles. The sandwich structure can be idealized by triangular and quadrilateral elements (CTRIA3, QUAD4). With the input data delivered from the PCOMP-Card the initial thickness and fiber orientation angle (figure 13) of each ply is

Control-Card	Description	Input Data
GRID	defines FEM nodes	grid coordinates
CTRIA	defines a triangular plate element	element nodes, material identification number
CQUAD4	defines a quadrilateral plate element	element nodes, material identification number
MAT8	defines the material proprty	material identification number, E1,E2 Youngs Moduli,NUE12 Poisson's Ratio, G12 Inplane Shear Modulus,RHO Mass Density
PCOMP	defines the materials of an n-ply composite laminate	propperty identification number, material ID of various plies, faller theory, initial thickness t_i and initial orientation Θ_i angle of various plies
CMAT	defines allowable stress, tension of an n-ply composite laminate	allowable strains in tension and compression for composite elements
DESV	defines the structural variables linked to a design variable	elements which belong to one design variable
BUCKFO	defines stability fields	elments which belong to a buckling field; buckling field definition nodes
CONBUCK	defines stability constraints	type number of stability module

Figure 11: Input cards for stability constraints

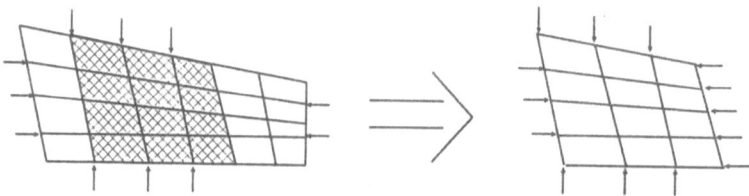

Figure 12: Substructures for local stability constraints

known and the initial distances z_{ref1} (figure 9) of the reference plane can be calculated. The material properties supplied by the MAT8-Card combined with the PCOMP-Card allow the calculation of the stiffness submatrices A,B and D for the various plies. The CMAT-Card supplies allowable tension, compression values to calculate the failure-coefficients (e.g. TSAI-WU failure theory) which are necessary to provide the static constraints for composite elements treated simultaneously to the other static constraints. With the variable linking (various elements are considered together to act as one group so that they all change their thickness or layer angle simultaneously) and the definition of buckling fields all necessary input data for the stability modules is available. In addition to the MAT8-card the youngs's modulus E_c, the shear modulus G_c and the height of the sandwich core must be delivered by the BUCKFO-Card.

3.2 STABILITY AND SENSITIVITY ANALYSIS

The necessary steps for the stability investigation of sandwich structures listed at the end of chapter 2.2 will now be discussed with respect to the integration in the program code of MBB-LAGRANGE. During the optimization loop the analysis module of MBB-LAGRANGE is activated by the selected optimization algorithm. The first major step in the analysis module is to calculate the displacement vector $u(x)$ solving equation

$$K(x)u(x) = f(x), \tag{17}$$

where the stiffness matrix $K(x)$, the displacement vector $u(x)$ and the load vector $f(x)$ is a function of the design variable vector x. With the help of the displacement vector $u(x)$ the static constraints can be calculated. In the event of chosing sandwich wrinkling constraints the following constraint equation must be taken into account

$$g_j(x) = 1 - \frac{\sigma_1(x, u(x))}{\sigma_{1cr}(x, u(x))} - max\left[0, \left(\frac{\sigma_2(x, u(x))}{\sigma_{2cr}(x, u(x))}\right)^3\right] \tag{18}$$

with

$$g_j(x) \geq 0, \qquad j = m, ..., n, \qquad \text{m,n addresses for wrinkling constraints} \tag{19}$$

and

$$\sigma_{1cr}, \sigma_{2cr} \leq 0, \qquad \sigma_1 \geq \sigma_2. \tag{20}$$

Therefore the critical face wrinkling stress σ_{1cr}, σ_{2cr} must be calculated by the use of equation(15), (16). To provide the principal stresses σ_1 and σ_2 for the two face sheets of each finite element the following equations are determined:

$$\sigma_1(x, u(x)) = \frac{\sigma_{xxm} + \sigma_{yym}}{2} + \sqrt{\left(\frac{\sigma_{xxm} - \sigma_{yym}}{2}\right)^2 + \tau_{xym}^2}, \tag{21}$$

$$\sigma_2(x, u(x)) = \frac{\sigma_{xxm} + \sigma_{yym}}{2} - \sqrt{\left(\frac{\sigma_{xxm} - \sigma_{yym}}{2}\right)^2 + \tau_{xym}^2}. \tag{22}$$

As well as the calculation of principal stress σ_1, σ_2, their direction angles Φ_1, Φ_2 can be evaluated by

$$\Phi_1(x, u(x)) = \frac{1}{2} \arctan \frac{2\tau_{xym}}{\sigma_{xxm} - \sigma_{yym}}, \tag{23}$$

$$\Phi_2(x, u(x)) = \Phi_1(x, u(x)) + \frac{\pi}{2}. \tag{24}$$

The stresses in a laminated sandwich face sheet are different for each layer as shown in figure 14. Therefore mean average values σ_{xxm}, σ_{yym} and τ_{xym} must be calculated by formula

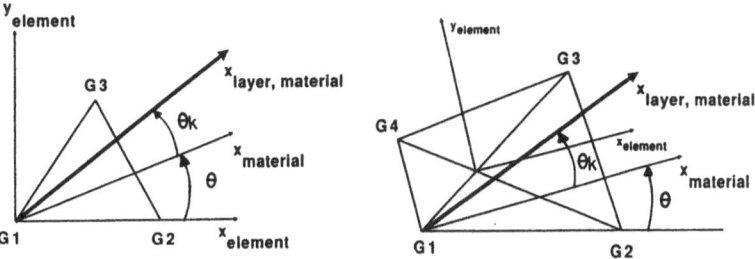

Figure 13: Material coordinate system

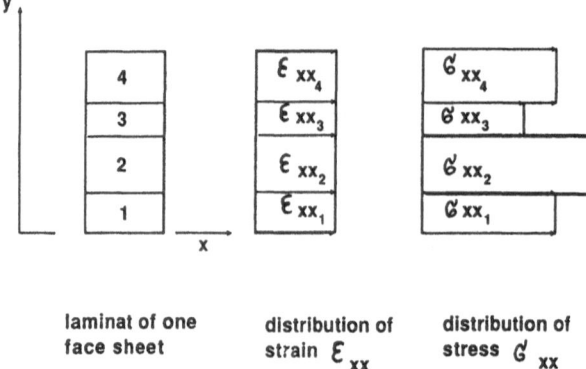

| laminat of one face sheet | distribution of strain ε_{xx} | distribution of stress σ_{xx} |

Figure 14: Strain- and stress distribution of a face sheet

$$\sigma_{xxm}(x, u(x)) = \frac{1}{t_f(x)} \sum_{k=1}^{\mu} t_k(x)\sigma_{xx,k}(u(x)), \tag{25}$$

$$\sigma_{yym}(x, u(x)) = \frac{1}{t_f(x)} \sum_{k=1}^{\mu} t_k(x)\sigma_{yy,k}(u(x)), \tag{26}$$

$$\tau_{xym}(x, u(x)) = \frac{1}{t_f(x)} \sum_{k=1}^{\mu} t_k(x)\tau_{xy,k}(u(x)). \tag{27}$$

As the stress vector $\underline{\sigma}_k = [\sigma_{xx}, \quad \sigma_{yy}, \quad \tau_{xy}]^T$ is calculated for each layer (illustrated with index k) in the element coordinate system (figure 13) by equation

$$\underline{\sigma}_k = \mathbf{Q}(x)\mathbf{B}u(x) \tag{28}$$

with

$$\mathbf{B} \quad := \quad \text{displacement strain matrix}$$

it must be noted that the material stiffness matrix $\mathbf{Q}$ results by the transformation

$$\mathbf{Q} = \mathbf{T}^T\overline{\mathbf{Q}}\mathbf{T} \tag{29}$$

with

$$\mathbf{T} = \begin{vmatrix} m^2 & n^2 & 2mn \\ n^2 & m^2 & -2mn \\ -mn & mn & (m^2 - n^2) \end{vmatrix} \tag{30}$$

$$m = \cos(\Theta_m + \Theta_k)$$
$$n = \sin(\Theta_m + \Theta_k)$$

and

$$\overline{Q} = \begin{vmatrix} \dfrac{E_1}{1-\mu_{12}\mu_{21}} & \dfrac{\mu_{21}E_1}{1-\mu_{12}\mu_{21}} & 0 \\[3mm] \dfrac{\mu_{21}E_1}{1-\mu_{12}\mu_{21}} & \dfrac{E_2}{1-\mu_{12}\mu_{21}} & 0 \\[3mm] 0 & 0 & G_{12} \end{vmatrix}. \tag{31}$$

Finally with the help of the direction angles Φ_1, Φ_2 the coefficients D_{11} and D_{22} acccording to equation (11) can now be calculated in these directions. At this point all values and equations are available for the wrinkling stress σ_{1cr}, σ_{2cr} and the constraint vector g_j ($j = 2*ielw$; ielw = number of wrinkling elements). After the calculation of the critical wrinkling stresses and the constraint by equation (18) the **Sensitivity Analysis** is carried out. This means that we require the derivative (=gradient) of the wrinkling constraints which can be fulfilled in three ways

- numerically,

- analytically,

- semi analytcally.

The numerical calculation of the gradient can be evaluated by

$$\frac{dg_j}{dx_i} = \frac{g_j((\underline{x}) + \Delta x_i \underline{e}_i) - g_j(\underline{x})}{\Delta x_i}. \tag{32}$$

For large scale problems with a large number of constraints the best time saving way is to solve this problem by the analytical formulation

$$\frac{dg_j}{dx_i} = \frac{\partial g_j}{\partial x_i} + \frac{\partial g_j}{\partial \underline{u}}\frac{\partial \underline{u}}{\partial x_i} \tag{33}$$

with the derivative of (17)

$$\frac{dg_j}{dx_i} = \frac{\partial g_j}{\partial x_i} + \frac{\partial g_j}{\partial \underline{u}}\mathbf{K}^{-1}\left[-\frac{\partial \mathbf{K}}{\partial x_i}\underline{u} + \frac{\partial \underline{f}}{\partial x_i}\right] \tag{34}$$

or in a shorter form

$$\frac{dg_j}{dx_i} = \alpha_{ij} + \underline{\beta}_{ij}\mathbf{K}\underline{\gamma}_{ij}. \tag{35}$$

As the derivative for the displacement vector $\frac{\partial \underline{u}}{\partial x_i}$ has already been calculated for the global static analysis by the term

$$\frac{\partial \underline{u}}{\partial x_i} = \mathbf{K}\left[-\frac{\partial \mathbf{K}}{\partial x_i}\underline{u} + \frac{\partial \underline{f}}{\partial x_i}\right] \tag{36}$$

only the scalar value α_{ij} (i index per design variable; j index per constraint) and the vector $\underline{\beta}_{ij}$ must be calculated to complete the sensitivity analysis in the wrinkling module of MBB-LAGRANGE. It must be noted, a very complex formulation was necessary to get the scalar value α_{ij} and vector $\underline{\beta}_{ij}$. This is not a trivial task. The terms

$$\alpha_{ij} = \frac{\partial g_j}{\partial \sigma_1}\frac{\partial \sigma_1}{\partial x_i} + \frac{\partial g_j}{\partial \sigma_2}\frac{\partial \sigma_2}{\partial x_i} + \frac{\partial g_j}{\partial \sigma_{1cr}}\frac{\partial \sigma_{1cr}}{\partial x_i} + \frac{\partial g_j}{\partial \sigma_{2cr}}\frac{\partial \sigma_{2cr}}{\partial x_i} \tag{37}$$

$$\underline{\beta}_{ij} = \frac{\partial g_j}{\partial \sigma_1}\frac{\partial \sigma_1}{\partial \underline{u}} + \frac{\partial g_j}{\partial \sigma_2}\frac{\partial \sigma_2}{\partial \underline{u}} + \frac{\partial g_j}{\partial \sigma_{1cr}}\frac{\partial \sigma_{1cr}}{\partial \underline{u}} + \frac{\partial g_j}{\partial \sigma_{2cr}}\frac{\partial \sigma_{2cr}}{\partial \underline{u}} \tag{38}$$

require the derivatives of equation (15), (16), (25), (26). Also must be taken into acount the dependency between the design variable vector x, the structural variable vector t (thickness of elements and layers), the reference plane z_{ref1}, z_{ref2} and the principle stress direction Φ_1, Φ_2 of each sandwich face sheet. With the wrinkling analysis and sensitivity as discussed in chapter 2 and 3 the optimization process leads to optimal designs with a minimum in the objective function by fulfilling the constraint requirements.

4. Example

As an example for the application of MBB-LAGRANGE on a real structure the horizontal stabilizer with endplates of a helicopter is investigated. The structure consists of an airfoil section like an airplane wing and endplates which act as vertical stabilizers. The design of one half of the airfoil section is shown in figure 15.

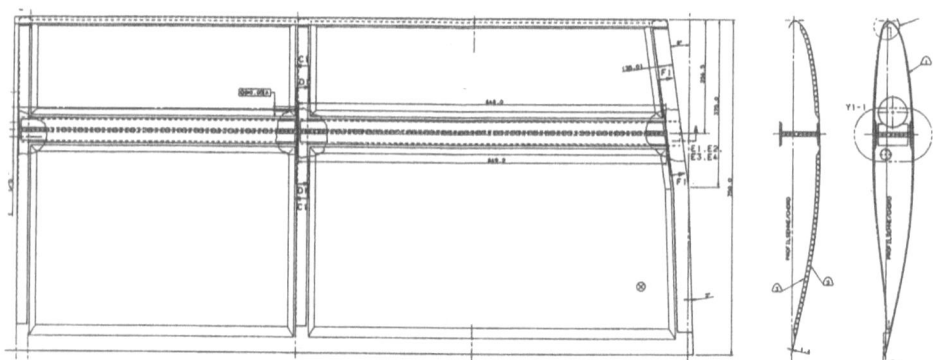

Figure 15: Horizontal stabilizer of a helicopter

The upper and lower panels of the airfoil section are sandwich plates with a honeycomb core and aramid fiber face sheets, The aerodynamic loads are transferred by these sandwich plates and ribs to a spar. This spar is an I-shaped bar with straps made from unidirectional carbon fiber reinforced material and a shear web which is a honeycomb sandwich with CFRP face sheets.

The endplates are sandwich plates of constant thickness consisting of aramid fiber reinforced face sheets and a honeycomb core. They are fixed to the airfoil section by screws.

As the structure is symmetrical, only one half of the structure was idealized. It is clamped in the plane of symmetry. Figure 16 shows the finite element idealization of one half of the stabilizer. For the idealization composite membrane elements are mainly used. The straps of the spar are represented by bars, some bars are used for the connection of the endplate to the airfoil section.

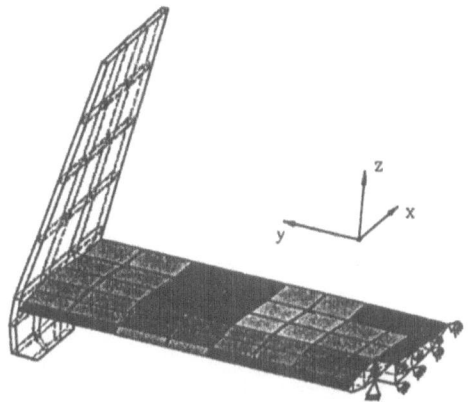

Figure 16: Finite element idealization of the horizontal stabilizer

To reduce the number of design variables and for manufacturing reasons, several sizing variables are linked together. In sum there are 144 design variables:

- 54 for the layer thicknesses of the upper panel (18 for every of the 0°, ±45° and 90° layers).
- 54 for the layer thicknesses of the lower panel (18 for every of the 0°, ±45° and 90° layers).
- 9 for the strap cross section area of the spar.
- 15 for the layer thicknesses of the spar shear web (5 for every of the 0°, ±45° and 90° layers).
- 12 for the layer thicknesses of the end plate face sheets (4 for every of the 0°, ±45° and 90° layers).

Figures 17 and 18 show the design variables of the different structural parts.

Three load cases define the loading of the stabilizer. One bending downward and one bending upward of the airfoil section and one loading of the endplates. The structure shall withstand these loadings with a factor of safety larger than 1.5. Sandwich wrinkling has to be considered as well as a composite failure criterion such as that from Tsai-Wu. Additionally to these constraints, a lower bound for the first eigenfrequency is given.

A further constraint is that the inner part of the upper sandwich panel must be walkable by a mechanic plus tools to enable maintainance of the tail rotor. From this requirement the lower bounds for the layer thicknesses in that sandwich area were fixed to 100% of the original value. For all other design variables the lower bound is defined to 10% and the upper bound to 400% of the original thicknesses.

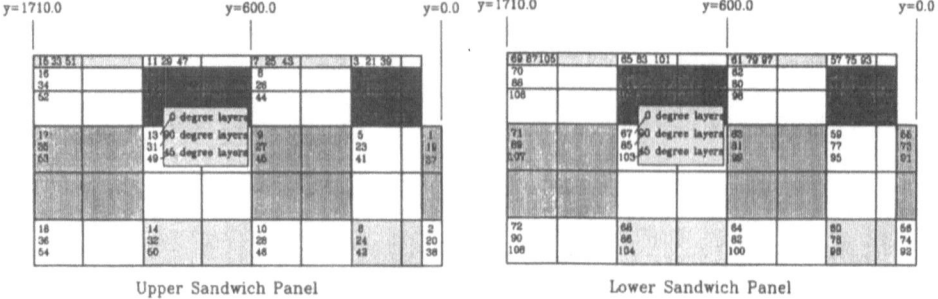

Figure 17: Design variables of the panels of the airfoil section

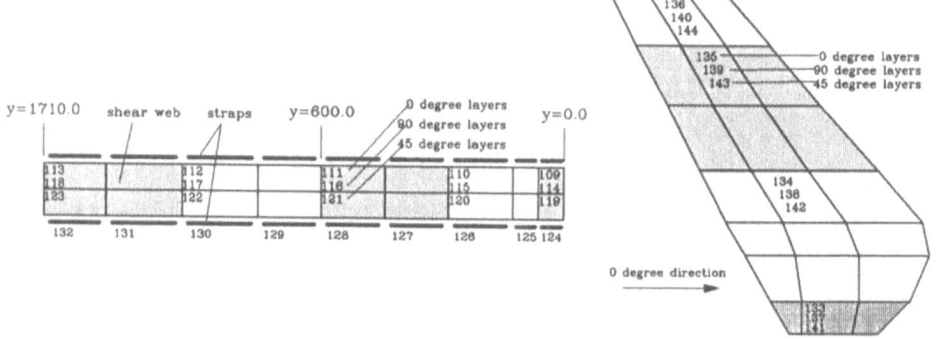

Figure 18: Design variables of the spar and the endplate

The whole finite element design model can be characterized by the following numbers:

Number of nodes	194
Number of elements	354
Number of DOFs	597
Number of sizing structural variables	1158
Number of design variables	144
Number of constraints	3439

Four questions should be answered by this investigation:

1. Is a weight reduction possible with the current design? How much does the first eigenfrequency decrease if only strength constraints are considered?

2. What is the weight saving, if a thicker aerodynamic profile of the airfoil section (15% instead of 12%) is used?

3. What design is more favourable: the current design with a spar and sandwich panels or a design without spar but with integrated stringers?

4. How much does the weight of the horizontal stabilizer increase, if the first eigenfrequency has to be raised.

To answer these questions different optimization runs are performed with the above described model. For the thicker aerodynamic profile, the coordinates of the nodes are were transformed. To investigate the design without a spar, the Young's modulus of the spar strap material was lowered by the factor 1000. Additionally the material of the 0° layers in the upper and lower panel face sheets were changed to carbon fiber reinforced material instead of aramid fiber reinforced material. Otherwise the required stiffness can not be reached and the design will be infeasible. The results are shown in figure 19.

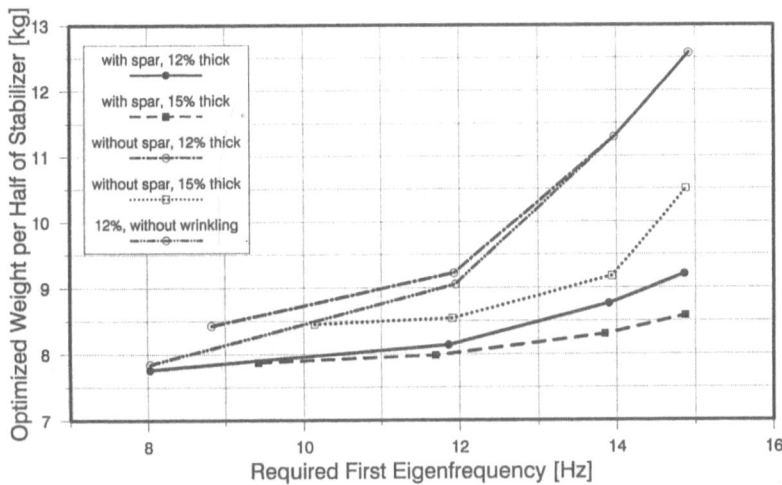

Figure 19: Minimum weight of the horizontal stabilizer versus the first eigenfrequency

The original design with spar has a structural weight of 10.6 kg (one half of the airfoil section plus one endplate) and a first eigenfrequency of 14 Hz. Considering only strength restrictions, the weight can be reduced by nearly 3 kg, but in this case the first eigenfrequency drops significantly. If the first eigenfrequency is held constant, then a weight reduction of about 2 kg is possible.

From figure 19 it can be clearly seen that the current design with a spar is far better than a design without a special spar. The reason for this is that the spar of this relatively small wing is located at the thickest location of the aerodynamic profile. As the first eigenfrequency is the design driver, the required stiffness can be reached more efficiently with a spar and concentrated straps than with unidirectional material spread over the whole sandwich panels. Moreover a spar with straps can carry a higher compression force than a sandwich panel, which is sensitive with respect to face sheet wrinkling.

Concerning the thickness of the airfoil section it can be stated that for low stiffness (e.g. a low first eigenfrequency) there is nearly no difference in the weight between a thin (12%) and a thicker (15%) profile. Only if a high stiffness is required (a high first eigenfrequency) a thicker aerodynamic profile is useful.

In figures 20 to 22 the optimized values of the design variables for the horizontal stabilizer (upper and lower sandwich panels and spar) are shown. These results belong to the current design with a spar and a required first eigenfrequency of 14.0 Hz (no drop of the eigenfrequency of the original design).

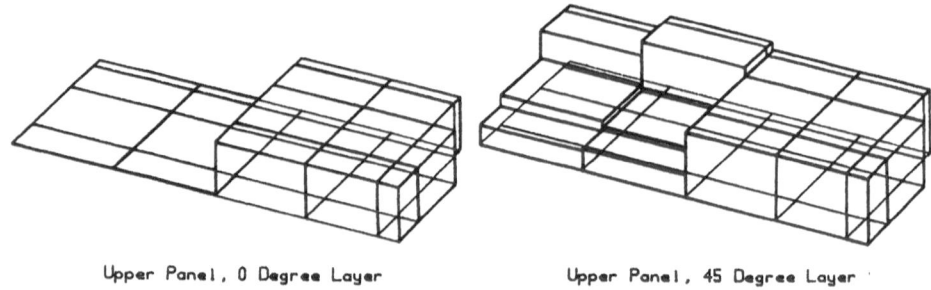

Upper Panel, 0 Degree Layer   Upper Panel, 45 Degree Layer

Figure 20: Optimized layer thicknesses for the upper panel

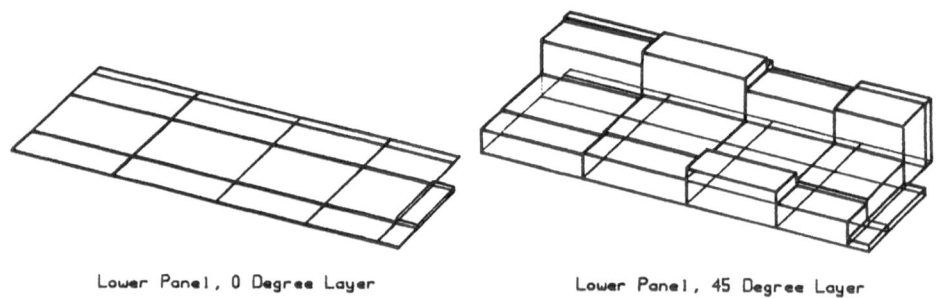

Lower Panel, 0 Degree Layer   Lower Panel, 45 Degree Layer

Figure 21: Optimized layer thicknesses for the lower panel

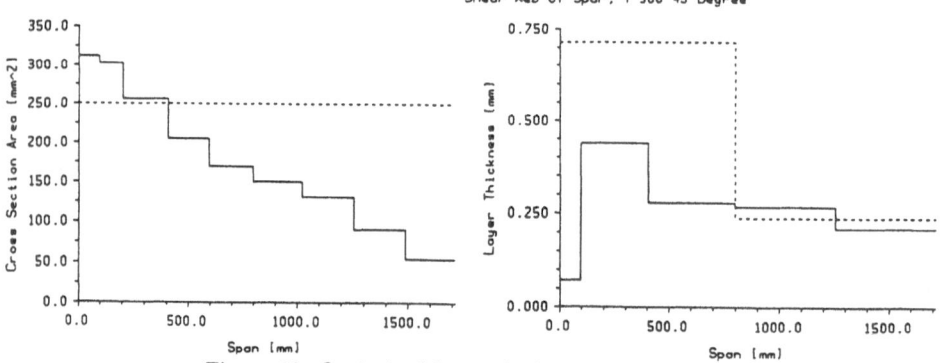

Figure 22: Optimized layer thicknesses for the spar

All 90° layers in the sandwich panels of the airfoil section are reduced to their minimum value of 10%. These layers are not necessary to transfer the loads. From figure 20 it can be seen that the inner part of the upper panel is dominated by the requirement of withstanding the maintainance loads. In the outer part only slightly more than the minimum thickness

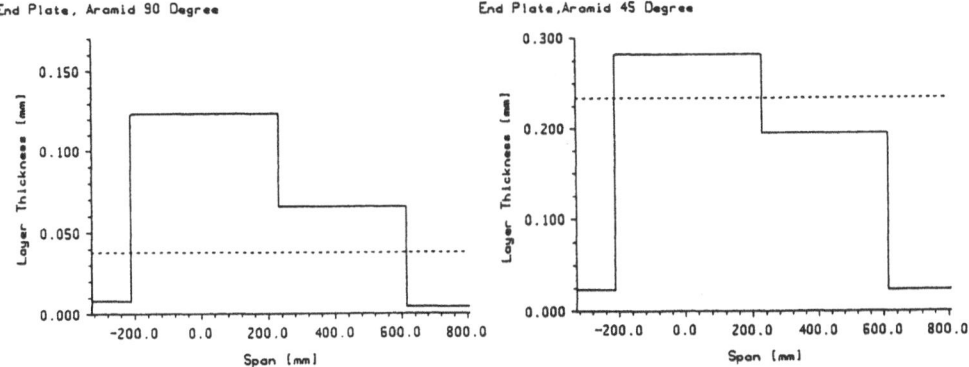

Figure 23: Optimized layer thicknesses for the endplate

is needed. To fulfill the frequency requirement nearly all of the bending stiffness is located in the spar. The upper and lower sandwich panels are only responsible for transferring the aerodynamic loads and for the torsional stiffness. For the first task relatively small layer thicknesses are sufficient. To reach the required torsional stiffness the front part of the wing, the so-called torsion box (e.g. shear web of the spar, front part of the upper and lower sandwich panel), is stiffened by ±45° layers.

The endplate can be considered as a wide bar which is clamped at approximately one fourth of its span. For strength and stiffness reasons no 0° layers (in direction of the aerodynamic profile) are necessary. The thicknesses of the 90° (perpendicular to the aerodynamic profil) and the ±45° layers are given in figures 22 and 23.

The convergence of the above described optimization problem depends on the finite element model and the design constraints. For the described investigations in maximum 15 iterations are allowed. In the case of the stabilizer with a spar these 15 iterations are always needed. Convergence then is not yet reached, but the design constraints are fulfilled up to 3%. This is enough for a practical application.

In the case of the stabilizer without spar, convergence is faster. In most cases, less than 12 iterations are sufficient to fulfill all requirements up to 0.5%.

All above investigations were performed with the optimization code MBB-LAGRANGE on an IBM 3090 XA mainframe. An optimization run with 15 iterations (16 function calls and 15 gradient calls) on this machine requires a total CPU time of 504 seconds (one reanalysis: 5 sec, one gradient: 22 sec).

5. Conclusion

In cooperation with an user team (Helicopter Department) of the program system MBB-LAGRANGE and the development team (IV-Systeme Technic) of MBB-LAGRANGE a further important constraint type for lightweight structures has been realized. As the first application examples demonstrate, very complex design studies can be performed to reach an optimal structure. The open architecture of MBB-LAGRANGE was again very helpful in realizing the additional requirements like the sandwich wrinkling constraint. Cooperation between the software specialist of the MBB-LAGRANGE team and the specialists from helicopter-structures department was necessary to obtain the understanding for the realizing of local stability constraints in the program code MBB-LAGRANGE. An important task in the future will be the incorperation of the analytical sensitivity analysis for layer angles in combination with local stability problems. As there were manufacturing restrictions for fixed layer angles in the shown example the wrinkling module still satisfies the actual requirements. But for other structures this can be a further step to reducing weight or fulfill the constraint requirements in a better way.

6. References

[1] Schnell W., Czerwenka G.,
Einführung in die Methoden des Leichtbaus II,
Bibliographisches Institut, Mannheim, 1970

[2] Benson A.S., Mayers J.,
General instability and face wrinkling of sandwich plates – unified theory and applications,
AIAA Journal, April 1967

[3] Sullins, Smith, Spier,
Manual for structural stability analysis of sandwich plates and shells,
NASA-Contractor report No. 1467, 1969

[4] Plantema F.J.,
Sandwich Construction, in: Airplane, Missile and Spacecraft Structures, Volume 3,
edited by N.J. Hoff, John Wiley & Sons, Inc., New York, 1966

[5] Kneppe G., Krammer J., Winkler F.,
Structural Optimization of Large Scale Problems Using MBB-LAGRANGE
5th World Congress and Exhibition on FEM Salzburg (Austria), 1987

[6] Kneppe G.,
MBB-LAGRANGE: Structural Optimization System for Space and Aircraft Structures
COMMETT Seminar Computer Aided Optimal Design, University of Bayreuth, FR Germany, 1990

[7] MBB-LAGRANGE,
MBB-LAGRANGE-Benutzerhandbuch; Version LAG09; MBB-Deutsche Aerospace intern Documentation

OPTIMIZATION IN STRUCTURAL DYNAMICS WITH APPLICATIONS

H. J. Baier
Dornier GmbH
799 Friedrichshafen, Germany

ABSTRACT. Structural optimization problems are discussed with special emphasis on consequences when requirements on structural dynamics are to be considered. Some remarks are made to sensitivity and perturbation analysis, and some specific algorithms for optimization problems in dynamics are addressed. Practical applications in spacecraft design and automotive parts with shape optimization are given and the formulation of mathematical dynamic model updating as a structural optimization problem is discussed with applications.

1. INTRODUCTION

In this paper, different topics and applications of structural optimization are discussed with problems containing requirements of structural dynamics. Apart from the fact that structural dynamic properties are more global properties of a structure (in contrast to stress/strength behaviour) which are influenced by stiffness and mass distribution, it should be noted that the solution process of such optimization problems is even more challenging due to the relatively high effort necessary to solve system equations (e. g. eigenvalue problems). This means that the requirement of a small number of iterations and small effort per iteration step (evaluation of system equations/sensitivities) is even more severe than in static problems. Moreover, the designers are usually less trained in 'thinking dynamics' than in 'thinking in static/stresses', i. e. support of structural optimization in the general design process will than likely have an even more significant effect. Since at least for the global modes such dynamic requirements influence the overall structural dimensions, such investigations and optimizations are to be carried out already in early design stages.

G. I. N. Rozvany (ed.), Optimization of Large Structural Systems, Vol. II, 973–985.

2. TWO INTRODUCTORY EXAMPLES

2.1. Optimization of Spacecraft Structure with Eigenfrequency Constraints

For a large and relatively heavy spacecraft, such as the multimirror X-ray satellite XMM (shown in figure 1), ensuring that a lower bound for the fundamental (bending) eigenfrequency is exceeded with minimum structural mass, is a design task to be solved already in earlier design stages. This is necessary to avoid severe dynamic interaction with the launcher, which would give rise to high dynamic loads for the satellite or causing difficulties for the launcher flight control. The corresponding statement as an optimization problem reads as follows:

- minimize: mass, dynamic response at specific degress of freedom;

- maximize: fundamental eigenfrequency;

- design variables: stiffness distribution (thicknesses, laminate build-up etc.)

- constraints: lower/upper bounds of design variables.

The problem is posed here as a multicriteria problem with different objective functions. Alternatively, in case of clearly specified bounds on the fundamental frequency (typically in the range of 8-10 Hz for bending models) and/or the dynamic response, these quantities could as well be constrained.

Figure 1 also shows the sandwich cylinder of the main structure of this satellite XMM with a truss-type structure having been considered as an alternative. Due to the satellite's height and its relatively large X-ray mirror masses of about 1 ton on top of the spacecraft, the fundamental eigenfrequency poses a basic design driver. Thus, optimization runs minimized structural mass while satisfying eigenfrequency constraints and lower bounds for the design variables which are the cross-sectional areas of the truss version and sandwich lay up for the cylindrical shell version. For each of both versions the mass could be improved by approximately 10 % compared to each starting design, and the cylindrical sandwich shell structure prooved superior compared to a truss type design. For the latter, masses and costs for the joints would have to be considered additionally. Thus, a clear and rational comparison between the two versions could be made on the basis of their individual optimal design.

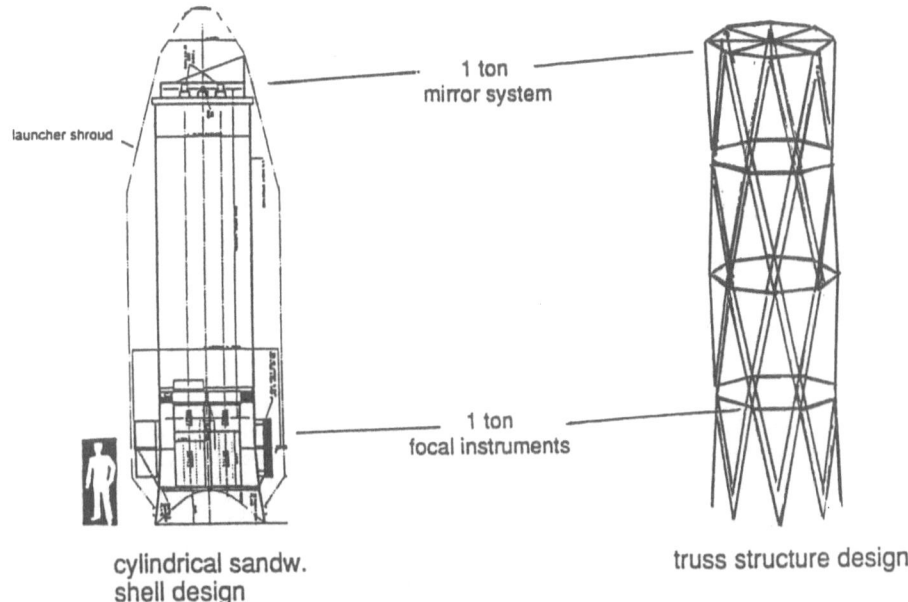

Figure 1: The future European X-ray satellite XMM

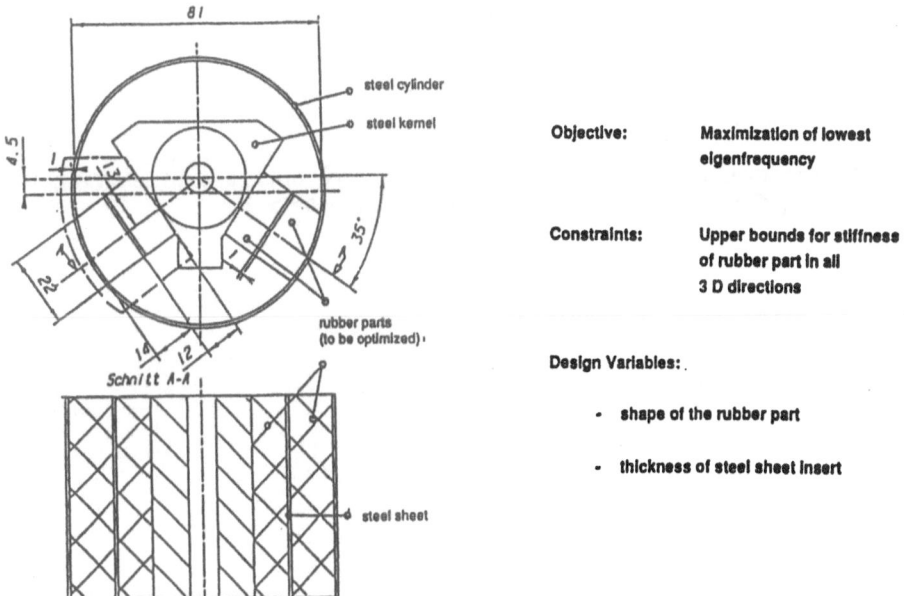

Objective: Maximization of lowest eigenfrequency

Constraints: Upper bounds for stiffness of rubber part in all 3 D directions

Design Variables:

- shape of the rubber part

- thickness of steel sheet insert

Figure 2: Rubber bushing of an automative support

2.2. Component Shape Optimization in Dynamics

A 3D shape optimization is requested in the design of rubber parts of an automobile bearing, figure 2. The objective is to maximize the fundamental eigenfrequency of such a rubber part with static stiffness properties to be held fixed.

For sensitivity analysis, the perturbation or semi-analytic approach as briefly described in the next chapter has been used. The optimization problem itself has been solved with the usable-feasible direction algorithm developed by Vanderplaats and implemented in CONMIN. For shape variation, the mesh variation and mesh updating approach as outlined in figure 3 and described in more detail in [1], [2] has been applied.

The resulting optimized shape of the rubber part is shown in figure 4. The related change of the frequency objective can be read from the diagram in figure 5. There are two jumps in the objective curve which indicate an approximation update for the frequency evaluation. This evaluation is based on a modal subspace approximation within each of the optimization cycles. The approximation subspace is defined by the 10 lowest eigenmodes evaluated a the beginning of each optimization cycle. Otherwise, eigenfrequency and sensitivity analysis has been applied as described in the following.

3. DYNAMIC STATE VARIABLES AND THEIR DERIVATIVES

Whatever optimization strategy is used, efficient determination of state variables and their derivatives with respect to optimization variables is essential for an overall efficient optimization package. Some remarks to this are made in the following.

3.1. Eigenvibration Data

The j-th eigenvalues w_j^2 and eigenvector ϕ_j are obtained by solving the eigenproblem for stiffness matrix $\underline{K}$ and mass matrix $\underline{M}$

$$(\underline{K} - w_j^2 \underline{M}) \phi_j = 0 \tag{1}$$

under the orthonomality condition

$$\phi_i^T \underline{M} \phi_j = \delta_{ij} \qquad i, j = 1, \ldots \tag{2a}$$

$$\phi_i^T \underline{K} \phi_j = \delta_{ij} w_j^2. \tag{2b}$$

For the repetitive solution of this eigenproblem e. g. by subspace iteration method, advantage should be taken from using eigenvectors of a previous optimization step as a starting vector of the new subspace iteration.

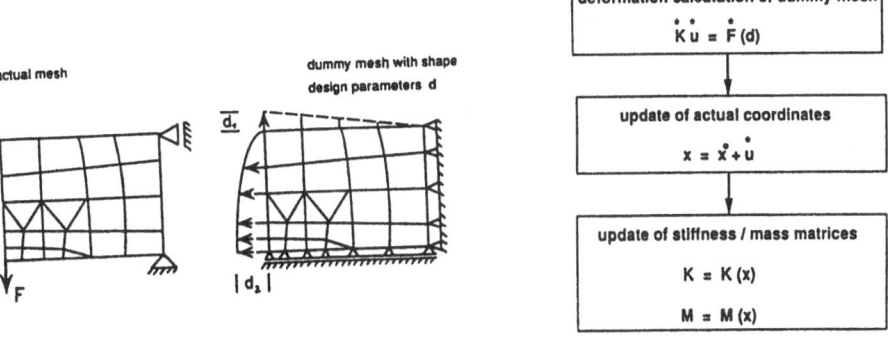

Figure 3: Concept of shape variation and mesh update

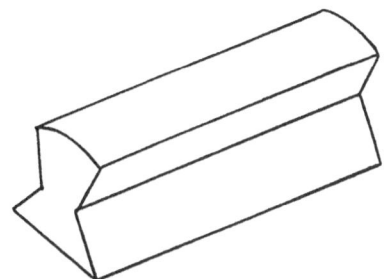

Figure 4: Optimized shape of rubber part

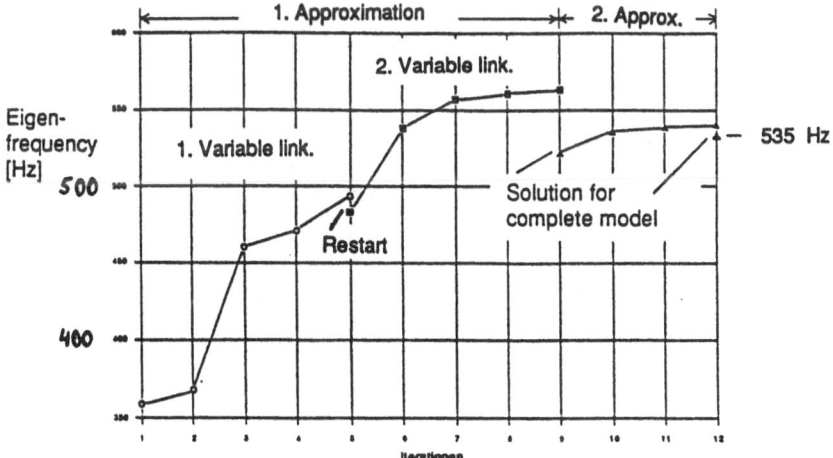

Figure 5: Eigenfrequency vs. optimization iteration number

Derivatives of the eigenvalue with respect to the i-th design variable x_i are obtained by

$$\frac{\partial \omega_j^2}{\partial x_i} = \phi_j^T \left(\frac{\partial}{\partial x_i} \underline{K} - \omega_j^2 \frac{\partial}{\partial x_i} \underline{M}\right) \phi_j \tag{3a}$$

and for the eigenvector for example by

$$\frac{\partial}{\partial x_i} \phi_j = \sum_k a_{ijk} \phi_k \tag{3b}$$

with a_{ijk} containing derivatives of stiffness and mass matrix as well as eigenvalues and modes.

Usually, $\frac{\partial}{\partial x_i} \underline{K}$ and the other matrix derivatives are approximated by finite differences ("semianalytic method"), allowing then relatively easy implementation into (fem)-computer codes.

Alternatively, the new mode ϕ_j' resulting from a modification in the structure can be assumed to be representable by a linear combination of a set of modes of the previous structure

$$\phi_j' = \sum_i c_i \phi_i \tag{4}$$

Substitution of (4) into the eigenvalue problem and left hand side multiplication by ϕ_j' results into a significantly smaller eigenvector problem in c_i, the size of which depends on the number of modes considered in (4).

In should be noted that these relations were still valid when working in a condensed system $\tilde{\phi}_j$

$$\phi_j = \underline{T} \, \tilde{\phi}_j \tag{5a}$$

$$\frac{\partial}{\partial x_i} \phi_j = \underline{T} \, \frac{\partial}{\partial x_i} \tilde{\phi}_i \tag{5b}$$

unless the transformation or condensation matrix $\underline{T}$ is influenced by the design variables x_i, which is usually not the case.

3.2. Response Variables and their Derivatives

Just as in "classical" dynamic response analysis, calculations via the modal space are usually also to be preferred in optimization. An exception might be in harmonic response without interest in or need of eigenvibration data. Then the response in physical degrees of freedom is obtained from

$$
\begin{bmatrix} \underline{K} - \Omega^2 \underline{M} & \Omega \underline{D} \\ - \Omega \underline{D} & K - \Omega^2 \underline{M} \end{bmatrix} \cdot \begin{bmatrix} U_c \\ U_s \end{bmatrix} = \begin{bmatrix} P_c \\ P_s \end{bmatrix} \tag{6}
$$

or briefly $$\underline{A} \cdot U = P \tag{7}$$

where the excitation is

$$P = P_c \cos \Omega t + P_s \sin \Omega t$$

and the steady state response

$$U = U_c \cos \Omega t + U_s \sin \Omega t.$$

The derivatives are obtained by differentiating (7)

$$\underline{A} \frac{\partial}{\partial x_i} U = (- \frac{\partial}{\partial x_i} \underline{A}) \cdot U \tag{8}$$

where derivation of $\underline{A}$ requires the derivatives of $\underline{K}$, $\underline{M}$ and $\underline{D}$, respectively.

In more general cases, the response or displacement vector U is obtained from

$$U(t) = \underline{\phi} \cdot Q(t) \tag{9}$$

with Q being the vector of modal degrees of freedom and the modal matrix $\underline{\phi}$ covering the "relevant" modes. The jth modal response Q_j is obtained from

$$[-\Omega_j^2 \underline{M}_{gen} + i\Omega_j \cdot \underline{D}_{gen} + \underline{K}_{gen}] \cdot Q_j = \underline{\phi}^T P_j \tag{10a}$$

or briefly $$\underline{B} \cdot Q_j = P_{gen} \tag{10b}$$

where the generalized matrices are diagonal or in the case of $\underline{D}_{gen}$ usually assumed to be diagonal.

Derivatives of U then are obtained from

$$\frac{\partial}{\partial x_i} U = \frac{\partial}{\partial x_i} \underline{\phi} \cdot Q + \underline{\phi} \cdot \frac{\partial}{\partial x_i} Q \tag{11}$$

where the derivatives of the modal response are calculated by

$$\underline{B} \cdot \frac{\partial}{\partial x_i} Q = (- \frac{\partial}{\partial x_i} \underline{B}) \cdot Q \tag{12}$$

if $\partial P_{gen}/\partial x_i = 0$. (Since $P_{gen} = \phi^T P$, it should be noted that $\partial P_{gen}/\partial x_i = 0$ in case of changes in the mode shapes, even if P is independent of x_i).

Usually some approximations are in order such as

$$[-\Omega_j^2 \underline{M}_{gen} + i\Omega_j \underline{D}_{gen} + \underline{K}_{gen} + \phi^T \underline{\Delta K} \phi] \cdot Q = \underline{\phi}^T P_j \tag{13}$$

if for example in the kth optimization step only stiffness parameters have been changed by ΔK. Then a differential quotient approximation could be made if derivatives are required. Also, the change in eigenmodes in equation (11) could be neglected for some iteration steps.

The general transient case, where the dynamic response is represented in the time space, in principle leads to an infinite number of constraints because of the time t acting as a continuous constraint function parameter. This infinite set can be discretized acording figure 6: in a first run, those time values t_{jk}^*, $k = 1,...,$ are determined, where the constraint function g_j is below some threshold value, i. e. where it is relatively critical. For these discrete times, constraint function values (and their derivatives) are then introduced into the constraint function (and derivatives) vector. These time values are to be updated during the optimization iteration, depending on the rate of change of response with respect to these values.

4. SPECIFIC STRATEGIES IN STRUCTURAL DYNAMICS OPTIMIZATION

Like in most structural optimization problems and especially when high numerical effort is envolved, it is most advisable to adapt a stepwise approach where in its first part a selection of variables and design change areas is done before the optimization is carried out. Such a stepwise approach is briefly outlined in the following.

4.1. A Stepwise Approach

The stepwise approach in problems with dynamics proceeds as follows

1. Determine the local distribution of strain and kinetic energy in the structure (also supported by graphical representations!), for the modes under consideration

- areas of high strain energy: strong influence of structural stiffness on eigenfrequency

- areas of high kinetic energy: strong influence of (structural) mass

2. Carry out a sensitivity or perturbation analysis in order to determine quantitatively the influence of the design variables on frequencies and modes.

3. Final selection of design variables according results of steps 1 and 2.

4. Solve a "classical" structural optimization problem or one of the alternative approximation problems such as briefly described in chapter 4.2.

5. Assessment of results, eventual repetition of steps 1/2/3/4.

4.2. Specific Approaches for Optimization in Structural Dynamics

In order to circumvent the considerable computational effort envolved, specific approaches are investigated by different researches for optimization problems in structural dynamics, even for the sake of accuracy in physics and/or optimization. Two typical classes are

- perturbation approaches of the left hand side of the system equations, combined with linearized optimization problems, see e. g. [5]

- control force approaches, specifically apt for optimization of discrete design elements (spring/damper systems) in a dynamically excited structure, see e. g. [3, 4].

In the perturbation approaches, perturbations such as $\underline{K} + \Delta\underline{K}$ and $\underline{\phi} + \Delta\underline{\phi}$ are introduced e. g. into the eigenproblem

$$\underline{K} + \Delta\underline{K} - (\lambda + \Delta\lambda)_j \, (\underline{M} + \Delta\underline{M}) \, (\phi_j + \Delta\phi_j) = 0 \tag{14}$$

By neglecting terms of higher order rather than of linear order, in a first step linear relations can be established to determine required changes $\Delta\underline{K}$, $\Delta\underline{M}$. In a second step, the new eigenvalues and modes are held fixed and related stiffness and mass variables are determined from the exact perturbation equation (14) by solving a nonlinear least squares problem. This is done in order to improve the physical accuracy, i. e. design and response variables then fit better together than in the first linearized step.

In the control force approaches, the discrete elements are substituted by control forces C which influence the response in the desired way. In parallel to this, stiffness (or also damping) values are determined

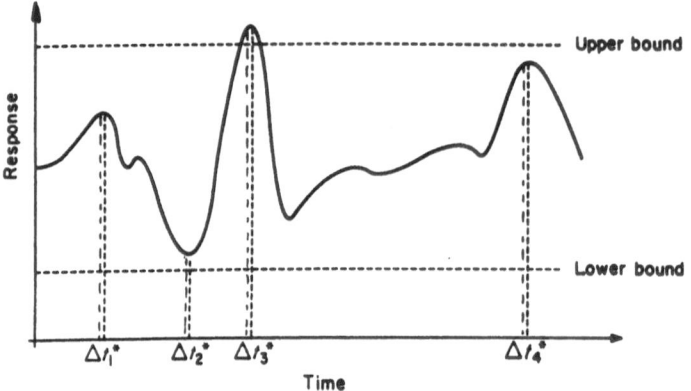

Figure 6: Approach of discretization of time-continuous constraint

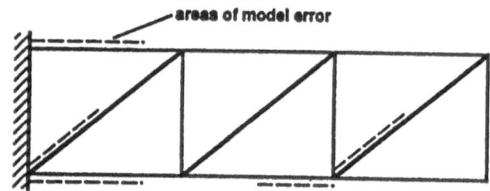

truss type structure with model errors

Eigenfrequencies f_k and modal assurance criterion (MAC)
before and after update

mode no k	Before Update		After Update	
	$\frac{f_k(Analysis)}{f_k(Test)} - 1$	MAC	$\frac{f_{k_A}}{f_{k_T}} - 1$	MAC
1	0.87	0.98	0.	0.997
2	0.09	0.99	0.01	0.994
3	0.02	0.85	0.	0.994
4	0.24	0.03	0.	0.999
5	0.04	0.07	0.01	0.980

Figure 7: Updating of mathematical dynamic models from test results

which generate these forces under this response. This leads e. g. to

$$\text{minimize } || \, C - \underline{K}(x) \cdot U \, || \tag{14a}$$

such that

$$\underline{M}_o \cdot \ddot{U} + \underline{D}_o + \underline{K}_o \, U = F - C \tag{14b}$$

and

$$g_j \, (U, \, C) \geqslant 0 \tag{14c}$$

with $\underline{K}(x)$ being the stiffness properties to be determined. By the objective (14a), a control force vector C is generated which can be matched by technically feasible (discrete) stiffness elements. Since the matrices of the left hand side in (14b) are held fixed, no kind of "matrix inversion" approaches to determine U for changing X are necessary. On the other side, (14) is now an optimization problem both in C and X.

In general it still has to be seen, whether such more specific (and with that also less general) approaches are competitive to the general "classical" structural mathematical optimization approaches, especially to those based on generating explicit approximation functions for the implicit objective and constraint functions.

5. MATHEMATICAL DYNAMIC MODEL UPDATING AS AN OPTIMIZATION PROBLEM

Dynamic mathematical model updating means the determination of model parameters in such a manner that the results from the updated model match as close and possible a set of measured experimental data. In modal analysis, these data of interest are frequencies and mode shapes. The updated model then can be used for further and even more accurate analyses. As described in the following, this update process can be treated as an optimization problem, see for example [6].

5.1. Model Updating as an Optimization Problem

The objective is to minimize the weighted differences of eigenfrequencies and mode shapes from analysis and test with minimum changes of the design parameters/update variables a.

The objective function $J(a) = J(a_1, a_2, \ldots a_1, \ldots a_L)$ consists of three terms.

$$\text{minimize } J(a) = J^f(a) + J^\phi(a) + J^a(a) \tag{15}$$

The first two terms $J^f(a)$ and $J^\phi(a)$ contain (u: update, m: measured)

1. weighted eigenfrequency differences ($f_k = \quad$)
2. weighted eigenvector differences.

$$J^f(a) = W^f \cdot \sum_{k=1}^{K} w_k^f \cdot [\frac{f_k^u(a) - f_k^m}{f_k^m}]^2 \tag{16}$$

$$J^\phi(a) = w^\phi \cdot \sum_{k=1}^{k} w_K^\phi \cdot \sum_{i=1}^{I} w_I^\phi \cdot [\frac{\phi_{ik}^u(a) - \phi_{ik}^m}{\phi_{max,k}^m}]^2 \tag{17}$$

For minimum changes of the design parameters a_i an additional third term $J^a(a)$ is added:

$$J^a(a) = w^a \cdot \sum_{i=1}^{L} w_i^a \cdot a_i^2 \tag{18}$$

As update variables may be selected the material properties (mass density δ and modulus of elasticity E) and the geometrical input parameters (cross sectional area A, moments of inertia for bending I_y and I_z and torsion I_i, ...), etc. The procedure also allows modifications of nodal point coordinates, thus even an update of the geometrical shape of the FE-model is possible. The set of these physical parameters has to be selected by the user and can be also lower and upper bounded.

Depending on the design parameter either the mass or stiffness matrix is modified or both matrices are changed simultaneously. In general the changes $\Delta M_i(a_1)$ and $\Delta K_i(a_i)$ depend nonlinearly on the different correction parameters $a_1, a_2, a_3, ..., a_1, ... a_L$:

$$M^u = M^i + \sum_i \Delta M_i(a_i) \tag{19a}$$

$$K^u = K^i + \sum_i \Delta K_i(a_i) \tag{19b}$$

For the solution of (15) and eventual constraints a modified feasible dirction method has been used.

Some typical examples and results before and after update are given in figure 7 with MAC desribing the parallelism of measured and analysed eigenvectors.

References

[1] Rajan, A.D., Belegundo, 'Shape optimal design using fictituous loads', AIAA Journal, Vol. 27, No. 1, 1989, pp 102-107

[2] Specht, B., Baier H., Shape optimization problems with constrained dynamic response', in Eschenauer, H., et.al., (eds), 'Engineering Optimization in Design Processes, Springer Verlag; 1991, pp 79-87

[3] Baier, Rausch, 'On right-hand side modifications for the solution of inverse problems in structural dynamic response', Mechanical Systems and Signal Processing, 1990, 4/3, pp 187-194

[4]   Wang, Pilkey W., Limiting performance characteristics of steady
      state systems, Journal of Applied Mechanics, Sep. 1975,
      pp 721-726

[5]   Bernitsus, M., Vang, B., 'Admissable large perturbations in
      structural redesign', AIAA Journal, Vol 29, No 1, 1990,
      pp 104-113

[6]   Eckert, L., Caesar, B., 'Model update under incomplete and noisy
      test data' Proc. Int. Forum Aeroelasticity, Aachen, Germany,
      1991

OPTIMIZATION IN COUPLED PROBLEMS

H. J. Baier
Dornier GmbH
799 Friedrichshafen, Germany

ABSTRACT. In coupled problems the influence of different disciplines such as mechanics, thermodynamics and control have to be considered. A complete coupled formulation can reasonable be solved only for smaller problems, and some approximations and decompositions are in order for larger practical problems. This is discussed with the thermomechanical analysis and optimization of plates. The optimization problems in controlled (smart) structures are addressed, and different decomposition strategies are outlined for optimizing light weight bridges.

1. INTRODUCTION

Coupled optimization problems are those where response quantities and design variables coming from different disciplines are envolved. Typical examples are:

- structure-thermal interaction problems, especially for thermally sensitive structures

- structure-fluid (acoustic) -interaction problems, to be found in hydromachines, airplanes or spacecraft launchers

- structure-control problems, again in aerospace structures and especially smart controlled structures.

We could differ between weakly, moderately and strongly coupled problems. In weakly coupled problems, the design and response variables coming from different disciplines can be treated very seperately in their related system equations. The response quantities determined then are introduced into the objective and constraint functions, and coupling occurs only on this level. More relevant are the moderately and strongly coupled problems, with the system equations

$$S_{A_i} (\{X\}_A, \{X\}_B, \{Y\}_A, \{Y\}_B) = 0 \quad , \; i = 1, \; .. \; n_A \qquad (1a)$$

$$S_{B_i} (\{X\}_A, \{X\}_B, \{Y\}_A, \{Y\}_B) = 0 \quad , \; i = 1, \; .. \; n_B \qquad (1b)$$

987

G. I. N. Rozvany (ed.), Optimization of Large Structural Systems, Vol. II, 987–996.
© 1993 Kluwer Academic Publishers.

where $\{X\}_A$, $\{Y\}_B$ and $\{Y\}_A$, $\{Y\}_B$ are the design variables and response variables of discipline/system A and B, respectively. In a moderately coupled problem, $\{Y\}_B = 0$ in S_A and $\{Y_A\} = 0$ in S_B, i.e. coupling occurs via the design variables, while the response variables of system A do not influence those of B and vice versa. For example, for a thermally loaded bending plate the plate thickness both influences the displacement and temperature distribution. In a strongly coupled problem, then also the displacement influences the temperature distribution e.g. when the plate is getting into contact with neighboured components. Some options to treat such problems are discussed in the following with a thermomechanical plate bending problem. General aspects are discussed for example in [1].

2. THERMO-MECHANICAL BENDING OF A PLATE

The problem of analyzing and optimizing thermally loaded plates (or shells) arises in different fields, such as in the stress analysis of the thermal protection system of a hypervelocity airplane or the deformation analysis of high precision (optical) reflectors. The plate thickness influences in both cases the mechanical deformation as well as the temperature distribution: the higher the plate thickness, the lower the bending deformation due to a <u>fixed</u> thermal gradient through the thickness, <u>but</u> usually also the higher the thermal gradient. So we have a typical moderately coupled problem where a design variable influences interacting responses from different disciplines. In principle, there are different options to deal with such problems.

1. keeping the input response variables (i.e. the temperature) fixed when optimizing the mechanical system, and after that updating of the temperatures with an eventual rerun of the mechanical optimization.

2. Varying structural parameters in thermal analysis and by that establishing sampling points for interpolation functions for approximate temperature variations vs. structural variations. These approximations are then used during structural optimization.

3. Determining the complete differential change of deformation $\{U\}$ with respect to structural variable change dx_i by

$$\frac{d}{dx_i} \{U\} = \frac{\partial}{\partial x_i} \{U\} + [B]_{UT} \frac{\partial}{\partial x_i} \{T\} \qquad (2)$$

with $\{T\}$ being the temperature field and

$$[B]_{UT} = \frac{\partial}{\partial T_j} \{U\} \qquad (3)$$

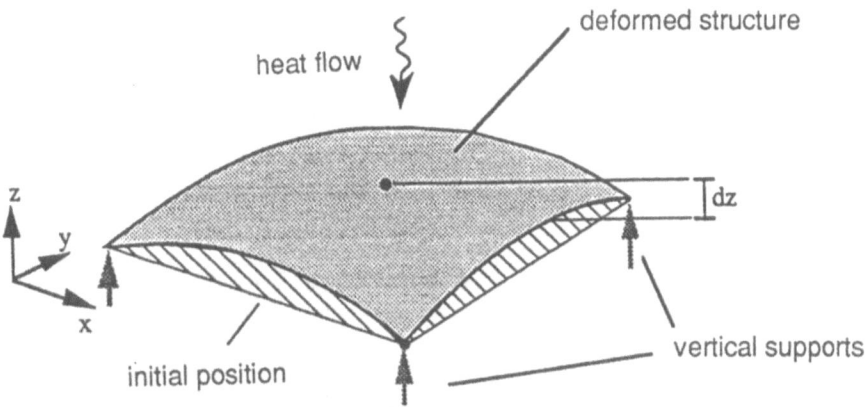

Figure 1:   Thermally loaded bending plate

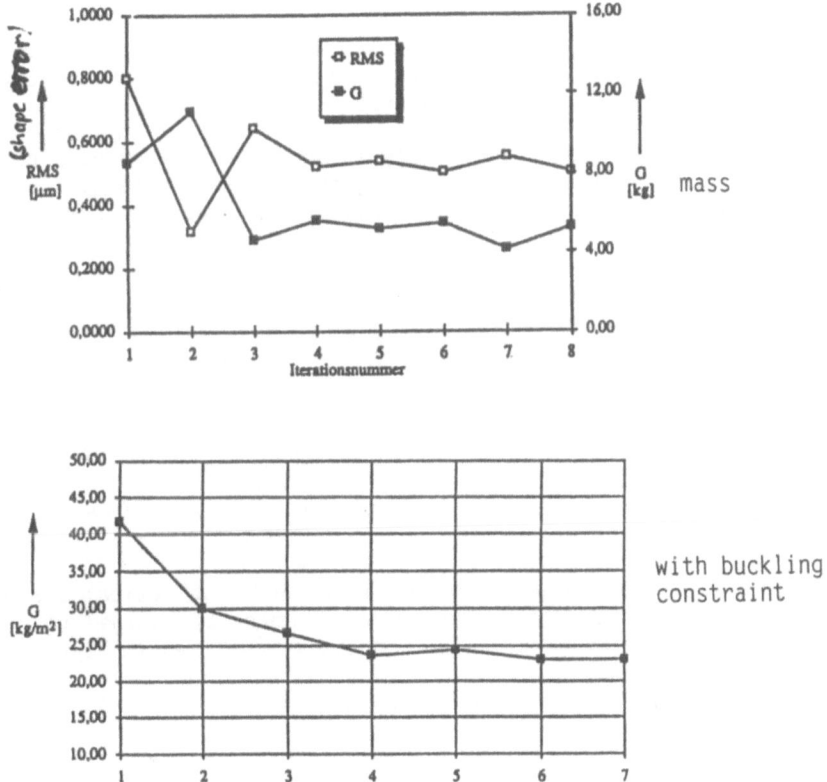

Figure 2:   Typical optimization results for thermally loaded plate

being the displacement vs. temperature sensitivity matrix. In (2) the complete sensitivity of the displacement vector due to a change dx_i also causing a differential change in the temperature is taken into account.

While the determination of $[B]_{UT}$ is relatively straight forward by

$$\frac{\partial}{\partial T_j} \{U\} = [K]^{-1} \cdot \frac{\partial}{\partial T_j} \{F\}_T \tag{4}$$

with $[K]$ being the stiffness matrix and $\{F\}_T$ the thermal load vector, the determination of $\partial\{T\}/\partial x_i$ is more complicated especially when also nonlinear thermal effects (especially radiation) are to be also considered. In that case a finite difference scheme is more appropriate.

Which of these possibilities is the best significantly depends on the kind of problem. While approach 1 is the most straight forward, it could nevertheless lead to convergence difficulties when significant changes in temperature occur due to changes in structural variables x_i. Approach 2 works well when the sampling points and the approximating interpolation functions are representative, and it also gives a lot of insight into the physical behaviour and is rather easily to implement. On the other side, for complicated behaviour a lot of sampling points might be necessary. Approach 3 is basically most apt to sensitivity and optimization analysis, but usually requires further software development and modifications e.g. for determining $[B]_{UT}$.

An example for thermomechanical analysis and optimization of a plate is given in figure 1. Apart from investigating thermal stressing of such plates, the main emphasis was in investigating components for precision (space) telescopes, where thermal deformation has to be minimized in addition to structural mass. Some results are shown in figure 2.

3. A SMART STRUCTURE OPTIMIZATION PROBLEM

3.1. General Description

In a smart (active, controlled, etc.) structure optimization problem a set of structural and control design variables is to be determined which minimize/maximal a set of performance criteria and satisfy a set of constraints.

Such smart structures range from active and adaptive optical structures to introduction of active damping into dynamically excited structures by counterbalancing actuators. The concept is outlined in figure 3.

Like in many other technical optimization problems there doesn't exist one and only one problem statement, but at best a typical or representative one. Concrete problem statements in a concrete development problem then are subsets or variations of a typical problem statement as outlined in the following.

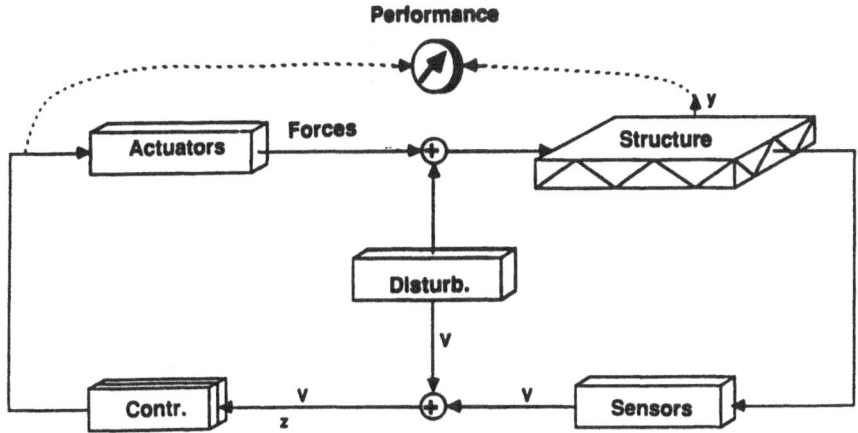

Figure 3: Concept of an active (smart) structure

3.2. Problem Statement

The formulation as a optimization problem means:

$$\text{minimize } f_1(\{X\}, \{C\}) + f_2(\{X\}, \{C\}) + f_3 \dots \tag{5}$$

$$\text{such that } \quad g_j(\{X\}, \{C\}) \geqslant 0, \ j = 1, \dots \tag{6}$$

and

$$[M(\{X\})] \cdot \{\ddot{U}\} + [D\{X\})] \{\dot{U}\} + [K(\{X\}] \cdot \{U\} = \tag{7}$$

$$= \{F\} - [T]\{C\}$$

$$(\text{e.g. } \{C\} = [G]\{E, U, \ddot{U}\}) \tag{8}$$

A set of <u>structural optimization variables</u> $\{X\}^T = \{X_1, X_2 .. X_n\}$ (such as cross-sectional areas, plate thicknesses, laminate built ups, etc.) and <u>control optimization variables</u> $\{C\}^T = \{C_1, C_2 .. C_m\}$ (such as control gains for displacement and/or velocities, control force amplitudes, etc...) optimize the objective functions f_1, f_2, f_3, .. and satisfy the constraints g_j and the system equations (7). So such kind of problems typically are multicriteria or vector optimization problems with objective functions on structural or total mass, shape deviation error for a precision structure, control energy, or robustness of the smart structure. While the other <u>objectives</u> are more or less straight forward, the criterion on robustness is a measure on the sensitivity of the smart structure with respect to variations in parameters such as stiffness, mass, damping etc. So this means that the smart structure has to be designed also such that the sensitivity is relatively small in order to achieve satisfactory overall performance even under varying structural parameters against those used in mathematical models, or due to degradation etc. With respect to structural design variables this means that they are selected e.g. such that a shift of a potentially critical eigenfrequency is small or goes into the uncritical direction, while for the controller some robustness its sensing/filtering, control force amplitudes and frequency content is to be achieved.

In its basic mathematical form, consideration of robustness means

$$\text{minimize } f = \sum_i [(\partial r/\partial p_i) \cdot \Delta p_i]^2 + \text{other objectives} \qquad (9)$$

where p_i is the i-th structural parameter which under a variation Δp_i causes a change in the response quantity r. These structural parameters could be also optimization variables. In case an optimization algorithm for (11) is used which requires gradients of the objectives with respect to the optimization variables, then (11) means that second derivatives of the response quantity r are to be determined. This could lead to a considerable numerical effort. So an alternative formulation for (3.2-7) could be

$$\text{minimize } f = (r_0 - r(X_1, \ .. \ X_u))^2 + \text{other objectives} \qquad (10)$$

with r_0 being a fixed value of the response quantity r.

The constraints as listed in (6) could typically relate to displacements or accelerations at the j-th degrees of freedom of the structure, eigenvalues of either the open or closed loop, and also to amplitudes of the control forces, etc. Of course, stress or reliability constraints are to be considered also, but this is usually better done in a separate design and optimization exercise.

The system equations (7) heavily influence the numerical algorithms, software and effort envolved. First of all, the structural variables {X} directly influence the system matrices, while the control forces {C} are treated as right hand side variables, with [T] being the transformation matrix which maps the control forces onto the degrees of freedom of the discretized model. The gains in the gain matrix [G] can be used as control optimization variables and can be applied to an error vector {E}, the displacement {U} or velocity {Ů} of the structure. In the latter two cases, this means an influence on the stiffness and damping matrices, respectively. This influence in general can be nonsymmetric, thus leading to nonsymmetric matrices and complex eigenvalue problems. Modal condensation and truncation also has to take into account a sufficient number of modes in order not to drop modes which could then the excited by the controller and might cause instability.

More details and applications are given e.g. in [2], [3].

4. DECOMPOSITION STRATEGIES

4.1. General Comments

Many real and larger scale problems have to be decomposed into sub-problems each of which can be treated with reasonable effort. This not only holds for problems coupling different disiplines, but also for "pure" structural optimization tasks. One can differ between

- process oriented decomposition, where the analysis, design and optimization process is decomposed into different subsequent steps, eventually including some iterations

- system oriented decomposition, where the system to be optimized is decomposed into proper subsystems which are treated seperately. Eventually, some iterations between inputs and outputs of the subsystems are carried out.

There are different developments under way in order to formalize such decomposition strategies, i.e. to formulate proper constraints for a subsystem in order to take into account effects of the other. Nevertheless, most decompositions in practise are carried out in an informal way by engineering judgment, an example of which is outlined in the following.

4.2. Decomposition Strategies in Optimization of Light Weight Bridges

Figure 4 shows the optimization task of a light weight bridge. This bridge has to span a gap of about 40-50 m and carry loads up to about 90 tons. Its cross-section is composed of two trapezoidal trackways made out of aluminum, with the top and bottom grider being strengthened and stiffned by carbon fibre composite layers.

For analysis and optimization, a global finite element model has been established which provides section forces under different loading conditions. These section forces then have been applied to the optimization models (mass minimization under strength, stability and fabrication constraints) for the top girder and side wall. The system equations have been based on classical mechanical plate equations with the section forces from finite element analysis as external loading.

So a process - oriented decomposition (global finite element model - local plate bending and stressing models) and system oriented decomposition (treating top grider, bottom girder and side walls seperately) has been used. After optimizing each subcomponent seperately with fixed section forces, the new design variables then are introduced into the global finite element model for a section force update and one additional decomposed optimization process.

Some results of this optimization problem for the top girder are given in figure 5.

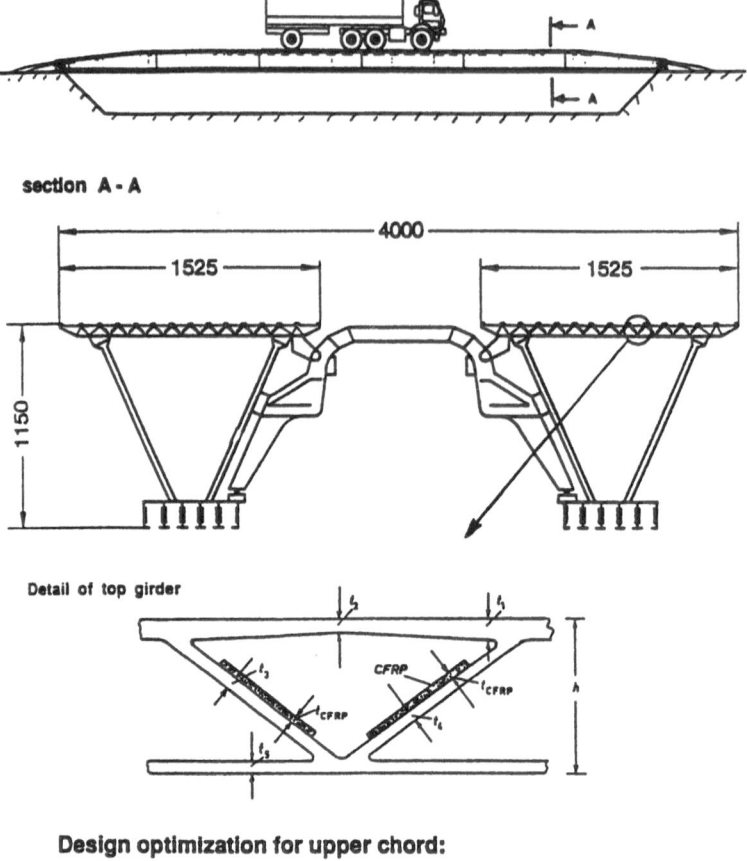

section A - A

Design optimization for upper chord:

    objective: weight (aluminum + CFRP)

    constraints: strength, fabrication

    design var.: thicknesses t (see above)

    alternative constraints: limit on probability of failure

Figure 4: Light weight aluminium/carbon-fibre bridge

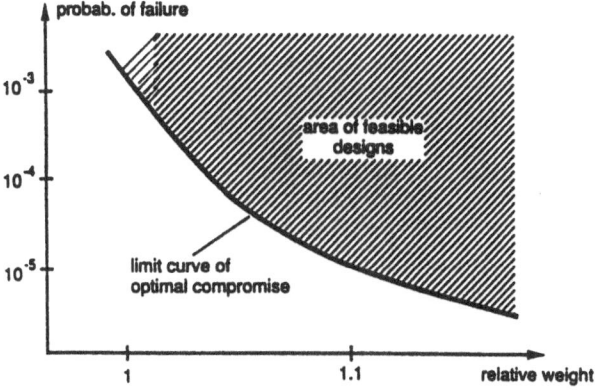

functional efficient boundary for probab. of failure vs. weight

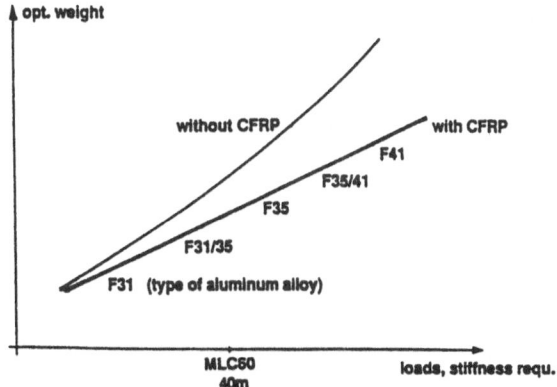

Optimal weight vs. increased requirements (loads, span) with and without CFRP

Figure 5: Typical results of top girder optimization problems

References

[1] Sobieski, J., "Multidisciplinary optimization of engineering systems", Bergmann H.W. (ed), Optimization Methods and Applications, Springer Verlag, Heidelberg, 1989

[2] Baier, H., "On shape control of precision structures: concepts, analysis and technology", Proceedings European Forum on Aeroelasticity and Structural Dynamics, Aachen, Germany, 1991

[3] Maurer, D., Charon, W., "Verification of performance of active structure after mechanical design", Proc. Europ. Forum Aeroelasticity and Structural Dynamics, Aachen, Germany, 1991

[4] Prenicker, M., Eschenauer, H., Post, U., "On a decomposition technique in design optimization process", in Eschenauer, H., Thierauf, G. (eds): Discretization Methods and Structural Optimization, Springer Verlag, Heidelberg, 1989

COMPOSITE LAMINATE AND SANDWICH OPTIMIZATION WITH APPLICATIONS

H. J. Baier
Dornier GmbH
799 Friedrichshafen, Germany

ABSTRACT. The design optimization problem of composite laminate structures is addressed with a decomposition into a structural and laminate
optimization process. Emphasis is given on the latter, with specifics
such as the determination of the fibre/matrix system in parallel to
fibre angles and lamina thicknesses. The nonconvexity of the problem is
highlighted. Practical examples are taken from high precision space
reflector design and the design of a large composite cylindrical shell
structure of the Ariane launcher.

1. INTRODUCTION

Composite materials have due to their excellent mass-specific stiffness
and strength an increasing field of applications not only for air- and
spacecraft, but also in other fields such as in automotive and sports
products. They are different from "conventional" and usually isotropic
metallic materials, because they can be tailored by the selection of
fibers- and matrix-systems and especially by the geometric variables of
fiber orientations and ply (or lamina) thicknesses. For those requirements composite optimization programmes are available, which determine
for minimum weight the orientations and ply thicknesses under
constraints like those on strength, stiffness, and thermal expansion.
But usually the material system (fibre/matrix combination) has to be
preselected and is held fixed during optimization.

The optimization problem gets even more challenging and interesting,
when the material system becomes an additional variable to be determined optimally from a material data bank. A formal treatment via
discrete optimization is usually out of scope because of lack of reliable and efficient algorithms. So some more heuristic approaches and
algorithms are presented, where in addition to the laminate lay up
(fibre angle and ply thickness), the type of lamina material is also
used as an optimization variable.

G. I. N. Rozvany (ed.), Optimization of Large Structural Systems, Vol. II, 997–1009.

Before discussing this in some detail, two typical composite structure applications will be presented briefly in order to show how the optimization of laminate and sandwich lay up fits into the overall composite structure design process.

2. COMPOSITE STRUCTURE EXAMPLES

Two typical composite structures are presented, namely a precision reflector, with severe requirements on stiffness and thermal expansion and a structural component of the Ariane 5 launcher the design of which is governed by stiffness and strength.

2.1. High Precision Reflector FIRST

A typical example of a high precision reflector for the Far Infrared and Submillimeter Space Telescope FIRST is shown in figure 1. It has to be stowed for launch and is then accurately deployed in orbit to a diameter of about 8 m. The total shape error (including those under environmental loads such as temperature gradients or dynamic excitations) must not exceed 10 μm, i.e. the root-mean-squares (rms) deviation of the actual shape from the theoretically required one (e.g. a paraboloid) must be smaller than this limit.

Because of the high ratio of strength and stiffness to mass and the very low coefficient of thermal expansion α_T, carbon fibre composites are extensively applied in such types of dimensionally stable structures. The corresponding laminate and sandwich lay up of the reflector shell thus is part of the design variables. A reasonable approach is to treat geometric and stiffness design variables in a structural optimization problem with the rms-error as the objective function to be minimized. In parallel but separately to this, the laminate and sandwich lay up is optimized under different criteria and constraints. Results of this then are introduced into the overall finite element model used for structural optimization. Through not strictly correct, this decomposition nevertheless is reasonable since it provides an efficient approach for the material definition.

For fibre laminates or a honeycomb core sandwiched between these laminates, the mass m and the weighted sum of the different coefficients of thermal expansion α_T, which occur because of anisotropic material properties, are taken as objectives and to be minimized:

$$f_2(x) = m\ (t_1,\ t_2,\ldots t_n), \tag{1a}$$

$$f_2(x) = \sum_j w_j\ \alpha_{Tj}^2\ (t_1,\ t_2,\ldots,t_n;\ \alpha_1,\ldots,\alpha_n). \tag{1b}$$

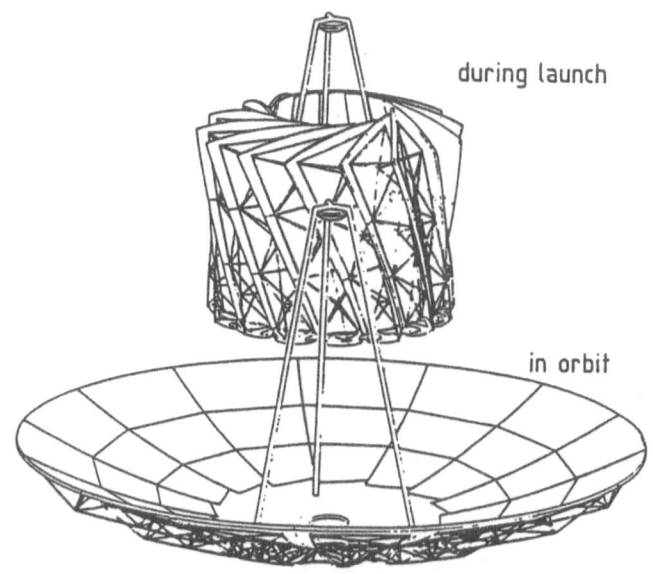

during launch

in orbit

Figure 1: The future European Far Infrared Telescope FIRST

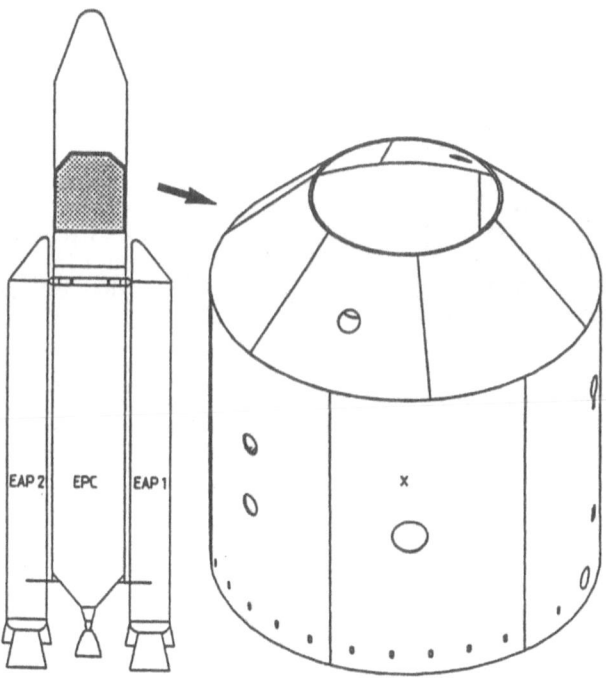

Figure 2: The SPELTRA structure of the future Ariane 5 launcher

The constraints are: failure criteria, stiffnesses, lower/
upper bounds on design variables;

The system equations are: laminate plate theory for plates consisting of stacked orthotropic plies.

In (1) the design variables t_i and α_i are the thickness and fibre angle
of the i-th lamina, respectively. The weighting factors w_j emphasize
the importance of low a_{Tj} in specific directions. Since the α_{Tj} may
become negative for carbon fibre laminates, it is important to use
square (or absolute) values within the objective function vector. This
design task is formulated as a multicriteria problem instead of con-
straining the α_T and minimizing mass because the selection of reasonable and feasible constraint levels for α_T might be problematic. On the
one hand small α_T are desirable while, on the other hand, constraint
levels which are too small could cause difficulties in finding feasible
variables or could even generate an empty feasible set.

Since in laminate plate theory fibre angles are introduced via trigonometric functions, the nonconvexity of the corresponding optimization
problems is often more prevalent than in many other types of structural
optimization in statics. The latter point is briefly addressed in chapter 4.3.

2.2. Launcher Shell Structure

Axisymmetric shells are often applied in launcher and satellites structures, for example as a satellite's central structure or booster structures tanks, launcher payload compartments etc. The shell structures
often consist of a honeycomb core sandwiched between composite material
facesheet, or a stringer stiffened thin shell, or a corrugated shell.
These structures should have a minimum weight and at the same time they
should satisfy stiffness, strength, as well as global and local
stability constraints. In some applications low or high temperatures
have to be taken into account. As an example the cylindrical and
conical upper shell of a launcher providing the interface to and carrying the satellites is shown in figure 2. This sandwich shell structure
SPELTRA with a diameter of more than 5 m and a height of about has to
carry loads which generate forces on the bottom section of N = 739 kN
and Q = 240 kN in the vertical and lateral direction, respectively, and
a bending moment of M = 3800 kNm. The structural mass shall be minimized.

The lay up of the facesheet as well as the core thickness of the sandwich are used as global design variables, in addition to some local
stiffeners.

A general approach to solve such optimization problems would be to use the relevant parameters of the mathematical finite element model of these structures as optimization variables. Mainly because of the considerable effort envolved it is often more practical to optimize the laminate or sandwich lay up seperately based on classical laminate plate theory with section forces from the structural finite element model taken as input data. The result of this optimization then can be either taken as a good starting vector for optimization on structural level, and in many cases is even sufficient to achieve a good if not optimal design. The main advantage of this decomposition as outlined in figure 3 is the relatively low numerical effort due to the use of classical laminate plate theory as system equations. This then allows in depth investigations and parametric optimizations, which in many cases are very valuable for the development process and an important prerequisite for a large scale optimization on structural level.

3. LAMINATE PLATE THEORY

The essential steps of laminate analysis comprise the builtup of the lamina properties dependent matrix relation between laminate strain and applied (section) forces, the determination of the stresses in each of the lamina, and eventual stiffness and thermal expansion properties. A concise textbook treatment is given e.g. in [1]. Since only relatively small matrices (typically of the order 6) are envolved in these system equations, the analysis effort is quite low. For convenience, this also allows the determination of sensitivities via finite differences without intolerable computational effort.

4. APPROACHES AND RESULTS IN LAMINATE OPTIMIZATION

In this chapter, some approaches and results for laminate optimization are discussed. First, the classical laminate optimization problem as implemented in the software package COOP [5] is outlined, and then the inclusion of material selection as additional design variable is given. Finally, some remarks on the nonconvexity of laminate optimization problems with consequences in local optima and difficulties in multicriteria problems are discussed.

4.1. The Standard Laminate Optimization Problem

The standard laminate optimization problem with fibre angle α_i and thickness t_i for each lamina of the laminate is given in figure 4. Typical objective is laminate weight while constraints on strength (e.g. Tsai-Wu failure criterion [2]), stiffness, hygrothermal expansion and bounds on design variable may exist. The resulting software package COOP (composite optimization) consists of the laminate analysis package LAMA and a feasible directions optimizer. Because laminate analysis and constraint function evaluations can be done rather quickly, the required gradients are determined via a finite difference scheme.

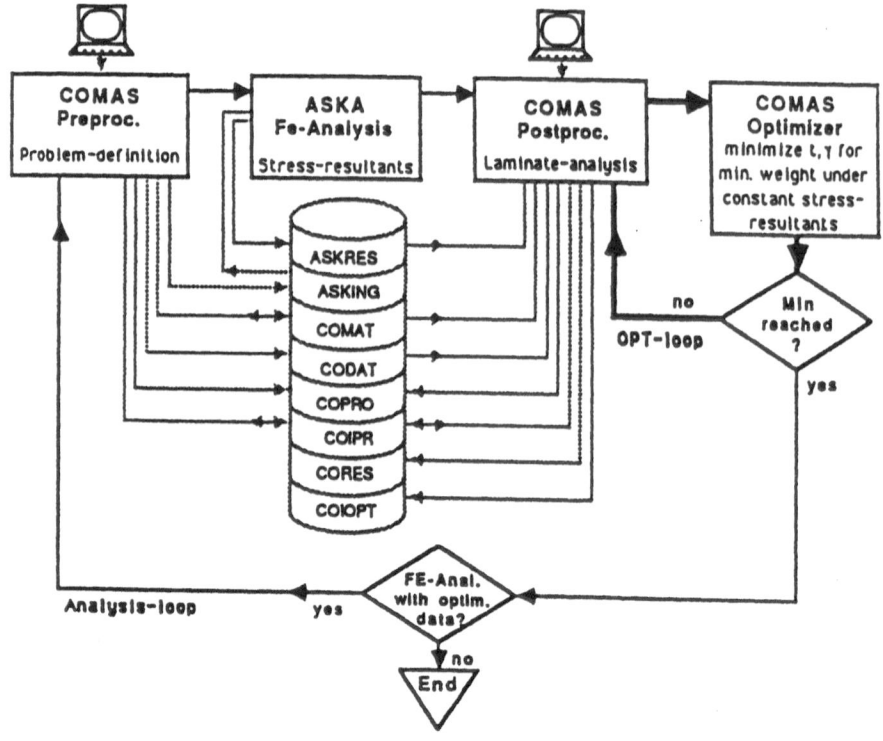

Figure 3: Decomposition of finite element structural analysis and
laminate optimization

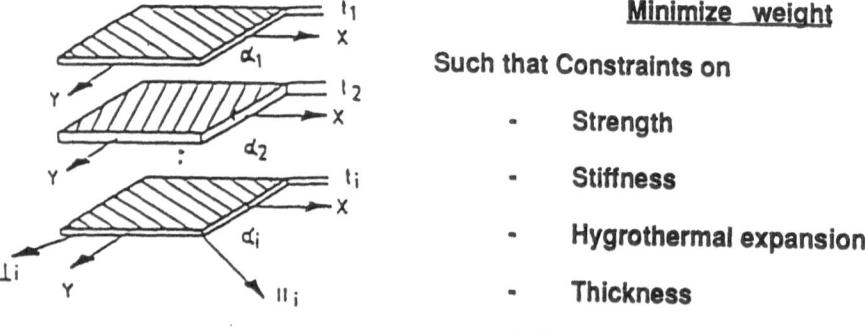

Figure 4: Basic problem statement of laminate optimization

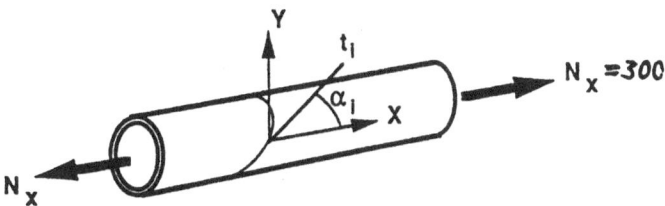

- **the weight is minimum**

- **Youngs Modulus $E_x = 100000$ N/mm^2**

- **Coefficient of thermal expansion $\alpha_x \leq 0.5 \cdot 10^{-6} \cdot C^{-1}$**

Figure 5:   Demonstration example for laminate optimization

```
*************************************************************
INPUT:                  GEOMETRY OF LAMINATE
*************************************************************

     NUMBER OF LAYERS:  6              LAMINATE IS: -  SYMMETRIC
                                                    -  BALANCED


     LAYER NR ! THICKNESS !   ANGLE   !        MATERIAL
              !   [MM]     !  [GRAD]   !
     _____!_____!_____!_____
              !           !           !
        1     !  0.100    !    50.0   ! CARBON FIBER T300-LY556
        2     !  0.100  S !   -50.0 M ! CARBON FIBER T300-LY556
        3     !  0.100    !     0.0 F ! CARBON FIBER T300-LY556

     _____  .  _____        LINE OF SYMMETRY      _____  .  _____
```

(comment)   S   =   SAME THICKNESS AS PRECEEDING LAYER
            M   =   MINUS FIBER ORIENTATION OF THE PRECEEDING LAYER
            F   =   FIXED VARIABLE

```
*************************************************************
               LOADSET(S) FOR CALCULATION
*************************************************************


     NUMBER OF LOADSET(S)    (MAX. 10):   ==> 1

     LOAD-SET NUMBER:  1

     N1 [N/MM] :    300
     N2 [N/MM] :
     N6 [N/MM] :
     M1 [N] :
     M2 [N] :
     M6 [N] :
     DELTA T [K] :
```

Figure 6:   Input parameters for laminate optimization package

```
**************************************************************
              OPTIMIZED GEOMETRY OF THE LAMINATE
************************************************************** .


     OBJECT FUNCTION:    -->    MINIMUM WEIGTH
     CONSTRAINTS:        -->    STRENGTH  TSAI-HILL
                         -->    COEFF. OF THERMAL EXP.
                         ALPHA X  -->    0.5000E-06
                         -->    STIFFNESS
                         EX       -->    0.1000E+06


     THICKNESS OF LAMINATE  [MM]            0.51
     WEIGTH  [KG/M**2]                      0.40058
```

PLY	!	PLY-ANGLE [GRAD]	!	PLY-THICKNESS [MM]
1	!	69.93	!	0.0392
2	!	-69.93	!	0.0392
3	!	0.00	!	0.1751

```
-------------------- LINE OF SYMMETRY --------------------


**************************************************************
         L A M I N A T E    P R O P E R T I E S
**************************************************************


YOUNG"S MODULI    [N/MM**2] :  EX = 0.9650E+05  EY =  0.4003E+05
SHEAR MODULUS     [N/MM**2] :  GXY = 0.8868E+04
POISSON"S RATIO            :  NXY = 0.1626E+00
COEFF. OF THERM. EXP.  [1/K] :  AX = 0.4806E-06  AY = 0.5311E-05

AVERAGE DENSITY:        [KG/M**3]       [G/CM**3]       [N*S**2/MM**4]
                        0.1580E+04      0.1580E+01       0.1580E-08
```

LAYER	!	RESERVE FACTOR (<1 MEANS FAILURE)
1	!	1.007
2	!	1.007
3	!	1.717
4	!	1.717
5	!	1.007
6	!	1.007

Figure 7: Essential output after laminate optimization runs

A simple application example for COOP is given in figure 5, with the required computer program input and the output for the optimized laminate given in figures 6 and 7, respectively. The user is prompted through the required input, and in case of other objectives than weight these can than be introduced via user subroutines.

4.2. Type of Material as Additional Optimization Variable

The use of material as an additional optimization variable formally leads to a discrete optimization problem. Since for nonlinear discrete optimization problems no general, efficient and reliable solution algorithm is available, a quasi-continuous problem is formulated as a first step and discretisation as a second as briefly described in the following.

The automatic material selection is algorithmically supported by first constructing continuous interpolation functions over the material data base (see for example table 1)and then using a "pointer variable" going through this interpolations in parallel to the geometric design variables as in figure 3. This then leads to a continuous nonlinear optimization problem. After solving this problem, the material with an actual "pointer variable" being most closely to the determined one is selected. Then, the optimization problem for minimum weight of a laminate reads:

$$\min\{ f(t_i, \rho_i) = \sum_{i=1}^{n} \rho_i \cdot t_i \mid g_j(t_i, \rho_i, \alpha_i) \leqslant 0; \ j=1,m\} \qquad (2)$$

(t_i = i-th ply thickness; ρ_i = i-th density; α_i = i-th ply orientation)
ρ is the "pointer variable" and t and α are the geometric variables for each of the n layers. Figure 8 shows the section-wise linearized strength of the data base in table 1. All other properties have to be arranged in the same manner. It is worthwhile to mention, that the here choosen "pointer variable" must not necessarily be the material density, but could be any other property, for instance Youngs modulus E. But for weight optimization the density as variable makes results directly interpretable.

In the following a simple example is given for a 2 layer composite with 2 load sets and the requirement: Find optimal thicknesses, angles and related materials out of databank, such that

- the weight is minimum
- stiffness E ⩾ 400000 N/mm
- strength criteria is satisfied under applied loads

Figure 9 shows the input (weight: 16.0 kg/m²) and results (weight: 5.23 kg/m²), which have been taken from [4].

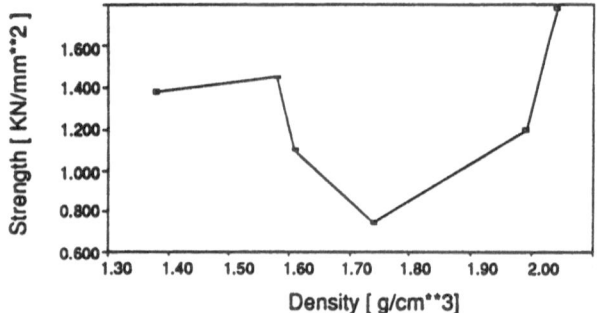

Figure 8: Linearized strength over density as "pointer variable"

Table 1: Except from composite materials data base:

		Aramid		CFRP		Glass	
		Kevlar49	T300	M-40	GY-70	E-Glas	S-Glas
ρ	[g/cm**3]	1.38	1.58	1.61	1.74	1.99	2.04
σ_{1Z}	[N/mm**2]	1380	1450	1100	750	1200	1780
σ_{1D}	[N/mm**2]	275	1400	1100	700	700	700
σ_{2Z}	[N/mm**2]	27	55	50	40	65	65
σ_{2D}	[N/mm**2]	138	170	150	130	150	192
τ_{12}	[N/mm**2]	44	90	75	70	62	62
E_1	[KN/mm**2]	75.8	135	220	290	83	53.8
E_2	[KN/mm**2]	5.5	10	7	5	46	13.4
G_{12}	[KN/mm**2]	2.0	5	5	5	13	4.46
ν	[-]	0.34	0.27	0.35	0.41	4.36	0.29
α_1	[1.e-6/K]	-4	-0.6	-0.8	-1	6	6
α_2	[1.e-6/K]	50	30	30	30	26	26

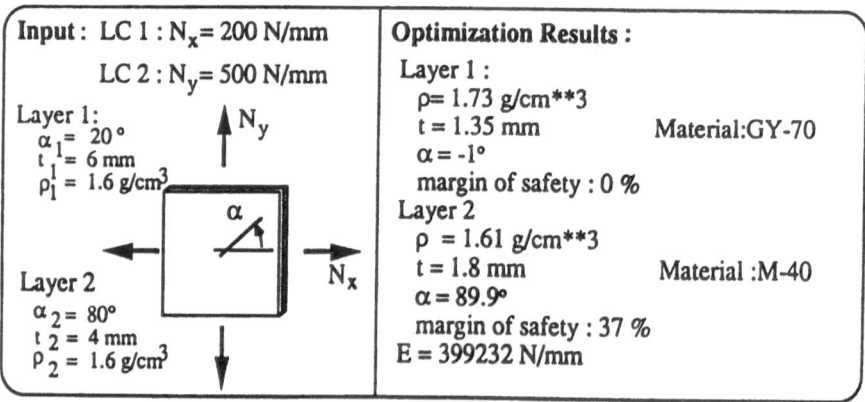

Figure 9: Demonstration example and optimization result with material selection

It is noticable that layer 1 with p = 1.73 g/cm³ is close to the material with the highest Youngs modulus and so contributes to the stiffness requirement. Layer 2 is in the neighbourhood of a material with a much smaller Youngs modulus but significant higher strength and therefore makes the composite stronger. The final discrete material decision then is (and should be!) made by the user. In that case, it is to use the high modulus GY-70 for layer 1 and higher strength M-40 for layer 2 with the corresponding optimal fibre orientations and lamina thicknesses.

4.3. Practical Consequenes of the Nonconvexity of the Problem

Fibre laminates, especially those based on carbon fibres are extensively applied to spacecraft structures because of their high stiffness and strength, their low specific mass, and their low coefficient of thermal expansion α_T. These properties mostly depend on the laminate structure, and a reasonable constraint limit is often difficult to specify, especially for α_T. Thus, a multicriteria formulation corresponds to a combined objective

$$p\,[f(x)] = m(x) + \sum_i w_i\,\alpha_{Ti}^2(x) \qquad (3)$$

where m is the mass and w_i stands for the weighting factors. Figure 10 gives a simple example. It considers a stack of two lamina where the fibre angle and the lamina thickness are the design variables. The objectives $f_1(x)$ = mass and $f_2(x) = \alpha_{T1} + \alpha_{T2}$ are outlined in figure 10. Here, the fibre angle ± α has been chosen as a parameter, and thicknesses are selected such that the failure criteria just take on critical values. It can be seen that minima are obtained for α = 0° and α = 45°, respectively. What is important to note is the nonconvexity of the functions, which is mainly due to the trigonometric functions of the fibre angles occuring in the system equations. So, if the combined objective function of (3) is used to determine the Pareto-optima describing the optimal trade-off between the objectives, the minima of p[f(x)] are either at α = 0° or α = 45° depending on the weighting factor. For that reason the Pareto-optima other than at α = 0° or α = 45° cannot be computed numerically in this nonconvex problem, whatever variation of w is used. This shows that the transformation of multicriteria optimization problems into scalar optimization problems of type (3) have to be performed carefully in nonconvex problems.

In addition to this difficulty in multi criteria problems, local optima in scalar optimization problems could be expected in laminate optimization more often than in other (static) structural optimization problems. This can be seen also from figure 11, where for thickness minimization of a single lamina under strength constraints the design space of thickness t and fibre angle α is shown together with the boundary of the feasibly region. Different local optima exist.

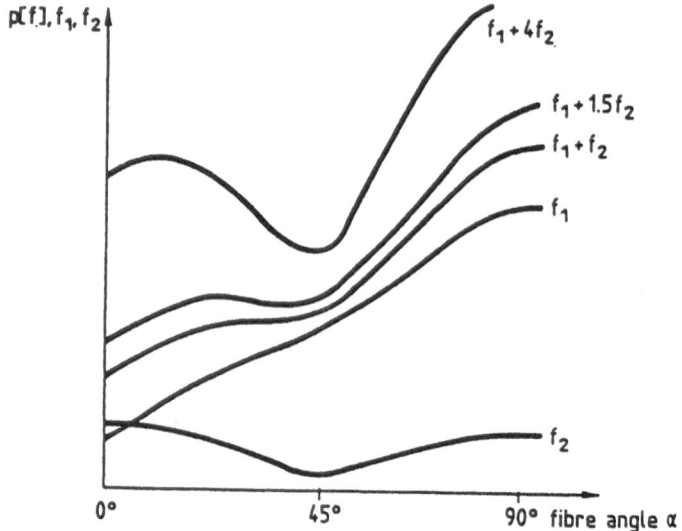

Objectives $f_1(x)$ (weight) and $f_2(x) = \sum |\alpha_T|_i$ as a function of the fiber angle α; preference objective $p[f] = f_1 + w\, f_2$;

Figure 10: **Nonconvex objective functions and their combination**

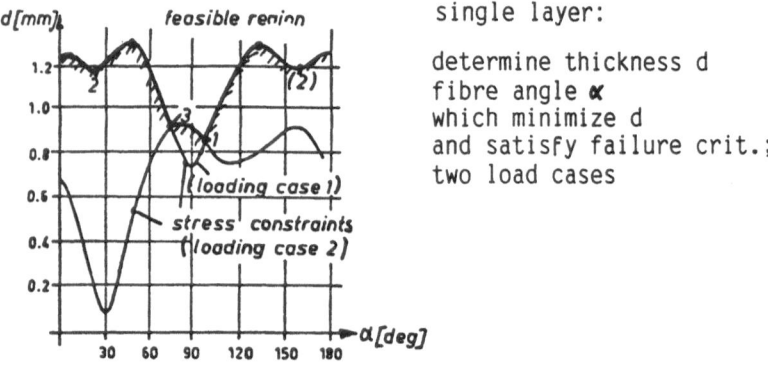

single layer:

determine thickness d
fibre angle α
which minimize d
and satisfy failure crit.;
two load cases

Figure 11: **A lamina optimization problem with local optima**

References

[1] Jones, R.M., Mechanics of Composite Materials, McGraw-Hill, New York, 1975

[2] Tsai, S.W., A survey of macroscopic failure criteria for composite materials, Journal of Reinforced Plastics and Composits, January 1984

[3] Baier, H., 'Multicriteria design of spacecraft structures with special emphasis on mass and stiffness', in H. Eschenauer, J. Koski, A. Osyczka (eds), Multicriteria Design Optimization, Springer Verlag, pp 244-260

[4] Seeßelberg, Ch., Helwig, G., Baier, H., 'Strategies of interactive structural optimization and composites material selection', in Eschenauer, H., Matthek, C., Othoff, N. (eds), Engineering Optimization in Design Processes, Springer Verlag, 1991, pp 133-142.

[5] Helwig, G., Becker, W., 'User manual for composite optimization package COOP', Report TMB/89, Dornier GmbH, Friedrichshafen, Germany

MULTIDISCIPLINARY OPTIMIZATION OF FIBER COMPOSITE AIRCRAFT STRUCTURES

H.A. ESCHENAUER, G. SCHUHMACHER
Research Laboratory for Applied Structural Optimization at the
Institute of Mechanics and Control Engineering, University of Siegen
5900 Siegen, Germany

W. HARTZHEIM
Deutsche Aerospace, MBB GmbH, Aircraft Division
Information Systems, Technical Application, CAE
P.O. Box 80 11 60
8000 Munich 80, Germany

ABSTRACT. The paper deals with the application of a method of structural optimization to find optimal layouts of fiber composite structures in aircraft construction. The goal of the optimization process is to minimize the structural weight and, simultaneously, to fulfill multidisciplinary constraints concerning failure criteria, stability requirements, dynamic responses, aeroelastic efficiencies and flutter speeds. Considering these requirements, the optimal values for the layer thicknesses and the fiber orientations of composite structures must be determined. This problem can efficiently be solved by structural optimization methods. The model formulation in structural optimization will be described in general as well as the special optimization model and the structural and sensitivity analysis for fiber composite structures. An application example demonstrates the performance of the whole procedure. This example shows that the structural optimization method is very useful to increase the efficiency of the design process as well as the quality and performance of aircraft structures.

1. Introduction

Due to their high weight-specific stiffness and strength, fiber composite materials are used in order to obtain light weight constructions. Therefore, they are extensively applied to the construction of air- and spacecrafts to achieve a low energy consumption. The different substructures of an aircraft, such as fuselage, wings, fins, rudders, etc. must satisfy especially rigorous and numerous demands with regard to strength, eigenfrequencies, aerodynamic efficiencies, flutter velocities etc. .

The various component properties depending on the stiffness and mass distribution are determined by material properties as well as the constructive layout of the structure. In most cases, fiber composite materials are laminates consisting of several single layers (Fig. 1.1). These layers, in turn, contain unidirectional fibers made of carbon, glass, boron, etc. which are embedded in a resin matrix.

1011

G. I. N. Rozvany (ed.), Optimization of Large Structural Systems, Vol. II, 1011–1022.
© 1993 *Kluwer Academic Publishers.*

1012

The stiffness and strength of the fibers are much higher than that of the resin matrix, so that the load carrying capacity is much higher in the fiber direction than perpendicular to it.

With the development of fiber composite structures, the thicknesses and fiber orientations of the single layers are available as design variables in addition to the shape of the component. Because of the multiple and often conflicting demands on the structure it is not possible, even with extensive engineering experience, to immediately find a design which satisfies all demands and simultaneously achieves a minimal structural weight. In order to find such a design, it is necessary to apply an iterative process, consisting of a continuous alternation between structural analysis and succeeding modification of the design. In many cases, this extensive iteration process is still carried out manually, and the construction is modified intuitively. The concept of structural optimization proved to be very useful to automate this iterative process and to improve its efficiency. Using an optimization procedure the component parameters are modified by an optimization algorithm until the desired objective (e.g. minimal structural weight) and any other demands or constraints are fulfilled. The optimization program LAGRANGE is such a procedure which has been developed by MBB, the Research Laboratory for Applied Structural Optimization at the University of Siegen, and several other university institutes since 1984 [2].

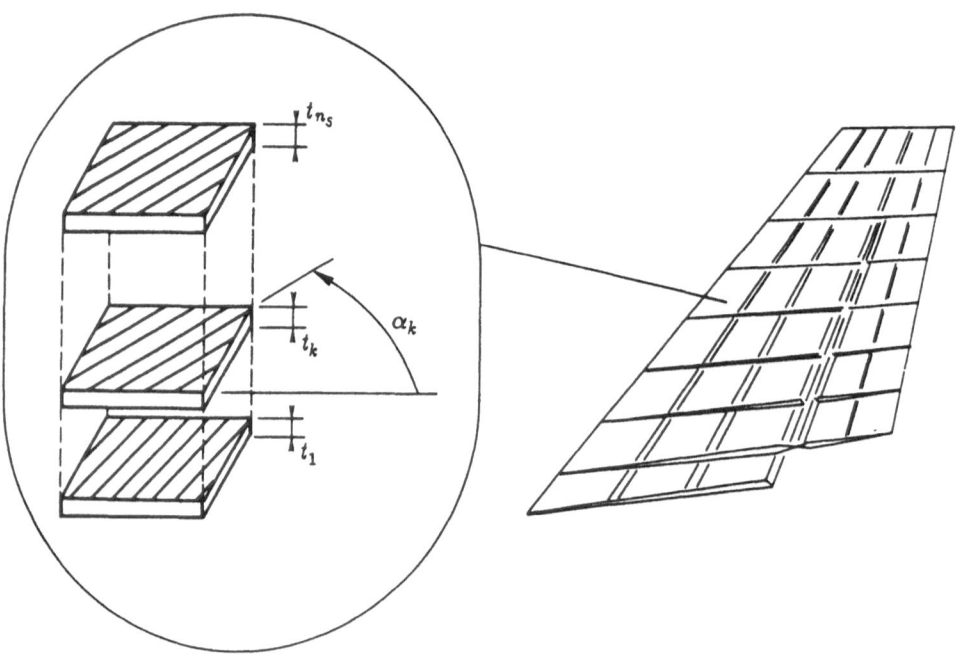

Fig. 1.1: *Aircraft fin made of fiber composite material*

Based upon the ability of determining optimal layer thicknesses of fiber composite structures, LAGRANGE was extended during the last two years in such a way that it can also calculate optimal fiber orientations in the single layers simultaneously, considering constraints on the static, dynamic and aeroelastic structural behaviour. This paper briefly describes the present state of the program as achieved by these extensions and gives a preview on future developments. In section 2 the different parts and the basic procedure of the structural optimization process are described. As major parts of the optimization loop, the optimization model as well as the structural and sensitivity analysis are then described in sections 3 and 4. All the mentioned parts are realized in the program LA-GRANGE. Using LAGRANGE, several fiber composite structures have been optimized. In order to demonstrate the performance of the procedure, the optimization results for a fin with rudder are presented.

2. Model formulation in structural optimization

In order to deal with optimization problems in the structural design process, a procedure following the "Three-Columns-Concept" [3] seems most suitable. Following this concept, it is most sensible to divide the problem into the "three columns":

I) Structural model,
II) Optimization model and
III) Optimization algorithm.

The **structural model** supplies a mathematical description of the physical behaviour of the structure using appropriate state variables (e.g. deformations, stresses, eigenvalues, etc.), which depend on the structural variables (dimensions of the component, cross-sections, ply thicknesses and angles, etc.). The Finite-Element-Method proved to be an especially efficient and universal method for the mechanical analysis of aircraft structures. However, any other analysis method could be applied in the optimization process as well.

An **optimization algorithm** is a mathematical procedure for solving a problem defined as follows:

$$Min \ \left\{ f(x) \ | \ h(x) = 0 \ ; \ g(x) \geq 0 \ ; \ x \in R^n \ ; \ x_l \leq x \leq x_u \right\} \quad , \quad (2.1)$$

$$X := \left\{ x \in R^n, \ x_l \leq x \leq x_u \ | \ h(x) = 0 \ ; \ g(x) \geq 0 \right\} \ , \quad (2.2)$$

with f objective function,
x vector of the n design variables,
h vector of the m_h equality constraints,
g vector of the m_g inequality constraints,
R^n n-dimensional, EUCLIDean space,
x_l, x_u lower and upper bounds of the design variables,
X feasible domain or design space.

With structural optimization problems there usually is a nonlinear relationship between the behaviour functions (objective and constraint functions) and the design variables. Because of this nonlinearity, the problem (2.1) cannot be solved explicitly but only by using an iterative, numerical process. For that purpose, several optimization algorithms with different solution strategies have been developed during the last decades. Experience shows that the efficiency and robustness of these algorithms often depends on the given problem. Therefore, it is advantageous to have more than one algorithm available to be able to choose the one which is most suitable for a given problem.

In order to solve a structural design problem by means of an optimization algorithm, it is necessary to describe the problem according to equation (2.1). This is carried out by the **optimization model** which is divided into a design- and an evaluation model. The design model defines the relationship between the structural variables and the design variables. Structural variables are physical parameters, for example cross-sections, fiber orientations, etc., of which the optimal values have to be determined, while the design variables represent the mathematical quantities which are processed by the optimization algorithm. By means of the evaluation model, it is stated which quantity of the structural model is to be minimized and which conditions or ultimate values are to be considered for the other state quantities. Thus, the optimization model is the mathematical description of the structural design task. For that reason, the optimization model is of special importance, since a truly optimal solution can only be achieved by

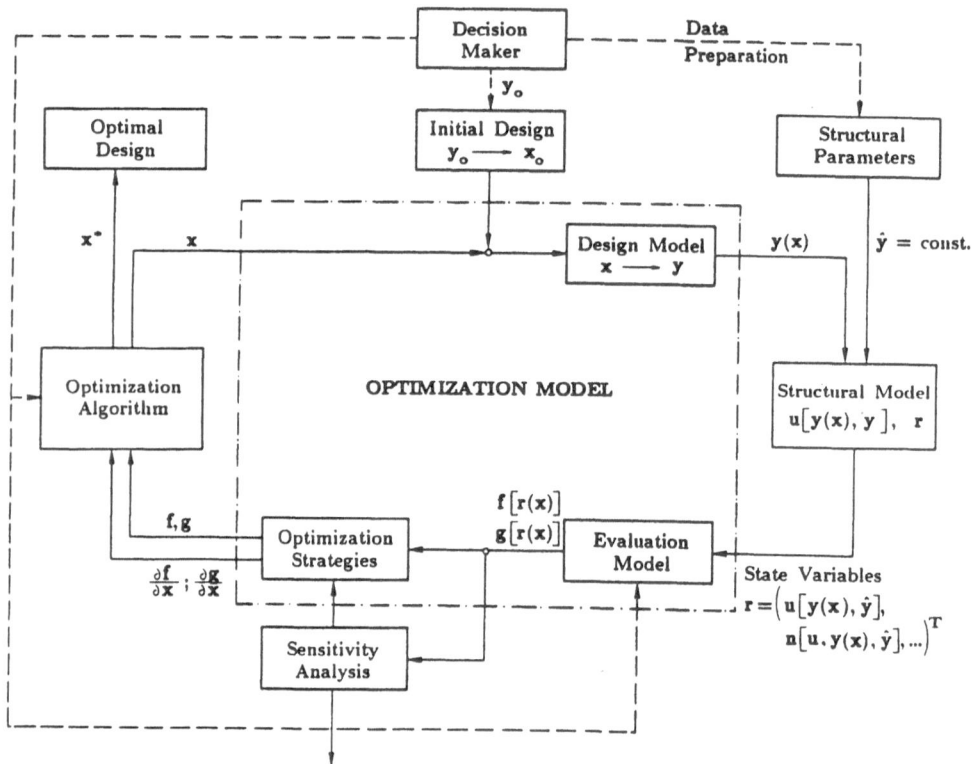

Fig. 2.1 : Optimization loop

a fully and accurately defined design task which considers all demands on the structure.

Fig. 2.1 shows the interaction of the "three columns" in the optimization loop. First, the decision maker has to describe the structural and optimization model for the special design problem. Based upon an initial design y_0 for the structural variables, the corresponding initial values x_0 for the design variables are determined. The design model then yields the variable subset of the structural parameters to be optimized. These, together with the constant structural parameters (material constants, nonvariable structural parameters), are taken to define a special design for which the state variables are calculated by the structural analysis. By means of the evaluation model the objective function and constraint values are calculated as one part of the input values for the optimization algorithm. In addition to the functional values, most optimization algorithms require the gradients of the behaviour functions with respect to the design variables which are evaluated by the sensitivity analysis. If a special optimization strategy is applied, for example a strategy for solving a multicriteria optimization problem, the behaviour functions and their derivatives are transformed into corresponding substitute values. Otherwise, they are directly transferred to the optimization algorithm. Using this information, the optimization algorithm calculates a new design variable vector and, thereby, one obtains a closed optimization loop. If the optimal design is achieved, which is indicated in the optimization algorithm by breaking-off criteria, the optimization loop is terminated.

3. Optimization Model

3.1. Design model

The design model describes the relationship between the structural variables and the design variables determined by the optimization algorithm. Apart from the shape of the structure, which is not adressed here, the thicknesses and fiber orientations of the single layers of a laminate are available as structural variables in the case of fiber composite structures. With the manufacturing process it is possible to produce variable layer thicknesses and fiber orientations by a regional application of single layers in different parts of the component. In order to fully utilize these manufacturing possibilities in the design model and to guarantee maximum flexibility, the general design model must also allow regionally variable layer thicknesses and fiber orientations. If the Finite-Element- Method is used for structural analysis, the FE-mesh already divides the structure into substructures in the form of finite elements. Within each finite element the fiber orientation and the layer thicknesses are defined independently. Thus, the finite element properties (thicknesses and fiber orientations) can be used for the design model formulation as principally independent structural variables determined by the design variables. (3.1.1) shows a way of describing these structural variables as a function of the design variables:

$$
\begin{bmatrix} t \\ \alpha \end{bmatrix} = \begin{bmatrix} A_t & 0 \\ 0 & A_\alpha \end{bmatrix} x + \begin{bmatrix} t^0 \\ \alpha^0 \end{bmatrix} ; \tag{3.1.1}
$$

with t vector of the layer thicknesses of all finite elements,
        α vector of the layer angles of all finite elements,
        A_t, A_α linking matrices of layer thicknesses and angles,
        x design variable vector,
        t^0, α^0 constant portions of the layer thicknesses and angles.

Arbitrary design models can be defined by the arrangement of the linking matrices A_t and A_α and the vectors of constants t^0 and α^0. It is also possible to link one common design variable with the structural variables of several elements in order to carry out a so-called "variable linking". On the other hand, a structural variable can maximally depend on one design variable which means that each row of the linking matrices contains maximally one coefficient different from zero. The coefficients of the linking matrices and the vectors of constants are chosen in such a way that the design variables take on the dimension "1" in the design space and are precisely "1" in the initial design. This standardization intends to avoid numerical difficulties in the optimization algorithm.

3.2 Evaluation model

The evaluation model serves to describe the requirements on the structure to be optimized. The special behaviour function, which should attain a minimal or maximal value in the optimal design, is chosen as the objective function. With aircraft design it is primarily important to find a design with a minimal structural weight. However, any other state quantity (e.g. costs, stress concentration, etc.) can be considered as objective as well. If there are several objective functions, the problem has to be solved by multicriteria optimization strategies, which are discussed in [1].

All nonobjective requirements on the structural behaviour are formulated as constraints which normally are upper and/or lower bounds on the corresponding state variables. For the design of aircraft structures many different types of design requirements have to be considered and, there still is quite a lot of work in order to combine *all* of the necessary analysis modules and optimization modules within one optimization program. Therefore, the following list contains only those constraint types, which can be considered in the procedure LAGRANGE:

a) strength (failure safety factor),
b) displacements,
c) stability (local buckling) [4,5],
d) dynamic quantities (eigenvalues, eigenvectors, transient- and frequency-response) [6,7],
e) aeroelastic efficiencies [9,11],
f) flutter speeds [9,10].

As an example for the mathematical formulation of the constraints, the flutter speed constraints will be described in the following.

Flutter speed constraints: Due to the interaction of the aerodynamic, the elastic and the inertia forces together with the structural deformation, an exchange of kinetic energy of the air flow and the elastic and kinetic energy of the structure occurs. Depending on the flow speed and the Mach number, energy is taken from the structure or added to it. An increase of the energy of the structure

causes unstable vibrations and, thereby, a destruction of the structure. In the critical case between damping and excitation there is no energy exchange, which
means that small disturbances lead to harmonic vibrations. Depending on the
stiffness and mass distribution of a structure, such a critical case occurs when
certain combinations of flow velocity and Mach number are given. The corresponding critical flow velocity is called flutter speed. Since no flutter case can be
admitted in the whole mission range of an aircraft, it must be required that for
a given altitude of the aircraft (determining the sonic speed) the smallest flutter
speed does not fall short of a certain limit given by the maximal flight speed
plus safety increase:

$$g = \frac{v_F}{v_{min}} - 1 \geq 0 , \qquad (3.2.1)$$

with v_F flutter speed,
         v_{min} maximum flight speed plus safety increase.

The limits for the flutter speed are taken from the so-called flight envelope
which depicts the mission range of an aircraft (maximal flight speeds dependent
on Mach number and altitude) [12].

By means of the constraints on the state variables presented by a) - f), many of
the most important requirements on aircrafts can be formulated. All of these
can be applied to the thickness optimization with LAGRANGE. The extension
of LAGRANGE to the simultaneous optimization of thicknesses and fiber orientations has already been carried out for the constraint types a,b,d (except for
transient- and frequency-response), e and f and will be completed in future.
All the different types of constraints and the objective function form the evaluation model, and together with the design model they completely describe the
optimization model and the design task.

4. Structural and Sensitivity Analysis

The task of the structural analysis is to calculate the state quantities of the structure required to determine the constraint- and objective values defined in the
evaluation model. As mentioned before, most optimization algorithms do not
only require the functional values of the behaviour functions but also their sensitivities with respect to the design variables. The calculation of these sensitivities
can be carried out analytically as well as numerically by means of simple differential quotients. Since in the aircraft design one has large scale design problems
with sometimes several hundreds of design variables, it is necessary, for the sake
of calculation effort and economy, to determine the sensitivities by the analytical
differentiation of all descriptive equations. The following depictions give a brief
survey of the structural and sensitivity analysis. A more detailed description can
be found in [15].
The analysis for the different types of state variables which is part of the optimization loop (Fig. 2.1), is based on the Finite-Element-Method [8]. It starts with
the computation of the element matrices (element-stiffness-matrices, el.-mass-

matrices,el.-load-vectors, strain-displacement-matrices, . .) which depend on the given structural variables y_ν of the current iteration ν (Fig. 4.1). From these the system-matrices can be assembled. The aerodynamic influence matrices C and A are externally calculated with an aerodynamics program. Then they are transformed in LAGRANGE in such a way that they correspond to the degrees of freedom of the finite element model.

Using the system-matrices, the necessary state variables r_ν are evaluated by solving the corresponding equations.

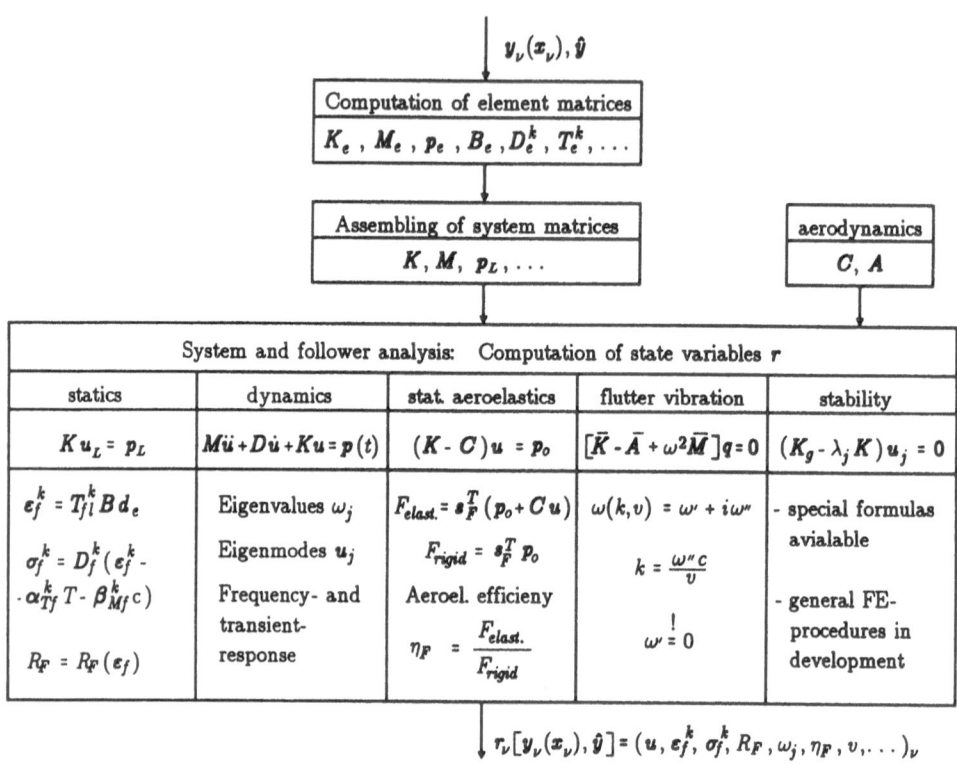

Fig. 4.1: FE-based structural analysis for the evaluation of the different state variables

The analytical evaluation of the constraint- and objective sensitivities requires the differentiation of the descriptive analysis equations. According to the chain-rule all different parts of the optimization loop contribute to the sensitivity analysis (Fig. 4.2). Because of the limited space available, Fig. 4.2 only depicts the sensitivity equations for the constraints on static and eigenvalue state variables, but the sensitivities for the other constraint types are also evaluated analytically.

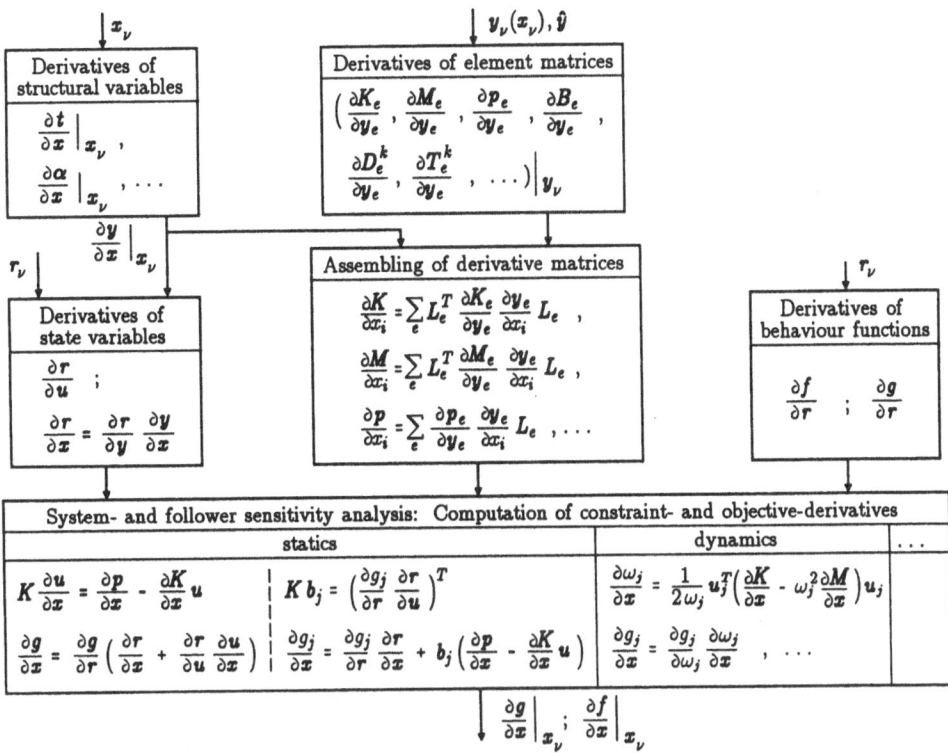

Fig. 4.2.: *Analytical evaluation of constraint sensitivities and objective sensitivities*

5. Application "Composite Fin"

All models and analysis procedures described in the sections before are realized in the optimization procedure LAGRANGE. Using LAGRANGE, the layer thicknesses and layer angles of a fin have been optimized considering constraints on failure safety, aeroelastic efficiencies and flutterspeeds. The objective of this examination was to demonstrate the efficiency of the optimization procedure and also the potential of possible weight savings. In the following the structural model and the optimization model as well as the optimization results will be described briefly.

Structural model: The fin can be subdivided into the main parts stabilizer, tip and rudder (Fig. 5.1). The cover skins of the fin are made of carbon fiber laminate with four different fiber orientations in the stabilizer and three in the rudder. The inner supporting structure is realized by an aluminium honeycomb core. The fin is supported at the connection points to the fuselage and the stiffness of the fuselage is modeled with a general stiffness element. As static load cases the aerodynamic forces of five different flight conditions are chosen. A detailed description of the aerodynamic model and the Finite Element model can be found in [13].

Design model: For the thickness variables, one independent design variable is defined for every layer in every finite element of the cover skins. With this, a variable linking is carried out according to the symmetry of the structure, i.e. the corresponding layer thicknesses of the left and right side of the fin are determined by common design variables. This condition of symmetry also applies to the design model of the fiber orientations. For the optimization of the fiber orientations, one design variable is assigned to each layer of the stabilizer and of the rudder. Altogether there are 102 design variables for the layer thicknesses and 7 for the fiber orientations.

Evaluation model: The structural weight of the fin is chosen as objective function to be minimized. The constraints include stress constraints for the 119 isotropic elements and 252 safety-factor constraints for the composite elements in each load case. In order to guarantee a sufficient performance of the flight-control-system, an aeroelastic lift efficiency of 0.8 for the fin and 0.5 for the rudder are required. Furthermore, the flutter speed must not be lower than 530 m/s when flying at sea level.

FINITE ELEM. MODEL:

Element types: ROD,
TRIA3, GENEL
CQUAD4, QUAD4

No. of elements :	188
No. of nodes :	118
Degr. of freedom:	354
Static load cases :	5

AERODYN. MODEL:

Panels :	88
Mach-numb.:	1.8
Dyn. press. :	91200 N/m²

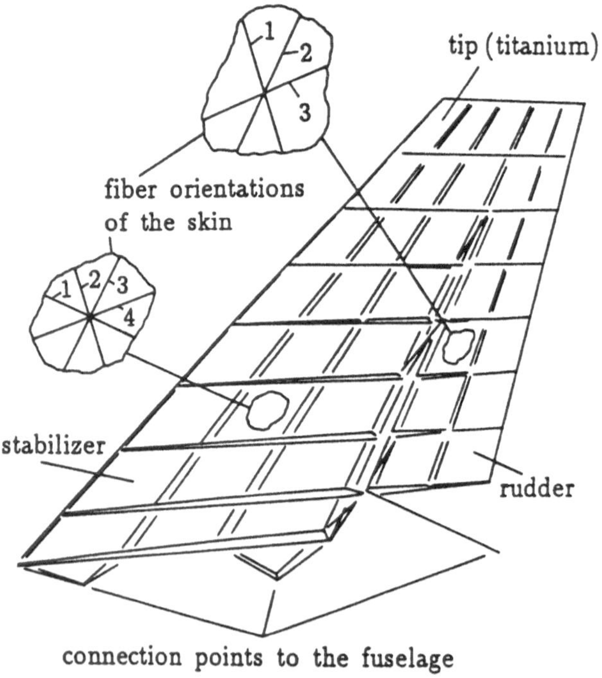

Fig. 5.1: Structural model of the fin (supporting structure: aluminium honeycomb core)

Initial design and optimization results: Different strength constraints as well as the efficiency and flutter constraints are violated in the initial design, whereas the optimized design meets all requirements defined in the evaluation model. The total weight of the structure consists of the following components:

concentrated masses (measuring instruments)	: 53.58 kg
interior structure and tip	: 45.43 kg
initial weight of the cover skins	: 53.98 kg (100.%)
optimal weight of the cover skins	: 34.67 kg (64.2%)

Thus, the optimization process yields a feasible design which additionally has a substantially reduced weight.

6. Conclusion

This paper presents a way of solving design tasks in aircraft construction using a structural optimization method. The simultaneous determination of optimal layer thicknesses and fiber orientations of fiber composite laminates is dealt with in particular. As design criteria, requirements on the static, dynamic, and aeroelastic behaviour of aircrafts are considered. The analysis procedures for the various state variables desribing this behaviour are based on the Finite-Element-Method. Because of the large-scale design problems in aircraft construction the sensitivities of the constraints and objective functions are evaluated analytically. All of the described models and analysis procedures are realized in the optimization program LAGRANGE. The application of this program demonstrates that the design process can be automated very efficiently by the structural optimization method. Another important advantage is the fact that the structural optimization enables to achieve technically optimal design. These advantages have to be payed for by an enormous effort for the development of an efficient and generally applicable optimization program. Nevertheless , this effort is profitable because of the great economical and technical profit which can be obtained by the application of an optimization procedure to the aircraft design.

In order to optimize real-life structures many important procedures and methods are combined in the optimization procedure LAGRANGE. Many further developments, however, must follow. For example, utilizing all design possibilities requires the consideration of shape design variables. Since an optimal design has to fulfill *all* demands on the structure simultaneously, suitable completions and extensions of the structural and sensitivity methods as well as the optimization models (local and global stability, heat transfer, thermal stresses, flight mechanics and control, manufacturing) are futhermore required [14].

For fiber composite materials in particular the characteristic possibilities and requirements of manufacturing must be included in the optimization process in order to guarantee that optimal designs can be produced efficiently by fully utilizing the design potential.

7. References

[1] Eschenauer, H.A.; Koski, J.; Osyczka, A.: Multicriteria Design Optimization. Berlin, Heidelberg, New York, London, Paris, Tokyo, Hong Kong: Springer-Verlag. 1990.

[2] Kneppe, G.: Multicriterion Optimal Layouts of Aircraft and Spacecraft Structures. In [1].

[3] Bremicker, M.; Eschenauer, H.A.; Post, P.U.: Optimization Procedure SAPOP - A General Tool for Multicriteria Structural Designs. In [1].

[4] Hörnlein, H.R.E.M.: Lokale Stabilität als Nebenbedingung in der Strukturoptimierung. COMETT-Seminar "Computerunterstützte Strukturoptimierung". 18.-22. Juni 1990, Thurnau/Bayreuth.

[5] Dobler, W.; Erl, P.; Rapp: Optimization of Sandwich Structures with Respect to Local Instabilities within MBB-LAGRANGE. Proceedings of the NATO Advanced Study Institute "Optimization of Large Structural Systems", Sept. 23 - Oct. 4 1991, Berchtesgaden, Germany.

[6] Ross, C.: Strukturoptimierung mit Nebenbedingungen aus der Dynamik. Dissertation, TU München. VDI-Fortschritt-Berichte, Reihe 20, Nr. 38, VDI-Verlag, Düsseldorf 1991.

[7] Pfeiffer, F.; Kneppe, G.; Ross, C.: Structural Optimization with Constraints from Dynamics in LAGRANGE. Third Air Force / NASA Symposium on recent Advances in Multidisciplinary Analysis and Optimization, September 24-26, 1990, San Francisco, CA. MBB-Report S-PUB-431.

[8] Bathe, K.J.: Finite-Elemente-Methoden. Deutsche Übersetzung von P. Zimmermann. Springer-Verlag, Berlin, Heidelberg, New York, Tokyo, 1986.

[9] Gödel, H.: Optimierung bezüglich Forderungen der stationären und instationären Aeroelastik. Kapitel 4. und 5. aus [12].

[10] Gödel, H.: Recent Developments in Structural Optimization with Respect to Dynamic and Aeroelasticity Problems. Proceedings of the "International forum on aeroelasticity and structural dynamics", June 3-6, 1991, Eurogress-Center, Aachen.

[11] Gödel, H.: Lösen großer linearer Gleichungssysteme. COMETT-Seminar "Computergestützte Strukturoptimierung", 26.- 30.9.1988, Bayreuth.

[12] N.N.: MBB-Lagrange, Theoriehandbuch, Version LAG09, 1990.

[13] Schneider, G.; Krammer, H.; Gödel, H.: Integrated Design Analysis and Optimization - Preliminary Results Fin Optimization. 70th AGARD SMP-Meeting, Sorrent, Italy, 1.-6. April 1990. MBB-Report o.: S-PUB-398.

[14] Sensburg, O.; Schweiger, J.; Gödel, H.; Lotze, A.: The Integration of structural optimization in the general design process for aircraft. 17th Congress International Council of the Aeronautical Sciencis (ICAS), September 9-14, 1990/Stockholm, Sweden. MBB-Report S-PUB-405.

[15] Eschenauer, H.A.; Schuhmacher, G.; Hartzheim, H.: Optimization of fiber composite structures in aircraft construction with the optimization procedure LAGRANGE. In: J. of. Computers & Structures (in review).

SHAPE OPTIMAL DESIGN OF AXISYMMETRIC SHELL STRUCTURES

C. A. MOTA SOARES*, J. INFANTE BARBOSA** and C. M. MOTA SOARES*
* CEMUL/IST, Av. Rovisco Pais, 1096 Lisboa Codex, PORTUGAL.
** ENIDH, Paço de Arcos, 2780 Oeiras, PORTUGAL.

ABSTRACT. In this section the sensitivity analysis for the optimization of axisymmetric shells subjected to static and dynamic constraints is presented. Thickness and shape design variables are considered. The model is based on a two node frustum conical finite element based on Love-Kirchhoff assumptions. The objective of the design is the minimization of the volume of the shell material or the maximization of the fundamental natural frequency or the minimization of the maximum stresses. The constraint functions are the displacements, stresses, enclosed volume of the structure, volume of shell material or the natural frequency of a specified mode shape. The sensitivities are calculated by analytical, semi-analytical and global finite difference techniques. The efficiency and accuracy of the models developed are discussed with reference to the applications.

1. INTRODUCTION

Structural optimization using finite element techniques requires the sequential use of structural and sensitivity analyses combined with a numerical optimizer. The success of the structural optimization process depends on the proper choices with respect to the finite element model, sensitivity analysis, objective function, constraints, design variables and method of solution of the nonlinear mathematical problem.

In this section the structural and sensitivity analyses of thin axisymmetric shell structures are presented using a frustum-cone finite element with 8 degrees of freedom, based on Love-Kirchhof assumptions. In the structural analysis program, the calculation of the sensitivities of displacements, stresses and natural frequency with respect to perturbations in the design variables are included. These sensitivities are evaluated using the semi-analytical method as presented by Gendong and Yingwei (1987) and Barthelemy et al. (1988). In this method the sensitivities of the stiffness or mass matrices or load vector, are obtained by a finite difference approach at element level considering a small perturbation of the design variables. Alternatively the sensitivities are evaluated analytically, using a symbolic manipulator to overcome the difficulty of obtaining the sensitivities when the design variables are nodal coordinates. It is also possible to calculate the sensitivities using a global finite difference technique. In this case two complete structural analyses for each design variable are necessary and they are

1023

G. I. N. Rozvany (ed.), Optimization of Large Structural Systems, Vol. II, 1023–1049.
© 1993 Kluwer Academic Publishers.

shown here only for comparison purposes.

The formulation presented is applied to the minimum weight design of thin axisymmetric shell structures subjected to constraints on displacements, stresses, natural frequency, volume of the shell material and enclosed volume of the structure. The minimization of maximum stresses or the maximization of a natural frequency of a specified mode shape can be also carried out. For static constraints the adjoint structure technique is used as presented in Haftka and Kamat (1987). The design variables are thicknesses and/or radial nodal coordinates. The ADS (Automated Design Synthesis) program of Vanderplaats (1984) is used to solve the nonlinear mathematical programming problem.

2. THIN AXISYMMETRIC SHELLS

2.1. KINEMATICS

Structural analysis of axisymmetric shells using finite element methods requires a discretized model where the complete shell can be idealized as a series of shell ring elements joined at their nodal point circles. Its behaviour will be characterized by the displacements of these nodal circles which are described in terms of a finite number of displacement variables or generalized displacements.

For an arbitrary shell the strain-displacement relations for small displacements in an orthogonal curvilinear system are given by (Kraus, 1967) :

$$\epsilon_i = \frac{\partial}{\partial \alpha_i}\left(\frac{u_i}{\sqrt{g_i}}\right) + \frac{1}{2\,g_i}\sum_{k=1}^{3}\frac{\partial g_i}{\partial \alpha_k}\frac{u_k}{\sqrt{g_k}} \qquad (i=1,2,3) \tag{1}$$

$$\gamma_{ij} = \frac{1}{\sqrt{g_i\,g_j}}\left[g_i\frac{\partial}{\partial \alpha_j}\left(\frac{u_i}{\sqrt{g_i}}\right) + g_j\frac{\partial}{\partial \alpha_i}\left(\frac{u_j}{\sqrt{g_j}}\right)\right] \qquad (i,\,j=1,2,3\;;\;i\neq j) \tag{2}$$

where ϵ_i are the normal strains, α_i the curvilinear coordinates, u_i the displacement components, g_i the first fundamental magnitudes and γ_{ij} the shear strains.

Considering the Love-Kirchhoff approximation of the theory of thin elastic shells which is based on the postulation that the shell is thin, the deflections of the shell are small, the transverse normal stress is negligible and normals to the reference surface of the shell remain normal to it and undergo no change in length during deformation. For a conical shell (Fig. 1) represented by its reference surface of revolution, the displacement components are represented as :

$$u_1 = U(S,\,\theta,\,\xi)\quad;\qquad u_2 = V(S,\,\theta,\,\xi)\quad;\qquad u_3 = W(S,\,\theta,\,\xi)$$

For this particular situation and assuming for thin shells $\xi/R_i \simeq 0$, where $1/R_i$ are the principal curvatures, equations (1) and (2) can be represented as :

$$\epsilon_{SS} = \frac{\partial U}{\partial S}$$
$$\epsilon_{\theta\theta} = \frac{1}{r}\left(\frac{\partial V}{\partial \theta} + U\cos\phi + W\sin\phi\right)$$
$$\epsilon_{\xi\xi} = \frac{\partial W}{\partial \xi} \tag{3}$$

$$\gamma_{S\theta} = \frac{\partial V}{\partial S} + \frac{1}{r}\left(\frac{\partial U}{\partial \theta} - V \cos \phi\right)$$

$$\gamma_{S\xi} = \frac{\partial W}{\partial S} + \frac{\partial U}{\partial \xi}$$

$$\gamma_{\theta\xi} = \frac{1}{r}\frac{\partial W}{\partial \theta} + r\frac{\partial}{\partial \xi}\left(\frac{V}{r}\right)$$

$$U = U(S, \theta, \xi)$$
$$V = V(S, \theta, \xi)$$
$$W = W(S, \theta, \xi)$$

where U, V, W are the components of the displacement vector of a spatial point and S, θ, ξ are, respectively, the coordinates along the meridian, parallel circle and normal to the reference surface of the shell (Fig. 1).

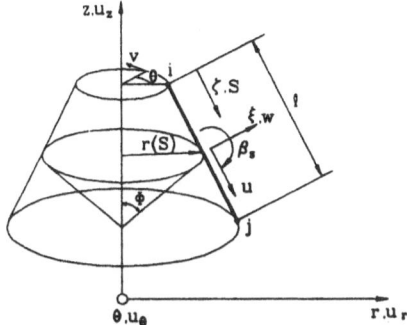

Fig. 1 - Frustum-Cone Finite Element. Geometry and Displacements.

Assuming the following displacement distribution :

$$U(S,\theta,\xi) = u(S,\theta) + \xi \frac{\partial U}{\partial \xi}\Big|_{\xi=0}$$

$$V(S,\theta,\xi) = v(S,\theta) + \xi \frac{\partial V}{\partial \xi}\Big|_{\xi=0} \qquad (4)$$

$$W(S,\theta,\xi) = w(S,\theta)$$

where $u(S,\theta)$, $v(S,\theta)$ and $w(S,\theta)$ represent the components of the displacement vector of a point on the reference middle surface of the shell and $\frac{\partial U}{\partial \xi}\Big|_{\xi=0}$ and $\frac{\partial V}{\partial \xi}\Big|_{\xi=0}$ represent the rotations of tangents to the reference surface oriented along the lines S and θ, respectively. Let $\frac{\partial U}{\partial \xi}\Big|_{\xi=0} = \beta_S(S,\theta)$ and $\frac{\partial V}{\partial \xi}\Big|_{\xi=0} = \beta_\theta(S,\theta)$ and using the Love-Kirchhoff assumptions $(\gamma_{S\xi} = \gamma_{\theta\xi} = 0)$, one obtains :

$$\beta_S = -\frac{\partial w}{\partial S}$$

$$\beta_\theta = \frac{v}{r}\sin\phi - \frac{1}{r}\frac{\partial w}{\partial \theta} \qquad (5)$$

The displacement vector $\underline{U} = [\ U \quad V \quad W\ ]^T$ of a given point (S, θ, ξ) can be expressed in terms of the displacement vector $U = [u, v, w]^T$ of the reference surface and the rotations of tangents (β_S, β_θ) of the same reference surface oriented along the parametric lines as :

$$
\underline{U} = \begin{bmatrix} u \\ v \\ w \end{bmatrix} + \begin{bmatrix} \xi & 0 \\ 0 & \xi \\ 0 & 0 \end{bmatrix} \begin{bmatrix} \beta_S \\ \beta_\theta \end{bmatrix}
\tag{6}
$$

Substituting relations (6) into the remainder of equations (3), the non-vanishing strains in the thin elastic shell are given by :

$$
\epsilon = \epsilon^\circ + \xi \chi
\tag{7}
$$

with:

$$
\epsilon = [\, \epsilon_{SS} \ \epsilon_{\theta\theta} \ \gamma_{S\theta} \,]^T \quad ; \quad \epsilon^\circ = [\, \epsilon^\circ_{SS} \ \epsilon^\circ_{\theta\theta} \ \gamma^\circ_{S\theta} \,]^T \quad ; \quad \chi = [\, \chi_{SS} \ \chi_{\theta\theta} \ \chi_{S\theta} \,]^T
$$

and:

$$
\epsilon^\circ = \Delta_m\, U
\tag{8}
$$

$$
\chi = \Delta_f\, U
\tag{9}
$$

where the operators Δ_m and Δ_f are, respectively :

$$
\Delta_m = \begin{bmatrix} \dfrac{\partial}{\partial S} & 0 & 0 \\[2mm] \dfrac{\cos\phi}{r} & \dfrac{1}{r}\dfrac{\partial}{\partial\theta} & \dfrac{\sin\phi}{r} \\[2mm] \dfrac{1}{r}\dfrac{\partial}{\partial\theta} & \left(\dfrac{\partial}{\partial S} - \dfrac{\cos\phi}{r}\right) & 0 \end{bmatrix}
$$

$$
\Delta_f = \begin{bmatrix} 0 & 0 & -\dfrac{\partial^2}{\partial S^2} \\[3mm] 0 & \dfrac{\sin\phi}{r^2}\dfrac{\partial}{\partial\theta} & -\dfrac{1}{r^2}\left(\dfrac{\partial^2}{\partial\theta^2} + r\cos\phi\,\dfrac{\partial}{\partial S}\right) \\[3mm] 0 & \dfrac{2\sin\phi}{r^2}\left(r\dfrac{\partial}{\partial S} - \cos\phi\right) & \dfrac{2}{r^2}\left(\cos\phi\,\dfrac{\partial}{\partial\theta} - r\dfrac{\partial^2}{\partial S\,\partial\theta}\right) \end{bmatrix}
$$

The quantities ϵ°_{SS} , $\epsilon^\circ_{\theta\theta}$, $\gamma^\circ_{S\theta}$ represent, respectively, the meridional, circumferential and shearing strains of the reference surface. The quantities χ_{SS} and $\chi_{\theta\theta}$ represent the changes in the curvature of the reference surface and $\chi_{S\theta}$ represents the torsion of the same surface during deformation.

Assuming that the displacements u, v, w can be expanded in Fourier series of the type :

$$
U = \sum_{n=0}^{N} \left(C_n\, U_n + \hat{C}_n\, \hat{U}_n \right)
\tag{10}
$$

where:

$$\mathbb{C}_n = \begin{bmatrix} \cos n\theta & 0 & 0 \\ 0 & \sin n\theta & 0 \\ 0 & 0 & \cos n\theta \end{bmatrix} \quad ; \quad \hat{\mathbb{C}}_n = \begin{bmatrix} \sin n\theta & 0 & 0 \\ 0 & \cos n\theta & 0 \\ 0 & 0 & \sin n\theta \end{bmatrix}$$

$$\mathbb{U}_n = [\, u_n \ v_n \ w_n \,]^T \quad ; \quad \hat{\mathbb{U}}_n = [\, \hat{u}_n \ \hat{v}_n \ \hat{w}_n \,]^T$$

The first and second terms of equation (10) represent these components of displacements which are, respectively, symmetric and antisymmetric with respect to the plane passing through $\theta=0$, u_n, v_n and w_n being the amplitudes of symmetric part and $\hat{u}_n$, $\hat{v}_n$ and $\hat{w}_n$ the amplitudes of antisymmetric part for the nth harmonic and N is the number of terms in the truncated Fourier series.

Since the angular dependence of displacement components is expressed in terms of trigonometric functions, the orthogonality properties of such functions yields a formulation of the problem as a series of uncoupled quasi two-dimensional problems, in which the displacement amplitudes u_n, v_n, w_n, $\hat{u}_n$, $\hat{v}_n$ and $\hat{w}_n$ are the unknowns. The meridional dependence of these displacements amplitudes along the frustum cone represented here only by the symmetric part, for sake of simplicity, can be assumed as :

$$\mathbb{U}_n = \mathsf{N} \, q_{e_n}^l \tag{11}$$

where :

$$\mathsf{N} = \begin{bmatrix} N_1 & 0 & 0 & 0 & N_2 & 0 & 0 & 0 \\ 0 & N_1 & 0 & 0 & 0 & N_2 & 0 & 0 \\ 0 & 0 & N_3 & N_4 & 0 & 0 & N_5 & N_6 \end{bmatrix} \quad ; \quad \begin{aligned} N_1 &= (1-\zeta) \\ N_2 &= \zeta \\ N_3 &= (1-3\,\zeta^2+2\,\zeta^3) \\ N_4 &= (\zeta-2\,\zeta^2+\zeta^3)\,\ell \\ N_5 &= (3\,\zeta^2-2\,\zeta^3) \\ N_6 &= (-\zeta^2+\zeta^3)\,\ell \end{aligned} \tag{12}$$

$\zeta = \dfrac{S}{\ell}$ - Coordinate over the length of the frustum-cone element.

$\ell = \sqrt{(r^j-r^i)^2+(z^j-z^i)^2}$ - Length of the frustum-cone element

$r = (1-\zeta)\,r^i+\zeta\,r^j$; r^i, r^j - Radial coordinates of nodes i and j

$q_{e_n}^l = [\, u_n^i \ v_n^i \ w_n^i \ \dfrac{dw_n^i}{dS} \ u_n^j \ v_n^j \ w_n^j \ \dfrac{dw_n^j}{dS} \,]^T$ - Vector of amplitudes displacement components in local coordinates

Substitution of equations (10) and (11) into equation (8) yields for the nth harmonic:

$$\epsilon_{m_n} = B_{m_n}^* \, q_{e_n}^l \tag{13}$$

being :

$$\epsilon_{m_n} = \begin{bmatrix} \epsilon_{SS_n}^\circ \\ \epsilon_{\theta\theta_n}^\circ \\ \gamma_{S\theta_n}^\circ \end{bmatrix} \quad ; \quad B_{m_n}^* = \begin{bmatrix} B_{11} & 0 & 0 & 0 & -B_{11} & 0 & 0 & 0 \\ B_{21} & B_{22} & B_{23} & B_{24} & B_{25} & B_{26} & B_{27} & B_{28} \\ B_{31} & B_{32} & 0 & 0 & B_{35} & B_{36} & 0 & 0 \end{bmatrix} \tag{14}$$

where $\epsilon_{SS_n}^{\circ}$, $\epsilon_{\theta\theta_n}^{\circ}$ and $\gamma_{S\theta_n}^{\circ}$ represent, respectively, the amplitudes for the nth harmonic of the meridional, circumferential and shearing strains of the reference surface. The $B_{m_n}^*$ are derived by applying the strain operator Δ_m to the displacement shape functions yielding :

$$B_{11} = -\tfrac{1}{\ell} \cos n\theta \quad ; \quad B_{21} = \tfrac{N_1}{r} \cos \phi \cos n\theta$$

$$B_{22} = \tfrac{N_1}{r} n \cos n\theta \quad ; \quad B_{23} = \tfrac{N_3}{r} \sin \phi \cos n\theta$$

$$B_{24} = \tfrac{N_4}{r} \sin \phi \cos n\theta \quad ; \quad B_{25} = \tfrac{N_2}{r} \cos \phi \cos n\theta$$

$$B_{26} = \tfrac{N_2}{r} n \cos n\theta \quad ; \quad B_{27} = \tfrac{N_5}{r} \sin \phi \cos n\theta \qquad (15)$$

$$B_{28} = \tfrac{N_6}{r} \sin \phi \cos n\theta \quad ; \quad B_{31} = -\tfrac{N_1}{r} n \sin n\theta$$

$$B_{35} = -\tfrac{N_2}{r} n \sin n\theta \quad ; \quad B_{32} = -(\tfrac{1}{\ell} + \tfrac{N_1}{r} \cos \phi) \sin n\theta$$

$$B_{36} = (\tfrac{1}{\ell} - \tfrac{N_2}{r} \cos \phi) \sin n\theta$$

Considering the transformation matrix L relating nodal local coordinates (S, θ, ξ) to nodal global coordinates (r, θ, z) (Fig. 1), one obtains the relationship between the displacements amplitudes in the local referential $(q_{e_n}^l)$ and the displacements amplitudes in the global referential (q_{e_n}) :

$$q_{e_n}^l = L \, q_{e_n} \qquad (16)$$

where :

$$q_{e_n} = [\, u_{r_n}^i \;\; u_{z_n}^i \;\; u_{\theta_n}^i \;\; \tfrac{dw_n^i}{dS} \;\; u_{r_n}^j \;\; u_{z_n}^j \;\; u_{\theta_n}^j \;\; \tfrac{dw_n^j}{dS} \,]^T$$

$$L = \begin{bmatrix} [\,A\,] & [\,0\,] \\ [\,0\,] & [\,A\,] \end{bmatrix}$$

$$[\,A\,] = \begin{bmatrix} \cos \phi & -\sin \phi & 0 & 0 \\ 0 & 0 & 1 & 0 \\ \sin \phi & \cos \phi & 0 & 0 \\ 0 & 0 & 0 & 1 \end{bmatrix} \quad ; \quad [\,0\,] = \begin{bmatrix} 0 & 0 & 0 & 0 \\ 0 & 0 & 0 & 0 \\ 0 & 0 & 0 & 0 \\ 0 & 0 & 0 & 0 \end{bmatrix}$$

Substituting equation (16) into the strain-displacement relation (eq. 13), one obtains the membrane terms of the strain-displacement in terms of the element degrees of freedom of the nth harmonic as:

$$\epsilon_{m_n} = B_{m_n} q_{e_n} \qquad (17)$$

where :

$$B_{m_n} = B^*_{m_n} \, L \tag{18}$$

For the bending terms the procedure is identical, yielding :

$$\chi_{f_n} = B^*_{f_n} \, q^l_{e_n} \tag{19}$$

where :

$$\chi_{f_n} = \begin{bmatrix} \chi_{SS_n} \\ \chi_{\theta\theta_n} \\ \chi_{S\theta_n} \end{bmatrix} \quad ; \quad B^*_{f_n} = \begin{bmatrix} 0 & 0 & B_{13} & B_{14} & 0 & 0 & -B_{13} & B_{18} \\ 0 & B_{22} & B_{23} & B_{24} & 0 & B_{26} & B_{27} & B_{28} \\ 0 & B_{32} & B_{33} & B_{34} & 0 & B_{36} & B_{37} & B_{38} \end{bmatrix}$$

and :

$$B_{13} = \frac{(6-12\zeta)}{\ell^2} \cos n\theta \qquad\qquad ; \; B_{14} = \frac{(4-6\zeta)}{\ell} \cos n\theta$$

$$B_{18} = \frac{(2-6\zeta)}{\ell} \cos n\theta \qquad\qquad ; \; B_{22} = \frac{N_1 \, n \, \sin\phi}{r^2} \cos n\theta$$

$$B_{23} = \left(\frac{n^2}{r^2} N_3 + \frac{\cos\phi}{\ell \, r}(6\zeta-6\zeta^2)\right)\cos n\theta \quad ; \; B_{24} = \left(\frac{n^2 N_4}{r^2} + \frac{(-1+4\zeta-3\zeta^2)\cos\phi}{r}\right)\cos n\theta$$

$$B_{26} = \frac{N_2 \, n \, \sin\phi}{r^2} \cos n\theta \qquad\qquad ; \; B_{27} = \left(\frac{n^2}{r^2} N_5 - \frac{\cos\phi}{\ell \, r}(6\zeta-6\zeta^2)\right)\cos n\theta$$

$$B_{28} = \left(\frac{n^2 N_6}{r^2} + \frac{(2\zeta-3\zeta^2)\cos\phi}{r}\right)\cos n\theta \; ; \; B_{32} = \left(-\frac{2\sin\phi}{r\,\ell} - \frac{2\,N_1\cos\phi\,\sin\phi}{r^2}\right)\sin n\theta$$

$$B_{33} = \left(-\frac{2n}{r^2} N_3 \cos\phi - \frac{2n}{\ell \, r}(6\zeta-6\zeta^2)\right)\sin n\theta$$

$$B_{36} = \left(\frac{2\sin\phi}{r\,\ell} - \frac{2N_2\sin\phi\,\cos\phi}{r^2}\right)\sin n\theta$$

$$B_{34} = \left(-\frac{2\,n\,N_4}{r^2}\cos\phi + \frac{2\,n\,(1-4\zeta+3\zeta^2)}{r}\right)\sin n\theta$$

$$B_{37} = \left(\frac{2n}{\ell \, r}(6\zeta-6\zeta^2) - \frac{2n}{r^2} N_5 \cos\phi\right)\sin n\theta$$

$$B_{38} = \left(\frac{2\,n\,(-2\zeta+3\zeta^2)}{r} - \frac{2\,n\,N_6}{r^2}\cos\phi\right)\sin n\theta$$

Similarly if one considers the transformation of coordinates (L) of the local referential (S, θ, ξ) to the global referential (r, θ, z), the changes of curvatures for the nth harmonic can be represented as :

$$\chi_{f_n} = B_{f_n} \, q_{e_n} \tag{20}$$

where :

$$B_{f_n} = B^*_{f_n} \, L \tag{21}$$

2.2. CONSTITUTIVE EQUATIONS

For a linear elastic solid, considering infinitesimal deformation and orthotropic materials the constitutive equation can be written as :

$$\tau = \mathcal{C} \, \varepsilon \tag{22}$$

where :

$$\tau = [\tau_{SS} \ \tau_{\theta\theta} \ \tau_{\xi\xi} \ \tau_{S\theta} \ \tau_{S\xi} \ \tau_{\theta\xi}]^T$$

$$\varepsilon = [\epsilon_{SS} \ \epsilon_{\theta\theta} \ \epsilon_{\xi\xi} \ \epsilon_{S\theta} \ \epsilon_{S\xi} \ \epsilon_{\theta\xi}]^T$$

and $\mathcal{C}$ is the constitutive matrix for the three-dimensional linear elastic solid.

As a consequence of the Love theory expressed by $\gamma_{S\xi}=\gamma_{\theta\xi}=\epsilon_{\xi\xi}=\tau_{\xi\xi}=0$, the system of stress-strain relations for thin orthotropic axisymmetric shells can be reduced to :

$$\tau = C \, \epsilon \tag{23}$$

where:

$$\tau = [\tau_{SS} \ \tau_{\theta\theta} \ \tau_{S\theta}]^T$$

$$C = \begin{bmatrix} E^*_S & \nu_{\theta S} E^*_\theta & 0 \\ \nu_{S\theta} E^*_\theta & E^*_\theta & 0 \\ 0 & 0 & G_{S\theta} \end{bmatrix}$$

$$E^*_\theta = \frac{E_\theta}{1 - \nu_{S\theta} \nu_{\theta S}} \ ; \qquad E^*_S = \frac{E_S}{1 - \nu_{S\theta} \nu_{\theta S}}$$

where E_S, E_θ, $G_{S\theta}$, $\nu_{S\theta}$, $\nu_{\theta S}$ are Young's moduli, shear modulus and Poisson's ratio for the material referred to the S and θ directions. Substitution of equations (7) in equations (23) yields :

$$\tau = C \, \epsilon^\circ + \xi \, C \, \chi \tag{24}$$

Integrating the stress distribution across the thickness of the shell by neglecting $\xi/R_i \simeq 0$ one obtains :

$$\tau = \frac{1}{h} \mathcal{N} + \xi \frac{12}{h^3} \mathcal{M} \tag{25}$$

where:

$$\mathcal{N} = [\mathcal{N}_{SS} \ \mathcal{N}_{\theta\theta} \ \mathcal{N}_{S\theta}]^T \qquad \text{(Membrane resultants)}$$

$$\mathcal{M} = [\; \mathcal{M}_{SS} \quad \mathcal{M}_{\theta\theta} \quad \mathcal{M}_{S\theta} \;]^{\mathrm{T}} \quad \text{(Bending moments)}$$

and :

$$\mathcal{N} = \int_{-h/2}^{h/2} \tau \; d\xi \; = D_m \sum_{n=0}^{N} \epsilon_{m_n} \tag{26}$$

$$\mathcal{M} = \int_{-h/2}^{h/2} \tau \; \xi \; d\xi \; = D_f \sum_{n=0}^{N} \chi_{f_n} \tag{27}$$

where D_m and D_f are the membrane and bending constitutive matrices given by :

$$D_m = h \; C \qquad ; \qquad D_f = \frac{h^3}{12} C$$

2.3. KINETIC ENERGY

The kinetic energy T^e of the eth element is given by the expression :

$$T^e = \frac{1}{2} \int \rho \; \underline{\dot{U}}^{\mathrm{T}} \underline{\dot{U}} \; d\Omega = \frac{1}{2} \int_0^{2\pi} \int_0^1 \int_{-h/2}^{h/2} \rho \; \underline{\dot{U}}^{\mathrm{T}} \underline{\dot{U}} \; \ell \; r \; d\xi \; d\zeta \; d\theta \tag{28}$$

where $\underline{\dot{U}} = \frac{d\underline{U}}{dt}$, $d\Omega$ is the elementary volume and ρ is the mass per unit of volume. Substituting equation (5) into equation (6) and integrating over the thickness of the element one obtains :

$$T^e = \frac{1}{2}\rho \int_0^{2\pi} \int_0^1 h \; \dot{U}^{\mathrm{T}} \dot{U} \; \ell \; r \; d\zeta \; d\theta \; + \; \frac{1}{2}\frac{\rho}{12} \int_0^{2\pi} \int_0^1 h^3 \; \kappa^{\mathrm{T}} \; \kappa \; \ell \; r \; d\zeta \; d\theta \tag{29}$$

where :

$$\dot{U} = \frac{d}{dt}[\; u \quad v \quad w\;]^{\mathrm{T}}$$

$$\kappa = \frac{d}{dt} [\; -\frac{\partial w}{\partial S} \quad \frac{1}{r}(v \sin\phi - \frac{\partial w}{\partial \theta}) \quad 0 \;]^{\mathrm{T}}$$

In equation (29) the first term represents the translational kinetic inertia and the second term represents the rotational kinetic inertia. Its important to note that the coupled terms of Fourier series of the type $(\sin m\theta \;\; \sin n\theta)$ and $(\cos m\theta \;\; \cos n\theta)$ for $m \neq n$ are not considered by making use of the orthogonality of the harmonic functions in the interval $0 \leq \theta \leq 2\pi$ and because the thickness is uniform along θ, $(h(\theta)=\text{constant})$, those terms are equal to zero resulting for the integration in θ :

$$\int_0^{2\pi} \cos n\theta \; \cos m\theta \; d\theta = \mathcal{A}_1$$

$$\int_0^{2\pi} \cos n\theta \; \sin m\theta \; d\theta = 0$$

$$\begin{array}{lll} \mathcal{A}_1 = \mathcal{A}_2 = 0 & \text{for } m \neq n & \\ \mathcal{A}_1 = 2\pi \; ; \; \mathcal{A}_2 = 0 & \text{for } m=n=0 & (30) \\ \mathcal{A}_1 = \mathcal{A}_2 = \pi & \text{for } m=n>0 & \end{array}$$

$$\int_0^{2\pi} \sin n\theta \; \sin m\theta \; d\theta = \mathcal{A}_2$$

Substituting equations (6), (10) and (11) in the equation (29) for the frustum-cone finite element one obtains :

$$T^e = \tfrac{1}{2}\rho \sum_{n=0}^{N} \int_0^{2\pi}\int_0^1 h \, (N_n^* L \, \dot{q}_{e_n})^T \, (N_n^* L \, \dot{q}_{e_n}) \, \ell \, r \, d\zeta \, d\theta \; +$$

$$+ \tfrac{1}{2}\tfrac{\rho}{12} \sum_{n=0}^{N} \int_0^{2\pi}\int_0^1 h^3 \, (R_n^* L \, \dot{q}_{e_n})^T \, (R_n^* L \, \dot{q}_{e_n}) \, \ell \, r \, d\zeta \, d\theta \qquad (31)$$

where :

$$\dot{q}_{e_n} = \frac{d}{dt}\, q_{e_n} \; ; \quad N_n^* = C_n N \; \text{ and } \; R_n^* = C_n R_n$$

in which :

$$
R_n =
\begin{bmatrix}
0 & 0 & \dfrac{-\partial N_3}{\partial S} & \dfrac{-\partial N_4}{\partial S} & 0 & 0 & \dfrac{-\partial N_5}{\partial S} & \dfrac{-\partial N_6}{\partial S} \\[2mm]
0 & \dfrac{N_1}{r}\sin\phi & \dfrac{n\,N_3}{r} & \dfrac{n\,\ell\,N_4}{r} & 0 & \dfrac{N_2}{r}\sin\phi & \dfrac{n\,N_5}{r} & \dfrac{n\,\ell\,N_6}{r} \\[2mm]
0 & 0 & 0 & 0 & 0 & 0 & 0 & 0
\end{bmatrix}
$$

The kinetic energy can be represented in a simplified form as :

$$T^e = \tfrac{1}{2} \sum_{n=0}^{N} \left(\dot{q}_{e_n}^T \left(M_{n_T}^e + M_{n_R}^e \right) \dot{q}_{e_n} \right) = \tfrac{1}{2} \sum_{n=0}^{N} \left(\dot{q}_{e_n}^T \, M_n^e \, \dot{q}_{e_n} \right) \qquad (32)$$

where M_n^e is the element mass matrix for the nth harmonic represented by :

$$M_n^e = \rho \int_0^{2\pi}\int_0^1 h \, (N_n^* L)^T \, (N_n^* L) \, \ell \, r \, d\zeta \, d\theta + \frac{\rho}{12} \int_0^{2\pi}\int_0^1 h^3 (R_n^* L)^T \, (R_n^* L) \, \ell \, r \, d\zeta \, d\theta \qquad (33)$$

The element mass matrices M_n^e are full (8x8) matrices which are evaluated taking into consideration the orthogonality properties of the trigonometric functions (eq. 30). The simplest mathematical model for inertia properties of structural elements is the lumped-mass representation. In this model the mass properties of the system are separated from the elastic properties and equivalent concentrated masses are placed at the nodal points to represent the inertia forces in the direction of the assumed element degree of freedom. These masses refer to both translation and rotational inertia of the element. This assumption excludes dynamic coupling between the element displacement and the resulting mass matrix is purely diagonal, thus implying a computer time reduction and a decrease in computer storage. The diagonal mass matrix is obtained from the full consistent mass matrix by adding the off diagonal terms to the appropriate diagonal elements (e.g. Desai and Abel, 1972).

2.4. STRAIN ENERGY

The strain energy of the εth element is represented by the expression :

$$E^e = \frac{1}{2} \int_\Omega \tau^T \epsilon \, d\Omega = \frac{1}{2} \int_0^{2\pi} \int_0^1 \int_{-h/2}^{h/2} \epsilon^T C \epsilon \, r \, \ell \, d\xi \, d\zeta \, d\theta \tag{34}$$

Introducing equation (7) in the above equation and taking advantage of the orthogonality properties of the trigonometric functions one obtains :

$$E^e = \frac{1}{2} \sum_{n=0}^N q_{e_n}^T K_n^e q_{e_n} \tag{35}$$

with :

$$K_n^e = K_{m_n}^e + K_{f_n}^e \tag{36}$$

$$K_{m_n}^e = \int_0^{2\pi} \int_0^1 B_{m_n}^T D_m B_{m_n} \, r \, \ell \, d\zeta \, d\theta \tag{37}$$

$$K_{f_n}^e = \int_0^{2\pi} \int_0^1 B_{f_n}^T D_f B_{f_n} \, r \, \ell \, d\zeta \, d\theta \tag{38}$$

where $K_{m_n}^e$ and $K_{f_n}^e$ are the membrane and bending terms of the element stiffness matrix for the nth harmonic. These matrices are evaluated analytically in the θ direction taking into consideration the orthogonality properties of the trigonometric functions. In the ζ direction the integration is carried out by Gaussian Quadrature formulae.

2.5. EXTERNAL WORK

Expanding the surface loads $p = [p_u \ p_\theta \ p_w]^T$ in Fourier series and using the assumption of orthogonality of the trigonometric functions on the interval $0 \le \theta \le 2\pi$, the external work becomes :

$$W^e = \int_0^{2\pi} \int_0^1 U^T \underline{p} \, r \, \ell \, d\zeta \, d\theta \tag{39}$$

where:

$$\underline{p} = \sum_{n=0}^N C_n \underline{p}^n \quad ; \quad \underline{p}^n = [\, p_u^n \ p_\theta^n \ p_w^n \,]^T$$

Assuming a linear dependence between the above magnitudes of the forces vector and the components of these vector on the nodes of the frustum-cone finite element, one obtains :

$$\underline{p}^n = \Re \, \underline{p}_{e_n}^l \tag{40}$$

where :

$$p^l_{e_n} = [\; p^n_{u_i} \;\; p^n_{\theta_i} \;\; p^n_{w_i} \;\; M^n_i \;\; p^n_{u_j} \;\; p^n_{\theta_j} \;\; p^n_{w_j} \;\; M^n_j \;]^T$$

$$\mathfrak{N} = \begin{bmatrix} N_1 & 0 & 0 & 0 & N_2 & 0 & 0 & 0 \\ 0 & N_1 & 0 & 0 & 0 & N_2 & 0 & 0 \\ 0 & 0 & N_1 & 0 & 0 & 0 & N_2 & 0 \end{bmatrix} \tag{41}$$

Considering the equations (6), (10), (11) and (16) and substituting in equation (39) one obtains :

$$W^e = \sum_{n=0}^{N} \int_0^{2\pi} \int_0^1 \left((C_n \, N \, L \, q_{e_n})^T \, C_n \, \mathfrak{N} \, p^l_{e_n} \right) r \, \ell \, d\zeta \, d\theta \tag{42}$$

Using $N^*_n = C_n \, N$ and making $\mathfrak{N}^*_n = C_n \, \mathfrak{N}$, yields :

$$W^e = \sum_{n=0}^{N} q^T_{e_n} \, Q^e_n \tag{43}$$

where :

$$Q^e_n = \int_0^{2\pi} \int_0^1 L^T \, N^{*T}_n \, \mathfrak{N}^*_n \, p^l_{e_n} \, r \, \ell \, d\zeta \, d\theta \tag{44}$$

is the load vector of the element for the nth harmonic which is consistent with the assumed displacement field used in deriving mass and stiffness matrices. Vectors Q^e_n are evaluated analytically in the θ direction taking into consideration the orthogonality properties of the trigonometric functions. In the ζ direction the integration may be carried out by Gaussian Quadrature formulae or alternatively by using a symbolic manipulator.

2.6. EQUATIONS OF MOTION

Problems of static equilibrium are governed by the variational principles such a minimum total potential energy, while the dynamic equilibrium are most simply formulated in terms of Hamilton's variational principle, defining a Lagrangian function through the expression :

$$\mathcal{L} = T^e - V^e \tag{45}$$

with :

$$V^e = E^e - W^e \tag{46}$$

Substituting the values of T^e, E^e and W^e given by the equations (32), (35) and (43) in the Lagrangian function $\mathcal{L}$, one obtains :

$$\mathcal{L} = \sum_{n=0}^{N} \left(\tfrac{1}{2} \left(\dot{q}^T_{e_n} \, M^e_n \, \dot{q}_{e_n} - q^T_{e_n} \, K^e_n \, q_{e_n} \right) + q^T_{e_n} \, Q^e_n \right) \tag{47}$$

Applying the appropriate Lagrange equations of the motion for equilibrium yields for the element and for the nth harmonic :

$$M_n^e \ddot{q}_{e_n} + K_n^e q_{e_n} = Q_n^e \qquad (48)$$

where:

$$\ddot{q}_{e_n} = \frac{d^2(q_{e_n})}{dt^2}$$

For the system these equations become :

$$M \ddot{q} + K q = Q \qquad (49)$$

where M, K, q and Q are, respectively, the mass and stiffness matrices, the displacement vector and the load vector of the nth harmonic.

Assuming free undamped vibration the static and dynamic equilibrium equations for the nth harmonic can be represented symbolically, respectively, as :

$$K q = Q \qquad (50)$$

$$K q = \omega^2 M q \qquad (51)$$

where ω is the natural frequency.

3. SENSITIVITY ANALYSIS

3.1. INTRODUCTION

The evaluation of sensitivities of structural response to changes in design variables is a crucial stage in the optimal design of complex structures, representing a major factor with regard to the computer time required for the optimization process. Hence it is important to have efficient techniques to calculate these derivatives. The simplest technique of evaluating sensitivities of response with respect to changes in design variables is through the finite difference approximation, here called global finite difference, which is computationally expensive or through the use of semi-analytical or analytical methods as described in the next section. These latter methods can both be applied through the direct or adjoint structure technique (Haftka and Kamat, 1987). In this section the formulation of the evaluation of sensitivities of axisymmetric shells is presented for the particular case of axisymmetric loading and boundary conditions. For this particular situation the element degrees of freedom of the frustum element are reduced from 8 to 6 degrees of freedom, since n=0 and $v_n = 0$.

More numerically based solutions are reported by Marcelin and Trompette (1988) using a finite element with a two node straight element and/or a three node parabolic element also based on Love-Kirchhoff shell theory and a semi-analytical method to evaluate the sensitivities. Other authors, such a Plaut et al (1984) and Chenais (1987), present alternative theories and models for the optimization of shell structures. More recently, Mehrez and Rousselet (1989) evaluate the analysis and optimization of shells of revolution using Koiter's model with the implementation of B-Splines for the middle surface and finite element for displacements.

3.2. ADJOINT STRUCTURE METHOD

3.2.1. *Statics*

A typical optimization constraint such as a limit on a displacement, stress component or effective stress can be represented by :

$$g_j = g_j (q ; b) \leq 0 \tag{52}$$

where b is the vector of design variables and $j \in (1, ..., m)$, m being the number of constraints. The sensitivity of the constraint g_j is :

$$\frac{dg_j}{db_i} = \frac{\partial g_j}{\partial b_i} + Z_j^T \frac{dq}{db_i} \tag{53}$$

where :

$$Z_j = \frac{\partial g_j}{\partial q} \tag{54}$$

is the vector of adjoint forces.

Thus in the method of the adjoint structure a virtual structure is defined that satisfies the equilibrium equation :

$$K \lambda_j = Z_j \tag{55}$$

where λ_j is the system adjoint degrees of freedom for the constraint g_j . The solution of the system equation (55) gives λ_j. It should be noted that the adjoint structure is identical to the real structure, but subjected to a different load. To increase computational efficiency, the already factorized form of the stiffness matrix should be used.

Considering the static equilibrium equations (50) and differentiating these with respect to a design variable b_i :

$$K \frac{dq}{db_i} = \frac{\partial Q}{\partial b_i} - \frac{\partial K}{\partial b_i} q \tag{56}$$

Premultiplying by Z_j^T one obtains :

$$Z_j^T \frac{dq}{db_i} = Z_j^T K^{-1} (\frac{\partial Q}{\partial b_i} - \frac{\partial K}{\partial b_i} q) \tag{57}$$

The inversion of the stiffness matrix K is easily avoided using the adjoint structure method through the solution of equation (55). Thus, the sensitivities given by equation (53) can be evaluated for the harmonic n=0, as :

$$\frac{dg_j}{db_i} = \frac{\partial g_j}{\partial b_i} + \lambda_j^T (\frac{\partial Q}{\partial b_i} - \frac{\partial K}{\partial b_i} q) \tag{58}$$

where $\frac{\partial K}{\partial b_i}$ is the sensitivity of the system stiffness matrix and $\frac{\partial Q}{\partial b_i}$ is the sensitivity of the system load vector. When the forces are independent of the design variables the sensitivity of the system load vector is zero and then equation (58) simplifies to :

$$\frac{dg_j}{db_i} = \frac{\partial g_j}{\partial b_i} - \lambda_j^T \frac{\partial K}{\partial b_i} q \qquad (59)$$

The term $\frac{\partial g_j}{\partial b_i}$ is usually zero or can easily be obtained.

3.2.2. Dynamics

Considering the mode of vibration q_k which corresponds to the natural frequency ω_k, the eigenvalue problem, equation (51), is represented for the system as :

$$K q_k = \omega_k^2 M q_k \qquad (60)$$

Differentiating the above equation with respect to a design variable b_i and premultiplying by q_k^T one obtains :

$$2 \omega_k \frac{\partial \omega_k}{\partial b_i} q_k^T M \; q_k = \; q_k^T \left(\left(\frac{\partial K}{\partial b_i} - \omega_k^2 \frac{\partial M}{\partial b_i} \right) q_k + \left(K - \omega_k^2 M \right) \frac{\partial q_k}{\partial b_i} \right) \qquad (61)$$

Considering the modal normalization $q_k^T M q_k = 1$, the sensitivity of the natural frequency corresponding to mode k with respect to changes in design variables is given by :

$$\frac{\partial \omega_k}{\partial b_i} = \frac{1}{2\omega_k} q_k^T \left(\frac{\partial K}{\partial b_i} - \omega_k^2 \frac{\partial M}{\partial b_i} \right) q_k \qquad (62)$$

where $\frac{\partial M}{\partial b_i}$ is the system mass sensitivity matrix. Thus to evaluate the sensitivity of natural frequencies with respect to changes in the design variables there is no need to define an adjoint structure.

3.3. SENSITIVITY ANALYSIS OF AXISYMMETRIC SHELLS

3.3.1. Analytical Method

The analytical derivative of the element stiffness matrix (eqs. 37 and 38) with respect to a variable b_i^* (not necessarily a design variable) can be represented in a symbolic form as :

$$\frac{\partial K^e}{\partial b_i^*} = 2\pi \int_0^1 \left\{ \left(\left(B^T D \frac{\partial B}{\partial b_i^*} \right)^T + \left(B^T D \frac{\partial B}{\partial b_i^*} \right) + \left(B^T \frac{\partial D}{\partial b_i^*} B \right) \right) r \, \ell \; + \right.$$

$$+ \left(B^T D B \right) \left(\ell \, \frac{\partial r}{\partial b_i^*} + r \, \frac{\partial \ell}{\partial b_i^*} \right) \Bigg\} \, d\zeta \tag{63}$$

where:

$$D = \begin{bmatrix} D_m & 0 \\ 0 & D_f \end{bmatrix} \qquad ; \qquad B = \begin{bmatrix} B_{m_n} \\ B_{f_n} \end{bmatrix}$$

The derivative of the element force vector is :

$$\frac{\partial Q^e}{\partial b_i^*} = 2\pi \int_0^1 \left\{ \left(\left(\frac{\partial L^T}{\partial b_i^*} N^{*T} \underline{p} \right) + \left(L^T \frac{\partial N^{*T}}{\partial b_i^*} \underline{p} \right) \left(L^T N^{*T} \frac{\partial \underline{p}}{\partial b_i^*} \right) \right) r\ell + \right.$$

$$\left. + \left(L^T N^{*T} \underline{p} \right) \left(\frac{\partial r}{\partial b_i^*} \ell + r \frac{\partial \ell}{\partial b_i^*} \right) \right\} d\zeta \tag{64}$$

The derivatives of the stiffness matrix are evaluated at each Gaussian point, separately for membrane and bending. When the design variables are radial coordinates, the derivatives of equation (63) and the integration of equations (64) and their derivatives $\partial Q^e / \partial b_i^*$ are carried out using the symbolic manipulator MATHEMATICA (Wolfram, 1988). Full details are presented by Barbosa (1990).

For the particular case of a frustum-cone finite element the sensitivities for the harmonic n=0 of stiffness matrix K^e or mass matrix M^e, are obtained easily when the design variables are thicknesses, using equations (33), (37) and (38). In fact the mass matrix M^e depends explicitly on the thickness while for the stiffness matrix K^e the dependence is only on constitutive matrices D_m and D_f. When the thickness is assumed constant within the element one obtains :

$$\frac{\partial K^e}{\partial h} = 2\pi \int_0^1 \left(B_m^T \frac{\partial D_m}{\partial h} B_m + B_f^T \frac{\partial D_f}{\partial h} B_f \right) r \, \ell \, d\zeta \tag{65}$$

$$\frac{\partial M^e}{\partial h} = 2\pi \int_0^1 \left(\left(\frac{1}{h} \right) \rho \, h \, (N^* L)^T (N^* L) + \left(\frac{3}{h} \right) \frac{\rho \, h^3}{12} (R^* L)^T (R^* L) \right) r \, \ell \, d\zeta \tag{66}$$

or :

$$\frac{\partial K^e}{\partial h} = \frac{1}{h} K_m^e + \frac{3}{h} K_f^e \tag{67}$$

$$\frac{\partial M^e}{\partial h} = \frac{1}{h} M_T^e + \frac{3}{h} M_R^e \tag{68}$$

Considering that the load vector Q^e is independent of the thickness, it becomes :

$$\frac{\partial Q^e}{\partial h} = 0 \tag{69}$$

For nodal coordinates or when the thickness distribution varies within the element, the shape of the model is related through the linking relation (Vanderplaats, 1984):

$$F = F^c + T b \tag{70}$$

where F is the vector of dependent variables (thicknesses and/or radial nodal coordinates of the finite element model), T the linking matrix which relates the vector of shape design variables b with the dependent variables and F^c a vector of constant terms.

The derivatives of the stiffness matrix with respect to a shape design variable required by equation (58) can be evaluated using relations (63) and the chain rule of differentiation as :

$$\frac{\partial K^e}{\partial b_i} = \frac{\partial K^e}{\partial b_j^*} \frac{\partial b_j^*}{\partial b_i} \qquad i = 1, n \quad ; \quad j = 1, 2 \tag{71}$$

where b_j^* is the value of the element nodal variable concerned, yielding :

$$\frac{\partial K^e}{\partial b_i} = \frac{\partial K^e}{\partial b_j^*} T_{ij}^e \qquad i = 1, n \quad ; \quad j = 1, 2 \tag{72}$$

where T_{ij}^e is related to the linking matrix T (eq. 70) through the topological finite element code procedure. The sensitivities of the element mass matrix or load vector are obtained in a similar way.

The sensitivities of a constraint function with respect to a design variable are evaluated at element level using equations (58) and (62), as :

$$\frac{dg_j}{db_i} = \frac{\partial g_j}{\partial b_i} + \sum_{\ell \epsilon E} \lambda_j^{(\ell)T} \left(\frac{\partial Q^{e(\ell)}}{\partial b_i} - \frac{\partial K^{e(\ell)}}{\partial b_i} q^{e(\ell)} \right) \tag{73}$$

$$\frac{dg_j}{db_i} = - \frac{1}{2\omega_o \omega_k} \sum_{\ell \epsilon E} q_k^{e(\ell)T} \left(\frac{\partial K^{e(\ell)}}{\partial b_i} - \omega_k^2 \frac{\partial M^{e(\ell)}}{\partial b_i} \right) q_k^{e(\ell)} \tag{74}$$

being E the set of elements which are affected by the design variable b_i, $\lambda_j^{(\ell)}$ the vector of the adjoint displacement of the element (ℓ), ω_k the natural frequency of the vibrating mode q_k and ω_o the limiting natural frequency.

3.3.2. Semi-Analytical Method

In this technique the vector of adjoint forces Z_j is obtained analytically and the gradients of equations (58) and (62), with terms of the type $\frac{\partial F}{\partial b_i}$, are evaluated by finite difference approximation, F being a function dependent on the design variables b.

The gradients $\frac{\partial F}{\partial b_i}$ can be evaluated through the approximations :

$$\frac{\partial F}{\partial b_i} \approx \frac{F(b + \Delta b) - F(b)}{\delta b_i} \qquad \text{(Forward difference - FFD)} \tag{75}$$

$$\frac{\partial F}{\partial b_i} \approx \frac{F(b+\triangle b)-F(b-\triangle b)}{2\delta b_i} \qquad \text{(Central difference - CFD)} \qquad (76)$$

where $\triangle b=[0, ..., \delta b_i, ..., 0]$ and δb_i is a small perturbation. It should be noted that to evaluate $F(b+\triangle b)$, the coordinates perturbations δR due to a design perturbation δb_i must be calculated, which is done through the linking relation.

With regard to shape design variables and considering the linking relation (70), the sensitivities of the element stiffness, mass or load vector can also be obtained analytically through:

$$\frac{\partial F^e}{\partial b_i} = \frac{\partial F^e}{\partial r_1^e}\frac{\partial r_1^e}{\partial b_i} + \frac{\partial F^e}{\partial r_2^e}\frac{\partial r_2^e}{\partial b_i} = \frac{\partial F^e}{\partial r_1^e} T_{i1}^e + \frac{\partial F^e}{\partial r_2^e} T_{i2}^e \qquad (77)$$

where F^e can be the stiffness and mass matrices or load vector of the eth element and r_j^e, z_j^e $(j=1,2)$ are the coordinates of the two ring nodes of the frustum conical element.

For the semi-analytical method the gradients with respect to changes in nodal radius are evaluated at element level through forward finite difference :

$$\frac{\partial K^e}{\partial b_i} \approx \frac{K^e(r_1^e+\delta r_1^e , z_1^e ; r_2^e+\delta r_2^e , z_2^e) - K^e(r_1^e , z_1^e ; r_2^e , z_2^e)}{\delta b_i} \qquad \text{a)}$$

$$\frac{\partial M^e}{\partial b_i} \approx \frac{M^e(r_1^e+\delta r_1^e , z_1^e ; r_2^e+\delta r_2^e , z_2^e) - M^e(r_1^e , z_1^e ; r_2^e , z_2^e)}{\delta b_i} \qquad \text{b)} \quad (78)$$

$$\frac{\partial Q^e}{\partial b_i} \approx \frac{Q^e(r_1^e+\delta r_1^e , z_1^e ; r_2^e+\delta r_2^e , z_2^e) - Q^e(r_1^e , z_1^e ; r_2^e , z_2^e)}{\delta b_i} \qquad \text{c)}$$

3.3.3. *Finite Difference Techniques*

Alternatively a global finite difference approach can be used. In this case the sensitivity of a constraint with respect to a change δb_i in a design variable is given by :

$$\frac{dg_j}{db_i} \approx \frac{g_j(b_1, ... , b_i+\delta b_i, ... , b_m) - g_j(b)}{\delta b_i} \qquad (79)$$

which needs one extra structural analysis for each design variable, increasing the computational effort.

3.3.4. *Limit on Displacements*

A constraint on a displacement is represented in normalized form by

$$g_j = \frac{q_f}{q_o} - 1 \leq 0 \qquad (80)$$

where q_f is the real generalized displacement corresponding to the system degree of freedom f and q_o is the maximum admissible generalized displacement. The vector of adjoint forces is :

$$Z_j = \left[\frac{\partial g_j}{\partial q_1} \cdots \frac{\partial g_j}{\partial q_f} \cdots \frac{\partial g_j}{\partial q_\eta} \right]^T = \left[0 \cdots \frac{1}{q_o} \cdots 0 \right]^T \tag{81}$$

where η is the total number of degrees of freedom. It should be noted that the adjoint structure is identical with the real structure and it is subjected to a force or moment of intensity $1/q_o$ on the corresponding degree of freedom where the displacement or rotation is limited.

3.3.5. Limit on Stresses

Limit on a stress component of the stress tensor or an effective stress is represented by :

$$g_j = \bar{\sigma}/\sigma_o - 1 \leq 0 \tag{82}$$

where σ_o is the maximum allowable stress, which may be different for tension and compression, and $\bar{\sigma}$ is the stress component or the effective stress. For instance for a Von Mises effective stress, $\bar{\sigma}$ is defined by the following relation :

$$\bar{\sigma} = \left[(\tau_{SS}^m)^2 + (\tau_{\theta\theta}^m)^2 - \tau_{SS}^m \tau_{\theta\theta}^m + 3\tau_{S\theta}^m + (\tau_{SS}^f)^2 + (\tau_{\theta\theta}^f)^2 - \tau_{SS}^f \tau_{\theta\theta}^f + 3\tau_{S\theta}^f \right]^{\frac{1}{2}} \tag{83}$$

where :

$$\sigma^m = \frac{1}{h} D_m B_m q_e = A q_e \quad \text{(Membrane and shear stresses)}$$

$$\sigma^f = \frac{6}{h^2} D_f B_f q_e = C q_e \quad \text{(Bending stresses)}$$

$$\sigma^m = \left[\tau_{SS}^m \ \tau_{\theta\theta}^m \ \tau_{S\theta}^m \right]^T ; \quad \sigma^f = \left[\tau_{SS}^f \ \tau_{\theta\theta}^f \ \tau_{S\theta}^f \right]^T$$

$$A = \left[A_1 \ A_2 \ A_3 \right]^T ; \quad C = \left[C_1 \ C_2 \ C_3 \right]^T$$

For a stress constraint at a Gaussian point of the element (ℓ), which corresponds to adjoint load $Z_j^{(\ell)}$, the system adjoint force vector is then :

$$Z_j^T = \left[0 \cdots Z_j^{(\ell)} \cdots 0 \right]^T \tag{84}$$

where :

$$Z_j^{(\ell)} = \frac{1}{\sigma_o \bar{\sigma}} \ [A_1^T A_1 + A_2^T A_2 - \tfrac{1}{2} A_1^T A_2 - \tfrac{1}{2} A_2^T A_1 + 3 A_3^T A_3 +$$

$$+ C_1^T C_1 + C_2^T C_2 - \tfrac{1}{2} C_1^T C_2 - \tfrac{1}{2} C_2^T C_1 + 3 C_3^T C_3] \ q_e^{(\ell)} \tag{85}$$

Thus, for a pointwise limit on a stress, such as defined by equation (83), the design sensitivity of stress to thickness variation, evaluated by equation (59), yields :

$$\frac{dg_j}{dh_i} = \frac{1}{\sigma_o \bar{\sigma} h^{(\ell)}} \left[(\tau^f_{ss})^2 + (\tau^f_{\theta\theta})^2 - \tau^f_{ss} \tau^f_{\theta\theta} + 3\tau^f_{s\theta} \right] - \lambda^T_j \frac{\partial K}{\partial h_i} q \tag{86}$$

where $h^{(\ell)}$ is the thickness of the element (ℓ), where the constraint g_j has been imposed.

3.3.6. *Limit on Natural Frequencies*

A constraint on the natural frequency of mode k can be easily evaluated once the eigenvalue problem, equation (62), is solved. There is no need to assemble the sensitivities of the system stiffness or mass matrices.

Following the procedure described for the static case, one obtains for a normalized constraint $g_j = 1 - \omega_k/\omega_o \leq 0$ and for each element:

$$\frac{dg_j}{db_i} = - \frac{1}{2\omega_o \omega_k} \sum_{\ell \in E} q_k^{T(\ell)} \left(\frac{\partial K^{(\ell)}}{\partial b_i^{(\ell)}} - \omega_k^2 \frac{\partial M^{(\ell)}}{\partial b_i^{(\ell)}} \right) q_k^{(\ell)} \tag{87}$$

where ω_o is the limiting natural frequency for mode k.

4. OPTIMAL DESIGN

The objective is the minimization of the volume of the material or the maximization of natural frequency or the minimization of the maximum stress. The problem is stated as :

$$\min V\ (b)\quad \text{or}\quad \max \omega_k\ (b)\qquad \text{or}\quad \min\ (\max \bar{\sigma}(b)) \tag{88}$$

subjected to :

$$g_j(b) \leq 0 \qquad j = 1, i$$

$$g_k(b) = 0 \qquad k = i+1,\ m \tag{89}$$

$$b_i^d \leq b_i \leq b_i^u \qquad i = 1, 2, \dots, n$$

where the objective function is the volume V of the material of the shell or the natural frequency ω_k or maximum stress $\bar{\sigma}$. Constraints g_j are inequality constraints (such as displacement or stress) and g_k are equality constraints (enclosed volume of the structure or the volume of the shell material), b_i^d and b_i^u are the lower and upper limiting bounds of the design variables and i, m, n are the number of inequality constraints, total number of constraints and total number of design variables, respectively.

The gradients of the constraints are evaluated using the analytical sensitivity analysis formulation for limits on displacements, stresses or natural frequencies when the design variables are thicknesses and the analytical, or alternatively the semi-analytical, formulation when design variables are the radial coordinates.

Bound formulation, Taylor and Bendsøe (1984), is used to solve min-max problem. The problem is restated as a simple minimization problem in terms of a bound β on the value of max $\overline{\sigma}$:

$$\begin{array}{l} \min \beta \\ (\beta - \epsilon \beta) < \max \overline{\sigma}(b) \leq \beta \end{array} \qquad (90)$$

where ϵ is defined by the user.
The optimal design problem is solved by the techniques of nonlinear mathematical programming described by Vanderplaats (1984) and the algorithms of the ADS program.

5. APPLICATIONS

A computer program for personal computers has been developed based on the formulation presented and using the modified feasible direction method of the ADS program for the optimization.

5.1. Supported Cylinder with End Shearing Force

The load, geometric and material properties are :
$Q = 1000$ N/m ; $\quad$ a $= 1$ m ; $\quad$ h $= 0.01$ m ; $\quad$ L $= 0.6$ m
$E = 200$ GPa (Young's modulus) $\quad$; $\quad \nu = 0.30$ (Poisson's coefficient)
A finite element model with 30 elements as been considered (Fig. 2). The design variable is the thickness of the cylinder.

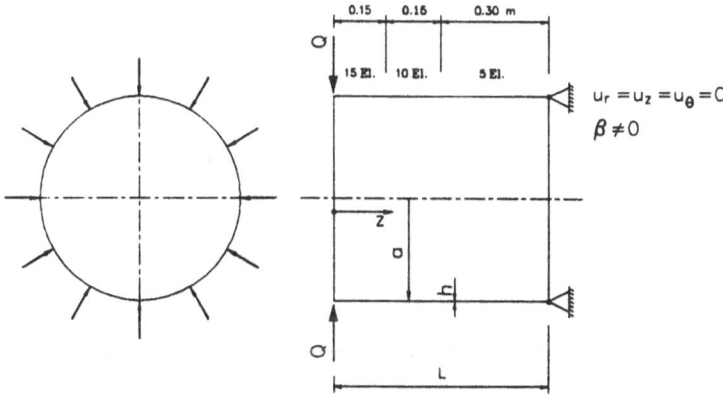

Fig. 2 Cylinder. Geometry and Load

The radial displacement distribution of the Love-Kirchhoff analytical solution (Kraus, 1967) is :

$$u_r = -\frac{Q}{2 \mu^3 D} e^{(-\mu z)} \cos(\mu z) \quad ; \quad \mu = \left(\frac{3 (1-\nu^2)}{a^2 h^2}\right) \quad ; \quad D = \frac{E h^3}{12 (1-\nu^2)}$$

which compares very favourably with the present numerical solution (Fig. 3).

The sensitivity distribution obtained using discrete finite element model with the analytical method has a very good agreement (Fig. 4) with the theoretic sensitivity obtained by differentiating the above expression using symbolic manipulator MATHEMATICA, yielding :

$$\frac{du_r}{dh} = \frac{dg}{dh} = \frac{P_1 \cos (P_2)}{P_3 \, h^{\left(\frac{5}{2}\right)} E} - \frac{P_4 \cos (P_2)}{P_3 \, h^3 \, E} - \frac{P_4 \sin (P_2)}{P_3 \, h^3 \, E}$$

where :

$$P_1 = 3.948 \, Q \, a^{\frac{3}{2}} \, (1-\nu^2)^{\frac{1}{4}} \qquad ; \qquad P_2 = \frac{1.316 \, z \, (1-\nu^2)^{\frac{1}{4}}}{(a \, h)^{\frac{1}{2}}}$$

$$P_3 = e^{\left(\frac{1.316 \, z \, (1-\nu^2)^{\frac{1}{4}}}{\sqrt{a} \, h}\right)} \qquad ; \qquad P_4 = 1.732 \, Q \, a \, z \, \sqrt{(1-\nu^2)}$$

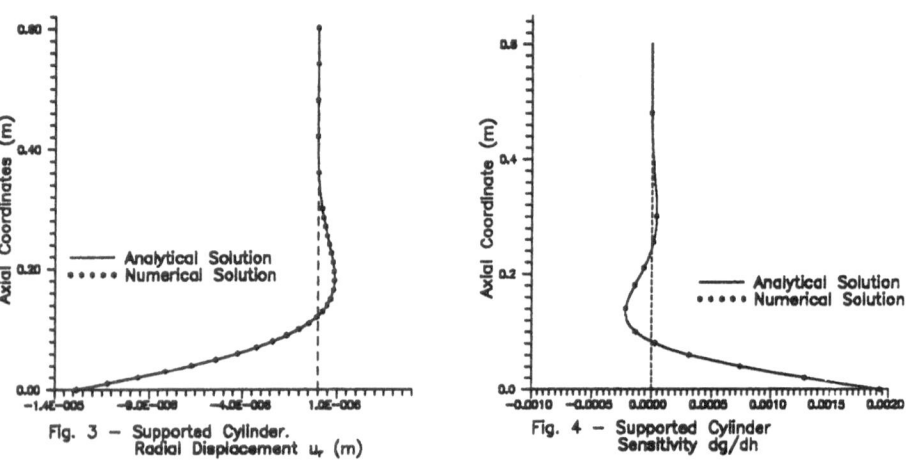

Fig. 3 — Supported Cylinder.
Radial Displacement u_r (m)

Fig. 4 — Supported Cylinder
Sensitivity dg/dh

5.2. Simply Supported Cone-Cylinder Connection with Internal Pressure

The geometric and material properties are:

R=1.0m (Cylinder radius), H=0.6m (Height of cylinder),
h=0.010m (thickness), E=200GPa, ν=0.3.

A finite element model with 50 elements as been considered (Fig. 5). The design variables are 6 radial coordinates (b_1, ... , b_6). Tables I and II shown the sensitivities for the initial design with respect to changes in the radial coordinates for a displacement radial constraint in the junction (r=1.0m , z=0.6m) and a meridional stress constraint τ_{SS} in the Gaussian point of the cylindrical element adjacent to the junction. The cone-cylinder connection is submitted to an internal pressure of p=0.2MPa. The global sensitivities are calculated using central finite difference (CFD) with a perturbation of Δb_i=0.001b_i.

From Tables I and II it is observed that the analytical sensitivities for shape design sensitivities compare very favourably with the global finite element sensitivities obtained with the same model. It is also seen that the semi-analytical sensitivities only compare favourably for very small perturbations in the design variables, being highly influenced by the perturbation used, due to the truncation on the finite difference method.

Table I : Sensitivities due to a radial displacement constraint

Source	Perturbation	Sensitivities					
		b_1	b_2	b_3	b_4	b_5	b_6
CFD	0.001b_i	-0.0430	0.6042	4.6708	-12.7946	0.0926	5.7383
Analytical		-0.0430	0.6041	4.6703	-12.7940	0.0922	5.7387
Semi-Analytical	0.000000001b_i	-0.0430	0.6042	4.6708	-12.7946	0.0922	5.7387
	0.0000001b_i	-0.0430	0.6042	4.6709	-12.7942	0.0928	5.7390
	0.00001b_i	-0.0430	0.6044	4.6811	-12.7507	0.1541	5.7679
	0.001b_i	-0.0428	0.6289	5.7062	-8.4132	6.2830	8.6519
	0.01b_i	-0.0410	0.8691	15.9369	29.7409	61.1342	34.3096

Table II : Sensitivities due to a meridional stress constraint

Source	Perturbation	Sensitivities					
		b_1	b_2	b_3	b_4	b_5	b_6
CFD	0.001b_i	-0.0077	0.8343	-6.5946	16.7119	-7.8595	-1.8374
Analytical		-0.0077	0.8342	-6.5940	16.7115	-7.8591	-1.8375
Semi-Analytical	0.000000001b_i	-0.0077	0.8343	-6.5946	16.7121	-7.8591	-1.8375
	0.0000001b_i	-0.0077	0.8343	-6.5947	16.7114	-7.8598	-1.8376
	0.00001b_i	-0.0077	0.8344	-6.6034	16.6458	-7.9269	-1.8488
	0.001b_i	-0.0077	0.8480	-7.4871	10.0850	-14.6369	-2.9660
	0.01b_i	-0.0077	0.9802	-16.3546	-48.1310	-74.5055	-12.912

For the initial design the maximum meridional stress is τ_{SS}=105.4 MPa at the cylindrical element adjacent to the junction. The model is optimized considering a radial deflection of u_{r_o}=0.3mm, a meridional stress limit of σ_o=120 MPa and the enclosed volume of the initial design at 2.932 m^3 (equality constraint).

The optimal design is obtained in 14 iterations, 34 function evaluations and 6 gradient evaluations, with a reduction of maximum stress of 68%. During the iteration process the enclosed volume of the initial design is an active constraint. For the optimal design the maximum meridional stress in the model decreases to τ_{SS}=33.5 MPa.

A CPU ratio of 2.5 is achieved between the analytical/semi-analytical evaluation of sensitivities. Table III shows the CPU ratio between semi-analytical method (SA) versus analytical method (AM), global central finite difference (CFD) and global forward finite difference (FFD).

The optimal design is also achieved through the method of modified feasible directions.

Table III : Evaluation of sensitivities

	AM / SA	FFD / SA	CFD / SA
CPU ratio	2.5	3.2	5.4

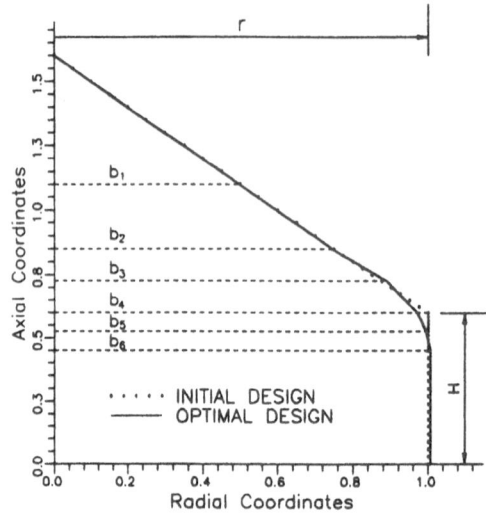

Fig.5 – Minimization of Maximum Stress

5.3. Conical Structure Clamped at Lower End

It is considered a conical shell clamped at lower end with geometry shown in Fig.6 and material properties E=200GPa (Young's modulus), $\nu = 0.15$ (Poisson's coefficient) and ρ=2410 Kg.m^{-3} (mass per unit volume). The structure has been modelled with 68 elements.

Table IV shows the sensitivities results for the initial design with a good agreement between the three methods.

Table IV : Sensitivities due to fundamental frequency

		Sensitivities	
Source	Perturbation	b_1	b_2
FFD	0.001b_i	-271.606	284.140
Analytical		-271.637	284.192
	0.000000001b_i	-271.720	284.192
	0.0000001b_i	-271.724	284.182
Semi-Analytical	0.00001b_i	-271.725	284.183
	0.01b_i	-272.109	285.353
	0.05b_i	-273.633	290.049

A two level optimization has been carried out. The objective of the design is the maximization of the fundamental frequency with a constraint of enclosed volume of the structure in the first level and a constraint of volume of the shell material in the second level.

In the first level the design variables are 2 radial coordinates b_1 and b_2. The optimal design (Fig.7) is obtained in 8 iterations, 41 function evaluations and 7 gradient evaluation, with a increase in natural frequency of 49%. During the iteration process the active constraint is the enclosed volume of the structure (equality constraint) that is imposed constant and equal to 59.157 m³. The natural frequency of the optimum solution is in agreement with a plot (Fig.8) of natural frequency versus radial coordinates carried out through 12 finite element analysis using constant enclosed volume (59.157 m³) for the structure .

In the 2nd level of optimization the design variables are 6 thicknesses, where the initial design is the final design of the first level of optimization. The constraint is the volume of the shell material that is imposed constant and equal to 3.706 m³. The optimal design (Fig.9) is obtained in 7 iterations, 27 function evaluations and 5 gradient evaluation, with a further increase in natural frequency of 24%.

The optimal designs are achieved through the method of modified feasible directions implemented in program ADS (Vanderplaats, 1987).

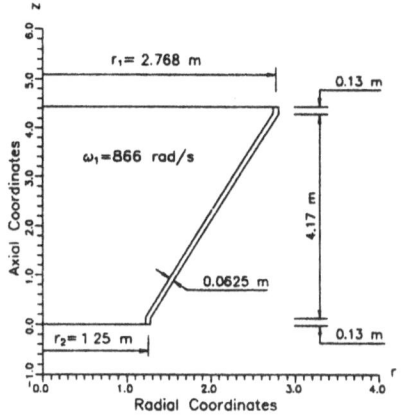

Fig.6 — Initial Design

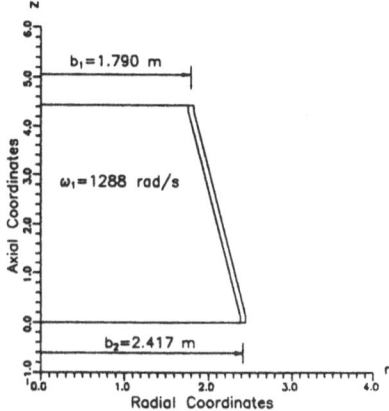

Fig.7 — Optimal Design (1st Level)

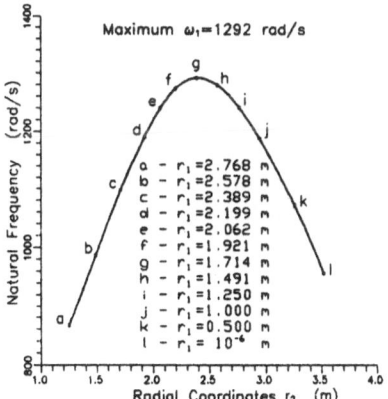

Fig.8 — FEM Analysis

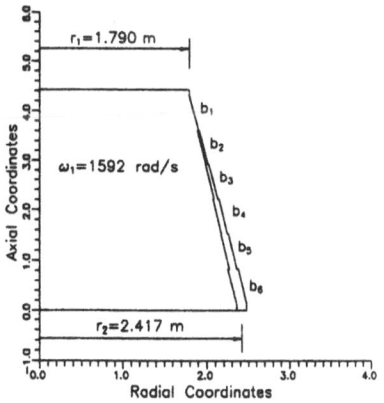

Fig.9 — Optimal Design (2nd Level)

5.4. Conclusions

The results presented in this section show that sensitivity analysis of statical and dynamic constraints of axisymmetric shells are efficiently and accurately obtained using the analytical method here described. When the design variables are thicknesses all the described techniques calculate the response sensitivities with accuracy.

From the observation of Tables I and II it can be concluded that analytical sensitivities for shape design are more accurate than the semi-analytical sensitivities. Hence analytical sensitivities should be recommended for shape optimization of axisymmetric type structures although they are more difficult to obtain and more expensive in terms of CPU time when compared with semi-analytical formulation.

Shell structures are very sensitive to small imperfections, consequently the semi-analytical techniques for sensitivities cannot be very accurate and it will require a very small perturbation to obtain acceptable results. However this perturbation can created problems of numerical stability.

The use of a symbolic manipulator to obtain sensitivities analytically is a very promising tool whenever it is required to obtain partial derivatives with respect to a specified variable. Limitations have been found by Barbosa (1990) in the use of the symbolic manipulator when the task is to integrate the element stiffness and mass matrices.

6. REFERENCES

Barbosa, J. I., (1990) 'Analytical Sensitivities for Axisymmetric Shells Using Symbolic Manipulator MATHEMATICA', CEMUL Report.

Barthelemy, B., Chon, C. T. and Haftka, R. T., (1988) 'Accuracy Problems Associated with Semi-Analytical Derivatives of Static Response', Journal of Finite Elements in Analysis and Design, Vol.4, pp. 249-265.

Chenais, D., (1987) 'Shape Optimization in Shell Theory : Design Sensitivity of the Continuous Problem', Engineering Optimization, 11, pp. 289-303.

Desai, C. S. and Abel, J. F., (1972) Introduction To The Finite Element Method, A Numerical Method For Engineering Analysis. Van Nostrand Reinhold Company, New York.

Gendong, C. and Yingwei, L., (1987) 'A New Computation Scheme for Sensitivity Analysis', Engineering Optimization., Vol. 12, pp. 219-234.

Haftka, R. T. and Kamat, M. P., (1987) 'Finite Elements in Structural Design', in C. A. Mota Soares (ed.), Computer Aided Optimal Design : Structural and Mechanical Systems , Springer-Verlag, Berlin, pp. 241-270.

Kraus, H., (1967) Thin Elastic Shells, John Wiley & Sons, Inc., New York

Marcelin, J. L. and Trompette Ph., (1988) 'Optimal Shape Design of Thin

Axisymmetric Shells', Engineering Optimization., Vol. 13, pp. 108-117.

Mehrez, S. and Rousselet, B., (1989) 'Analysis and Optimization of a Shell of Revolution', in C. A. Brebbia and S. Hernandez (eds.), Computer Aided Optimum Design of Structures : Applications. Computational Mechanics Publications, Springer-Verlag, pp. 123-133.

Plaut, R. H., Johnson, L. W. and Parbery, R., (1984) 'Optimal Form of Shallow Shells with Circular Boundary', Transactions of the ASME, Vol. 51, pp. 526-538.

Taylor, J. E. and Bendsøe, M. P., (1984) "An Interpretation for Min-Max Structural Design Problems Including a Method for Relaxing Constraints", Int. J. Solids Structures, Vol. 20, No 4, pp. 301-314.

Thambiratnam, D. P., Thevendran, V. and Lee, S. L., (1989) 'Computer Aided Optimum Design of Structures for Vibration Isolation', in C. A. Brebbia and S. Hernandez (eds.) Computer Aided Optimum Design of Structures: Recent Advances, , Computational Mechanics Publications, Springer-Verlag, pp. 49-59.

Vanderplaats, G. N., (1984) Numerical Optimization Techniques for Engineering Design, McGraw-Hill, New York.

Vanderplaats, G. N., (1987) ADS - A Fortran Program for Automated Design Synthesis, Version 2.01, Engineering Design Optimization, Inc., St. Barbara, California.

Wolfram, S., (1988) MATHEMATICA - A System for Doing Mathematics by Computer, Addison - Wesley Publishing Company, Inc.

APPLICATION OF ANALITICAL MODELS FOR THE OPTIMIZATION OF LARGE STRUCTURAL SYSTEMS

N.V.BANICHUK

USSR Academy of Sciences, Moscow USSR

ABSTRACT. The paper deals with continues analytical models used for analysis and optimization of dynamical characteristics of plate-like large space structures. The models are supposed to be adequate to thin-walled structures made from locally orthotropic mate rials. The influence of anisotropic tailoring on the basic integral characteristics investigated. The orientations of the axes of the orthotropy are assumed as unknown design variables and are found to maximize fundamental eigenfrequencies of bending vibrations. Computed optimum designs are compared with conventional layouts.

INTRODUCTION

Analytical models are used for analysis and optimization of mechanical characteristics of large structural systems such as plate-like large space structures. These models give in some cases the possibility to find out the effective design solution and to determine the basic properties of the structures. Applications of analytical models give also the opportunity to compare the advantages and shortcoming of various structural and optimization

1051

G. I. N. Rozvany (ed.), Optimization of Large Structural Systems, Vol. II, 1051–1072.
© 1993 *Kluwer Academic Publishers.*

concepts and to evaluate the required characteristics. Considered models are suppo sed to be adequate to thin-walled continuous structures made from elastic locally orthotropic materials. Determining the best orientations of the axes of orthotropy for two-dimensional structures was investigated by Banichuk (1978,1979) analytically and numerically. The books of the author and Kobelev and Rikards (1983,1988) include other references. Similar research has been carried out by Pedersen (1989,1990).Applied aspects of the problem of determining the best orientations of the orthotropy axes are very important. Thus we need to decide how to effectively exploit the anisotropic properties of materials and anisotropic schemes application is known as aeroelastic tailoring.N.J.Krone (1975) may be was the first who performed study on divergence elimination with advanced composites.Perspective approach to increasing the divergence speed of the swept wing consists in proper selection of ply angle and laminate thickness dis tribution. Practical application of the approach owe their success to researches the simplified analysis published by Weisshaar (1980).Various applications of anisotropic schemes and materials were discussed by Hertz, Shirk, Ricketts, Weisshaar (1982),Frolov,Shanygin (1984);V.V.Kobelev, Larichev (1988). The papers of Birjuk, Sharanjuk (1984), Birjuk, Epurash and the author presents results derived by finite element techniques and concerning the optimum orientation of anisotropic axes for wing skins. The wings are characterized by small aspectratio and their skins consist of a set of orthotropic panels. The best orientation of the orthotropic axes is determined in these papers such that one of the aeroelastic characteristics (divergence or flutter speed, integral stiffness, aileron efficien cy) is maximized.In this paper the smallest nonzero eigenfrequency of unrestrained plate-like structures will be selected as a measure of structural stiffness (dunamical rigidity) and maximized with the help of finding an optimum distribution of the orthotropic axes orientations. The optimal designs will be obtained with appli cation the technique of successive optimization and finite element method and presented for several types of plate-like large space structures and it's dynamical properties will be discussed briefly.

1.Optimization problem.

Modern trends of developing space structures consist in improving of weight index and essential inlargement of their sizes. Extreme light-weight design must be required for economical reasons, where the necessary transport energy for establishing the geostationary space structures is dominant. Large space structures [LSS] generally exhibit a high degree of flexibility. The lowest frequency of natural vibrations of LSS depends inversely on its size and become close to zero. One proposal for reducing the

effect of structural flexibility and to comply the
stringent requirements expressed by frequency and stiffness
constraints has been to optimize structural performance
in terms of stiffness to mass ratio. Adequiate cha
racteristic of dynamical stiffness of the unrestrained
structure is the smallest nonzero [fundamental]
eigenfrequency of free elastic vibrations. Selection of
the fundamental frequency as a measure of structural
stiffness is reasonable because this frequency is a minimal
nonzero diagonal element of stiffness matrix of finite-
element model. Choice of the efficient structural scheme is
the most important way for increasing of fundamental
frequency. To describe the mechanical behavior of such
complicated built-up structures as regular and nonregular
huge space trusses, platforms and antennas we can use
dynamically similar models. Because of structu ral
complexity the analytical models with distributed locally
ani- sotropic characteristics must be applied for effective
modeling.We keep in mind such distributed parameter models
as beam, plate and shell-like models. Modelling of a
variety of structures can be based on application of
locally orthotropic schemes. These schemes permit effective
force transmission modelling. Note particularly that the
variations of anisotropic internal structure of the system and connected variations of force transmission
directions give the possibility to improve dynamical
characteristics of LSS. The possibilities for control of

mechanical characteristics lead to statement of optimal structural design problem. Optimization of the structure consists in determining the best orientation of the axes of orthotropy for analytical structural models. The best ori entation of the orthotropic axes is determined such that the fun- damental frequency is maximized.

$$J_* = \max_{\alpha} \omega^2 = \max_{\alpha} \min_{\underline{u}} (V/T) \qquad (1.1)$$

$$(\overline{u}, \overline{u}_r^{-l}) = 0, \quad l = 1,2,\ldots n_r \qquad (1.2)$$

where ω – the smallest eigenfrequency of free vibrations, $\overline{u}$ – vector of amplitude displacements, $\omega^2 T$ – amplitude kinetic energy, V – amplitude expression of potential energy. We assume that the elastic medium of the plate-like structure is locally orthotropic and denote by ξ and η the axes of orthotropy. At the point with coordinate $(x,y) \in \Omega$ (the domain ocupied by orthotropic material) the di- rection of the $\xi - \eta$ axes of orthotropy, relative to the $x - y$ axes is given by the angle $\alpha(x,y)$, where α is the angle between the x and ξ axes. The plate is supposed to be unrestrained. The spectrum of free vibrations has zero frequencies, corresponding ri gid body displacements and rotations. For determining non-zero eigenfrequencies of free vibrations with the help of variational principles it is necessary to take into account the orthogonality conditions for elastic modes $\overline{u}$ and rigid body modes $\overline{u}_r^{-l}$. For given function $\alpha = \alpha(x,y)$ the smallest nonzero eigenfrequency ω is calcu lated according to Rayleigh's variational principle. The minimum of V/T with respect to $\bar{u}$ computed in the class of smooth functions $\bar{u}(x,y)$ that satisfy the orthogonality conditions (2). The boundary conditions for free edges of the plate are a "natural" conditions for the Rayleigh quotient V/T (Rayleigh functional), so there is no need to satisfy these conditions a priory. If the minimum of the Rayleigh quotient V/T is found in the class of functions satisfying the orthogonality constraints (2), then the minimizing function $\bar{u}(x,y)$ automatically satisfies the boundary conditions (absence of forces and moments at the free edges of the plate). The external maximum with respect to α is determined on the set of arbitrary distributions of orthotropy angles α. This formulation of the optimization problem permits us to account for design and technological limitation. Therefore, solution of the problem is of substa- ntial interest in technological applications. Finding an optimum distribution of the orthotropic axes orientations permits us to determine the optimum direction of structural reinforcement and to formulate an opinion with regard to traditionally accepted structural designs. Even when it is difficult to construct a structure having the optimum anisotropy, it is useful to have a theoretical solytion to the optimization problem, to clarify limitations and to apply quasi-optimal reinforcement designs. To numerically solve the optimal design problem, we apply the technique of successive optimization, which

was described in the book of Banichuk (1983) and use expli
cit expression for gradients of Rayleigh quotient with res
pect to design variables.The algorithm consists of obtain
ing successive approximations to the optimal solution and is
founded on the idea of "small" perturbations of the design
function $a(x,y)$ with a solution of the "direct" problem
of finding $\bar{u}$ by the finite-element method with the values
of the function a^k regarded as fixed (here the superscript k
denote the k-th approximation). For every distri bution of
displace ment function $\bar{u}$ was calculated by the finite-
element method. The matrices of rigidity and mass have been
constr ucted with the help of one and the same
approximation of displacement function. The
expressionsfor Rayleigh quotient gradients have been
derived taking into account the possibility of appearance
of not only simple but multiple eigenfrequencies. If two
lowest frequ encies come close to each other and double
eigenfrequency appears then modified expression for
gradient is taken. This expression ta kes into account two
linear independent vibrational modes. Note that for many
cases of optimal design of vibrating plate-like LSS it has
been observed that points of discrete spectrum frequently
come close to each other and ultiply eigenfrequencies
appear.This phenomenon creates real difficulties in
designing structures with optimal stiffness properties.
Alarge number of research works on the subject of

structural optimization has as its objective a study
different aspects of cases with multiple eigenvalues
(see,for example,Haug, Choiand Komkov (1986). Note also
that the method of decomposition was applied for direct
analysis of vibrations to reduce the volume of
computational and increase the precision. So the original
eigenvalue problem for rectangular plates was reduced to
solving of four or five problems for 1/4 plate section with
corresponding boundary conditions. The original
eigenvalue problems for square and hexagon plates were
reduced to the problems for 1/8 and 1/12 plate sections.
All computations were performed in assistance with
A.A.Barsuk, L.R.Trifanova, N.R.Trifanova,A.V.Sharanjuk.

2.Optimal structures for bending vibrations.

Describe basic relations necessary for solution of optimal
design problems with application of local-orthotropic plate
models (see Banichuk, Barsuk, Trifanova, Sharanjuk (1991)).
Amplitude expressions for potential energy and kinetic
energy for thin elastic plates are given by.

$$V = \frac{1}{2} \int_{\Omega} \{ D_{11} w_{xx}^2 + D_{12} w_{xx} w_{yy} + D_{22} w_{yy}^2$$
$$+ D_{16} w_{xx} w_{xy} + D_{26} w_{yy} w_{xy} + D_{66} w_{xy}^2 \} d\Omega \;) \tag{2.1}$$

$$\omega^2 T = \frac{1}{2} \rho \omega^2 \int_{\Omega} w^2 \, d\Omega \tag{2.2}$$

where $W(x,y)$ - amplitude latteral deflection of the plate, ω
-frequency of free vibrations, Ω - domain in plane xy.The
values of the bending orthotropic constants

$D^o, D^o_{\xi\xi}, D^o_{\eta\eta}, D^o_{\xi\eta}$ are known. In the fixed $x-y$ coordinate system, the bending elastic moduli $D_{11}, \ldots, D_{66}$ are related to the assigned constants $D^o, D^o_{\xi\xi}, D^o_{\eta\eta}, D^o_{\xi\eta}$ in the $\xi\eta$ system by the well-known transformation formulas.

$$D_{11}(\alpha) = \frac{1}{2}(D^o_{\xi\xi}+D^o_{\eta\eta})+\frac{1}{2}(D^o_{\xi\xi}-D^o_{\eta\eta})Cos2\alpha-\frac{1}{4}D^o_t Sin^2 2\alpha \qquad (2.3)$$

$$D_{22}(\alpha) = \frac{1}{2}(D^o_{\xi\xi}+D^o_{\eta\eta})-\frac{1}{2}(D^o_{\xi\xi}-D^o_{\eta\eta})Cos2\alpha-\frac{1}{4}D^o_t Sin^2 2\alpha$$

$$D_{12}(\alpha)=2D^o+\frac{1}{2}D^o Sin^2 2\alpha$$

$$D_{16}(\alpha)=(D^o_{\xi\xi}-D^o_{\eta\eta}+D^o_t Cos\ 2\alpha)Sin\ 2\alpha$$

$$D_{26}(\alpha)=(D^o_{\xi\xi}-D^o_{\eta\eta}-D^o_t Cos\ 2\alpha)Sin\ 2\alpha$$

$$D_{66}(\alpha)=4D^o_{\xi\eta}+D^o_t Sin^2 2\alpha, \quad D^o_t=D^o_{\xi\xi}+D^o_{\eta\eta}-2D^o-4D^o_{\xi\eta}$$

The spectrum of free bending vibrations (determined with the help of Rayleigh variational principle) has three zero frequencies, corresponding rigid body vertical displacement and rotations with respect to axes x and y. The orthogonality conditions for elastic modes and rigid body modes have the form.

$$\int_\Omega w\ d\Omega = 0 \qquad (2.4)$$

$$\int_\Omega x\ wd\Omega = 0 \qquad (2.5)$$

$$\int_\Omega y\ wd\Omega = 0 \qquad (2.6)$$

Condition (6) means the absence of latteral external forces

acting in normal toplate direction,and conditions (7),(8) mean the absence of external torques.Computation were performed for square,rectangular and hexagonal plates.Inall cases we assume $D_{\xi\xi}^{o}=1$, $D_{\xi\xi}^{o} = 0.33$, $D_{\xi\eta}^{o}= 0.16$, $D^{o} = 0.08$ (glasslike texture).Present the result of computation for square plate $(-0.5 \leqslant x \leqslant 0.5, -0.5 \leqslant x \leqslant 0,5)$ of unit area. Numerically computed optimum distribu tion shown in Fig.1 for the angle $a(x,y)$ of inclination of the axes of orthotropy are slightly change from initial distribution for $a(x,y)$ and correspond to the value of the functional $J_{*} = \omega_{1} = 3.527$.

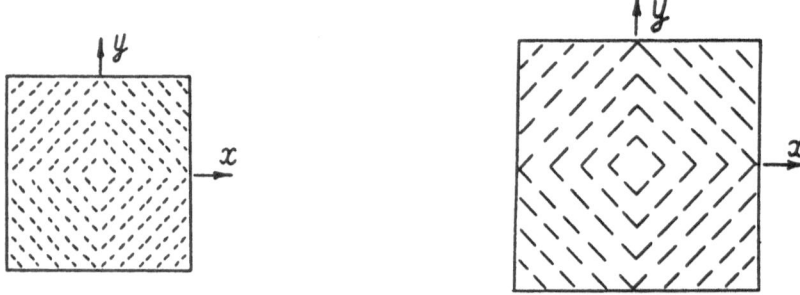

Fig 1 Optimal design of square   Fig.2 Conventional design plate for bending vibrations.   sguare plate.

Initial approximation for $a = a^{o}(x,y)$ shown in Fig.2 corresponds to the smallest nonzero eigenfrequency $\omega_{1}^{o} = 3.501$

 So the small optimization gain is explained by succes sful choice of initial approximation for design variable, which canbe taken as quasioptimal solution. For the smallest nonzero frequency the vibrating mode is odd function of x and y, $w(x,y)=-w(-x,y)$, $w(x,y)=-w(x,-y)$. To compare the optimal distribution with the conventional layouts we computed the frequencies for the plate shown in Figs.2-4 and displayed them in Table 1.

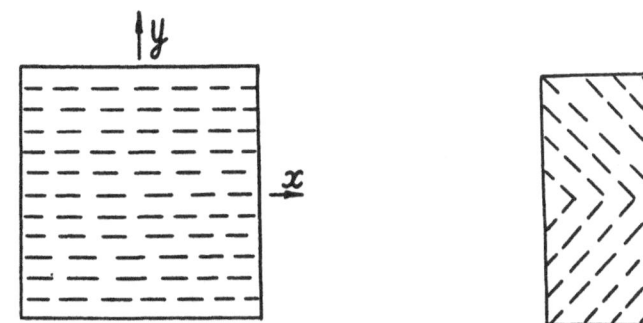

Fig.3 Conventional design for Fig.4 Conventional design

for square plate. square plate.

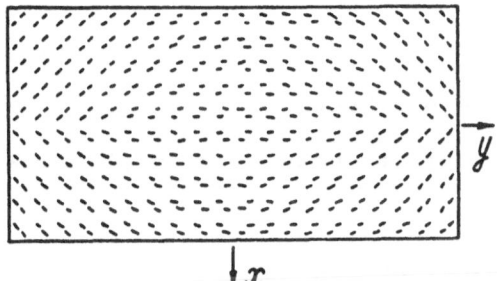

Fig.5 Optimal design for rectangular plate for bending vibrations Table 1. Eigenfrequencies of bending vibrations for square plates with optimal andconventional designs.

	Fig.1	Fig.2	Fig.3	Fig.4
w_1	3.527	3.501	2.636	2.923
	(1)	(1)	(1)	(1)

ω_2	3.814	3.825	3.670	3.863
	(1)	(1)	(1)	(1)
ω_3	4.796	4.857	6.412	5.008
	(1)	(1)	(1)	(1)
ω_4	7.360	7.395	6.481	7.612
	(2)	(2)	(1)	(2)
ω_5	13.13	12.75	8.250	13.09
	(2)	(2)	(1)	(1)
ω_6	14.22	14.32	10.25	13.23
	(1)	(1)	(1)	(1)
ω_7	14.71	14.61	12.91	13.98
	(1)	(1)	(1)	(1)
ω_8	17.40	16.78	13.02	18.91
	(1)	(1)	(1)	(1)

First column of the Table 1 corresponds to the optimal plate shown in Fig.1.Second, third and fourth columns correspond to the plates presented in Figs.2-4.The numbers in round brackets indicate the multiplicity of the eigenfrequencies.Comparison of presented values ω_1 for nonoptimal plates with the optimal values ω_1 shows lerge possibilities of increasing the dynamical rigidity with the help of optimal anisotropic structure. Present the results of calculation for the rectangular plate $(-0.25 \leqslant x \leqslant 0.25, -1 \leqslant y \leqslant 1)$ of unit area. Optimal distribution of orthotropy axes orientation is shown in Fig.5. Finding of optimal solution was realized for the plate with initial orthotropy angle distribution $\overset{o}{\alpha}(x,y)$ shown in Fig.6, which corresponds to smallest frequency $\omega_1=2.86$.The optimal functional $J_*=\omega_1=3.04$ is double for optimal distribution. The first eigenfrequency and the highest eigenfrequencies of the optimal plate bending vibrations are displayed in the first column of the Table 2.For comparison the frequency spectrums were computed also for the plates with nonptimal orthotropy axes orientations shown in Figs.7,8.The computed values are represented in the second and third columns of the Table 2.Interesting to hote that vibrating eigenmade, corresponding the smallest nonzero eigenfrequency,

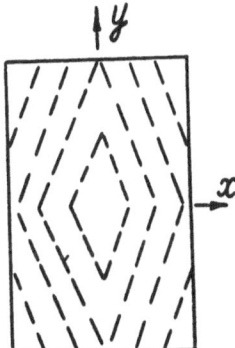

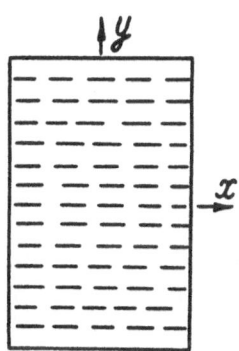

Fig.6 Conventional design of of rectangular plate.

Fig.7 Conventional desing rectangular plate.

is unsymmetric function of x and $y(w(x,y)=-w(-x,y),$ $w(x,y)=-w(x,-y))$ for the plate with homogeneous angle distribution $\alpha(x,y)=\pi/2$ and the eigenmode is symmetric function $(w(x,y)=w(-x,y),\ w(x,y)=w(x,-y))$ for the plate with angle distribution $\alpha(x,y)=0$. Comparison of the optimal values of smallest nonzero frequencies for square

(ω_1=3.527) and rectangular (ω_1=3.035) plaes shows that
dynamic stiffness depends essentially on structural shape.
Table 2. Eigenfrequencies of bending vibrations for rectang
ular plates with optimal and conventional designs.

	Fig.5	Fig.7	Fig.8
ω_1	3.035	1.838	2.628
	(2)	(1)	(1)
ω_2	6.885	2.582	3.179
	(1)	(1)	(1)
ω_3	7.750	5.088	6.208
	(1)	(1)	(1)
ω_4	8.142	5.585	7.404
	(1)	(1)	(1)
ω_5	9.616	9.452	8.656
	(1)	(1)	(1)
ω_6	11.99	9.985	9.265
	(1)	(1)	(1)
ω_7	18.92	12.74	11.88

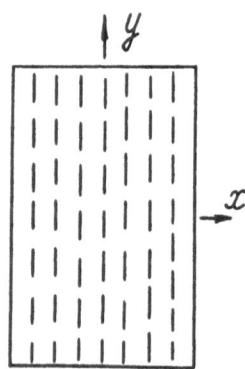

Fig.8 Conventional design of rectangular plate.

Distinctive peculiarity of unrestrained plate-like structures con- sists in the increase of the smallest nonzero frequency of free bending vibrations with the invreasing of structural symmetry.Note that for clamped plate-like structures increasing of the symmetry cause the reduction of the smallest nonzero eigenfrequency (isope- rimetric theorem). Optimal distibution of orthotropy axes orientation is shown in Fig.9 for hexagan plate. This distribution cor- responds to the smallest eigenfrequency $J_* = \omega_1 = 4.045.$ Note that the maximized eigenfrequency is double. The frequency spectrum of the plate with optimal distribution of orthotropy axes orientation is displaced in the first column of the Table 3.Next columns correspond to the plates with nonoptimal layouts shown in Figs.10-12.

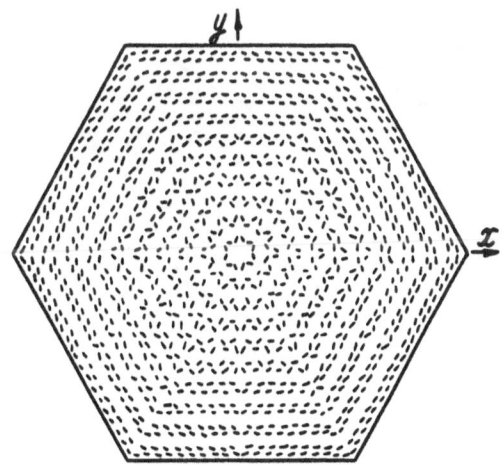

Fig.9 Optimal design of hexagan plate for bending vibrations.

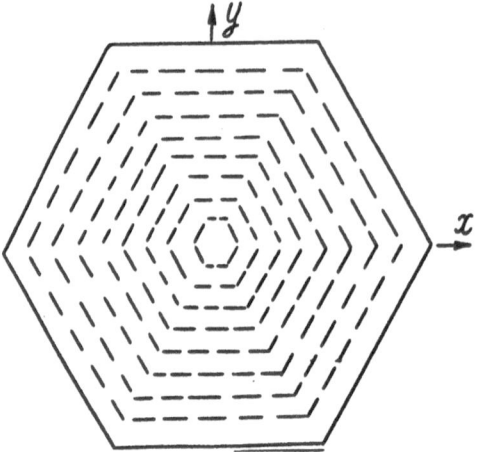

Fig.10 Conventional desing of hexagan plate.

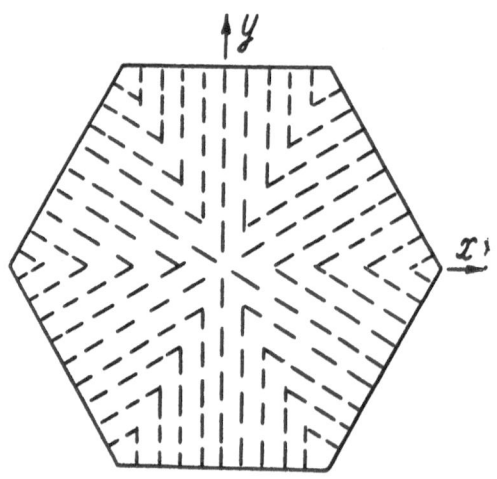

Fig.11 Conventional desing of hexagan plate.

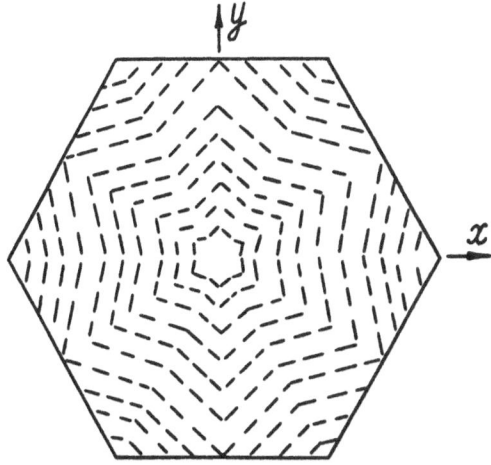

Fig.12 Conventional desing of hexagan plate.

Table 3. Eigenfrequencies of bendig vibrations for hexagan
plates with optimal and conventional designs.

	Fig.9	Fig.10	Fig.11	Fig.12
w_1	4.045	3.844	3.472	3.654
	(2)	(2)	(2)	(2)
w_2	6.558	6.554	5.263	5.939
	(1)	(1)	(1)	(1)
w_3	8.866	7.684	6.628	9.017
	(1)	(1)	(1)	(1)
w_4	10.56	11.51	8.076	9.417
	(1)	(1)	(1)	(1)
w_5	13.03	12.42	12.26	12.34
	(2)	(2)	(2)	(2)
w_6	16.72	16.50	13.45	15.27
	(2)	(2)	(2)	(2)

w_7	21.97	22.26	20.15	21.00
	(1)	(1)	(1)	(1)
w_8	26.35	23.51	22.43	22.62
	(2)	(2)	(1)	(1)

3.Certain conclusions and remarks.

The problems discussed above are connected with an important class of problems of optimization of the internal structure of elastic bodies.Particular attention is paid to the optimum orientation of the anisotropy axes.Analysis of the stationary orientation schemes of the elastic moduli and comparison with the conventional layouts have been performed. Other difficulties of solution are connected with the appearance of repeated fundamental eigenfrequencies for optimum.It is also important to note the distinctive peculiarity of unrestrained plate-like structures consisted in the increase of the smallest nonzero frequency of free bending with the increasing of the structural symmetry. Application of the numerical methods of solution is not simple due to the existence of a large number of local maxima. Comparison of presentad values of eigenfrequencies for nonoptimal plates with the optimal eigenfrequencies values shows lerge possibilities of increasing the dynamical rigidity with the help of optimal anisotropic structure.However,the numerical results discussud in the paper demonstrate the considerable effects which may be attained by means of optimization of anisotropic properties. The solutions of optimization problems presented in this paper are based on the purely phenomenological approach and on the equations of the theory of elasticity of anisotropic bodies.The elastic moduli occurring in these equations are assumed to be known (from experiments).A more detailed analysis of the problems of optimization of constructions made of anisotropic structures and composite materials indicates the possibility of the prospective utilization of deformation and fracture mechanics based on macrostructural properties.In this approach the distributed mechanical characteristics of continious structure depend on the mechanical and deometrical characteristics of the micro-structural elements and other microstructural parameters.The advantages of such approaches lie the possibility (see Bolotin(1966)) of connecting the deformation and strength problems of elastic bodies,predicting the mechanical properties of structures on the basis of the mechanical properties of their components,solving the problems of optimum design of structures.The governing function of such processes of optimization of anisotropic properties may be represented by certain distributed parameters of microstructure.In fact,viewing the elastic moduli D_{ij} as certain averaged properties depending on the microstructural parameters (concentration of structural masses,dimensions, position and orientation of structural element or reinforcing elements,etc.),the distributed moduli may be assumed as the governing magnitude.Such an approach to optimization problems enables us to take into account varions structural and technological limitations and,as a result,the solution may answer the question of practical interest. Acknowledgement The author is indebted to A.A.Barsuk, A.V.Sharanjuk, L.R.Trifanova and N.R.Trifanova for assistance in computations and discussion of the results.

References.

Banichuk,N.V.(1978):A certain extremal problem for a system with distributed parameters and determination the optimal properties of an elastic medium.Dokl.Akad.Nauk.SSSR, vol.242, No 5, p.1042-1045.

Banichuk,N.V.(1979):Optimization of anisotropic properties of deformable media in the framework of two-dimensional theory of elasticity. Izv.Akad.Nauk SSSR, MTT, No 1, p.71-77.

Banichuk,N.V.(1983):Problems and methods of optimal structural design.Plenum Press,New-York.

Banichuk,N.V.; Birjuk,V.I.; Epurash,D.M. (1984) :Onoptimization problem for constructive power schemes with the help of anisotropic models (in Russian).Scientific Notes, TsAGI,vol.XV,No 2, p.134-138.

Banichuk,N.V.; Kobelev,V.V.; Rikards,R.B. (1988) : Optimization of structural elements made of composite materials.Mashinostroenie, Moscow.

Banichuk,N.V.;Barsuk,A.A.;Trifanova,L.R.; Sharanjuk,A.V.
(1991):Modeling of large unrestrained anisotropic structures.
Proc.of Gagarin scientific seminars on cosmonautics and
aviation (in Russians),Nauka,Moscow.

Birjuk,V.I.;Sharanjuk,A.V.(1984):Aeroelastic optimization
of constructive force schemes with the help of anisotropic
models (in Russian).Scientific Note, TsAGI, vol.XV, No 6,
p.77-84.

Bolotin,V.V.(1966):Plane problem of the theory of
elasticity for elements made of reinforced materials (in
Russian).Sb.Rasc%ety na Pro%cnost,Mashinostroenie,12.

Frolov,V.M.,Shanygin,A.N.,(1984):Application of composites
to anisotropic swept forward wing with aeroelastic and
strength properties consideration (in Russian).Mech of
composite materials,No 2,p.353-359.

Haug,E.J.;Choi,K.K.;Komkov,V.(1986):Design sensitivity
analysis of structural systems.Academic Press.

Hertz T.J.,Shirk M.H.,Ricketts R.H.,Weisshaar,T.A.(1982):
On the track of practical forward-swept wing.Astronautics
and Aeronautics,Jan., p.40-52.

Kobelev,V.V.;Larichev,A.D.(1988):Divergence of
composite thin-walled beams of closed cross-section in gas
flow (in Russian).Scientific Notes, TsAGI, vol.XIX.

Krone,N.J.(1975):Divergence elimination with advanced
composites, AIAA Paper,No 75-1009, Aug.

Pedersen,P.(1989):On optimal orientation of orthotropic
materias.Struct.Optim.1, 101-106.

Pedersen,P.(1990):Bounds on elastic energy in solids of orthotropic materials.Struct.Optim.2,55-63.

Weishaar,T.A.(1980):Divergence of forward swept composite wings.Journ.of Aircraft.vol.17, June, p.442-448.

Weishaar T.A.(1980);Aeroelastic tailoring of forwart swert composite wing.AIAA paper No 80-0795, May.

OPTIMIZATION OF LARGE SCALE SYSTEMS IN ELASTICITY

E. Schnack and G. Iancu
Institute of Solid Mechanics
Karlsruhe University
P.O. Box 6980
D - W 7500 Karlsruhe 1
Germany

ABSTRACT. Using the present procedure optimal stress designs have been computed in the last thirteen years for two-dimensional, axisymmetric and whole three-dimensional structures. Constraints on the variation of the boundary are considered in form of linear equations. The sensitivity behaviour for elastic structures can be predicted from the physical background which gives information about the influence of geometrical properties on the stress field at each point on a traction-free surface. In comparison with sensitivity methods, irregular boundaries can be avoided here by a priori introducing a smoothing operator.

The iterative nongradient procedure is realized in two- and three-dimensional software packages which give the possibility for example to analyze practical mechanical engineering problems like rotor masts of helicopters or connecting rods treated as a three-dimensional structure.

1. Introduction

The finite element formulation has been used by many researchers for shape optimization. Nonlinear programming with sensitivities obtained by implicitly differentiating the discretized equations has been used by Zienkiewicz and Campbell [46], Francavilla, Ramakrishnan and Zienkiewicz [07], Ramakrishnan and Francavilla [23] and Kristensen and Madsen [18] to solve this problem in two dimensions. The papers of Pedersen and Laursen [22], Zhang and Beckers [44] and Trompette and Marcelin [41] treat shape optimization of axisymmetric structures in a similar manner. Aspects associated with three-dimensional structures are discussed in this context by Botkin, Yang and Benett [03], Imam [14], and Kodiyalam and Vanderplaats [17]. A detailed description about computation of structural response using the FE based discrete approach and numerical problems associated with this are presented by Haftka [08], Wang, Sun and Gallagher [42] and Haftka and Barthelemy [09].

Variational equations as such derived in the papers of Choi and Haug [04], Chun and Haug [05,06], Zolesio [45] and in the book of Haugh, Choi and Komkov [10] take advantage of the variational character

G. I. N. Rozvany (ed.), Optimization of Large Structural Systems, Vol. II, 1073–1086.
© 1993 Kluwer Academic Publishers.

of FE formulation. A comparison between the discrete and continuum approach can be found for instance in [44].

Beside the classical gradient methods of mathematical programming, we have the possibility to develop a nongradient strategy of the feasible direction type for minimization of stress concentrations. The research work has shown that this is possible for two-dimensional, axisymmetric and general three-dimensional problems. For previous works and actual state of research on this topic, see Schnack [26-34], Schnack and Iancu [35-37], Schnack and Spörl [38], Schnack, Spörl and Iancu [39], Iancu and Schnack [12-13], Iancu [11] and Spörl [40].

Attention has also been paid to shape optimization with the boundary element approach. Mota Soares et al. [20,21], Rodriguez and Mota Soares [24] and Rodriguez [25] use a variational approach to optimize the shape of shafts. The paper of Kane and Saigal [15] is devoted to the sensititivity analysis by differentiating the discretized BE-equations for two-dimensional problems. Shape sensitivity using Analytical differentiation of the boundary integral equation was formulated for three-dimensional linearly elastic structures by Barone and Yang [01] and by Zhang and Mukherjee [43] for the plane case.

Contributions about adaptive meshing in context of shape optimization can be found in the works of Benett and Botkin [02], Kikuchi et al. [16] and Leal [19].

2. Analysis

Several problems of engineering are formulated for weight and cost minimization of structures involving beam, truss and plate elements with constraints on the design variables, displacements, stresses or eigen-

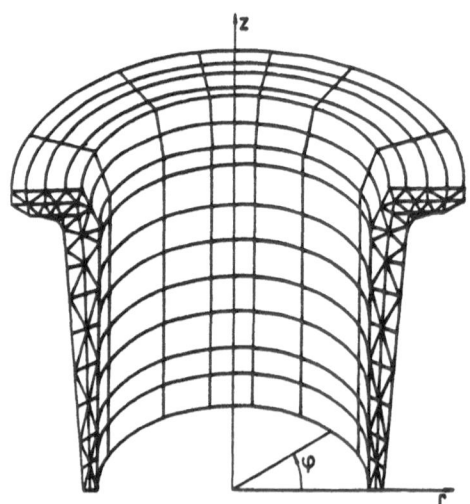

Figure 1. Axisymmetric structure - Rotor mast.

values. These are called sizing problems because the design variables: cross sectional dimensions, area's moments of inertia, moments of re-

sistance as well as the physical constraints are defined on a given domain. In the linear case one can derive analytical expressions for partial derivatives of the problem functions without any difficulties.

Because shape optimization problems with objective function or
constraints depending on state of system do not fall in this category,
they require a more complex treatement. An example is shape optimization with the objective minimization of stress concentration of the rotor mast of a helicopter shown in Figure 1. An axial section of this
axisymmetric structural component together with the boundary conditions, design boundary and variation domain are given in Figure 2. Here
denotes Γ design boundary and Γ* variation domain. In Figure 3 is shown
a typical FE mesh for the starting design. This problem has been treated in the case of axisymmetric loading, also in [44]. In the present
lecture we compare two optimal solutions for nonaxisymmetric loading.
They have been obtained in [41] using the augmented Lagrangian multiplier together with the DFP method of nonlinear programming and in [13]
by a nongradient method.

The optimal solution with the nongradient strategy is shown in Figure 4(a), while the optimal design with the DFP-Method from [41] is
demonstrated in Figure 4(b). A comparison of the stress distributions
of the starting design and of the solutions from [41] and [13] is shown
in Figure 5.

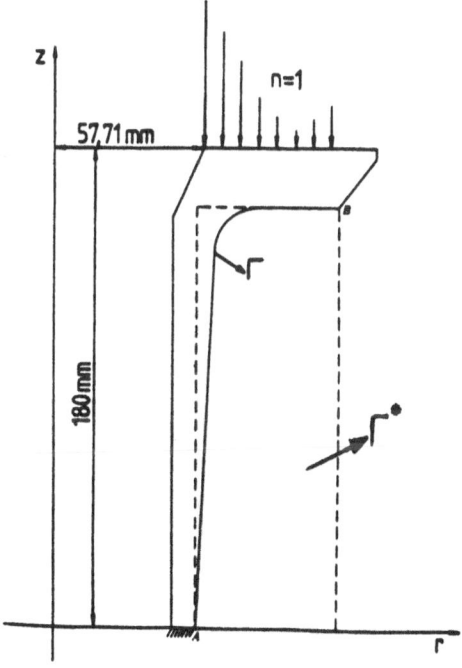

Figure 2. Axial section of the rotor mast.

It can be seen that in the critical area, the stress peak is rapidly reduced by the nongradient strategy. The design variable of such

a problem is the shape. This will be described in the following by the vector of design variables b. The vector b appears in the functions which describe the physical state of structure such as the stress components B in two ways: explicitly and implicitly through the displacement vector u(b):

$$B(b, u(b)) \qquad\qquad b \in \mathbb{R}^n. \qquad\qquad (1)$$

The minimization of stress concentrations can be written as follows:

$$\min \ f(B(b, u(b))) \qquad b \in \mathbb{R}^n. \qquad\qquad (2)$$

$$g_i(b) \le 0. \qquad\qquad (3)$$

$$h_j(B(b, u(b))) \le 0, \qquad\qquad (4)$$

with $i = 1(1)k, \ j = 1(1)\ell.$

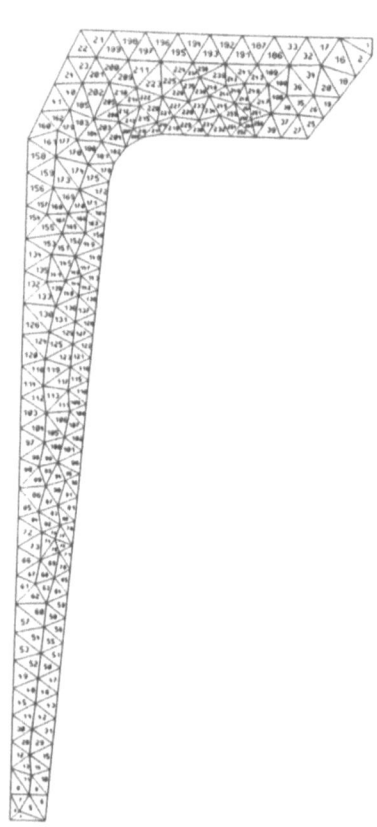

Figure 3. Starting Design.

The function f is defined as the maximum von Mises stress value $\bar{\sigma}_\mu$ of all M loading cases in a subdomain of the boundary value problem Ω^*:

$$f: \text{Max } \bar{\sigma}, \text{ in } \Omega^* \subset \Omega \text{ and for } \mu = 1(1)M. \tag{5}$$

The geometrical and physical constraints denoted by g and h respectively mean:

$$g: \quad \Gamma \subset \Gamma^*. \tag{6}$$

$$h: \quad \bar{\sigma}_\mu \leq \sigma \text{ in } \Omega. \tag{7}$$

In the inequality (7) σ denotes an upper stress bound. Because the shape optimization problem (5-7) is in general nondifferentiable, we have to transform it, if we want to use mathematical programming procedures for the solution. This can be done by many methods. In all these cases we have to compute the sensitivities of functions of type B.

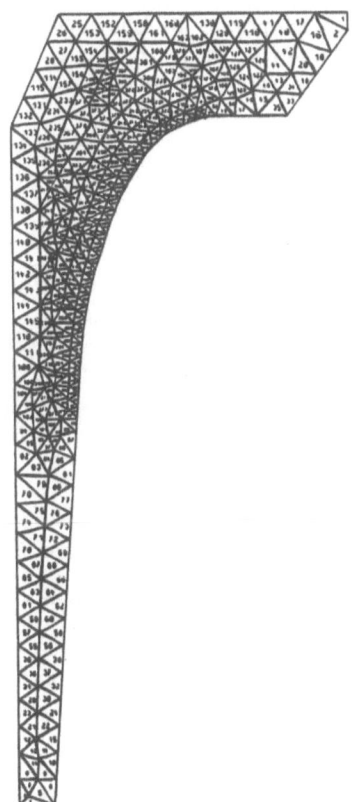

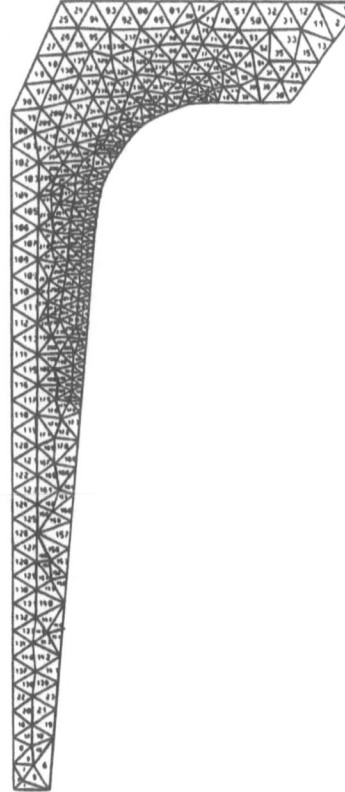

Figure 4(a): Optimal design with the nongradient method from [13]

Fig. 4(b): Optimal design with the DFP-method from [41]

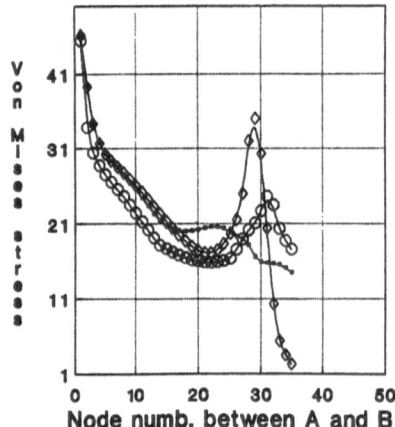

Figure 5. Stress distributions on Γ for: - -◆- starting design
- -◇- DFP-method from [41]
- -—- nongradient method from [13]

We can also predict pointwise the stress response to shape variation at a traction free boundary using the monotonicity theorem. The
nongradient procedure for minimization of stress concentration has the
following form:

$$b^T = \left[x_1, \ldots, x_{NB}, \ y_1, \ldots, y_{NB}, \ z_1, \ldots, z_{NB}\right], \tag{8}$$

$$\min \max \left(\bar{\sigma}^{\mu}_1(b), \ldots, \bar{\sigma}^{\mu}_{NB}(b)\right), \tag{9}$$

$$\bar{\sigma}^{\mu}_i(b) - \sigma \leq 0, \tag{10}$$

$$Ab - B \leq 0, \tag{11}$$

with $\mu = 1(1)M$ and $i = 1(1)N$.

with NB number of nodal points on the design boundary.
The transition function f_j which describes the iteration rule for
changing the nodal point coordinates on the optimizing boundary is defined on a physical basis:

• A geometrical perturbation on the design boundary Γ produces a rapid
  fade away of the von Mises stress $\bar{\sigma}$ in the neighbourhood of the perturbation.

• From the monotonicity relation of the two principal stresses with respect to the corresponding normal curvatures in the principal stress
  directions we can derive a relation for the control of the von Mises
  stress on Γ. We have pointwise:

$$\delta\bar{\sigma}^{\mu} = K_{w}^{s}\delta h_{n},$$ (12)

with K_{w}^{s} as weighted surface curvature which depends generally on stress state and geometrical data on the surface, and δh_{n} as the perturbation in the normal direction of the boundary Γ.

- By increasing the minimum effective stress, the maximum effective stress in the direct neighbourhood can also be reduced:

$$b_{j} = f_{j}(b_{j-1}, v_{j}) \quad \text{explicit (for tetrahedron elements).}$$ (13)

$$x_{j}^{i} = x_{j-1}^{i} + v_{j}^{i} \cdot \frac{1}{N_{i}} \sum_{k=1}^{N_{i}} \bar{a}_{j-1}^{i,k},$$ (14)

$$y_{j}^{i} = y_{j-1}^{i} + v_{j}^{i} \cdot \frac{1}{N_{i}} \sum_{k=1}^{N_{i}} \bar{b}_{j-1}^{i,k},$$ (15)

$$z_{j}^{i} = z_{j-1}^{i} + v_{j}^{i} \cdot \frac{1}{N_{i}} \sum_{k=1}^{N_{i}} \bar{c}_{j-1}^{i,k},$$ (16)

for $i = 1(1)NB$.

with $\bar{a}_{j-1}^{i,k}$, $\bar{b}_{j-1}^{i,k}$ and $\bar{c}_{j-1}^{i,k}$ coefficients of the Hessian form for each of the N_{i} surface triangles abutting the node i. Because the magnitude of shifting v_{j} is controlled by an arithmetical smoothing algorithm, we have at the nodal point i:

$$v_{j}^{i} = v_{j}^{i*} \xi_{j-1}^{i} \left(1 - \frac{s^{i}}{M_{s}}\right), \quad s^{i} = 1(1)M_{s},$$ (17)

with M_{s} = number of smoothing zones,

v_{j}^{i*} = magnitude of shifting at a point with maximum or minimum stress.

Making the approximation that each of two principal stresses is a linear function of the normal curvature we get for the sign of shiftig ξ_{j-1}^{i}:

$$\xi_{j-1}^{i} = \text{sgn}\,(\sigma_{1,j-1}^{i} + \sigma_{2,j-1}^{i})\,\text{sgn}(\bar{\sigma}_{j-1}^{i} - \overset{\approx}{\sigma}_{j-1}^{i}),$$ (18)

for $i = 1(1)NB$,

where σ_1^i and σ_2^i denote the two principal stresses at the nodal point i and $\overset{\alpha}{\sigma}$ the average von Mises stress on the design boundary.

As a result, we have a discrete, dynamic optimization problem with the following 'cost function' g_j:

$$g_j := (\bar{\sigma})_j^{max} - (\bar{\sigma})_{j-1}^{max}. \tag{19}$$

$$(\bar{\sigma})_j^{max} = Max \ ((\bar{\sigma}_1^\mu)_j, \ldots, (\bar{\sigma}_{NB}^\mu)_j), \ \mu = 1(1)M. \tag{20}$$

Problem:

$$min \sum_{j=1}^{l} g_j(b_{j-1}, v_j), \tag{21}$$

$$b \ \epsilon \ \Xi.$$

$$v_j \ \epsilon \ \Omega_j(b_{j-1}), \tag{22}$$

with $\Omega_j(b_{j-1})$ as feasible control space.

$$\left(x_j^i, y_j^i, z_j^i\right) \ \epsilon \ \bar{\Omega}_j^* \subset \mathbb{R}^n, \tag{23}$$

with $n = 3$, $i = 1(1)NB$, $j = 1(1)l$,

$\bar{\Omega}^*$ closed set: $\bar{\Omega}^* = \Omega \cup \Gamma_- \cup \Gamma_+$,

state space Ξ: $\Xi \subset \mathbb{R}^{3NB}$.

Ξ is compact, as a product of compact sets from geometrical constraints:

$$v_j^i \ \epsilon \ \left[-d^i\left(\Gamma_{d, j-1}, \Gamma_-\right), \ d^i\left(\Gamma_{d, j-1}, \Gamma_+\right)\right] = \Omega_j^i(b_{j-1}). \tag{24}$$

For the decision space it follows:

$$\Omega_j(b_{j-1}) = \prod_{i=1}^{NB} \Omega_j^i(b_{j-1}) \subset \mathbb{R}^{NB}. \tag{25}$$

With the theorems of Tychonoff (Ω_j^i is compact) and Weierstrass and making the supposition that the objective function from the dynamic optimization problem is continuous, we have the existence of the solution.

3. 3D-Test Problem

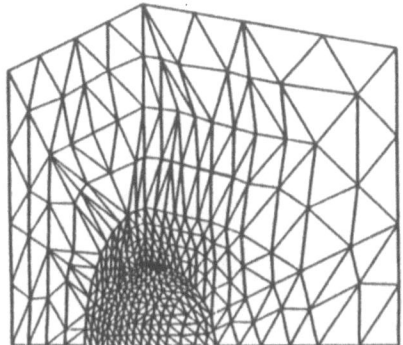

Figure 6. FE-discretization of body with spherical hole.

An example of application of the nongradient procedure for three-dimensional structures is the stress optimal shape of a cavity in a large elastic domain (see [11] and Figure 6). The starting geometry for this problem (see Figure 7) was a spherical hole:

$$a : b = a : c = 1, \tag{26}$$

with the maximum stress $3.36\ \sigma_{Ox}$.

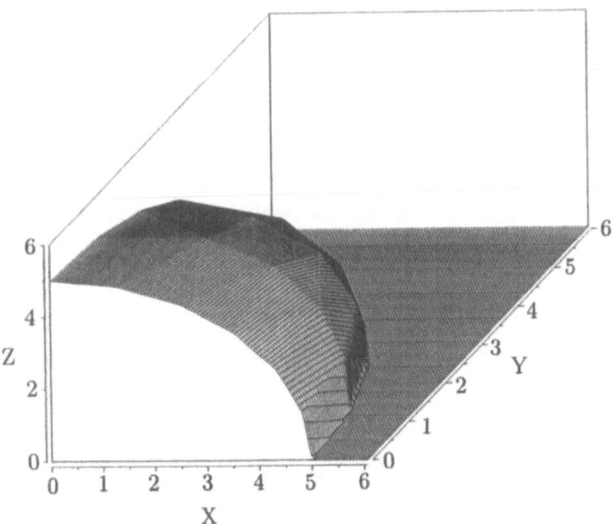

Figure 7. Shape of the starting surface.

The loading case is:

$$\sigma_{Ox} : \sigma_{Oy} : \sigma_{Oz} = 1 : 2 : 2. \tag{27}$$

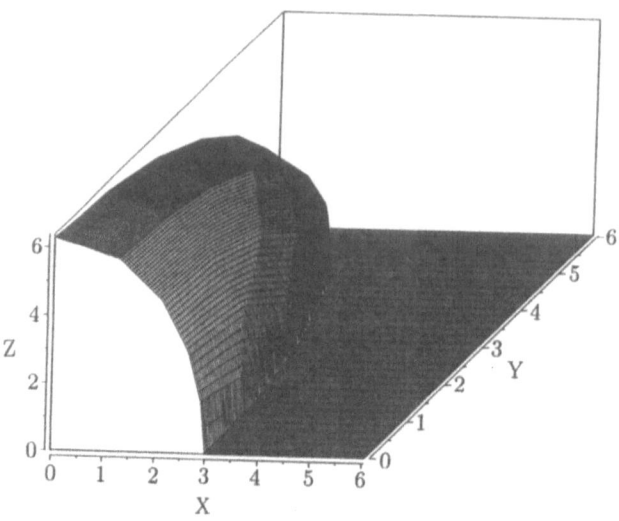

Figure 8. Optimal shape of cavity in a large elastic domain.

After 24 iteration steps we obtain an axisymmetric ellipsoid with:

$$b : a = c : a = 2.160. \tag{28}$$

The maximum stress value is reduced here to $2.70 \, \sigma_{Ox}$. This shape is shown in Figure 8. The proof of optimality for the cavity problem can be givenanalytically with the maximum priciple. The result is:

$$b: a = c : a = 2.614, \tag{29}$$

and a constant von Mises stress on Γ of $2.5 \, \sigma_{Ox}$.

The examples of the last chapter show the high performance of the nongradient strategies in optimizing notch problems of elasticity.

4. References

01. Barone, M.R. and Yang, R.J. (1989) 'A Boundary Element Approach for Recovery of Shape Sensitivities in Three-Dimensional Elastic Solids', Computer Methods in Applied Mechanics and Engng. 74, 69-82.
02. Benett, J.A. and Botkin, M.E. (1983) 'Shape Optimization of Two-Dimensonal Structures with Geometric Problem Description and Adaptive Mesh Refinement', AIAA Journal.

03. Botkin, M.E., Yang, R.J. and Benett, J.A. (1985) 'Shape Optimization of Three-Dimensional Stamped and Solid Automotive Components', Paper presented at the International Symposium on Optimum Shape, General Motors Research Labs, Warren, Michigan.

04. Choi, K.K. and Haug, E.J. (1983) 'Shape Design Sensitivity Analysis of Elastic Structures', J. of Structural Mech. 11/2, 231-269.

05. Chun, Y.W. and Haug, E.J. (1978) 'Two-Dimensional Shape Optimal Design', Int. J. Num. Meth. Engng. 13, 311-336.

06. Chun, Y.W. and Haug, E.J. (1983) 'Shape Optimization of a Solid of Revolution', J. of Engng. Mech. 109/1, 30-46.

07. Francavilla, A., Ramakrishnan, C.V. and Zienkiewicz, O.C. (1975) 'Optimization of Shape to Minimize Stress Concentration', J. of Strain Analysis 10/2, 63-69.

08. Haftka, R.T. (1987) 'Finite Elements in Optimal Structural Design', in C.A. Mota Soares (ed.), Computer Aided Optimal Design: Structural and Mechanical Systems, Springer-Verlag, Berlin and New York, pp. 271-297.

09. Haftka, R.T. and Barthelemy, B. (1989) 'On the Accuracy of Shape Sensitivity', in C.A. Brebbia and S. Hernandez (eds.), Computer Aided Optimum Design of Structures: Recent Advances, Springer-Verlag, Berlin, Heidelberg and New York, pp. 327-336.

10. Haug, E.J., Choi, R.K. and Komkov, V. (1986) 'Design Sensitivity Analysis of Structural Systems', in W. F. Ames (ed.), Mathematical Science and Engineering, Academic Press, Orlando, San Diego and New York.

11. Iancu, G. (1991) 'Optimierung von Spannungskonzentrationen bei dreidimensionalen elastischen Strukturen', Doctoral Thesis, Karlsruhe University.

12. Iancu, G. and Schnack, E. (1989) 'Knowledge-Based Shape Optimization', in C.A. Brebbia and S. Hernandez (eds.), Computer Aided Optimum Design of Structures: Recent Advances , Proceedings of the First Conference on Computer Aided Optimum Design of Structures (CAOD) - OPTI 89, Southampton, United Kingdom, 1989, Springer-Verlag, Berlin, Heidelberg and New York, pp. 71-83.

13. Iancu, G. and Schnack, E. (1990) 'Shape Optimization Scheme for Large Scale Structures', in Proceedings of the Second World Congress on Computational Mechanics, 27 - 31 August 1990, Stuttgart, Germany, International Association of Computational Mechanics, pp. 216-219.

14. Imam, M.H. (1982) 'Three-Dimensional Shape Optimization', Int. J. of Num. Meth. Engng. 18, 661-673.

15. Kane, J. and Saigal, S. (1988) 'Design-Sensitivity Analysis of Solids Using BEM', J. of Engng. Mech. 114/10, 1703-1722.

16. Kikuchi, N., Chung, K.Y., Torigaki, T. and Taylor, J.E. (1986), 'Adaptive Finite Element Methods for Shape Optimization of Linearly Elastic Structures', Comp. Meth. in Appl. Mech. and Engng. 57, 67-89.

17. Kodiyalam, S. and Vanderplaats, G.N. (1989) 'Shape Optimization of Three-Dimensional Continuum Structures via Force Approximation Techniques', AIAA Journal 27/9, 1256-1263.

18. Kristensen, E.S. and Madsen, N.F. (1976) 'On the Optimum Shape of Fillets in Plates Subjected to Multiple In-Plane Loading Cases',

Int. J. Num. Meth. Engng. 10, 1007-1019.

19. Leal, R.P. (1985) 'Boundary Elements in Bidimensional Elasticity', Master Sc. Thesis, Technical University of Lisbon.

20. Mota Soares, C.A, Rodrigues, H.C., Oliveira Faria, L.M. and Haug, E.J. (1984) 'Optimization of the Geometry of Shafts Using Boundary Elements', ASME J. of Mechanisms, Transmissions and Automation in Design 106, 199-203.

21. Mota Soares, C.A., Rodriguez, H.C., Oliviera Faria, L.M. and Haug, E.J. (1985) 'Boundary Elements in Shape Optimal design of Shafts', in J.S. Gero (ed.), Optimization in Computer Aided Design, pp. 155-175, North-Holland.

22. Pedersen, P. and Laursen, L.L. (1982/83) 'Design for Minimum Stress Concentration by Finite Element Elements and Linear Programming', J. of Struct. Mech. 10/4, 375-391.

23. Ramakrishnan, C.V. and Francavilla, A. (1974/75) 'Structural Shape Optimization Using Penalty Functions', J. of Struct. Mech. 3/4, 403-422.

24. Rodriguez, H.C. and Mota Soares, C.A. (1983) 'Shape Optimization of Shafts', in Proceedings of the Third National Congress of Theoretical and Applied Mechanics, Lisbon, Portugal.

25. Rodriguez, H.C. (1984) 'Shape Optimization of Shafts Using Boundary Elements', Master Sc. Thesis, Technical University of Lisbon.

26. Schnack, E. (1977) 'Ein Iterationsverfahren zur Optimierung von Spannungskonzentrationen', Habilitationsschrift, University of Kaiserslautern.

27. Schnack, E. (1978) 'Ein Iterationsverfahren zur Optimierung von Kerboberflächen', VDI-Forschungsheft 589, VDI-Verlag, Düsseldorf.

28. Schnack, E. (1979) 'An Optimization Procedure for Stress Concentrations by the Finite Element Technique', Int. J. Num. Meth. Engng. 14/1, 115-124.

29. Schnack, E. (1980) 'Optimierung von Spannungskonzentrationen bei Viellastbeanspruchung', ZAMM 60, T151-T152.

30. Schnack, E. (1982) 'Optimal Designing of Notched Structures without Gradient Computation', in J.P. Barbary and L. Le Letty (eds.), Control of Distributed Parameter Systems, Proceedings of the 3rd IFAC-Symposium, Toulouse, France, 1982, Pergamon Press, Oxford, New York, Toronto, Sydney, Paris and Frankfurt, pp. 365-369.

31. Schnack, E. (1985) 'Computer Simulation of an Experimental Method for Notch-Shape-Optimization', in J. Burger and Y. Janny (eds.), Simulation in Engineering Sciences, Proceedings of the Int. Symp. of IMACS, Nantes, France, 1983, Tome 2, Elsevier Science Publishers B.V. (North Holland), Amsterdam, The Netherlands, pp. 269-275.

32. Schnack, E. (1985) 'Local Effects of Geometry Variation in the Analysis of Structures', in P. Ladevèze (ed.), Studies in Applied Mechanics 12: Local Effects in the Analysis of Structures, Elsevier Science Publisher, Amsterdam, pp. 325-342.

33. Schnack, E. (1986) 'Free Boundary Value Problems in Elastostatics', in R.P. Shaw, J. Periaux, A. Chaudouet, J. Wu, C. Marino and C.A. Brebbia (eds.), Innovative Numerical Methods in Engineering, Proceedings of the 4th Int. Symp. on Numerical Methods in Engineering, Atlanta, Georgia/USA, 1986, Computational Mechanics Publications Southampton, Springer-Verlag, Berlin, Heidelberg and New York, pp.

435-440.

34. Schnack, E. (1988) 'A Method of Feasible Direction with FEM for Shape Optimization', in G.I.N. Rozvany and B.L. Karihaloo (eds.), Structural Optimization, invited lecture: Proceedings of the IUTAM-Symp. on Structural Optimization, Melbourne, Australia, 1988, Kluwer Academic Publishers, Dordrecht, Boston and London, pp. 299-306.

35. Schnack, E. and Iancu, G. (1989) 'Control of the von Mises Stress with Dynamic Programming', in H.A. Eschenauer and G. Thierauf (eds.), Proceedings of the GAMM-Seminar, Vol. 43, GAMM-Seminar on Discretization Methods and Structural Optimization - Procedures and Applications, Siegen, Federal Republic of Germany, 1988, Springer-Verlag, Berlin and Heidelberg, pp. 154-161.

36. Schnack, E. and Iancu. G. (1989) 'Shape Design of Elastostatics Structures Based on Local Perturbation Analysis', Structural Optimization 1, 117-125.

37. Schnack, E. and Iancu, G. (1989) 'Non-Linear Programming Applicable for the Control of Elastic Structures', in A. El Jai and M. Amouroux (eds.), Preprints of the 5th IFAC Symposium on Control of Distributed Parameter Systems, Proceedings of the 5th IFAC Symposium on Control of Distributed Parameters, Perpignan, France, 1989, Institut de Science et de Génie des Matériaux et Procédés (CNRS), Groupe d'Automatique, Université de Perpignan, pp. 163-168.

38. Schnack, E. and Spörl, U. (1986) 'A Mechanical Dynamic Programming Algorithm for Structure Optimization', Int. J. Num. Meth. Engng. 23/11, 1985-2004.

39. Schnack, E. Spörl, U. and Iancu, G. (1988) 'Gradientless Shape Optimization with FEM', VDI Forschungsheft 647/88, 1-44.

40. Spörl, U. (1985) 'Spannungsoptimale Auslegung elastischer Strukturen', Doctoral Thesis, Karlsruhe University.

41. Trompette, Ph. and Marcelin, J.L. (1986) 'On the Choice of the Objectives in Shape Optimization', in C.A. Mota Soares (ed.), Computer Aided Optimal Design: Structural and Mechanical Systems, Springer-Verlag, Berlin and New York, pp. 247-261.

42. Wang, S.-Y., Sun, Y. and Gallagher, R.H. (1985) 'Sensitivity Analysis in Shape Optimization of Continuum Structures', Computer & Structures, 20/5, 855-867.

43. Zhang, Q. and Mukherjee, S. (1990) 'Design Sensitivity Coefficients for Linear Elasticity Problems by Boundary Element Methods', in G. Kuhn and H. Mang (eds.), Proceedings of the IUTAM/IACM Symposium on Discretized Methods in Structural Mechanics, Vienna, Austria, Springer-Verlag, Berlin and Heidelberg, pp. 283-289.

44. Zhang, W.H. and Beckers, P (1989) 'Comparison of Different Sensitivity Analysis Approaches for Structural Shape Optimization', in C.A. Brebbia and S. Hernandez (eds.), Computer Aided Optimum Design of Structures: Recent Advances, Springer-Verlag, Berlin, Heidelberg and New York, pp. 346-356.

45. Zolesio, J.-P. (1981) 'The Material Derivative (or speed) Method for Shape Optimization of Distributed Parameter Structures', in E.J. Haug and J. Cea (eds.), Zuithoff and Noordhoff, Alphen aan den Rhijn, pp. 1089-1151.

46. Zienkiewicz, O.C. and Campbell, J.S. (1973) 'Shape Optimization and Sequential Linear Programming', in R.H. Gallagher and O.C. Zien-

kiewicz (eds.), Optimum Structural Design, John Wiley & Sons, London, New-York, and Sydney.

MINIMUM WEIGHT DESIGN OF STRUCTURES UNDER NON-CONSERVATIVE FORCES

W. Gutkowski *, O. Mahrenholtz **, M. Pyrz *
* Institute of Fundamental Technological Research, Warsaw, Poland
** Technical University of Hamburg–Harburg, Germany

ABSTRACT. The paper deals with a method of determining minimum weight design of a structure subjected to nonconservative forces with stability constraints. The relatively simple solution algorithm based on Kuhn–Tucker necessary conditions is proposed. The procedure of scaling design variables for required critical loading and normalization condition for eigenvectors are introduced at each iterative step to overcome difficulties due to the high sensitivity fo the problem. The numerical example of Beck's column shows efficiency of the proposed method which in this case gives the highest so far known value of the critical force with no additional constraints added on equality of two first critical loads.

1 Introduction

Since about half a century a lot of attention has been paid to dynamic stability. The most elaborated problem so far is the problem of a column fixed at one end and loaded by circulatory load at the other end. The problem is known as Beck's column. Works in the field of optimization of this kind of structures have been carried out in two main directions. The first one is related to variational principles for nonconservative stability problems e.g. Claudon (1975), Hanaoka, Washizu (1980), Seguchi, Kojima (1988), Tada et al. (1989). The second direction has been aimed on direct analysis of various system parameters influence on optimum critical load e.g. Bogacz, Mahrenholtz (1986), Mahrenholtz, Bogacz (1988).

G. I. N. Rozvany (ed.), Optimization of Large Structural Systems, Vol. II, 1087–1099.
© 1993 Kluwer Academic Publishers.

The optimum solution of the problem in question is very sensitive to variation of design variables. This is one of the reasons that so far proposed algorithms for their solution e.g. by Hanaoka, Washizu (1980) or Tada et al. (1989) include some rather complex procedures difficult to incorporate in an algorithm which might be useful for a broader class of structures subjected to follower forces.

The object of the paper is to give a consistent and relatively simple method for optimum design of an arbitrary elastic structure subjected to nonconservative forces. The classical FEM procedures are applied with additional matrix arising from the fact that nonconservative forces are applied to the structure.

The presented method is based on Kuhn–Tucker necessary conditions for an optimum problem proposed in Gutkowski et al. (1990). It is free of approximations assumed in Optimality Criteria approach and nonlinear programming technique which consists in separability condition and linearization of some constraints e.g. Berke, Khot (1986). The proposed method may be easily applied together with a standard software of FEM. The solution is obtained by an iterative process carried out on a system of equations of motion and optimality conditions.

Two relatively simple procedures introduced in the proposed method are helping to overcome the main difficulties arising from the high sensitivity of the problem. The first one is scaling. The second one is the normalization relation for eigenvectors. The relation is derived from the necessary conditions for the optimum problem. Both procedures are applied at each iterative step of the proposed algorithm which gives the highest so far known value of critical force for Beck's column.

2 Preliminary remarks

The classical FEM formulation of a linear dynamic structural problem leads to the known relations obtained under assumption of harmonic motion

$$(K - \omega^2 M)u = 0 \quad , \tag{1}$$

where K–global stiffness matrix of the structure; M–global mass matrix; ω–natural frequency, u–displacements of nodal points.

On the other hand the initial stability problem for a structure subjected to static conservative load is described by the relation

$$(K - pK_\sigma)u = 0 \tag{2}$$

where $\mathbf{K}_\sigma$ is the initial stress matrix (geometric matrix) and p is the loading multiplier.

It has been shown in Hanaoka, Washizu (1980) that the conservative force P may be seen as a vector sum of two mutually perpendicular forces V and F (Fig. 1). The force in V (linear approximation equal to P) may be regarded as a conservative force. The component $F = \vartheta P$ can be seen as an additional force applied to the structural node. In the case of finite elements with a rotational degree of freedom ϑ is well defined as a rotation of the node to which P is applied. In the case of trusses ϑ has to be evaluated as rigid rotation of the truss member along which the nonconservative force is acting. Under these assumptions the principle of virtual work in FEM leads to an additional loading vector composed of nonconservative forces multiplied by angles of their rotation during the process of deformation (Hanaoka, Washizu (1980)). This vector then may be represented by a product of a asymmetric matarix $\mathbf{F}$ and displacement vector $\mathbf{u}$. This way we arrive at the system of equations in the following form

$$(\mathbf{K} - \omega^2\mathbf{M} - p\mathbf{H})\mathbf{u} = 0 \tag{3}$$

where

$$\mathbf{H} = \mathbf{K}_\sigma + \gamma\mathbf{F} \tag{4}$$

and γ is a factor taking into account the fact that P may not exactly follow the angle of rotation ϑ but only its γ part. In other words $\gamma = 0$ is equivalent to the problem of a structure under conservative forces. The force $\mathbf{F}$ as well as $\mathbf{K}_\sigma$ are constant in the problem. The variation of the axial force is obtained by multiplication of $\mathbf{F}$ by multiplier p. The known necessary condition for nontrivial solutions of the system of equations (3) is

$$Det(\mathbf{K} - \omega^2\mathbf{M} - p\mathbf{H}) = 0 \quad . \tag{5}$$

In the cases of natural frequency or stability the determinant for (1) and (2) gives polynomials as a function of one variable ω^2 or p. The zero value of the determinant in these cases results in distinct values for ω^2 or p. In the present problem the condition (5) represents then a set of curves $D(\omega^2, p)$ along which the determinant is zero with p and ω^2 being real. Beyond these curves the roots of the polynomial (5) either do not exist or some of them are complex.

With the assumption of harmonic motion of the system the force P related to two distinct frequencies joining in a double roots is a critical force. This is due to the fact that the double root solution possesses only one eigenvector such that the amplitude assumed in the dynamic analysis is growing linearly with respect to time.

The nature of these curves has so far not been fully recognized. However, observing $p - \omega^2$ curves for the first four eigenvalues obtained in listed papers three cases of relations between the critical multiplier p_{cr} and ω_i^2 may be distinguished (Fig. 2). For all these cases two common stability constraints may be concluded:

$$\omega_{i+1}^2 \geq \omega_i^2; \ p_{cr} \geq p . \tag{6}$$

3 Formulation of the problem for minimum weight structure

Assume a given layout of the middle surface of a structure (beam, plate, shell) together with matrices $\mathbf{K}_\sigma$ and $\mathbf{F}$ and multiplier p_0. In other words, a given nonconservative system of external loads which magnitude is expressed by multiplier p_0.

Under the above assumption the formulation of the problem in question is as follows:

The cost function is the volume of the material,

$$W = \mathbf{A}^\mathbf{T}\mathbf{L} \rightarrow \min , \tag{7}$$

where $\mathbf{A}$ is a vector of design variables A_j representing the cross section of a beam or the thickness of a plate or shell, $\mathbf{L}$ denotes the vector of length of beams or areas of finite elements.

Let us now specify the equality constraints of our problem. First of all we have to fulfill the equation of motion by (3)

$$(\mathbf{K} - \omega^2\mathbf{M} - p_0\mathbf{H})\mathbf{u} = 0 . \tag{8}$$

Next is the normalization condition for eigenvector u e.g.:

$$(\mathbf{u}^T\mathbf{M}\mathbf{u} - 1) = 0 . \tag{9}$$

The stiffness matrix $\mathbf{K}$ contains the flexural rigidities of particular finite elements. Denoting it by I_j we assume its relation to design variable A in the following form:

$$I_j = rA_j^q ;$$ (10)

and for plates $r = 1/12(1 - \nu^2)$ and $q = 3$.

Let us now define inequality constraints. We start with the relation between two successive eigenfrequencies discussed above:

$$\omega_{i+1}^2 - \omega_i^2 \geq 0 .$$ (11)

The critical multipliers p_{cr}^k should be larger than the given value p_0. Then, the appropriate inequality constraint has the form

$$p_{cr}^k - p_0 \geq 0 .$$ (12)

The last inequality constraint considered here is related to the minimum value of available design variable A_j. Denoting it by A_j^{min} we have

$$A_j - A_j^{min} \geq 0 .$$ (13)

The above relation also implies for stiffness I_j

$$I_j - r(A_j^{min})^q \geq 0 .$$ (14)

Having cost function and constraints we can write down the Lagrangian for the problem in question:

$$
\begin{aligned}
L = \ & -\ \mathbf{A}^T \mathbf{L} + \sum_{i=1}^n \lambda_i^{\theta} [(\mathbf{K} - \omega^2 \mathbf{M} - p_0 \mathbf{H}) \mathbf{u})]_{(i)} \\
& + \sum_{i=1}^{n-1} \lambda_i^{\omega} (\omega_{i+1}^2 - \omega_i^2) + \lambda^0 (\mathbf{u}^T \mathbf{M} \mathbf{u} - 1) + \\
& + \sum_{k=1}^{n/2} \lambda_k^p (p_{cr}^k - p_0) + \sum_{j=1}^m \lambda_j^c (A_j - A_j^{min}) \quad , \quad \lambda_i^{\omega}, \lambda_k^p, \lambda_j^c \geq 0 ,
\end{aligned}
$$ (15)

where $\lambda_i^\theta, \lambda^0, \lambda_i^\omega, \lambda_k^p, \lambda_j^c$ are Lagrange multipliers associated respectively with equality constraints (equilibrium and normalization conditions) and inequality constraints imposed on eigenfrequencies, axial force and minimum value of design variable A_j and (i) denotes the i-th equation.

Equality (10) is substituted into stiffness matrix $\mathbf{K}$. This way, design variables are cross section areas A_j only.

Let us assume that $A_j, \mathbf{u}, \omega_i^2$ and p_{cr} are independent variables. Then, Kuhn–Tucker necessary conditions for our problem take after some transformations the following form:

$$(\mathbf{K} - \omega_0^2\mathbf{M} - p_0\mathbf{H})\mathbf{u} = 0,$$

$$\lambda_i^\theta(\mathbf{K} - \omega_0^2\mathbf{M} - p_0\mathbf{H})_{(i)} = 0, \qquad\qquad\qquad i = 1, 2, \ldots, n,$$

$$-L_jA_j + \sum_{i=1}^n \lambda_i^\theta[(q\mathbf{K}_{[j]}] - \omega_0^2\mathbf{M}_{[j]})\mathbf{u}]_{(i)} + \lambda_j^cA_j = 0, \quad j = 1, 2, \ldots, m, \tag{16}$$

$$\sum_{i=1}^n \lambda_i^\theta(\mathbf{M}\mathbf{u})_{(i)} = 0, \quad \lambda_j^c(A_j - A_j^{min}) = 0, \quad \lambda_j^c \geq 0 \quad ,$$

where $[j]$ denotes elements of a matrix depending only on the j-th design variable A_j and ω_0 corresponds to the force multiplier p_0. It should be noted that due to the fact that $(\mathbf{K} - \omega_0^2\mathbf{M} + p_0\mathbf{H})$ is an asymmetric matrix $\mathbf{u}$ and λ^θ are right and left eigenvectors and they are not equal like in the case of a symmetric matrix. The further discussion is devoted to the solution of the above system of equations and inequalities given in (16). The main difficulty consists here in high sensitivity of the critical force for minimum weight structure with variation of design variables. Two subsequent sections of the paper are devoted to two procedures helping to overcome these difficulties.

4 Scaling

Let us assume for A_j some arbitrary values. Substituting them into (5) we can find eigenfrequencies and eigenvectors for chosen values of multiplier p and then for the critical, minimal force multiplier p_{cr}^{min}. The latter one will in general differ from p_0, it means from the value for which we are designing a given structure. However, we can easily change the value of A_j by scaling it with a number obtained from the ratio p_0/p_{cr}^{min}. This allows to get new values of A_j for critical multiplier $p_0 = p_{cr}^{min}$ without need of resolving the eigenproblem.

The scaling procedure is as follows. We multiply (3) by coefficient $s = p_0/p_{cr}^{min}$. It means that just by multiplication of A_j by $s^{1/q}$ we get new values of design variables for which $p_0 = p_{cr}^{min}$.

5   Normalization condition for eigenvectors

As mentioned above right and left eigenvectors respectively $\mathbf{u}$ and λ^θ are involved in the problem. From $(16)_3$, it may be seen that their normalization conditions may not be independent.

Let us recall that with (9) we have already chosen a certain normalization condition for the eigenvector $\mathbf{u}$. We have to find then normalization for the left eigenvector which would match our problem.

Let us assume $A_j > A_j^{min}$, then $\lambda_j^c = 0$. Now we add $(16)_3$ for $j = 1, 2, \ldots, p$. The result of summation is the following:

$$-W + \sum_{i=1}^{n} \lambda_i^\theta [(q\mathbf{K} - \omega_0^2 \mathbf{M})\mathbf{u}]_{(i)} = 0 \ . \tag{17}$$

The above equation together with $(16)_4$ give

$$W = q \sum_{i=1}^{n} \lambda_i^\theta (\mathbf{K}\mathbf{u})_{(i)} \ , \tag{18}$$

relating magnitudes of both eigenvectors to the cost function and power of q specified in (10).

6   The solution algorithm

The presented problem is solved by successive approximation solution of a set of equations and inequalities (16) arising from Kuhn–Tucker necessary conditions for optimum problem. In order to make the solution more stable two procedures mentioned in paragraphs 4 and 5 are introduced at each iterative step.

With the above remarks the following algorithm is proposed:

Step 1.   Assume an arbitrary vector $\mathbf{A}(0)$ of design variables

$A_j(0)$.

Step 2. $\nu := 0$ (ν is an iteration counter).

Step 3.   For given $\mathbf{A}(\nu)$ find by successive bisections $p_{cr}^{min}; \omega_{cr}^{min};$

$\mathbf{u}; \lambda^\theta$ with increasing values of multiplier p.

Step 4.   Scale design variables

$$A_j(\nu) := A_j(\nu)\sqrt[q]{s(\nu)}$$

where

$$s(\nu) = \frac{p_0}{p_{cr}^{min}(\nu)}$$

Step 5.   Normalize eigenvector λ^θ according to (18)

$$W = q\sum_{i=1}^{n} \lambda_i^\theta (\mathbf{Ku})_{(i)}$$

(a normalization for eigenvector $\mathbf{u}$ included in the
applied standard numerical eigenproblem procedure)

Step 6.   Find residual $D_j(\nu)$ from $(16)_3$

$$-L_j A_j(\nu) + \sum_{i=1}^{n} \lambda_i^\theta(\nu)\{[q\mathbf{K}(\nu)_{[j]} - \omega_{cr}^2(\nu)\mathbf{M}(\nu)_{[j]}]\mathbf{u}(\nu)\}_{(i)} = D_j(\nu),$$
$$j = 1, 2, \ldots, p;$$

if $|D_j(\nu)| \ge C_j$ (C_j – given small number) go to step 7
else go to step 9.

Step 7. $\nu := \nu + 1$.

Step 8. $A_j(\nu + 1) = A_j(\nu)[1 + \beta(\nu)\dfrac{D_j(\nu)}{[\sum\limits_{j} D_j^2(\nu)]^{1/2}}]$

go to step 3. (where $\beta(\nu)$ is an accelerating coefficient)

Step 9.   Stop.

7   Numerical example

An elastic column clamped at one end and subjected to a tangential compressive follower force P at the other end is considered (Beck's column). The problem is to determine the mass distribution of this cantilever which for a given load multiplier p_0 minimizes the weight of the structure and satisfies the condition of dynamic stability. The numerical example is recalculated so that the total mass equals 1 and is compared with the results from the literature.

The column is divided into 10 elements and it is assumed that all cross sections along the column are geometrically similar and depend on a one nondimensional design variable m – the linear mass density i.e. $A_j = m_j A_0$, $I_j = m_j^2 I_0$, where A_j is the cross–sectional area, I_j is the moment of inertia and A_0, I_0 are reference values. We assume the following dimensionless quantities

$$\lambda = \frac{\omega^2 \rho A_0 L^4}{E I_0}, \qquad p = \frac{P L^2}{E I_0},$$

where ω is eigenfrequency, ρ–mass density, L–length of the column.

The final results are presented in Fig. 3. The results have been scaled so that the total mass equals 1. The optimal column is characterized by the critical load $p_{cr} = 92,56$ which represents an increase in p_{cr} of 462% over the uniform column and is the highest known value in the literature. The best results calculated so far, i.e. $p_{cr} = 83.53$ by Hanaoka, Washizu (1980) and $p_{cr} = 90.80$ by Tada et al. (1989), have been obtained on the way of the force maximization with the constraint on mass equals 1.

Fig. 4 and 5 represent the characteristic curves of the optimal design and it can be seen that the first and the second eigenvalues join in a double root at 92.56 as well as the third and the fourth at the same value of p_{cr}. The characteristic curves are found to be very sensitive to small variations of design variables. The second and the third eigenvalues have a tendency to join but they do not for obtained values of m and there is a small distance of about 72 between them.

It is worth to note that no additional constraints on equality of p_{cr}^1 and p_{cr}^2 have been added in the optimization process and the equality of two first critical loads is the natural result of the applied numerical procedure only.

8 Conclusions

The paper presents a method of determining minimum weight of a structure subjected to nonconservative forces. The proposed solution algorithm is free of any additional assumptions and constraints applied in reference papers. Two procedures of scaling and normalization condition for eigenvectors are proposed. Applied at each iterative step they help to minimize the numerical difficulties due to the high sensitivity of the problem. The numerical example of Beck's column shows efficiency of the proposed method which in this case gives the highest so far known value of the critical force.

References

1. Berke, L.; Khot, N.S. 1986: Structural optimization using optimality criteria. In: Mota Soares C.A. (ed.) Computer aided optimal design: structural and mechanical systems., 1, pp. 235–269, Troja, Portugal.

2. Bogacz, R.; Mahrenholtz, O. 1986: On stability of a column under circulatory load. Arch. Mech. 38 (3), pp. 281–288.

3. Claudon, J.L. 1975: Characteristic curves and optimum design of two structures subjected to circulatory loads. J. de Mechanique, 14 (3), pp. 531–543.

4. Gutkowski, W.; Bauer, J.; Iwanow, Z. 1990: Explicit formulation of Kuhn–Tucker necessary conditions in structural optimization. Comp. and Structures. 37 (5), pp. 753–758.

5. Hanaoka, M.; Washizu, K. 1980: Optimum design of Beck's column. Comp. and Structures. 11, pp. 473–480.

6. Mahrenholtz, O.; Bogacz, R. 1988: On the optimal design of columns subjected to circulatory loads. In: Structural Optimization. Proceedings of the IUTAM Symposium on Structural Optimization. pp. 9–13, Melbourne.

7. Seguchi, Y.; Kojima, S. 1988: On the shape determination of nonconservative system: a case of column under follower force. In: Structural Optimization. Proceedings of the IUTAM Symposium on Structural Optimization. pp. 315–322, Melbourne.

8. Tada, Y.; Matsumoto, R.; Oku, A. 1989: Shape determination of nonconservative structural systems. In: Brebbia, C.A.; Hernandez, S. (eds.) Proc. 1-st. Int. Conference. Computer aided optimum design of structures. Recent advances. pp. 13–21, Southampton, Berlin, Springer.

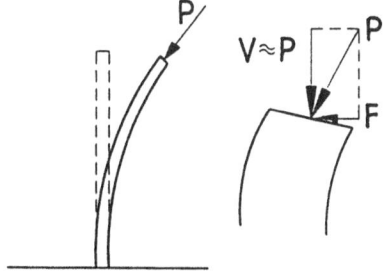

Figure 1: Beck's column

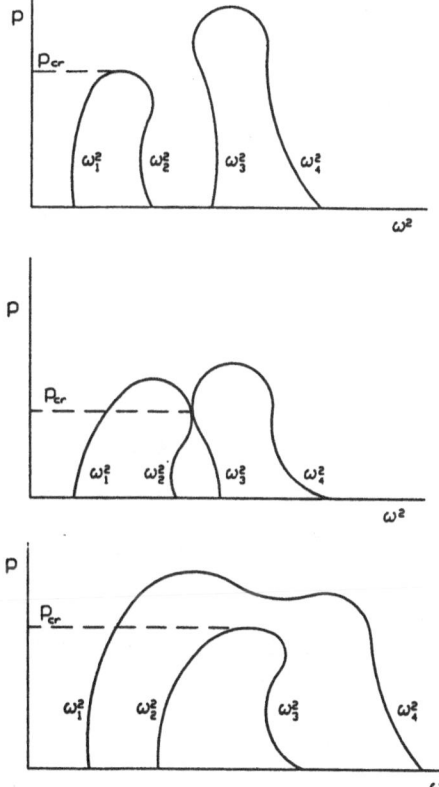

Figure 2: Three cases for p_{cr} related to the inequality $\omega_{i+1}^2 \geq \omega_i^2$.

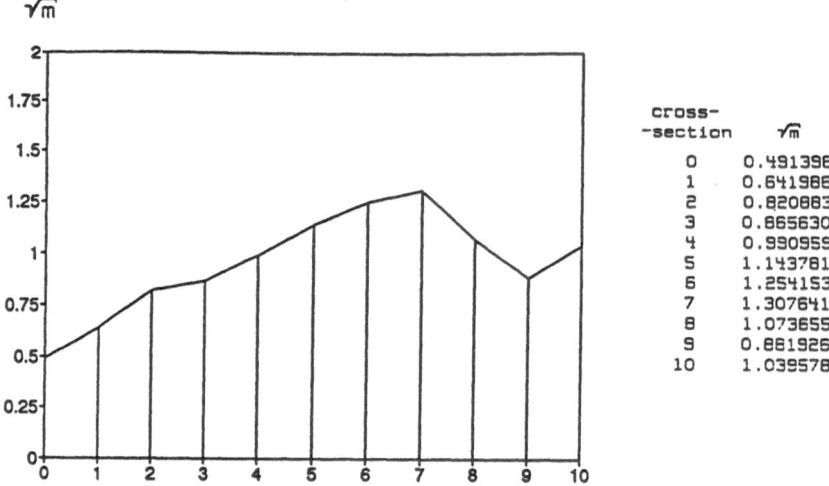

Figure 3: Mass distribution along the column length.

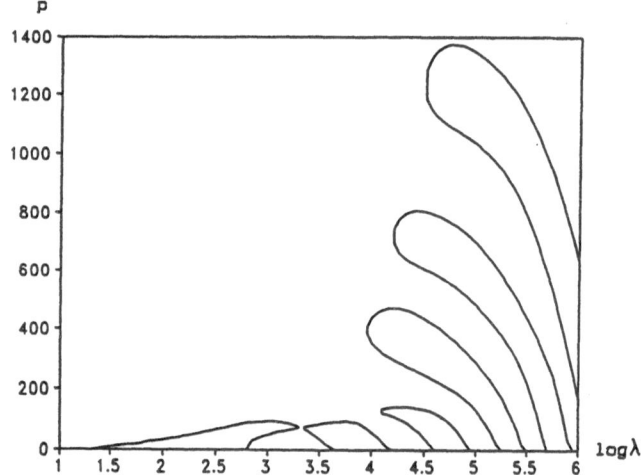

Figure 4: Characteristic curves $p - -log\lambda$ for optimum solution

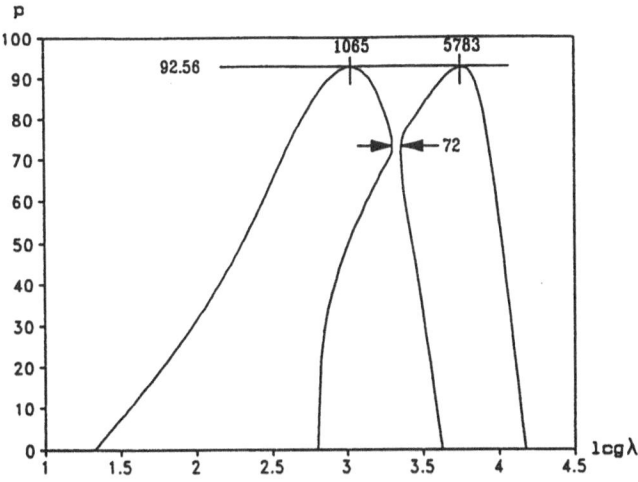

Figure 5: First characteristic curves $p - -log\lambda$ for optimum solution

MULTI-MODEL OPTIMIZATION OF LARGE SCALE STRUCTURES

P. MORELLE
and
V. BRAIBANT
Samtech, Bd. Frère-Orban, 25, 400 Liège, Belgium

Summary

The paper is devoted to the structural analysis and optimization of large scale problems when using the *SAMCEF* finite element code and eventually a C.A.D. package to describe the geometry of the structure.

We consider here two classes of problems:

■ sizing of thin walled structures;
■ shape optimization.

We shall first show that a realistic (large scale) engineering problem has to be defined as a multi-model optimization problem.

Then, we shall show how this fact has an influence on the optimization code and the way it has to be used by the users, practically speaking.

After that, we shall continue to have a look on the impact of the problem size on the optimization process with special attention to:

❚ definition of design variables, restrictions, etc.;
❚ resolution strategies;
❚ use of modern computers (vectorization, parallelism).

To close the discussion, some examples of applications will be given, related to the *OPTI* module of the *SAMCEF* package [4].

G. I. N. Rozvany (ed.), Optimization of Large Structural Systems, Vol. II, 1101–1114.
© 1993 *Kluwer Academic Publishers.*

1. Multi-model optimization concept ─────────────────

■ The present application of large scale optimization codes are mainly related to the design of mobile structures such that:

▮ airplanes;
▮ space launchers, shuttles, etc.

■ but also:

▮ sport equipment (tennis, golf, ...);
▮ motorcars, etc.;

for which a weight reduction is of highest importance. Very oftently, the designer is looking for a design which can minimize the weight in the same time that it maximize the stiffness, and those two objective functions are obviously conflictual.

Such a design problem has to take into account a lot of aspects which are classified as a multidiscipline design problem. We can for example distinguish:

▮ the structural analysis;
▮ the aerodynamic behaviour;
▮ the user's point of view for what concern comfort, space, etc.;
▮ the manufacturing point of view for what concern the method of fabrication;
▮ the quality insurance problem;
▮ etc.

In the field of optimization problems, all those aspects furnish a lot of restrictions or constraint functions which define the design space into which we try to optimize one or several objective functions with respect to an initial choice of design variables.

If we restrict ourself to the design restrictions related to structural analysis (which have also some impacts on the other category of analysis), it is clear that it is impossible to formulate the problem by using one single structural modelization. If we consider for example the behaviour of an airplane during its commercial exploitation:

▮ the airplane can be flying (no support) or posed on his landing gear (simply supported structure);
▮ we have to take account of a lot of possible behaviours: static, dynamic, aero-elastic, fatigue, etc.

so that a realistic large-scale problem naturally leads to the definition of a multi-model optimization problem. That means that several finite element analysis (using eventually several finite element models) have to be performed simply to evaluate the behaviour of the structure.

In what follows, we consider the class of optimization codes which solve the problem by using convex linearization methods (a good description can be found in [1] or [2]) or related algorithms like SVANBERG's method of moving asymptotes [3] etc., which use only first order derivatives (also called sensitivities).

That means that at each iteration, an explicit approximation of the real (non explicit) problem is builded and solved. The set of design variables which are optimal for the current approximated problem furnish then the basis for the next iteration, until the variation between two successive optimization becomes very small. The true optimum has then been reached.

This class of methods has the wonderful property to convergence quickly (the number of iteration is oftently lower than 10) and independently of the problem size. It has been used in the field of design and shape optimization in the *OPTI* module of the *SAMCEF* finite element system [4].

An other interesting property of this algorithm is that it can automatically solve some conflictual situations by using a relaxation strategy of the imposed bounds.

2. Impact of the multi-model aspects on the optimization procedure

When trying to solve a multi-model optimization problem, the most important remark concerns the distinction between successive and simultaneous multi-model optimization.

In the first case, which could be called the "naive" approach, a first optimization step takes into account the first finite element model with all its related constraints. The second step is devoted to the second model etc. until all the models have been taken into account.

It is clear that at the end of this looping process, the restrictions related to the first model are no more satisfied so that a second loop is necessary, ...with no guarantee of convergence !!!...

In the second case, all the models are taken into account at each optimization step so that all the restrictions relative to the various behaviours are incorporated simultaneously into the problem and satisfied at the optimum. This last method is the only one which can be used for realistic problems.

In order to be able to furnish this possibility, an optimization code must be built on the basis of the following architecture illustrated on Figure 1.

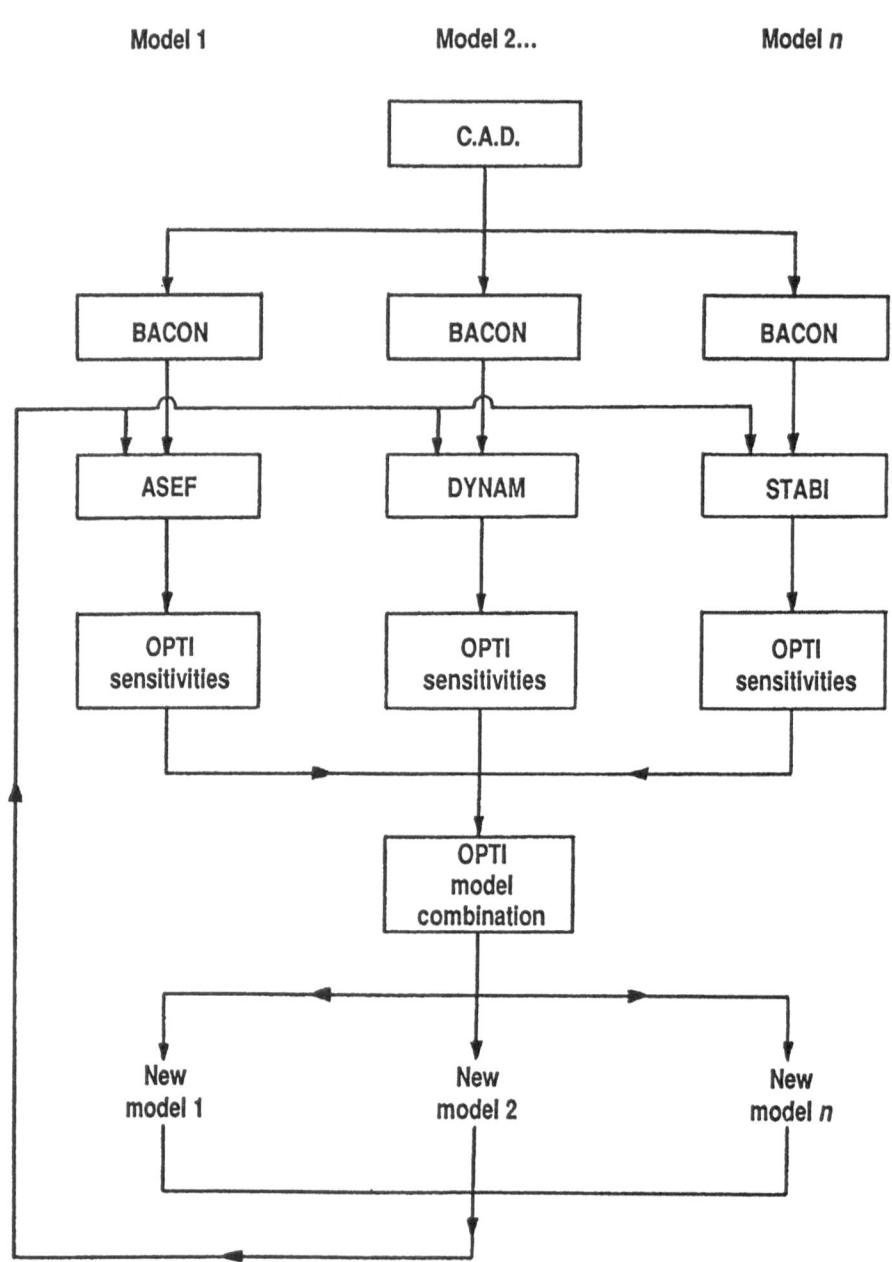

Figure 1

Of course, this architecture has an impact on the looping process and the flow of information which is to be interchanged between the modules. A good automatic procedure has to be furnished by the software developper in order to simplify this transmission of informations.

But the most important consequence of such an architecture is related to the organization of the design office itself: it is clear that now, the engineers responsible for the aerodynamic problems and those working on structural aspects can now work TOGETHER and design the structure IN THE SAME TIME to lead to an optimized structure or in other words to the best compromise between all of those aspects.

Unfortunately, it is not so easy to change completely the organization of a company and the way it works till 10 or 20 years...

The consequence is that, as it can be oftently seen that in real life applications, the first method described here (successive optimization) is used mainly because of "historical" reasons and the difficulty to change completely the organization of the work at a so high level...

Furthermore, in order to be used for real life applications, some other requirements exist related to the exchange of informations between the different teams working on a common project:

▮ the datas have to be defined at a very high level of description that means independently of the finite elements themselves. The C.A.D. level is surely the easiest that can be used for this purpose. The problem here is that several C.A.D. models have to be defined to describe an engineering problem. It is then necessary to manage the relations between those various description... A beautiful problem of interfacing !!!

▮ to define a design variable with the help of a C.A.D. package, it is assumed that the initial C.A.D. model can be easily modified or in other terms that the C.A.D. model can be parametized. This is far to be the case in several of the present C.A.D. packages, but a lot of progress have been made in this field...

▮ last but not least, the internal flow of information inside the optimization module has to be organized in a very efficient way. The modern solutions use a Data Base structure to manage and store the information.

All this description must be seen as the ideal situation of the design office of tomorrow. On the basis of an existing system like *SAMCEF*, the software developpers have now to increase the possibilities of interfacing CAD and FEM systems as well as FEM analysis and optimization modules. In the current release of the *SAMCEF* system, it exists a lot of interfaces which allow to exchange datas between the *SAMCEF* preprocessor and CAD packages like *SUPERTAB*, *EUCLIDE*, *CATIA* etc., using for example an *IGES* file. We are on the way...

3. Impact of the problem size on the data definition in sizing

At the beginning of automatic sizing, the whole optimization problem was defined at the finite element level. That means that the user had to say:

"The first design variable is the thickness of the element number 12",

and this for each element and design variable. The pre-historical period of informatic was like that !

But this kind of data definition and of design variable description is not only a waste of time and energy: it also leads to non-industrial, non usable optimum because the fact that such a "free" variation of thickness from one element to the other leads to a thickness distribution which is completely stupid at the practical point of view.

So came the idea of "linking" between some groups of elements: this idea which is still the standard possibility available in the *SAMCEF-OPTI* software [4] consists in the definition of sets of elements. At each set is linked a design variable which is a multiplier of the initial thickness distribution as illustrated at Figure 2.

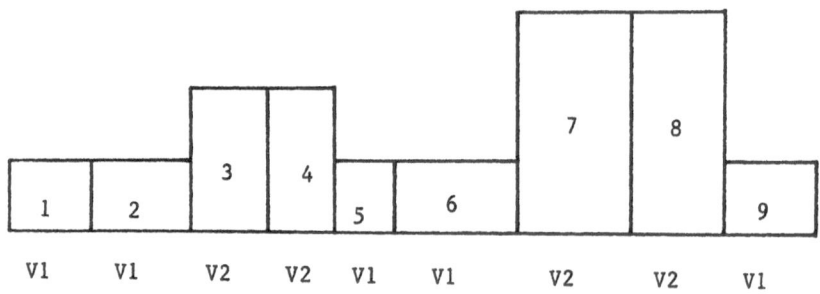

```
set 1 : elements 1, 2, 5, 6, 9
set 2 : elements 3, 4, 7, 8
```

Figure 2

It is even possible to define a complex variation (using B-splines for exam- ples) so that a complicated BUT SMOOTH variation of thickness can be obtained on a given domain. Furthermore, some "tangency" conditions can be defined for two adjacent domains so that no discontinuity of thickness is observed.

In such a situation, the sizing problem become very close to a shape optimization problem: it is only a particular case in which the shape of the structure is changed in one direction only.

The use of C.A.D. tools to define a sizing problem is, in our opinion, the best way for large-scale problems definition. It has the advantages of a reduction of the number of design variables and to lead to industrially applicable structures.

If for example, it is requested to have a constant thickness on a given part of a structure, the solution is to define a set containing all the finite elements belonging to this part, to give a first constant value for the thickness of all the elements belonging to the set and then to link one design variable to this initial value.

When a linearly varying thickness distribution is feasible at an industrial point of view, it is possible to define an initial distribution on the basis of such a variation so that the optimal distribution will also be linearly variable.

However, some disadvantages are still inherent to this method of linking:

■ first, it is not possible in the previous example to modify the initial linear distribution: the modifications of slope are then limited. In fact, it could be interesting to link more than one design variable to the initial distribution;

■ second, the definition of design variable is still made at the finite element level and not at the C.A.D. description level. This could be a very heavy task for a large-scale problem.

For those reasons, a next step has to be done and it will be made in the direction of this C.A.D. description of the structure. Using C.A.D. models, it is possible to define domains, (more or less) independently of the finite element definition. On a given domain, it can now be said that for example (see Figure 3):

"The thickness definition on the domain 1 is a bi-linear function governed by design variables v1 to v3"

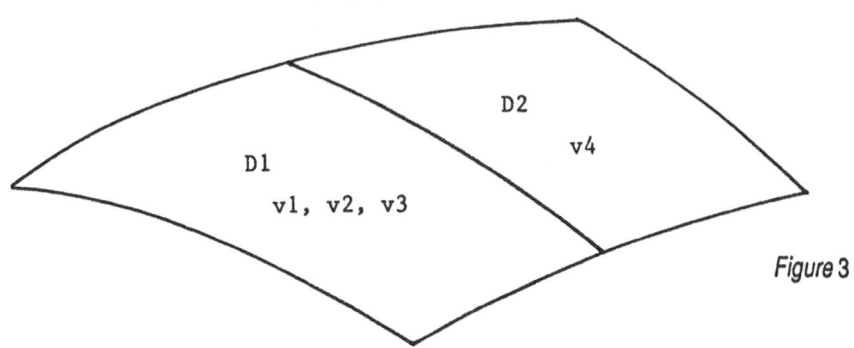

Figure 3

4. Intelligent selection of active restrictions ———————

Until here, we have not yet mentioned the problem of C.P.U. time related to large scale problem resolution. This is because, in our opinion, the perfor- mances of modern super-computer become sufficiently high to allow an effective resolution of this class of problems for industry.

However, this category of computers is still restricted to very big companies or research centres and is not yet available for smaller industries: in this case, the use of work-stations become more and more popular so that, despite the fact that the speed of computation is also drastically increasing for this class of computers, it remains necessary to look for an economy of C.P.U. time.

One classical way to decrease the computation time is to use what have been called a "strategy for active constraint reduction".

The principle is very simple and comes from the fact that generally, for a large scale problem that means in our opinion (for classical super-computers):

∎ 1000 design variables;
∎ 10.000 restrictions.

only a few of all those restrictions are potentially active. A great number have an initial value which is very far from the imposed bounds and will keep such a value after optimization. They do not act on the optimal state so that they can be ignored for the present iteration.

The problem is that if the user select a too small number of restrictions, at the next iteration, a great number of those constraints will be violated and this has generally a very bad impact on the convergence rate. In other words, we have increased the C.P.U. time when trying to decrease it !...

However, this problem becomes less and less severe when we come in the vicinity of the optimum because the variations between successive states become very small at this moment.

In *SAMCEF* [4], the following classical solution (see for example [5]) has been adopted. Let us consider a set of restrictions denoted:

$$\underline{h}_i \leq h_i(x) \leq \overline{h}_i \qquad (4.1)$$

A constraint is potentially active for the current iteration if:

∎ it violated some bound at the initial state;

■ or if one of the following inequalities is satisfied:

$$\frac{\bar{h}_i - h_i}{\bar{h}_i - \underline{h}_i} < CA \qquad \frac{h_i - \underline{h}_i}{\bar{h}_i - \underline{h}_i} < CA \qquad (4.2\text{-}4.3)$$

where the CA constant can be choosed by the user. Its value can be fixed between 0 and 1.

■ If $CA = 1$, (4.2) (4.3) lead to:

$$\underline{h}_i < h_i < \bar{h}_i \qquad (4.4)$$

that means that all the constraints are potentially active and no constraint reduction occurs.

■ At the contrary, if $CA = 0$ (limit value), (4.2)(4.3) lead to:

$$\underline{h}_i > h_i > \bar{h}_i \qquad (4.5)$$

that means that no constraints are active any more !

The result is that the CA parameter acts exactly as a filter allowing to sup- pres the constraints whose value are very small in comparison with the imposed bounds at the beginning of the present iteration.

To modify automatically the value of CA during the iterative process, a second relation is added to (4.2) (4.3):

$$(CA)_{i+1} = \frac{(CA)_i}{CB} ; (CA)_{i+1} > CC \qquad (4.6)$$

where, again, CB and CC are selected by the user. In that case, we must have CB greater than 1 (to have a decreasing value for CA) and CC is the minimum value for the CA constant.

So what should be the strategy used for practical, real life applications ?

If the initial design is infeasible (some constraints are violated), the variation of design variables will normally be large during the first iterations because the system first try to find a feasible state before trying to optimize anything. The initial value for CA must then be very close (or equal) to one because the set of active constraints will normally change from one iteration to the other.

If the initial design is feasible, convergence is normally faster so that it is possible to begin with a smaller value of CA, for example 0.8.

At the end of the optimization loop (that means when the objective function changes very few from one iteration to the other), CA can be decreased to smaller values. This can be done automatically by using the CB, CC parameters.

One possible improvement of this technique is to use Artificial Intelligence methods to develop an automatic "learning" data-base related to the choice of the CA, CB, CC constants. This could be done efficiently for a given cate- gory of problems. The difficulty is then to characterized the problem category using A.I. techniques...

5. Use of super-computers to compute sensitivities and optimal state

Let us focus on computation of sensitivities in static problems. If we use the classical semi-analytical approach to compute the sensitivities, we have to compute:

$$\frac{dK}{dx} \cong \frac{K(x+dx) - K(x)}{\partial x} \tag{5.1}$$

$$\frac{dq}{dx} = K^{-1}\bar{g} \; ; \bar{g} = \frac{dg}{dx} - \frac{dK}{dx}q \tag{5.2}$$

$$\text{if } dc = B(q) \; ; \frac{dc}{dx} = \frac{\partial B}{\partial x} + \frac{\partial B}{\partial q}\frac{dq}{dx} \tag{5.3}$$

What is it possible to do in order to take advantage of vectorized computers or of parallel architectures ? This is what we are going to discuss now.

At the informatic level, the computation of (5.3) involves the following tasks:

task 1   : for each finite element, computation of the sensitivities of the ele- mentary stiffness (and eventually mass, stress, etc.) matrices;

task 2   : assembly of the pseudo loads $\bar{g}$;

task 3   : computation of the nodal displacements sensitivities;

task 4   : computation of the functions sensitivities.

Obviously, task 1 can only benefits of a multi-processor architecture: in fact, one processor for each finite element. All the matrices can then be computed in parallel: this is the idea of a finite element machine allowing to generate elementary matrices all together.

For what concern the other tasks, we have not here to invert the structural stiffness matrix (if it is possible to read it from the analysis module). However, a lot of vectorized operations, that means products:

vector * vector

matrix * vector

have to be performed. They can be optimized to work on vectorized machines in order to obtain a highly efficient code.

6. Example of application of multi-model optimization _____

6.1 Design of a railway carriage

This first example is only described to illustrate the necessity of a multi-model formulation.

It concerns a medium scale application of the *SAMCEF – OPTI* module to the optimization of a railway carriage. This is a structure composed of shell elements to modelize the panels and an assembly of beams to modelize the frame. The finite element model is illustrated at Figure 4. This medium scale problem has 1500 degrees of freedom, 30 design variables and 284 restrictions.

The objective function is the structural weight and the design variables are the thicknesses of the sizes, the floor and the roof panels for the shells, and also the cross sections and inertia of some beams.

The multi-model problem comes from the fact that several structural situations have to be taken into account:

1st model : *the vehicle is travelling* and static behaviour is taken into account. Several loading cases are related to the same finite element model: service loads, traction and compression loads, extreme loads. The constraint is an upper bound on the stress level and on some displacements;

2nd model : *the vehicle is lifted* (for repairing purpose). The constraints here are again stress level and maximum displacement;

3rd model : *travelling vehicle:* eigenfrequency analysis. It is requested that no eigenfrequency appear in the interval 9-12 Hz.

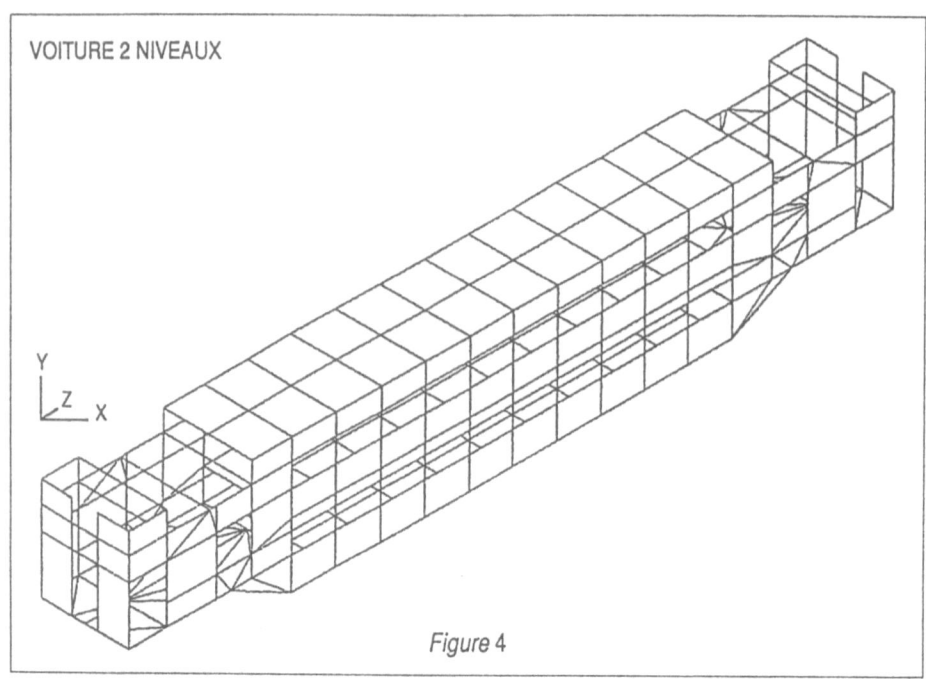

VOITURE 2 NIVEAUX

Figure 4

6.2 Optimization of the front casing of *ARIANE* 5 launcher

We describe here the optimization of a part of the *ARIANE* 5 launcher: it is illustrated at Figure 5. The finite element model is composed of 7000 degrees of freedom and 655 elements. It uses beam, shell and volume elements. The problem has been solved in *SAMTECH* on a *VAX* 3600 machine. For this class of computer, it can be called a real life application. However, *AEROSPATIALE* which is responsible of the project solves with *OPTI* problems which are five time bigger on *IBM* 3090 VM/SP mainframes.

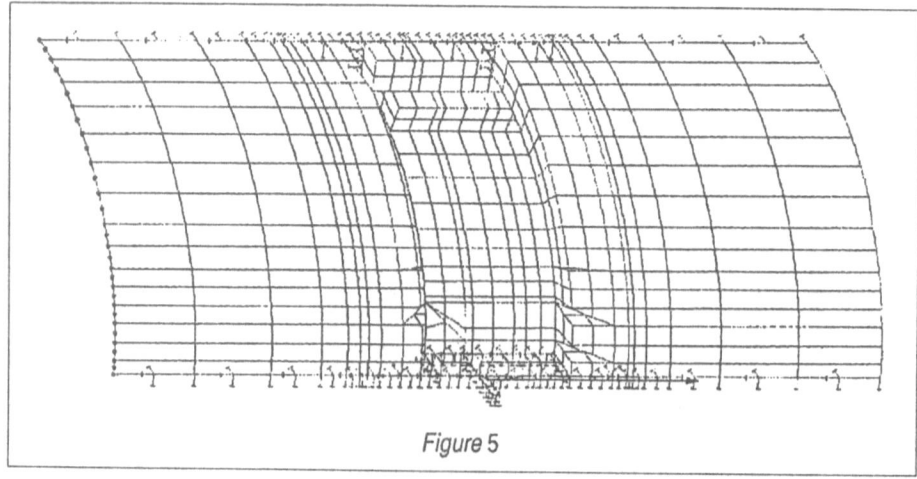

Figure 5

In fact, the cost of the resolution is mainly related to the number of design variables and constraints than to the number of finite elements. Here, the design variables are again thickness distribution of the shells and beam properties (cross section and inertia). In our model, there were 110 design variables.

The objective function was the structural weight.

For what concern the constraints, we had to take into account several finite element models with, for each of them, different fixations. We had:

1ˢᵗ model	: restrictions on the stress level in each finite elements.
2ⁿᵈ and 3ᵈ models	: restriction of the structural stiffness. This is done by imposing an upper bound on the displacement of some points of the structure when some loads are applied. To kind of limit conditions had to be considered.
4ᵗʰ and 5ᵗʰ models	: restrictions on the flux of stresses flowing from this part of the launcher to the neighbourhood. Again, two situations have to be considered with different fixations.

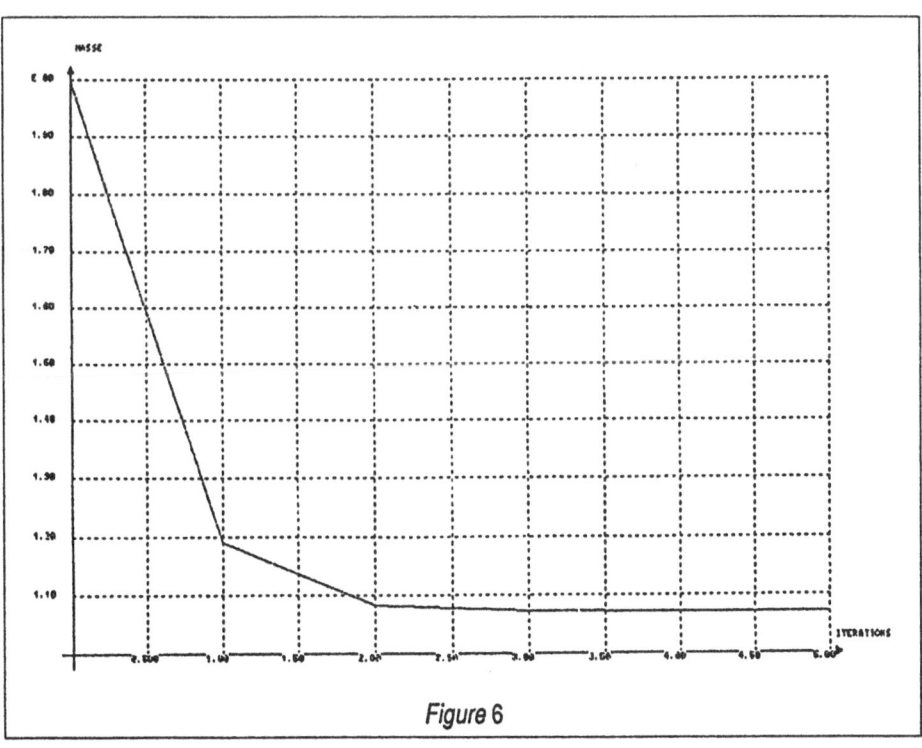

Figure 6

There were a total number of 700 restrictions.

The convergence curve for the structural weight is illustrated at Figure 6. It can be seen that optimum is reached after 5 iterations. This is due to the fact that:

∎ the initial design was feasible;
∎ all the restrictions concern the static behaviour;

all circumstances which are favourable for an increase of the convergence rate.

7. Conclusions

It has been shown that the size of the problem in structural optimization has not only an impact on the CPU time, but that it has also some consequences:

∎ on the way the code architecture is organized;
∎ on the way the users can use the code efficiently.

Multi-model optimization is a must. It is the only correct way to solve real life problems of design structures: in *SAMCEF*, the same kind of approach is used for sizing and for shape optimization.

A lot of progress have still to be done especially, as it has been described in the paper:

∎ to increase the efficiency of data definition using CAD entities;
∎ to take advantage of the parallel and vectorized computers.

References

[1] Cl. FLEURY and V. BRAIBANT, *Convex approximation strategies in structural synthesis*, in "Computer Aided Optimal Design" Vol. 3, NATO/NASA Lect. notes, pp. 156-175, July 1989, Troia (Portugal).

[2] Cl. FLEURY, *First and second order convex approximation strategies in structural optimization*, Structural Optimization 1, pp. 3-10, (1989).

[3] K. SVANBERG, *Method of moving asymptotes — a new method for structural optimization*, Int. J. Num. Meth. Eng., 24, pp. 359-373, (1987).

[4] *OPTI V.4.1 User's Manual*: SAMTECH Editor, (1991).

[5] B. PRASAD, *Approximation, adaptation and automation concepts for large scale structural optimization*, Engineering Optimization, Vol. 6, Num 3., (1983).

The Analytic Solution of the Structural Analysis Problem and Its Use in Structural Synthesis

M. B. FUCHS
Department of Solid Mechanics, Materials and Structures
Faculty of Engineering, Tel-Aviv University
Ramat-Aviv, Tel-Aviv 69978, Israel

ABSTRACT. This author has presented the analytic solution of the linear structural analysis problem. It was established that the internal forces, the nodal displacements and the inverse of the system stiffness matrix can be expressed as the ratio of multilinear polynomials in terms of the unimodal stiffnesses of the elements. Instead of performing a classical analysis via a displacement or a force method, for given values of the rigidities of the elements, one can construct directly the solution, be it an internal force or a nodal displacement, with the element stiffnesses as explicit parameters. What looks as Eldorado for structural engineers is hampered by the inordinate length of the exact expressions. However, researchers can now contemplate for the first time the exact form of the expressions for which they have been devicing approximate ones. This may pave the way to approximations based on structural considerations rather than on mathematical artifices. This paper is a compendium of the theory underlying the exact analytic expressions and will illustrate the process by means of some treated examples. An approximate model, which emulates the construct of the exact formulation, will also be presented. As will be shown, the method applies equally to trussed and to framed structures.

1. The 3-bar Truss

We will choose an unusual route and present the method for assembling the expression for the internal forces in a truss prior to discussing the theoretical aspects of the technique. Consider the familiar 3-bar truss in Fig. 1. The structure is composed of three elements of rigidities EA_j (j=1,2,3) and has two nodal degrees of freedom u_i (i=1,2). The truss is subjected to two external loads p_i at the free node. The nodal equilibrium equations in terms of the internal forces t_j are

G. I. N. Rozvany (ed.), Optimization of Large Structural Systems, Vol. II, 1115–1133.
© 1993 *Kluwer Academic Publishers.*

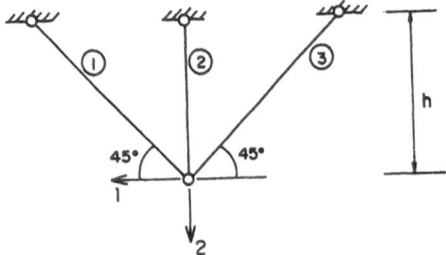

Figure 1. The 3-bar truss

$$\frac{1}{2}\begin{bmatrix} -\sqrt{2} & 0 & \sqrt{2} \\ \sqrt{2} & 2 & \sqrt{2} \end{bmatrix}\begin{Bmatrix} t_1 \\ t_2 \\ t_3 \end{Bmatrix} = \begin{Bmatrix} p_1 \\ p_2 \end{Bmatrix} \tag{1}$$

where the coefficients matrix is the statics matrix Q. We have here 2 equations in 3 unknowns and the structure has thus a statical redundancy of degree 1. The method for constructing the solution hinges on the number of stable statically determinate substructures which are embedded in the original structure. For every stable substructure, denoted by subscript m, one computes the determinant of the corresponding statics matrix $|Q_m|$ and the internal forces t_{jm} in the substructure when it is subjected to the applied loads. For a given determinate configuration, the forces in the missing bars are obviously zero. With the above definitions, the force t_j in member j of the structure is

$$t_j = \frac{\displaystyle\sum_m |Q_m|^2 \, t_{jm} \, \pi_m}{\displaystyle\sum_m |Q_m|^2 \, \pi_m} \tag{2}$$

where the summation is taken over all the stable substructures, and π_m is the product of the stiffnesses s_j $(=E_j A_j/L_j)$ of the bars of the substructure, with E_j, A_j and L_j being respectively Young's modulus, the cross-sectional area and the length of element j.

Selecting in turn columns (2,3), (1,3) and (1,2) of the Statics matrix we can write the equilibrium equations for the determinate subtructures m=1,2,3 respectively in terms of the corresponding internal forces. Note, the substructures are identified by their missing bar (m).

$$\frac{1}{2} \begin{bmatrix} 0 & \sqrt{2} \\ 2 & \sqrt{2} \end{bmatrix} \begin{Bmatrix} t_{21} \\ t_{31} \end{Bmatrix} = \begin{Bmatrix} p_1 \\ p_2 \end{Bmatrix} \tag{3a}$$

$$\frac{1}{2} \begin{bmatrix} -\sqrt{2} & \sqrt{2} \\ \sqrt{2} & \sqrt{2} \end{bmatrix} \begin{Bmatrix} t_{12} \\ t_{32} \end{Bmatrix} = \begin{Bmatrix} p_1 \\ p_2 \end{Bmatrix} \tag{3b}$$

$$\frac{1}{2} \begin{bmatrix} -\sqrt{2} & 0 \\ \sqrt{2} & 2 \end{bmatrix} \begin{Bmatrix} t_{13} \\ t_{23} \end{Bmatrix} = \begin{Bmatrix} p_1 \\ p_2 \end{Bmatrix} \tag{3c}$$

with $t_{11} = t_{22} = t_{33} = 0$.

The t_{jm} forces, the determinants $|Q_m|$ and their squares are summarized in Table 1.

m	1	2	3
t_{1m}	0	$\sqrt{2}/2 \, (p_1 + p_2)$	$\sqrt{2} \, p_1$
t_{2m}	$p_1 + p_2$	0	$p_2 - p_1$
t_{3m}	$-\sqrt{2} \, p_1$	$\sqrt{2}/2 \, (p_2 - p_1)$	0
$|Q_m|$	$\sqrt{2}/2$	1	$\sqrt{2}/2$
$|Q_m|^2$	1/2	1	1/2

Table 1: Three-bar truss; forces and determinants of statically determinate substructures

We are now ready to write the analytic expression for the internal forces. Noting that the product of stiffness are for the three substructures

$$\pi_1 = (EA_2/L_2)(EA_3/L_3) = \frac{\sqrt{2}}{2} (E/h)^2 A_2 A_3$$

$$\pi_2 = (EA_1/L_1)(EA_3/L_3) = \frac{1}{2}(E/h)^2 A_1 A_3 \tag{4}$$

$$\pi_3 = (EA_1/L_1)(EA_2/L_2) = \frac{\sqrt{2}}{2}(E/h)^2 A_1 A_2$$

and introducing these expressions, in conjuction of the values in Table 1, in (2) we obtain the analytic solution for the internal forces in the truss

$$t_1 = \frac{(p_1+p_2) A_1 A_3 + \sqrt{2} p_1 A_1 A_2}{A_2 A_3 + \sqrt{2} A_1 A_3 + A_1 A_2}$$

$$t_2 = \frac{(p_1+p_2) A_2 A_3 + (p_2-p_1) A_1 A_2}{A_2 A_3 + \sqrt{2} A_1 A_3 + A_1 A_2} \tag{5}$$

$$t_3 = \frac{-\sqrt{2} p_1 A_2 A_3 + (p_2-p_1) A_1 A_3}{A_2 A_3 + \sqrt{2} A_1 A_3 + A_1 A_2}.$$

These expressions are explicit in the member rigidities. In the case of the 3-bar truss they are also explicit in terms of the applied loads. For more complex structures the external loads are usually treated numerically. It is clear that with the aid of (2) the construction of the analytic expressions is straightforward. We will show in the following section a relatively simple way to derive the basic construct of (2).

2. Analytic Representation of Member Forces

One approach to derive the analytic representation of the member forces is rooted in the exact univariate model given by Fuchs and Steinberg (1991) for redundant trusses. The univariate model gives the value of the internal force in bar j when the stiffness in bar i varies. This was done along the following lines.

Let s_i and t_i be the stiffness and the internal force of a typical bar i of a redundant truss. Let k denote any arbitrary bar of the structure except bar i. We disconnect bar i from one of its nodes thus freeing the redundant force t_i which is now applied in opposite directions at the cut section. The structure, which may still be redundant, is now subjected to the external loads and to the equal and opposite forces t_j which are considered as additional external loads. The relative displacement δ at the cut section is obtained via the unit-load method

$$\delta = \left(\sum_k n_k^2/s_k + 1/s_i\right) t_i + \sum_k N_k n_k/s_k \tag{6}$$

where n_k and N_k are the forces in bar k due to unit and opposite loads at the cut section and to the applied loads respectively. Since the structure may still be statically redundant, these forces depend on the stiffnesses s_k of the structure, but not on stiffness s_i. Note, $n_i = 1$ and $N_i = 0$. Compatibility of deformations requires $\delta = 0$, and consequently (6) leads to an equation of the form

$$t_i = \frac{s_i}{c + d\, s_i} \qquad (7)$$

where c and d are constants depending on the stiffnesses s_k and are independent of s_i. Applying the principle of superposition to the released structure we find that the force in an arbitrary bar j (j is one of the bars of the k set) is equal to the sum of the force in j due to the external loads and the force in j due to the equal and opposite loads t_i

$$t_j = N_j + n_j\, t_i. \qquad (8)$$

Introducing (7) in (8) yields

$$t_j = \frac{a_j + b_j\, s_i}{c + d\, s_i} \qquad (9)$$

where the constants a_j and b_j are also independent of s_i. The subscript j indicates that these constants vary with j whereas c and d are common to all the forces t_j. Equation (9) is called the univariate model, and as stated earlier, it constitutes the basis for an exact expression for an arbitrary force t_j explicitly in terms of all the stiffnesses of the structure.

The question which arises now is: how do the a_j, b_j, c and d parameters in (9) depend on the stiffnesses s_k? What (9) tells us is that no matter which stiffness we choose to vary, it will always appear linearly both in the numerator and in the denominator of (9). This being the case we can fairly surmise (Fuchs, 1992) that the generalization of (9) takes the form

$$t_j = \frac{A_{jo} + \sum_m A_{jm}\pi_m}{B_o + \sum_n B_n \pi_n} \qquad (10)$$

where the A's and B's are constants, and the π's are some products of the stiffnesses of the structure. In the case of the 3-bar truss for instance, these products can be selected from $s_1 s_2 s_3$, $s_2 s_3$, $s_1 s_3$, $s_1 s_2$, s_1, s_2, or s_3. Setting two stiffnesses to constant values will always yield linear expressions in the third stiffness, both in the numerator and in the denominator of (10), which is in line with (9). A term of the form $s_1 s_3^2$ must be ruled out since by fixing s_1 and s_2 we would get a non-linear term in s_3, in contradiction with (9). The number of terms of the m and n summations is at this point unknown.

The forces in linear trusses are scaling invariant. When we multiply all the rigidities by a constant factor, the internal forces are not modified. In other words, t_j is homogenous of degree zero in the stiffnesses (assuming that we keep the geometry constant). This immediately implies that the constants A_{jo} and and B_o must be zero. It also demands that all the π's be the products of a same number of stiffnesses. Any other combination would render (10) non homogeneous. Falling back on the 3-bar truss example, we can either have terms in $s_1 s_2 s_3$, or $s_2 s_3$ $s_1 s_3$ $s_1 s_2$ or s_1 s_2 s_3. Let us assume that we have products of O stiffnesses in each of the π_s.

Now, Eq. (10) should be valid for all values of the stiffnesses, as long as the structure remains stable. It includes in particular the solution for statically determinate substructures. Let M be the number of elements of the structure and let N be the number of unconstrained nodal degrees of freedom. Its degree of statical redundancy is R (=M-N). In the 3-bar truss M=3, N=2 and R=1. A determinate substructure has N non-zero stiffnesses and R zero stiffnesses. Assuming that O<N, by requiring that R components are null, more than one π will be different from zero. For the 3-bar truss we would have terms in s_1, s_2 and s_3. Setting s_1 to zero we find that the member forces are a function of s_2 and s_3. Consequently, the forces t_j in a statically determinate structure would be some function of the stiffnesses, which is unacceptable. In a statically determinate structure the forces can be computed purely on the basis of statics. On the other hand, if O>N, by setting R redundant components to zero, all the π's will be identically zero, a possibility which must equally be ruled out since it would give an undetermined solution in a statically determinate structure. Considering again the 3-bar truss, the only term in (10) would be $s_1 s_2 s_3$ which is zero for any substructure.

We are thus left with the only remaining option, namely O=N (terms in $s_2 s_3$, $s_1 s_3$, $s_2 s_3$ for the 3-bar truss). Both the numerator and the denominator of (10) are multilinear homogeneous polynomials of degree N in the stiffnesses. How many terms in each polynomial? At this stage we will assume that we have all possible combinations of N rigidities out of M, in both polynomials

$$t_j = \frac{\sum\limits_{m} A_{jm} \, t_{jm} \, \pi_m}{\sum\limits_{m} B_m \, \pi_m} \tag{11}$$

with

$$\pi_m = s_p s_q s_r \cdots \quad \text{(N stiffnesses)} \tag{12}$$

where the number of terms in the polynomials is equal to M!/N!R!. For structures with R=1 (as in the 3-bar truss) the number of terms is equal to the number of bars. In all other cases the number of terms is unfortunately very much larger than the number of bars.

Equation (11) can be further simplified by considering the member forces in substructure k for instance. These are obtained by setting its R redundants to zero in (11). Consequently, all the π's but π_k disappear. We are left with one term both in the numerator and in the denominator of (11). Since π_k cancels out, we obtain the following expression for the force t_{jk} in an arbitrary component j of the determinate substructure k

$$t_{jk} = A_{jk} / B_k \qquad (13)$$

which as expected is independent of the stiffnesses. Equation (13) can now be used to express A_{jk} in terms of the corresponding B_k, thus yielding the final form for the member forces in a redundant truss

$$t_j = \frac{\displaystyle\sum_m B_m\, t_{jm}\, \pi_m}{\displaystyle\sum_m B_m\, \pi_m}. \qquad (14)$$

Comparing (14) with (2) we note that both expressions are identical provided that one can show that B_m is the square of the determinant of the corresponding statics matrix. This was shown in Fuchs (1991c) by employing theorems on determinants attributed to Binet and Cauchy and writing the explicit expression for the inverse of the stiffness matrix of a structure.

3. The Explicit Inverse of the Stiffness Matrix

Consider the field equations governing the analysis of linear elastic trusses:

Statics:	$Q\, t = p$	(15a)
Elasticity:	$S\, e = t$	(15b)
Kinematics:	$R\, u = e$	(15c)

where Q is the (NxM) statics matrix, t are the member forces, S is the (MxM) unassembled (diagonal) element stiffness matrix ($S_{ij} = \delta_{ij} s_j$, where δ_{ij} is the Kronecker delta), e is the M-vector of element deformations, R is the (MxN) Kinematics matrix and u is the N-vector of nodal displacaments. The Statics-Kinematics duality (SKD) gives $R = Q^T$.

Backsubstitution of (15b) and (15c) into (15a) yields the nodal equilibrium equations of the Displacement method

$$K\, u = p \qquad (16)$$

where the (reduced) system stiffness matrix K is given in the congruent product form

$$K = Q\, S\, R \qquad (17)$$

Premultiplying both sides of (16) by the inverse of the stiffness matrix gives the nodal displacements

$$u = K^{-1} p \tag{18}$$

The inverse of the stiffness matrix is equal to the ratio of the adjoint of the stiffness matrix to the determinant of the stiffness matrix

$$K^{-1} = \frac{adj \ K}{|K|} \tag{19}$$

An explicit expression for the determinant of K is readily found. Since K (17) is a product of three matrices (note, Q and R are respectively of size NxM and MxN), we can use a Binet and Cauchy theorem (Aitken, 1959) to write its determinant as follows

$$|Q \ S \ R| = \sum_m |Q_m| \ |S_m| \ |R_m| \tag{20}$$

where Q_m is typically a NxN submatrix of Q, and S_m and R_m are corresponding NxN sub-matrices of S and R respectively. The number of terms ie equal to the number of different Q_m matrices which are present in Q. It will be recognized that the terms in the RHS of (20) are the determinants of the stiffness matrices of the statically determinate substructures. The index m runs over all the statically determinate and stable substructures which can be derived from the original structure (unstable substructures have zero Q_m determinants). We find here that the determinant of the stiffness matrix of a structure is equal to the sum of the determinants of the stiffness matrices of all its statically determinate stable substructures.

$$|K| = \sum_m |K_m| \tag{21}$$

It will also be noted that S_m being diagonal, its determinant is equal to the product of its diagonal terms, i.e.,

$$|S_m| = \pi_m \tag{22}$$

Since the determinant of a matrix is equal to the determinant of its transpose, (20) with (22) gives the following explicit expression for the determinant of the stiffness matrix

$$|K| = \sum_m |Q_m|^2 \ \pi_m \tag{23}$$

We now need an expression for adj K (19). The adjoint of the stiffness matrix will be computed via the (N-1) compound of the stiffness matrix, denoted by $K^{(N-1)}$. This matrix is formed by all the minors of order (N-1) which can be obtained from K. The row and column position of a given minor in the compound matrix is determined by the rows and columns of the minor in K, in increasing order, row-wise and column-wise. Both the adjoint and the (N-1) compound of **K** are matrices of order NxN and are composed of elements which are minors of order (N-1) of **K**. They differ in two respects: the rows and columns of both matrices are in inverse order and the two matrices also differ by the sign of their elements. The adjoint is composed of signed minors, or cofactors, whereas the compound is populated by (unsigned) minors. It is easy to verify (Fuchs, 1991c) that the adjoint can be obtained by pre- and post-multiplying the (N-1) compound with the signed inversion matrix $\check{I}$

$$\text{adj } \mathbf{K} = \check{I} \, \mathbf{K}^{(N-1)} \, \check{I}^T \tag{24}$$

where the inversion matrix $\check{I}$ has alternatively the values 1 and -1 on its secondary diagonal, and zero entries elsewhere.

Having written the adjoint in terms of the (N-1) compound of **K** we will rely again on Binet and Cauchy and employ their remarkable theorem on product of compound matrices: the k-compound of a product of matrices is equal to the product of the k-compounds of these matrices, in the same order. This theorem holds also for rectangular matrices. Consequently, (24) with (17) becomes

$$\text{adj } \mathbf{K} = \check{I} \, \mathbf{Q}^{(N-1)} \, \mathbf{S}^{(N-1)} \, \mathbf{R}^{(N-1)} \, \check{I}^T \tag{25}$$

Note, the (N-1) compounds of **Q**, **S** and **R** are of the respective order of NxC_{N-1}^M, $C_{N-1}^M x C_{N-1}^M$ and $C_{N-1}^M x N$, where

$$C_{N-1}^M = \frac{M!}{(N-1)! \, (R+1)!} \tag{26}$$

represents the set of unique combinations of (N-1) different columns which can be selected out of the M columns of the Statics matrix.

Let a_n be the n-th column of $\mathbf{Q}^{(N-1)}$ (also the n-th row of $\mathbf{R}^{(N-1)}$ and let μ_n be entry (n,n) of $\mathbf{S}^{(N-1)}$. Note, μ_n is the product of (N-1) stiffnesses. It can be shown that the adjoint of **K** (25) can also be written as

$$\text{adj } \mathbf{K} = \sum_n \mu_n \, a_n a_n^T \tag{27}$$

where the summation index n is carried over all the column indeces of the (N -1) compound of **Q**. With the notation

$$A_n = a_n a_n^T \tag{28}$$

we obtain after introduction of (27) in (19) the following expression for the inverse of the stiffness matrix, explicit in terms of the stiffnesses,

$$\mathbf{K}^{-1} = \frac{1}{|\mathbf{K}|} \sum_n \mu_n \mathbf{A}_n \tag{29}$$

where $|\mathbf{K}|$ is given by (23). The number of generic matrices $\mathbf{A}_n$ in the numerator is given by (26), each generic matrix is constant and depends on the geometry only and the μ_n's are products of (N-1) stiffnesses.

Returning to the 3-bar truss example, to obtain the explicit inverse of the stiffness matrix we have to compute the (N-1) compounds of $\mathbf{Q}$ and $\mathbf{S}$ of the structure. The number of nodal displacements is N=2, consequently we need the 1-compounds of the above matrices. By definition, the 1-compound of a matrix is a new matrix populated by all minors of order 1 of the original matrix. Minors of order 1 are simply the entries of the matrix. For instance, the minor of $\mathbf{Q}$ using row 1 and column 1 is $Q_{11} = \sqrt{2}/2$. Its position in the compound matrix is (1,1).

Very conveniently, we note that the 1-compounds of $\mathbf{Q}$ and $\mathbf{S}$ are equal to $\mathbf{Q}$ and $\mathbf{S}$ respectively. Note, this is true only for 1-compounds. In general, the (N-1) compound of $\mathbf{Q}$, for instance, is a matrix with N rows but with a much larger number of columns than M. Performing, the row inversion in (25) on $\mathbf{Q}^{(1)}$, yields

$$\mathbf{I}\ \mathbf{Q}^{(1)} = \frac{1}{2} \begin{bmatrix} -\sqrt{2} & -2 & -\sqrt{2} \\ \sqrt{2} & 0 & -\sqrt{2} \end{bmatrix} \tag{30}$$

Building the generic matrices in (28) by employing the 3 columns in (30) gives us the following expression for the inverse of the stiffness matrix of the 3-bar truss (29)

$$\mathbf{K}^{-1} = \frac{1}{|\mathbf{K}|} \left\{ \frac{s_1}{2} \begin{bmatrix} 1 & -1 \\ -1 & 1 \end{bmatrix} + s_2 \begin{bmatrix} 1 & 0 \\ 0 & 0 \end{bmatrix} + \frac{s_3}{2} \begin{bmatrix} 1 & 1 \\ 1 & 1 \end{bmatrix} \right\} \tag{31}$$

where the determinant of the stiffness matrix is computed from (23) with the values in Table 1,

$$|\mathbf{K}| = \frac{s_2 s_3}{2} + s_1 s_3 + \frac{s_1 s_2}{2} \tag{32}$$

The nodal displacements are obtained by post-multiplying both sides of (31) by the applied loads vector $\mathbf{p}$

$$\mathbf{u} = \frac{1}{|\mathbf{K}|} \sum_n \mu_n \mathbf{A}_n \mathbf{p} \tag{33}$$

Note, the number of terms in the numerator (dummy index n) is equal to the number of columns in the (N-1) compound of $\mathbf{Q}$. The member forces are similarly computed by premultiplying both sides of (33) by $\mathbf{SR}$

$$t = \frac{1}{|K|} \left(\sum_n \mu_n \, S \, R \, A_n \, p \right) \tag{34}$$

It was shown in Fuchs (1991c) that the above equation reduces to

$$t = \sum_m \frac{|K_m|}{|K|} \, t_m \tag{35}$$

where subscript m runs over all the stable statically determinate substructures, $|K_m|$ is the determinant of the stiffness matrix of substructure m, and t_m are the member forces in substructure m when subjected independently to the external loads. This remarkable result merits a verbal restatement

"The member forces in a redundant structure are equal to the weighted sum of the member forces in all the embedded statically determinate and stable substructures when subjected to the external loads. The weighting factors are the ratios of the determinants of the stiffness matrices of the substructures to that of the original structure".

It is clear that by using (23) in (35), we recover the alternate expression in (2) for the member forces in the truss.

Considering again the 3-bar truss example, (33) yields the nodal displacements

$$u = \frac{1}{|K|} \left(\frac{s_1}{2} \begin{Bmatrix} p_1 - p_2 \\ -p_1 + p_2 \end{Bmatrix} + s_2 \begin{Bmatrix} p_1 \\ 0 \end{Bmatrix} + \frac{s_3}{2} \begin{Bmatrix} p_1 + p_2 \\ p_1 + p_2 \end{Bmatrix} \right) \tag{36}$$

Multiplying the displacements by SR yields the member forces

$$t = \frac{1}{|K|} \begin{Bmatrix} \frac{\sqrt{2}}{2} (p_1 + p_2) \, s_1 s_3 + \frac{\sqrt{2}}{2} p_1 \, s_1 s_2 \\ \frac{1}{2} (p_1 + p_2) \, s_2 s_3 + \frac{1}{2} (p_2 - p_1) \, s_1 s_2 \\ -\frac{\sqrt{2}}{2} p_1 \, s_2 s_3 + \frac{\sqrt{2}}{2} (p_2 - p_1) \, s_1 s_3 \end{Bmatrix} \tag{37}$$

with $|K|$ given in (32). It will be verified that these member forces are identical to those given in (2) in conjunction of the values in Table 1.

To summarize the results obtained so far, we have derived for linear elastic trusses explicit exact expressions for the inverse of the stiffness matrix, the nodal displacements and the member forces. In the context of this paper, the latter are of particular interest. Before proceeding we will show in the next section that the above results can be generalized to the case of linear elastic frames composed of uniform prismatic members.

4. Extension to Framed Structures

Traditionally, the structural analysis of trusses and frames, especially in a force method approach, are somehow different. Similarly, in structural optimization, we have techniques and properties relating to the design of trusses, which do not readily generalize to the case of frames. This is somehow intriguing since the same field equations (15) apply both to trusses and to frames. However, what sets the truss apart from framed structures is the fact that in the former the stiffness matrix in deformation coordinates, S, is diagonal. For frames this matrix is only block-diagonal. It all has to do with bending deformation. Extensional elements (and torsional elements in 3D space) have one mode of deformation. Bending deformation is bimodal. Consequently, a bending element stiffness matrix in deformation coordinates is a 2x2 matrix. Hence, S has also 2x2 bending matrices along its diagonal. In the derivation of the present theory we have implicitly taken advantage of S being diagonal. This has allowed us to formulate the determinant of S_m (22) as the product of its diagonal elements. When S is no longer diagonal the analytic form of the right-hand side of (22) may not be that simple to obtain. In order to incorporate frames in the present theory we will write the field equations (15) in modal deformation coordinates, for which S becomes diagonal. We will deal with 2D structures although the method is valid for 3D structures as well.

Consider a linear elastic uniform prismatic element. In 2D-space the element deforms in two orthogonal deformation patterns: extensional and bending. The extensional deformation is characterized by one parameter, the longitudinal extention δ. The bending action is determined by two deformation quantities, the end-rotations θ_A and θ_B with respect to the chord

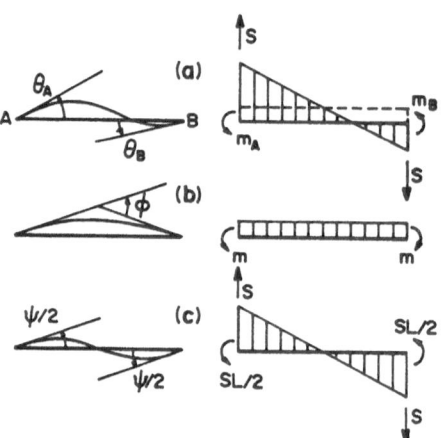

Figure 2. Modal deformations of bending element

AB of the element (Fig. 2). In these coordinates the 3x3 element stiffness matrix is block-diagonal

$$\begin{Bmatrix} n \\ m_A \\ m_B \end{Bmatrix} = \begin{bmatrix} EA/L & 0 & 0 \\ 0 & 4EI/L & 2EI/L \\ 0 & 2EI/L & 4EI/L \end{bmatrix} \begin{Bmatrix} \delta \\ \theta_A \\ \theta_B \end{Bmatrix} \tag{38}$$

where n is the axial force and m_A and m_B are the end moments. Fuchs (1991a) has shown that in modal deformation coordinates, δ for the extensional mode and ϕ and ψ for the bending modes the above equations take the uncoupled form

$$\begin{Bmatrix} n \\ m \\ sL/2 \end{Bmatrix} = \begin{bmatrix} EA/L & 0 & 0 \\ 0 & EI/L & 0 \\ 0 & 0 & 3EI/L \end{bmatrix} \begin{Bmatrix} \delta \\ \phi \\ \psi \end{Bmatrix} \tag{39}$$

with

$$\begin{Bmatrix} \phi \\ \psi \end{Bmatrix} = \begin{Bmatrix} \theta_A - \theta_B \\ \theta_A + \theta_B \end{Bmatrix} \tag{40a}$$

$$\begin{Bmatrix} m \\ sL/2 \end{Bmatrix} = \begin{Bmatrix} (m_A - m_B)/2 \\ (m_A + m_B)/2 \end{Bmatrix} \tag{40b}$$

The physical interpretation of the modal deformation coordinates, ϕ and ψ, is given in Fig. 2. The corresponding force quantites are, m, the average bending moment on the element and, sL/2, where s is the constant shear force. Consequently, a 2D uniform prismatic element can be viewed as the sum of its three unimodal components: an axial element (δ,n) which carries the normal load, a moment element (ϕ,m) which carries the average moment in pure bending, and a shear element $(\psi,sL/2)$ which carries the differential moments and the shear force in "pure" shear.

In 3D space the prismatic element is equivalent to six unimodal components attached in parallel: axial, moment and shear in two orthogonal planes, and torsion with stiffness GJ/L where G is the shear modulus and J is the torsional constant. In these modal deformation coordinates, all structures composed of prismatic elements have diagonal stiffness matrices in equations (15b). Consequently, the present theory applies also to frames. A statically determinate substructure of a frame is obtained by removing R unimodal components, be they axial, moment or shear components. The analytic representation of a member force (2) in a frame (n, m or sL/2) will include products of N stiffnesses of the type EA/L, EI/L, and 3EI/L.

5. Approximate Models for the Internal Forces

The straightforward use of the analytic expressions for the analysis and design of structures must unfortunately be ruled out. The number of terms of the polynomials render them simply intractable for common engineering structures. Take for instance the internal forces. The number of combinations of structures composed of N components out of a total of M is

$$C_N^M = \frac{M!}{N! \, R!} \tag{41}$$

which is a formidable number. Now, even assuming that most of them are unstable, it seems hopeless to track down the stable ones. However, the present theory is not merely of academic nature. After all, for the first time engineers can visualize the construct of the exact solution of the structural analysis problem. In this vein, we will present here an approximate model based on the construct of the exact expression and some preliminary results which were reported in Fuchs and Maslovitz (1991).

Consider the analytic representation for the member forces in a structure (14). Every term in the polynomials corresponds to a statically determinate substructure. We will construct an approximation for the member forces by using a truncated set of the terms in (14). Note, since every t_m satisfies equilibrium, the truncated set satisfies equilibrium for any nuber of selected terms. What we are loosing by dropping terms is compatibility of deformations. To ensure overall compatibilty we need the full set of terms, and then the constants B_m are given by

$$B_m = \left| Q_m \right|^2 \tag{42}$$

which yields expression (2). In the truncated set, if we take the B_m's as parameters and adjust them in order to satisfy compatibility of deformations at preselected "anchor" points in the design space, (14) becomes an approximate model for the member forces. The model is exact at the anchor points and for the substructures which were retained in the truncated form of (14). Elsewhere, the model constitutes an approximation.

Following the procedure used in the classical Force method, by eliminating the nodal displacements from the Kinematics equations (15c) one obtains R compatibility conditions in the form of R homogeneous linear equations in the M modal deformations of the components of the structure

$$\sum_{j=1}^{M} G_{ij} \, e_j = 0 \qquad i=1,\dots,R \tag{43}$$

where the G_{ij}'s are constants which can be computed from matrix R. If we select P anchor points, at which we require exact results, we will be writing the R conditions (43) at those points, thus producing a total of RP equations. Since one B coefficient is arbitrary, one will need RP+1 terms in the truncated expressions for such a model.

The expression for the modal deformations, e_j, is obtained by dividing both sides of Eq. (14) by the stiffness s_j of the corresponding element, for all the components of the structure

$$e_j = \frac{\sum_{m=1}^{PR+1} B_m \, t_{jm} \, \pi_{m/s_j}}{\sum_{m=1}^{PR+1} B_m \, \pi_m} \qquad j=1,...,M \qquad (44)$$

Introducing (44) in (43), and multiplying throughout by the denominator of (14), which is common to all the terms, yields after grouping the expressions by B_m, R compatibility equations, linear in the PR coefficients

$$\sum_{m=1}^{PR+1} D_{im} \, B_m = 0 \qquad i=1,...,R \qquad (45)$$

with

$$D_{im} = \sum_{j=1}^{M} G_{ij} \, t_{jm} \, \pi_m \, / \, s_j \qquad (46)$$

We note that the coefficients in Eq. (46) depend on the unimodal stiffnesses of the structure, $D_{im}(s)$, via the products π_m/s_j. If we evaluate these coefficients at an arbitrary design point s_p we obtain R conditions in order for the model of Eq. (14) to be exact at that point. By writing the compatibility conditions for P distinct points in the design space, we obtain, with $B_1=1$, PR+1 linear equations in the PR+1 B_m constants

$$\sum_{m=1}^{PR+1} D_{im}(s_p) \, B_m = 0 \qquad i=1,...,R; \quad p=1,...,P$$

$$B_1 = 1 \qquad (47)$$

It will be recognized that the model is scaling invariant, a basic property of linear structures, in modal coordinates. Such models were used to interpolate the member forces between anchor points located along a line in the analysis space spanned by the unimodal components for various structures. From limited numerical experimentation some common characteristics have emerged. They will be delineated in the simple case of a portal frame.

7. A Numerical Example

Consider the portal frame reported in Vanderplaats and Salajegheh (1989) which is subjected to two lateral and equal loads at the free nodes (Fig. 3). The left column and the horizontal

1130

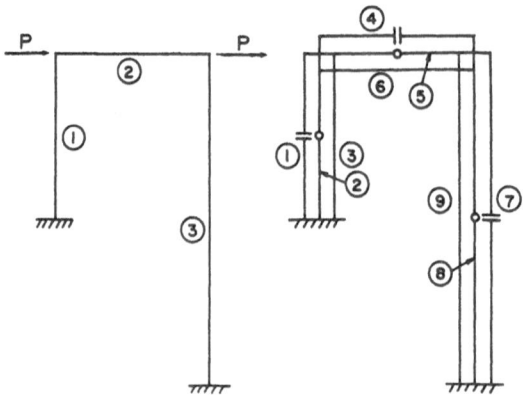

Figure 3. The portal frame

beam are of length L and the right column has a length of 2L. Decomposing every element into its unimodal components - extension, moment, and shear - we obtain a structure with M=9 components, N=6 nodal displacements and a redundancy of R=M-N=3. Every determinate substructure of the portal frame will be composed of 6 unimodal components. Using Eq. (41) we find that there are 84 different combinations of 6 components out of a total of

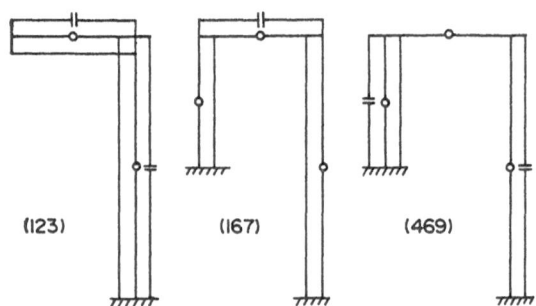

Figure 4. Sample of stable substructures of portal frame

9. However, only 46 of them yield stable configurations. In Fig. 4   we show a sample of stable substructures. The missing redundant components are given in paranthesis. Note, moment and shear components are depicted respectively with a shear and a moment release at mid-span. It can be shown that such bending elements are structurally equivalent to the unimodal bending components. Extensional components are represented by a simple line.

Consequently, the exact expressions of the internal forces in the frame will be based on 46 B_m coefficients. Setting $B_1=1$ we note that by enforcing compatibility of deformations at P=15 distinct points in the analysis space the PR=45 equations (47) will give us the coefficients of the exact model. For our purposes we will construct an approximate model which will be enforced to be exact at selected points along a line connecting the stiffnesses x_0 and

x_n in the analysis space which is spanned over the unimodal components. The values of

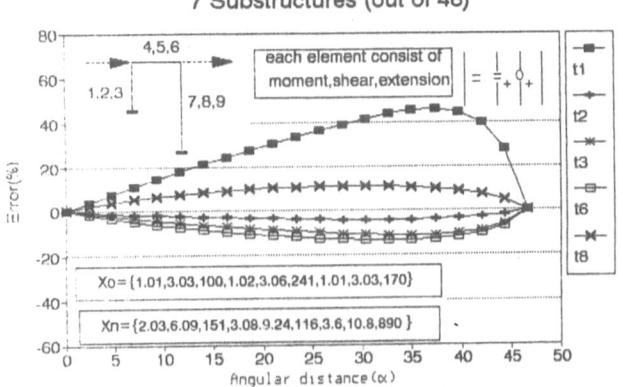

Figure 5. A two-point model for the portal frame

these design points can be seen in Fig. 5. For the left column of the frame, at design point x_0 for instance, we have $s_1=(EI/L)=1.01$ and $s_3=(EA/L)=100$, where L=3. The applied loads are set equal to 1. The approximate model was compared to the exact results for design points s on the line joining the two points

$$s = (1-\lambda)\, x_0 + \lambda\, x_1 \qquad (48)$$

where λ ranges from 0 to 1. The error ϵ_j for a typical internal force t_j is given by

$$\epsilon_j = (t_j^a - t_j)\, /\, t_j \qquad (49)$$

where t_j^a is the approximate value of the internal force.

In the analysis space, spanned over the unimodal stiffnesses, the approximate model is scaling invariant. To filter the scaling effect, the error ϵ in the unimodal member forces at s was plotted against the angular distance α between the initial design x_0 and s

$$\alpha = \text{arc cos}\ \frac{x_0 \cdot s}{\|x_0\|\ \|s\|} \qquad (50)$$

where the numerator is the scalar product of vectors s_0 and s, and the denominator is the product of their Euclidean Norms. The error ϵ_j for a typical internal force t_j is given by

$$\epsilon_j = (t_j^a - t_j)\, /\, t_j \qquad (50)$$

where t_j^a is the approximate value of the internal force.

We have considered models which are exact at one point (x_0), two points $(x_0$ and $x_1)$ and at three points, the two extreme points and an intermediate point half-way on the α scale. This calls for approximate (truncated) models which are based on respectively, 4, 7 and 10 statically determinate substructures (3 compatibility equations at each anchor point) and which have a corresponding number of B_m coefficients. The substructures were selected randomly. The one-point model, not reported herein, gave rather poor results. The two-point model, shown in Fig. 4, has in general error levels of the order of 10%, but for the axial force in the left column were the error reaches 50%. The three-point model depicted in Fig. 5 is excellent with errors of a fraction of a percent over the entire α range. It should be noted that an angular distance of 45° (the distance between x_0 and x_n) represents a major redistribution of rigidities.

Similar results have been obtained in other cases. In particular, Fuchs and Maslovitz (1991) have reported an example of a three-point model for an arch truss. Employing 15 substructures out of a total of 985 stable configurations, all member forces had error levels

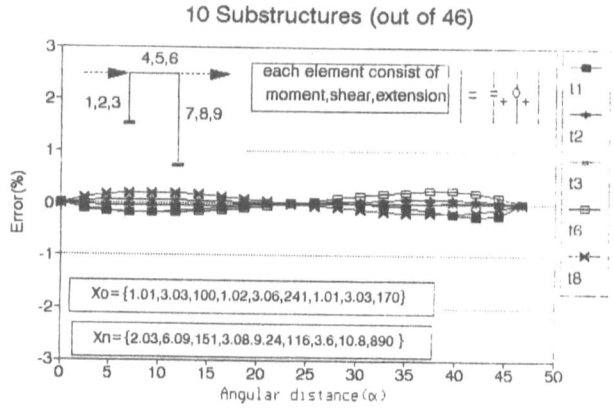

Figure 6. A three-point model for the portal frame

of less than 0.5% for a stiffness redistribution range of 30°.

8. Conclusions

This paper has presented an overview of the analytic expressions for the inverse of the stiffness matrix, the nodal displacements and the internal forces in linear elastic redundant structures. The inverse of the stiffness matrix and the nodal displacements were obtained by using Binet and Cauchy's theorem on product of compound matrices. The formula for the internal forces was derived from principles of structural mechanics. The approach was shown to apply to all framed structures via the unimodal stiffnesses of its elements. Departing from the exact formulation, approximate models were constructed which are exact at preselected points along a line in the analysis space. In the presented example as well as in other cases, three-point models gave excellent results.

The paper also advocates the use of ratios of multilinear polynomials as an alternative to the Taylor expansion-based approximations. What has been shown is manifestly not sufficient to convince investigators of the pertinence of approximations using ratios of truncated multilinear polynomials. It is hoped that it will nevertheless trigger a research effort for an in-depth check of this new alley for structural reanalysis.

9. References

Abu Kassim, A.M. and Topping, B.H.V. (1985) 'Static Reanalysis of Structures: A Review', in Civil-Comp, Proc.Sec. Int. Conf. Civ. and Str. Eng. Comp., London, (Ed. B.H.V. Topping), 2, Civil-Comp Press, Edinburgh GB, 137-146.

Aitkin, A.C. (1956) 'Determinants and Matrices', Oliver and Boyd, London.

Arora, J.S. (1976) 'Survey of Structural Reanalysis Techniques', J. Str. Div., ASCE, 102, 783-802.

Fuchs, M.B. (1991a) 'Unimodal Beam Elements', Int. J. Solids Struct., 27(5), 533-545.

Fuchs, M.B. (1991b) 'Analytic Representation of Member Forces in Linear Elastic Redundant Trusses', Int. J. Solids Struct., (accepted).

Fuchs, M.B. (1991c) 'The Explicit Inverse of the Stiffness Matrix', (submitted).

Fuchs, M.B. (1992) 'The Explicit Expression of the Internal Forces in Prismatic Membered Structures', Mechanics of Structures and Machines, 20(2).

Fuchs, M.B. and Steinberg, Y. (1991) 'An Efficient Approximate Analysis Method Based on an Exact Univariate Model for the Element Loads', Structural Optimization, 3(1), 107-114.

Fuchs, M.B. and Maslovitz, G. (1991) 'Predicting the Stress Resultants in Framed Structures Based on the Exact Analytic Expressions', 23nd AIAA/ASME/ASCE/AHS/ASC Structures, Structural Dynamics and Materials Conference, 8-10 April 1991, Baltimore, Maryland.

Vanderplaats, G.N. and Salajegheh, E. (1989) 'A New Approximation Method for Stress Constraints in Structural Synthesis', AIAA J., 27, 352-358.

LARGE SCALE STRUCTURAL OPTIMIZATION WITH NONLINEAR GOAL PROGRAMMING

By:

Mohamed E. M. El-Sayed and T. S. Jang

Department of Mechanical Engineering, Florida International University

Miami, Florida 33199

ABSTRACT

Traditional optimization models require the formulation of a single objective function. While many problems can be modeled in this form, many real situations involve not only multiple objectives but also multiple, conflicting objectives. In this paper a multiple objective goal structural optimization approach is presented using a nonlinear goal programming technique. To demonstrate the approach, the formulation and solution of some truss problems using nonlinear goal programming are presented.

INTRODUCTION

Mathematical programming is becoming a very important tool in industrial and engineering design. The application of optimization techniques to structural problems has continued to increase and to gain considerable interest since the pioneering work by Schmit [1], in 1960. One of the major difficulties in applying traditional mathematical programming techniques to design problems, however, is matching the model with reality. Real-life design problems occur with conflicting and multiple objectives rather than with a single one. It is often difficult to define the problems exactly in mathematical programming traditional form. There are, however, techniques that are capable of handling weights and priority factors for conflicting and multi-objective

1135

G. I. N. Rozvany (ed.), Optimization of Large Structural Systems, Vol. II, 1135–1153.

optimization. These techniques are called linear and nonlinear goal programming [2,3]. In a goal programming (GP) formulation, the design goals are defined and weights and priority factors are assigned to each. The algorithm then attempts to minimize the sum of the deviational variables with weights and priority factors, starting with the highest priority goal, which makes goal programming an ideal real life design tool.

The initial development of the goal programming concept was made in 1961 by Charnes and Cooper [4]. In 1965 Ijiri [5], further developed the linear goal programming (LGP). He presented the definition of "preemptive priority factors" to treat the multiple conflicting goals according to their ranked importance. He also assigned weights to the goals of the same priority level. In the late 1960's and early 1970's, the LGP began to receive wide acclaim as a decision-making tool. Veikko Jaaskelainen [6] developed a computer code for the multi-phase simplex method in 1969. The application of LGP to many areas was conducted in the 1970's. Lee [7], illustrated the application of LGP and contributed to the continuous growth in the use of LGP. He also developed a computer program using the revised simplex method for the solution of LGP problems.

In the late 1970's and mid 1980's the emphasis was more evenly shared between applications for management [8], marketing [9], and industrial engineering [10]. The development of goal programming models basically consisted of improving the solution procedures [11]. There was also some efforts to apply multi-objective criteria to structural optimization problems [12]. Recently, Rao [13] applied fuzzy goal programming to structural systems. El-sayed et al [14,15], used successive linearization method to apply LGP techniques to large scale optimization problems. As indicated In references

[14,15], for large scale structural optimization problems with high nonlinearity, successive LGP may not converge to the optimum design.

In this paper the general formulation of the structural optimization problem into a nonlinear goal programming (NLGP) form is presented. The resulting NLGP problem is then solved using unconstrained nonlinear optimization program.

FORMULATION OF THE STRUCTURAL OPTIMIZATION PROBLEM INTO NONLINEAR GOAL PROGRAMMING MODEL

The general mathematical programming model for the optimization problem can be expressed as follows:

Find $\qquad \mathbf{x} = (x_1, x_2, x_3, \ldots, x_n)$

to minimize $\qquad F(\mathbf{x})$

subject to $\qquad G_i(\mathbf{x}) \leq 0 \qquad$ (for i=1,2, $\cdots$,l) $\hspace{3cm}$ (1)

and $\qquad \mathbf{x}^L \leq \mathbf{x} \leq \mathbf{x}^U$

where $F(\mathbf{x})$ is the objective function and $G_i(\mathbf{x})$ represents the inequality constraint function including the equality constraints. The elements of the vector $\mathbf{x}$ are the design variable.

For structural optimization problems, the constraint function $G_i(\mathbf{x})$ may be a function of stress, displacement, and natural frequency. The elements of the vector $\mathbf{x}$ correspond to the size of the structural members. The variables $\mathbf{x}^L$ and $\mathbf{x}^U$ represent the lower and upper bound on the design variables, respectively.

The minimum weight structural optimization problem with stress, displacement, and frequency requirements can be expressed as :

Find $\quad \mathbf{x} = (x_1, x_2, x_3, \ldots, x_n)$

to minimize $\quad W(\mathbf{x})$

subject to $\quad \sigma_i / \sigma_a - 1 \leq 0$ $\qquad\qquad$ (2)

$\qquad\qquad\quad u_j / u_a - 1 \leq 0$

$\qquad\qquad\quad f / f_a - 1 \leq 0$

and $\qquad\qquad x^L \leq x \leq x^U$

Following [16], the standard form of the nonlinear goal programming model is:

Find $\quad \mathbf{x} = (x_1, x_2, x_3, \ldots, x_n)$

to minimize $\quad \mathbf{z} = (f_1(\mathbf{d}^-, \mathbf{d}^+), f_2(\mathbf{d}^-, \mathbf{d}^+), \ldots, f_k(\mathbf{d}^-, \mathbf{d}^+), \ldots, f_K(\mathbf{d}^-, \mathbf{d}^+))$

subject to $\quad g_i(\mathbf{x}) + d_i^- - d_i^+ = b_i \qquad$ (for i=1,2, $\cdots$,I) $\qquad\qquad$ (3)

and $\qquad\quad d_i^-, d_i^+ \geq 0$

where $\mathbf{z}$ represents an achievement vector, structured as an ordered set such that a preemptive priority structure is maintained. The dimension of $\mathbf{z}$ represents the number of preemptive priority levels which is equal to or less than the number of objectives. The value of $\mathbf{z}$ will be equal to zero vector if all the objectives meet their aspiration levels.

$\qquad g_i(\mathbf{x}) + d_i^- - d_i^+ = b_i \qquad$ (for i=1,2, $\cdots$,I)

represent the design objectives where d_i^- measures the negative deviation from the aspired level for for objective i and d_i^+ measures the positive deviation.

$f_k(\mathbf{d^-}, \mathbf{d^+})$ is a weighted linear function of the deviation variables $\mathbf{d^-}, \mathbf{d^+}$. Each

function can be represented as the weighted sum of negative and the positive deviation variables for all the objective contained in a priority level P_k with weights w_n and μ_n as:

$$f_k(\mathbf{d^-}, \mathbf{d^+}) = \sum_{j \varepsilon P_k} (w_j\, d_j^- + \mu_j\, d_j^+) \quad \text{(for } k = 1, 2, \ldots, K,) \tag{4}$$

The nonlinear goal programming model for the structural optimization problem using equation (2) and (3), for weight, stress, displacement, and frequency requirements can be expressed as :

Find $\quad \mathbf{x} = (x_1, x_2, x_3, \ldots\ldots, x_n)$

to minimize $\quad \mathbf{z} = (\, f_1(\mathbf{d^-}, \mathbf{d^+}),\ f_2(\mathbf{d^-}, \mathbf{d^+}), \ldots, f_k(\mathbf{d^-}, \mathbf{d^+}), \ldots, f_K(\mathbf{d^-}, \mathbf{d^+})\,)$

subject to $\quad W/W_a - 1 + d_1^- - d_1^+ = 0,$

$$\sigma_i\, /\, \sigma_a\ -1 + d_i^- - d_i^+ = 0 \tag{5}$$

$$u_j\, /\, u_a - 1 + d_j^- - d_j^+ = 0$$

$$f/\, f_a - 1 + d_l^- - d_l^+ = 0$$

and $\quad x^L \leq x \leq x^U,$

$$d_1^-, d_1^+, d_i^-, d_i^+, d_j^-, d_j^+, d_l^-, d_l^+ \geq 0,$$

where W is the total weight of the structure, σ_i is the i th stress objective, u_j is the j th displacement objective, and f is the natural frequency. W_a, σ_a, u_a, and f_a represent the target weight, allowable stress, allowable displacement, and frequency, respectively.

NONLINEAR GOAL PROGRAMMING SEARCH ALGORITHM

A variety of methods can be used to solve nonlinear optimization problems. The success of a method over other methods to find a solution may be relative to the particular problem. A straightforward approach for solving the nonlinear goal structural optimization problem of equation (5) is to use a zero-order optimization method. In the following the Hooke-Jeeves pattern search method is used for its simplicity and robustness.

The nonlinear goal programming algorithm developed in [16], was modified for the structural optimization applications and linked with a finite element routine. The NLGP algorithm first minimizes, as nearly as possible, the objectives with the highest priority level. It then proceed to satisfy the objectives of the next priority level, as nearly as possible, without degrading the achievement of any objective in a higher priority level. This process is continued until all priority levels have been considered.

At each priority level the search is terminated when the difference between present and previous achievement function value becomes sufficiently small. The value of z will be equal to zero vector if all the objectives meet their aspiration levels. The value of z_k will be positive if one or more objectives in priority level k are not met.

TEST CASES

To demonstrate the application of the nonlinear goal programming to design optimization problems some numerical test cases are conducted using the 10 member planar truss Figure (1), to demonstrate the effect of the priority levels and weights on the structural optimization problem. The 200 member planar truss Figure (1), is used to demonstrate the applicability of the method large scale structural optimization problems.

The 10 Member Planar Truss:

The design and load data for the 10 member planar truss are given in Table (1).

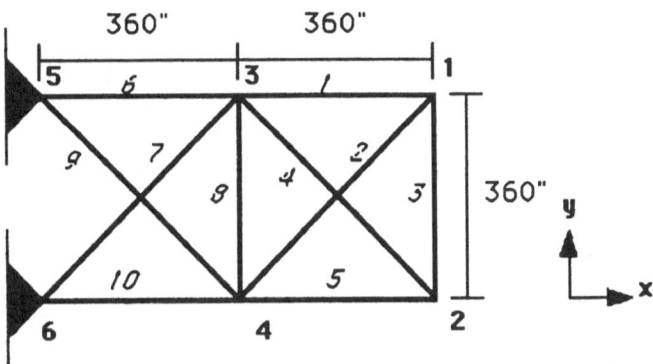

Figure (1) 10 Member Planar Truss

Table (1) Design and Load Data for 10 Member Planar Truss

Modulus of elasticity = 10^4 ksi

Material specific weight = 0.10 lb/in^3

Lower limit on cross-sectional areas = 0.10 in^2

Initial value of design variable   = 10 in^2

Stress limit = $\pm$ 25 ksi

Number of loading conditions = 2

Load Data

Loading Condition	Node	Load component (kips) in Direction		
		x	y	z
1	2	0.0	-100.0	0.0
	4	0.0	-100.0	0.0
2	1	0.0	50.0	0.0
	2	0.0	-150.0	0.0
	3	0.0	50.0	0.0
	4	0.0	-150.0	0.0

Three different cases of nonlinear goal structural optimization problems are considered. The first case the represents the minimum weight design with stress constraints only. The second case represents the minimum weight design with stress and displacement constraints with different priority levels. The third case represents the minimum weight design with stress, displacement and frequency constraints with different priority levels and weights. In all cases the maximum positive or negative deviation was limited to be less than 0.01. The stresses, displacements and frequencies were obtained using a finite element analysis program. A Vax 8650 computer was used to solve the structural optimization problems. In the following we discuss the three case and the obtained results.

Case I : In this case, the minimum weight design with stress constraints only is performed using the developed NLGP technique. Two loads of 100 kips in the negative y direction are applied at nodes 2 and 4 as in loading condition (1). The structural optimization problem was solved using NLGP technique with deviational variables only. The results for the optimum cross-sectional area of each member, and the optimum weight are compared with the results obtained from a traditional mathematical programming algorithm [19], in Table (2).

Table (2) Results for Case I of 10 Member Truss

Optimum Cross-Sectional Area in in^2.

Member Number	Ref. [19]	NLGP
1	7.9379	7.9375
2	0.1000	0.1039
3	8.0621	8.0625
4	3.9379	3.9375
5	0.1000	0.1000
6	0.1000	0.1000
7	5.7447	5.7461
8	5.5690	5.5703
9	5.5690	5.5703
10	0.1000	0.1000
Optimum Weight (lb)	1593.18	1593.51
Total CPU (Sec.) on Vax 8650		35.42

Case II : In this case, the structural optimization is performed for loading condition (2) with stress and displacement constraints of ± 2.0 in. at nodal points 1 and 2. The NLGP algorithm was used to solve two structural optimization problems. The first problem was solved with deviational variables only. The second problem was solved with priority priority level 1 for the weight and displacement constraints and priority level 2 for the stress constraints. The results for the two optimization problems are compared with the results obtained from a traditional mathematical programming algorithm [19], in Table (3).

Table (3) Results for Case II of 10 Member Truss

Optimum Cross-Sectional Area in in^2.

Member Number	Ref. [19]	NLGP	NLGP (P)
1	23.564	24.300	29.344
2	0.1	0.1	0.1
3	25.277	23.000	24.875
4	14.343	13.900	15.25
5	0.1	0.1	0.1
6	1.9698	1.9890	1.1625
7	12.401	13.200	5.0938
8	12.850	13.900	20.906
9	20.285	19.750	21.875
10	0.1	0.1	0.1
Optimum Weight (lb)	4676.13	4673.30	4992.41
Total CPU (Sec.) on Vax 8650		34.17	86.30

Case III : This case is the same as case II except that a natural frequency
constraint of 22. Hz was imposed on the structure. The NLGP algorithm was
used to solve three structural optimization problems. The first problem was
solved with deviational variables only. The second problem was solved with
priority level 1 for the weight, displacement and frequency constraints and
priority level 2 for the stress constraints. For the third problem priority level 1
was given to the weight, displacement and frequency constraints and a
weighting factor of 3 was given to the frequency constraint. Priority level 2 was
given to the stress constraints. The results for the three optimization problems
are compared with the results obtained from a traditional mathematical
programming algorithm [20], in Table (4).

Table (4) Results for Case III of 10 Member Truss

Optimum Cross-Sectional Area in in^2.

Member Number	GRGA [20]	NLGP	NLGP (P)	NLGP (P&W)
1	24.8602	25.1614	24.5356	7.928
2	0.1	0.1	0.12	0.1
3	25.9762	23.780	24.893	10.398
4	13.1350	13.218	13.348	4.5563
5	0.1	0.12	0.12	0.1
6	1.9727	1.9888	1.16	0.5063
7	13.1924	13.664	8.911	7.9281
8	15.3214	15.899	15.753	4.5563
9	17.5268	18.011	18.726	5.8156
10	0.1	0.1	0.1	0.1
Optimum Weight (lb)	4730.29	4744.41	4524.49	1785.97
Total CPU (Sec.) on Vax 8650		165.65	151.38	105.26

The 200 Member Plane Truss:

The geometry and dimensions of a 200 member plane truss are shown in Figure (2). This structure has 77 joints and 150 degrees of freedom. The structure is designed to withstand 3 loading conditions. Due to the symmetry, the number of design variables is reduced to 96 variables. Table (5) gives the design data of the structure. The structural optimization problems considered is subjected to weight, stress, and displacement goal constraints.

The results for the structural optimization problem is compared with the results obtained from a traditional mathematical programming algorithm [19], in Table (6). The iteration history of this problem is given in Table (7).

Table (5) Design Data for 200 Member Plane Truss

Modulus of elasticity = 30000 ksi

Material specific weight = 0.283 lb/in^3

Lower limit on cross-sectional areas = 0.1 in^2

Initial value of design variable   = 1 in^2

Stress limit = $\pm$ 30 ksi

Displacement limits at all nodes and in x and y directions = 0.5 in

---

Loading Condition 1. One kip acting in position x direction at node
points 1,6,15,20,29,34,43,48,57,62,71.

Loading Condition 2. 10 kips acting in position y direction at node
points 1,2,3,4,5,6,7,8,9,10,12,14,15,16,
17,18,19,20,22,24,..........,71,72,73,74,75.

Loading Condition 3.   Loading Conditions 1 and 2 acting together

---

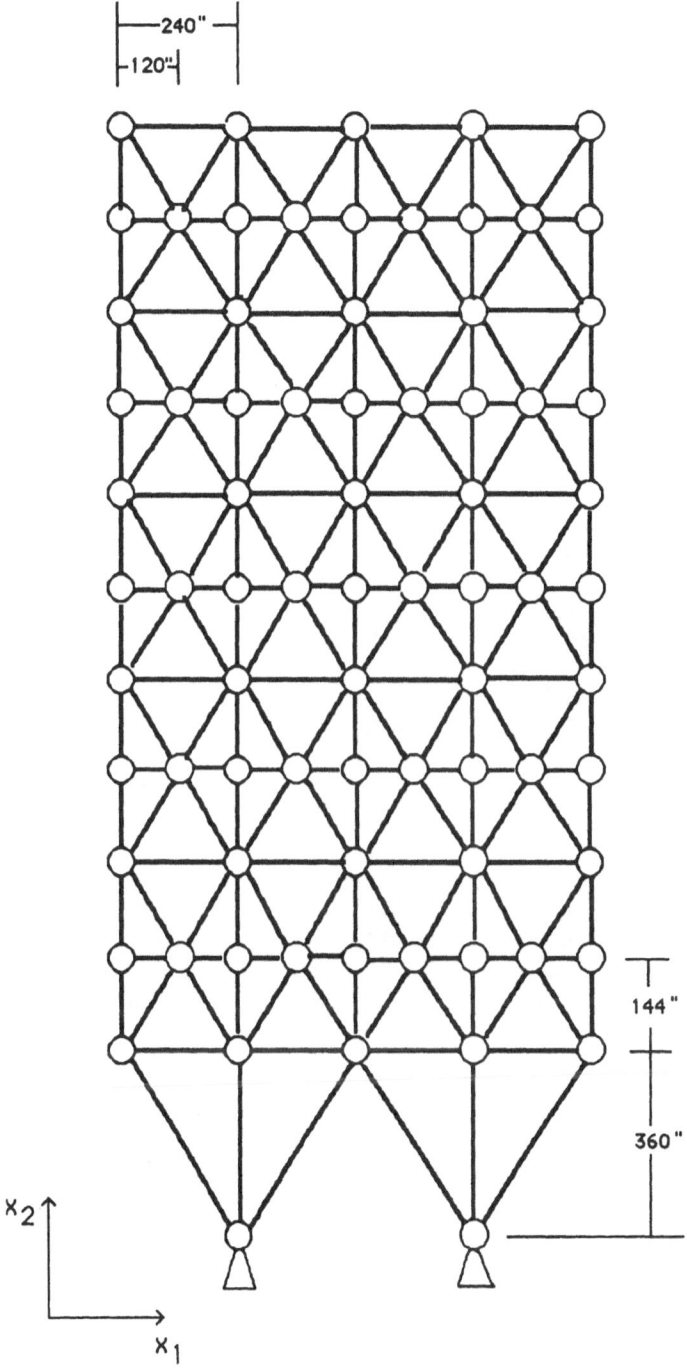

—240"—

—120"—

144"

360"

x_2

x_1

Figure (2) 200 Member Planar Truss

Table (6) Results for 200 Member Truss

Optimum Cross-Sectional Area in in^2.

Member Number	NLGP-HJ	Ref. [19]
1,4	0.2656	0.1878
2,3	0.1250	0.1
5,17	3.5	4.7832
6,16	0.375	0.1703
7,15	0.1	0.1
8,14	1.7719	2.3462
9,13	0.125	0.1876
10,12	0.5625	0.1
11	1.4125	2.8809
18,25,56, 63,94,101, 132,139, 170,177	0.1211	0.1
19,20, 23,24	0.1719	0.1
21,22	0.3125	0.1
26,38	6.625	6.7767
27,37	0.1	0.1
28,36	0.4125	0.2361
29,35	3.1563	3.3133
30,34	0.8125	0.1732
31,33	0.225	0.2227
32	3.375	4.1473
39,42	0.1	0.1
40,41	0.1	0.1
43,55	7.9375	8.1292
44,54	0.1875	0.2476
45,53	0.1	0.1
46,52	4.8125	4.4206
47,51	0.25	0.2802
48,50	0.5	0.2673
49	3.625	4.7929
57,58, 61,62	0.125	0.1
59,60	0.225	0.1002
64,76	9.5	9.3889
65,75	0.1	0.1
66,74	0.25	0.3362
67,73	6.0	5.0733

Table (6) (Continued)

Member Number	NLGP-HJ	Ref. [19]
68,72	0.5	0.3008
69,71	0.25	0.3096
70	4.875	5.5744
77,80	0.375	0.4967
78,79	0.5	0.3865
81,93	9.8125	9.5196
82,92	0.75	0.9366
83,91	0.1719	0.1
84,90	7.0	6.2617
85,89	0.25	0.3508
86,88	0.5	0.4835
87	5.5625	5.8679
95,96, 99,100	0.1	0.1
97,98	0.1	0.1
102,114	11.25	10.48
103,113	0.25	0.1108
104,112	0.9531	1.0313
105,111	7.5	6.8203
106,110	0.5	0.5012
107,109	0.25	0.3754
108	6.4375	6.4768
115,118	0.7813	1.9807
116,117	0.5	1.4784
119,131	10.75	9.1546
120,130	1.3438	3.1979
121,129	0.125	0.1
122,128	10.0938	9.0271
123,127	0.5	0.2074
124,126	0.5	0.9717
125	6.75	6.5338
133,134, 137,138	0.1	0.1
135,136	0.1	0.1219
140,152	11.75	9.9624
141,151	0.25	0.1341
142,150	1.75	3.3
143,149	11.0	9.5771
144,148	0.5	0.9814
145,147	0.5	0.2269
146	8.9375	7.0561

Table (6) (Continued)

Member Number	NLGP-HJ	Ref. [19]
153,156	1.75	2.55
154,155	1.0	0.6074
157,169	9.5	7.5376
158,168	3.0	4.1216
159,167	0.1	0.1
160,166	13.25	13.329
161,165	1.125	1.8691
162,164	0.5	0.3045
163	9.8125	7.4246
171,172, 175,176	0.8438	0.1
173,174	0.75	0.1
178,190	10.5	8.2183
179,189	0.5313	0.1
180,188	3.75	4.1916
181,187	14.25	13.833
182,186	0.75	0.3354
183,185	1.5	1.9082
184	10.0	7.884
191,194	6.5	5.8649
192,193	4.25	3.4248
195,200	11.5	10.656
196,199	14.0	17.777
197,198	7.0	7.714
Optimum Weight (lb)	28917.5	28963.0
No. of Iterations	62	
Total CPU Time (sec), Vax 8650	2626.72	
No. of Function Evaluations	10190	
Max. Positive Deviation	2.59×10^{-2}	
Max. Constraint Violation		$1. \times 10^{-3}$

Table (7) Iteration History Data for
200 Member Truss

Iteration Number	Weight (lb) NLGP
0	9963.4
1	39535.3
5	55905.0
10	29284.6
15	31814.9
20	29316.2
22	29847.4
25	29224.7
30	29002.9
35	29009.1
40	28931.7
44	28920.7
50	28916.3
55	28913.9
62	28917.5

CONCLUSION

A method for solving large scale structural optimization problems using nonlinear goal programming is developed. The method uses any unconstrained nonlinear multivariable optimization algorithm to minimize the achievement function at each priority level. At the end of the optimization process the achievement vector indicates the degree to which the desired aspiration levels are met.

The test cases performed, using the developed algorithm, demonstrated that classical structural optimization problems can be solved with same efficiency using the nonlinear goal programming formulation. By using different priority levels and weighting factors different designs can be achieved. The resulting optimum designs, however, depend on the priority levels and the weighting factors used. Therefore, careful selection for the priority levels and the weighting factors is of great importance for achieving the required design.

The greatest advantage of using nonlinear goal programming in solving structural optimization problems is that the method allows the design engineer to model the problem realistically. Another advantage of using nonlinear goal programming with unconstrained search methods is that the solution to the optimization problem can always be obtained. This is due to the fact that unconstrained search methods do not require feasibility.

REFERENCES

[1]   Schmit, L. A., "Structural Design by Systematic Synthesis," Proceedings of the 2nd Conference on Electronic Computation, ASCE, New York, 1960, pp. 105-122.

[2]   Schniederjans, M. J. and Kwak, N. K., " An Alternative Solution Method for Goal Programming: a Tutorial," J. of Operational Research Society, Vol. 33, pp. 247-251, 1982.

[3]   Ignizio, J. P., " Generalized Goal Programming: An Overview," Computer and Operational Research, Vol. 10, No. 4, pp. 277-289, 1983.

[4]   Charnes, A. and Cooper, W. W., Management Models and the Industrial Applications of Linear Programming, Vols. 1 and 2, John Wiley and Sons, New York, 1961.

[5]   Ijiri, Y., Management Goals and Accounting for Control, Rand McNally, Chicago, 1965.

[6]   Jaaskelainen, V. " Accounting and Mathematical Programming

[7]   Lee, S. M., Goal Programming for Decision Analysis, Auerback Publishers, Philadelphia, 1972.

[8]   Fisk, J. C., "A Goal Programming Model for Output Planning," Decision Sciences, Vol. 10, No. 4, pp. 593-603, 1979.

[9]   Wilson, J. M., "The Handling of Goals in Marketing Problems," Management Decision, Vol. 13, No. 3, pp. 175-180, 1975.

[10] Kwak, N. K. and Schniederjans, M. J., " A Goal Programming Model for Improved Transportation Problem Solutions," Omega, Vol. 7, No. 4, pp. 367-370, 1979.

[11]   Ignizio, J. P., Linear Programming in Single and Multiple Objective
       Systems, 1982, Prentice-Hall, Inc., Englewood Cliffs, New Jersey.

[12]   Rao, S.S., "Multiobjective Optimization in Structural Design with
       Uncertain Parameters and Stochastic Processes," AIAA J., Vol. 22, NO.
       11, pp. 1670-1678, 1984.

[13]   Rao, S. S., "Multi-Objective Optimizations of Fuzzy Structural Systems,"
       Int. J. for Numerical Methods in Engineering, Vol. 24, pp. 1157-1171,
       1987.

[14]   El-Sayed, M. E., Ridgely, B. J., and Sandgren, E., "Nonlinear Structural
       Optimizations Using Linear Goal Programming," Computers and
       Structures, Vol. 32, No. 1, pp. 69-73, 1989.

[15]   El-Sayed, M. E. and Jang, T. S., "Large Scale Structural Optimizations
       Using with Linear Goal Programming," Proceedings of the ASME
       Design Automation Conference, Chicago, 1990.

[16]   Ignizio, J. P., Elchak, T. L., and Yang, T., "Advances in Multicriteria
       Engineering Design," U. S. Army Final Report, DASG 60-77-C-0078,
       1979.

[17]   Powell, M. J. D., " An Efficient Method for Finding the Minimum of a
       Function of Several Variables without Calculating Derivatives," Computer
       J., Vol. 7, No. 4, pp. 155-162, 1964.

[18]   Vanderplaats, G. N., Numerical Optimization Techniques for Engineering
       Design: with Applications, McGraw-Hill, New York, 1984.

[19]   Haug, E. J. and Arora, J. S., Applied Optimal Design, John Wiley and
       Sons, New York, 1979.

[20]   Abadie, J. and Carpenter, J., " The Generalization of the Wolfe Reduced
       Gradient Method to the Case of Nonlinear Constraints," in Optimization,
       Fletcher,R.(ed), Academic Press, New York, pp. 37-47, 1969.

Isoperimetric inequalities in stability problems

V.V.Kobelev

Institute for Problems in Mechanics, Russian Academy of Science.
117526, Moscow, Russia.

summary. The isoperimetric inequalities for the problem of determining the shape of a compressed column that has minimum weight and can withstand a given load, without loss of stability are treated. Two new isoperimetric inequalities, arizing in the problems of stability optimization, are proved: 1) the optimal column has the largest buckling load among all the columns with the same weight and clamped ends; 2) the optimal generalized Pfluger column has the largest buckling load among all generalized Pfluger columns with the same weight.

1.ISOPERIMETRIC INEQUALITIES IN A BIMODAL STABILITY PROBLEM.

Stability optimization of columns loaded by dead axial force.
Optimization problem for a column, loaded by axial forces, which direction and value remain constant, is studied in this part. A rigorous mathematical justification of optimality of solution, which is found in (Clausen, [1851]), was given in (Tadjbakhsh, Keller [1962]). They obtained an analytic solution for two types of boundary conditions: clamping/hinging and clamping/clamping, and proved corresponding isoperimetric inequalities. However, it was shown in (Olhoff, Ramussen, [1977]), that the solution, obtained for a rod clamped at both ends is invalid, since this rod has a different buckling form (*bimodal case*) and the critical load is less then expected in (Tadjbakhsh, Keller [1962]). The analytical solution of the bimodal optimization problem for a rod, clamped at both ends, was obtained independently in (Seiranyan [1984], Masur [1984]). The corresponding isoperimetric inequalities are formulated below.

The variation principle for the stability of a rod in the dimensionless form can be written as

$$\Lambda_o[\hbar] = \mathcal{R}[M,\hbar] = \min_m \mathcal{R}[m,\hbar]$$

1155

$$(1.1) \qquad \mathcal{R}[m,\hbar] = \left[\int_{-1/2}^{1/2} m'^2(x)\ dx \right] \left[\int_{-1/2}^{1/2} \hbar^{-n}\ m^2(x)\ dx \right]^{-1}$$

The admissible functions $m(x)$ in (1.1) are all functions, having piecewise continuous first derivatives, satisfying definite boundary conditions. The function $\hbar = \hbar(x) > 0$ designates the non - dimensional cross-sectional area. The function $\hbar^n(x)$ is directly proportional to the bending rigidity of the rod, and will be referred to as the non - dimensional rigidity. The exponent in $\hbar^n$ takes the integer values 1,2,3. The cases n=1,2, 3 correspond, respectively, to light - core sandwiches of constant width and depth and to solid rectangular sections of constant width.
The boundary conditions for the rod which is clamped at x=-1/2 and free at x=1/2 are the following

$$(1.2) \qquad m'(-1/2) = 0, \qquad\qquad\qquad m(1/2) = 0.$$

If the rod is clamped at x=-1/2 and hinged at x=1/2, the boundary conditions assume the form

$$(1.3) \qquad m'(-1/2) + m(-1/2) = 0, \qquad\qquad m(1/2) = 0.$$

If the ends x=-1/2 and x=1/2 are clamped, the boundary conditions are

$$(1.4) \qquad m(1/2) = m(-1/2) + m'(-1/2), \qquad m'(-1/2) = m'(1/2).$$

The Euler equation for variational problem (1.1) is

$$(1.5) \qquad m'' + \Lambda\, \hbar^{-n}\, m = 0, \qquad\qquad\qquad \text{if } -1/2 < x < 1/2$$

Eigenvalue problem (1.5) with boundary conditions (1.2), or (1.3) or (1.4) is self-adjoint. The conditions (1.2) and (1.3) are of Sturm type, and there exist an infinite set of eigenvalues, all eigenvalues are real and positive and can be arranged as a monotonic sequence, and each eigenvalue is simple.
The conditions (1.4) are of mixed type, and the eigenvalues are real and positive and can be arranged as two sequences

$$\Lambda_1 \; < \; \Lambda_3 \; < \; \ldots \; < \; \Lambda_{2n-1} \; < \; \Lambda_{2n+1} < \; \Lambda_{2n+3} < \; \ldots \; ,$$

$$\Lambda_2 \; < \; \Lambda_4 \; < \; \ldots \; < \; \Lambda_{2n} \; < \; \Lambda_{2n+2} < \; \Lambda_{2n+4} < \; \ldots$$

such that $\Lambda_{2n-1} \; \leq \; \Lambda_{2n} \; < \; \Lambda_{2n+1}$. Thus, the eigenvalues of boundary value problem are simple or double. Consider hereafter the columns with equal weight

$$(1.6) \qquad \int_{-1/2}^{1/2} \hbar(x) \; dx \; = \; 1.$$

The column with the thickness distribution $\hbar_0(x)$, which satisfies the equations

$$(1.7) \qquad M_{10}^2 = C \hbar_0^{n+1}, \qquad M_{10}'' + \Lambda_{10} \hbar_0^{-n} \; M_{10} = 0,$$

and boundary conditions (1.2) or (1.3), is called *optimal* for the boundary conditions of Sturm type. Here C is a Lagrange multiplier. The following isoperimetric inequality was stated in (Tadjbakhsh, Celler, [1962]): *If the eigenvalues (1.1) are simple, the optimal column has the largest buckling load among all the columns with the same weight*

$$(1.8) \qquad \Lambda_1 \leq \Lambda_{10},$$

and the equality sign holds only for the optimal column.
The similar statement can be proved for the case of clamped ends. Namely, there exist the functions M_{10}, M_{20} and $\hbar_0$, such that

$$(1.9) \; M_{10}^2 + M_{20}^2 = C \hbar_0^{n+1}, \qquad\qquad \Lambda_{10} = \Lambda_{20},$$

$$\Lambda_{10} = \int_{-1/2}^{1/2} M_{10}'^2(x) \; dx \; , \qquad M_{10}'' + \Lambda_{10} \hbar_0^{-n} \; M_{i0} \equiv 0, \quad i = 1, 2.$$

Function $\hbar_0(x)$ satisfies isoperimetric condition (1.5), while $M_{10}(x)$ and $M_{20}(x)$ satisfy the boundary conditions (1.4) and condition

$$(1.10) \qquad \int_{-1/2}^{1/2} \Lambda_o^{-n} M_{io} M_{jo} dx = \delta_{ij}, \quad i,j = 1,2.$$

The column with the thickness distribution $\Lambda_o(x)$ is optimal and isoperimetric inequality (1.7) can be stated in this case too. The functions $M_{10}(x)$ and $M_{20}(x)$ are the first and the second buckling modes of the optimal column respectively.

Consider an arbitrary distribution of thickness $\Lambda(x) > 0$. Because $\Lambda_1 \le \Lambda_2$, variational principle for eigenvalues gives, that

$$(1.11) \qquad \Lambda_1 = \frac{\Lambda_1 + \Lambda_1}{2} \le \frac{1}{2} \left[\frac{\int_{-1/2}^{1/2} M_a'^2 dx}{\int_{-1/2}^{1/2} \Lambda^{-n} M_a^2 dx} + \frac{\int_{-1/2}^{1/2} M_b'^2 dx}{\int_{-1/2}^{1/2} \Lambda^{-n} M_b^2 dx} \right]$$

Here M_a and M_b are the admissible functions, satisfying orthogonality condition. We can choose, particularly,

$$M_a(x) = \cos \alpha_o \; M_{10}(x) + \sin \alpha_o \; M_{20}(x),$$

$$(1.12)$$

$$M_b(x) = -\sin \alpha_o \; M_{10}(x) + \cos \alpha_o \; M_{20}(x).$$

where α_o - is a root of the equation

$$2 \sin 2\alpha_o \int_{-1/2}^{1/2} \Lambda^{-n} M_{10} M_{20} dx = \cos 2\alpha_o \int_{-1/2}^{1/2} \Lambda^{-n} (M_{20}^2 - M_{10}^2) \, dx$$

One can show, that

$$\int_{-1/2}^{1/2} \Lambda^{-n} M_a \, dx = \int_{-1/2}^{1/2} \Lambda^{-n} M_b \, dx = \frac{1}{2} \int_{-1/2}^{1/2} \Lambda^{-n} \left[M_{10}^2 + M_{20}^2 \right] dx$$

$$\int_{-1/2}^{1/2} \left[M_a'^2 + M_b'^2 \right] dx = \int_{-1/2}^{1/2} \left[M_{10}'^2 + M_{20}'^2 \right] dx.$$

Substitution of M_a and M_b into (1.11) shows, that

$$(1.13) \qquad \Lambda_1 \le \frac{\int_{-1/2}^{1/2} (M_{10}'^2 + M_{20}'^2) \, dx}{\int_{-1/2}^{1/2} \Lambda^{-n} (M_{10}^2 + M_{20}^2) \, dx} \le$$

$$\le \frac{\int_{-1/2}^{1/2} (M_{10}'^2 + M_{20}'^2) \, dx}{\int_{-1/2}^{1/2} \Lambda_o^{-n} (M_{10}^2 + M_{20}^2) \, dx} = \frac{\Lambda_{10} + \Lambda_{20}}{2} = \Lambda_{10}$$

To prove the second inequality the optimality condition (1.9) is used. The nominators of fractions to the left and right of the third inequality sign are identical, and the inequality is equivalent to

$$(1.14) \quad \int_{-1/2}^{1/2} \hbar_0^{-n} \left[M_{10}^2 + M_{20}^2 \right] dx \quad \leq \quad \int_{-1/2}^{1/2} \hbar^{-n} \left[M_{10}^2 + M_{20}^2 \right] dx.$$

Substitution of the optimality condition $M_{10}^2 + M_{20}^2 = C\hbar^{n+1}$ reduces the unproved jet inequality (1.14) to

$$(1.15) \quad \int_{-1/2}^{1/2} \hbar_0 \, dx \leq \int_{-1/2}^{1/2} \hbar_0^{n+1} \, \hbar^{-n} \, dx.$$

The inequality (1.15) follows immediately from Hölder inequality (Beckenbach, Bellman [1971]):

$$(1.16) \quad \int \xi \zeta \, dx \leq \left[\int \xi^p \, dx \right]^{1/p} \left[\int \zeta^q \, dx \right]^{1/q}, \quad \frac{1}{p} + \frac{1}{q} = 1,$$

after substitution

$$\xi = \hbar_0 \, \hbar^{-1/p}, \quad \zeta = \hbar^{1/p}, \quad p = \frac{n+1}{n}, \quad q = n+1.$$

To complete the proof of the isoperimetric inequality (1.13), note, that the optimal column, which satisfies the equations (1.9), really exists. Thus, we proved that *the optimal column has the largest buckling load among all the columns with the same weight and clamped ends.*

2. ISOPERIMETRIC INEQUALITY IN A STABILITY PROBLEM FOR A CONSERVATIVE SYSTEM OF THE SECOND KIND.

A concept of conservative systems of the second kind. This chapter deals with the problem of optimal tapering of a conservative system of the second kind, namely generalized Pfluger column. The conservative systems of the second kind are truly nonselfadjoint and nonconservative systems in a classical sense, but selfadjoint in a generalized sense. These systems buckle by divergence, do possess a generalized conservation theorem and a generalized Rayleigh's quotient. The optimization problems for generalized

Pfluger column with pinned-pinned or sliding ends, subjected to distributed compressive follower forces are considered. It is shown, that by means of a special transformation of independent variable, the problem is reduced to a classical conservative bifurcation problem for the column, loaded by the axial distributed load. The optimal solutions for some load distributions are found analytically.

The shape of columns, for which the buckling load is largest among all columns of given length and volume, for various boundary conditions and a "dead" conservative load and the corresponding isoperimetric inequalities were discussed above. The equations of stability are reduced to secondary-order selfadjoint equations with homogeneous, but generally mixed boundary conditions. Due to the conservative character of the load, the stability problem was analyzed by means of Euler static method.

The general nonconservative systems are described by nonselfadjoint equations of motion and loose the stability by bifurcation (divergence) and/or flutter. The stability of such systems is studied by dynamic method. Optimization of nonconservative systems is surveyed by Weisshaar and Plaut (1981) and Gajewski and Zyczkowski (1988).

However, certain nonselfadjoint systems have only bifurcation type of instability, despite of the presence of nonconservative forces, Leipholz (1974a, 1974b). Such a system is called a conservative system of the second kind. This system is selfadjoint in a generalized sense with respect to an assigned operator, there exist a Lyapunov functional and a generalized Rayleigh quotient.

The purpose of this paper is the study of optimization problem for conservative systems of the second kind. For certain optimization problems the analytical solutions are obtained and the corresponding isoperimetrical inequalities are proved.

Conservative system of the second kind: generalized Pfluger column. The column under non-uniform tangential follower force with simply-supported or guided ends is called generalized Pfluger column. Stability of the generalized Pfluger column is described by the eigenvalue problem governed by the differential equation

(2.1) $-\mu \, \omega^2 \, w + (\, EJ \, w'' \,)'' + P \, f(x) \, w'' = 0$, $-1/2 \leq x \leq 1/2$.

and the boundary conditions

(2.2) $w = EJ\ w'' = 0$

for a pinned end, or

(2.3) $w' = (EJ\ w'')' = 0$

for a guided end. Here $w = w(x)$ - is the lateral deflection of the column, ω - frequency of free vibration, $\mu = \mu(x)$ - linear density of the column, $f = f(x)$ is axial force in the column, $P > 0$ is a load parameter. The column with guided ends admits a translatory isometric mode, $EJ=EJ(x)$ - is a bending stiffness.

The operator $f(x)(d^2/dx^2)$ is not selfadjoint (in classical sense) with respect to the prescribed boundary conditions. However, the operator

(2.4) $\mathcal{A} = -\mu\ \omega^2 + \dfrac{d^2}{dx^2}(\ EJ\ \dfrac{d^2}{dx^2}) + P\ f\ \dfrac{d^2}{dx^2}$

is selfadjoint in generalized sense, Leipholz (1974a), with respect to the operator

(2.5) $\mathcal{B} = EJ\ \dfrac{d^2}{dx^2}.$

The generalized selfadjointness implies

$$\int\limits_{-1/2}^{1/2} \mathcal{A}u\ \mathcal{B}v\ dx = \int\limits_{-1/2}^{1/2} \mathcal{A}v\ \mathcal{B}u\ dx,$$

for all admissible functions $u(x)$ and $v(x)$, satisfying all boundary conditions prescribed for $w(x)$.

Because the operator $\mathcal{L}$ is selfadjoint in generalized sense, the generalized Pfluger column can only have bifurcational mode of instability (divergence) and Euler method can be applied to study the stability. From the stability theory, there exists a generalized Rayleigh quotient for approximating the lowest buckling load, and this lowest buckling load is the minimum of the generalized Rayleigh quotient in the space of admissible functions:

(2.6) $P = \min\limits_{w}\ \left[\int\limits_{-1/2}^{1/2} (EJ\ w'')'^2 dx\right]\left[\int\limits_{-1/2}^{1/2} EJ\ f\ w''^2\ dx\right]^{-1}$

In terms of bending moment $m(x) = EJ\ w''$ the equation (2.1) assumes the form

(2.7) $m'' + P f \dfrac{m}{E J} = 0$.

The boundary conditions for equation (2.7) in the case of guided ends are written in the form $m'=0$, while for the pinned ends $m=0$. The equation (2.7) and the corresponding Rayleigh quotient could be written in the dimensionless form.

For this purpose introduce the function $\hbar = \hbar(x) > 0$, designating the dimensionless cross-sectional area. The function $\hbar^n$ is directly proportional to the bending stiffness of the column. The dimensionless critical parameter λ is directly proportional to p and is a principal positive eigenvalue of the equation

(2.8) $m'' + \lambda f \hbar^{-n} m = 0$

or, equivalent, a minimum of generalized Rayleigh quotient

(2.9) $\lambda[\hbar] = \min\limits_{m} \left[\displaystyle\int_{-1/2}^{1/2} m'^2 dx \right] \left[\displaystyle\int_{-1/2}^{1/2} f \hbar^{-n} m^2 dx \right]^{-1}$

in the space of all admissible functions $m=m(x)$, i.e. functions which are once diffentiable and do satisfy the corresponding boundary conditions.

Stability optimization of Pfluger column. Consider now a problem of maximization of critical buckling load. The design variable is the dimensionless cross-sectional area $\hbar = \hbar(x)$, and the shape of the cross-section is prescribed. Assume as the design objective the critical buckling load $\lambda = \lambda[\hbar]$. The volume of the column is constrained:

(2.10) $\displaystyle\int_{-1/2}^{1/2} \hbar \, dx = 1$.

Consider thereafter the boundary conditions in form: $m(-1/2)=m(1/2)=0$ (pinned ends) or $m'(-1/2)=m'(1/2)=0$ (guided ends) or $m(-1/2)=m'(1/2)=0$ (one pinned end and one guided end). The boundary conditions are of Sturm type, so the principal eigenvalue is simple. The necessary optimality condition is obtained using method of Lagrange multipliers, (Banichuk [1990]):

(2.11) $f \hbar^{-1-\alpha} m^2 = C$.

Isoperimetric inequality. Prove the isoperimetric inequality for the buckling load of generalized Pfluger column. Namely, there exist the functions M_o and Λ_o, such that

$$(2.12) \quad f\,M_o^2 = C\Lambda_o^{n+1}, \quad \lambda_o = \int_{-1/2}^{1/2} M_o'^2(x)\,dx\,, \quad M_o'' + \lambda_o f\,\Lambda_o^{-n}\,M_o = \emptyset.$$

Function $\Lambda_o(x)$ satisfies isoperime t ric condition, while $M_o(x)$ satisfies the boundary conditions. and condition

$$(2.13) \quad \int_{-1/2}^{1/2} f\,\Lambda_o^{-n}\,M_o^2\,dx = 1.$$

The column with the thickness distribution $\Lambda_o(x)$ is optimal and isoperimetric inequality can be stated. The function $M_o(x)$ represents the buckling mode of the optimal column.

Consider now an arbitrary distribution of thickness $\Lambda(x) > \emptyset$. Variational principle for eigenvalues gives, that

$$(2.14) \quad \lambda \le \frac{\displaystyle\int_{-1/2}^{1/2} M_o'^2\,dx}{\displaystyle\int_{-1/2}^{1/2} f\Lambda^{-n}\,M_o^2\,dx} \le \frac{\displaystyle\int_{-1/2}^{1/2} M_o'^2\,dx}{\displaystyle\int_{-1/2}^{1/2} f\Lambda_o^{-n}\,M_o^2\,dx} = \lambda_o$$

To prove the second inequality the optimality condition (1.9) is used. The nominators of fractions to the left and right of the third inequality sign are identical, and the inequality is equivalent to

$$(2.15) \quad \int_{-1/2}^{1/2} f\,\Lambda_o^{-n}\,M_o^2\,dx \le \int_{-1/2}^{1/2} f\,\Lambda^{-n}\,M_o^2\,dx\,.$$

Substitution of the optimality condition (2.12) reduces the unproved jet inequality (2.15) to (1.15), which follows immediately from Hölder inequality. Thus, we proved that *the optimal generalized Pfluger column has the largest buckling load among all generalized Pfluger columns with the same weight.*

Acknowledgement. This work was supported by AvHumboldt Foundation.

BIBLIOGRAPHY

N.V. Banichuk, 1990. *Introduction to optimization of structures.* Springer - Verlag, Berlin, Heidelberg, N.-Y.

E.F. Beckenbach, R.Bellman, 1971. *Inequalities.* Springer - Verlag, Berlin, Heidelberg, N.-Y.

T. Clausen, 1853. Über die Form architektonischer Saules. *Bulletin physico - mathematiques et Astronumiques.* Tome I, 1849-1853, 279 - 294, StPetersburg.

A. Gajewski, M. Zyczkowski, 1988. *Optimal Structural Design under Stability Constraints.* Kluwer Academic Publishers. Dordrecht.

H.H.E. Leipholz 1974a. On Conservative Elastic Systems of the First and Second Kind. *Ingenieur-Archiv,* Vol. 43, pp.255-271.

H.H.E. Leipholz 1974b. On a Generalization of the Concept of Self-Adjointness and of Rayleigh's Quotient. *Mechanics Research Communications.* Vol. 1, pp. 67-72.

E.F. Masur, 1984. Optimal structural design under multiple eigenvalue constraints. *Int. J. Solids & Structures.* 20, No 3, 211 - 231.

N. Olhoff, S.H. Ramussen, 1977. On single and bimodal buckling loads of clamped columns. *Int. J. Solids & Structures.* 13, No 7, 605 - 614.

A. P. Seiranyan, 1984. On a problem of Lagrange. *Mech. of Solids (MTT),* 19, No 2, 101-111.

I. Tadjbakhsh, J.B. Keller, 1962. Strongest columns and isoperimetric inequalities for eigenvalues. *Trans. ASME, ser. E, J. Appl. Mech.* 29, No 1, 159-164.

T.A. Weisshaar, R.H. Plaut, 1981. Structural Optimization Under Nonconservative loading.- In *Optimization of Distributed Parameter Structures.* Eds. E.J. Haug, J. Cea. Sijthoff and Noordhoff, Alphen aan den Rijn. p. 843-864.

OCTOPUS — A TOOL FOR DISTRIBUTED OPTIMIZATION OF MULTI-DISCIPLINARY OBJECTIVES

B. Esping, P. Clarin and O. Romell
ALFGAM Optimering AB,
Roslagsvägen 101,
S-104 05 Stockholm,
Sweden

Abstract

OCTOPUS is a tool where objective and constraint functions from different types of disciplines can be combined and treated simultaneously. The system is especially designed for computer networks. Evaluation of functions and gradients for each discipline can be executed on separate computers, the values are transferred to a central optimizer once per iteration. The communication may be administered from a separate 'chief' computer which can be of rather small size (workstation) even if the analyses are of super computer size.

The communication between the different computers is handled through small ascii files, this makes the concept very flexible and enables communication between all types of computers that can exchange ascii files. OCTOPUS is an open system where new types of analyses (jobs) can be included during the process. Jobs can also be temporarily removed and reinserted later, killed or put in passive analysis mode, i.e. pure analysis with no gradient evaluation. Jobs may be entered in parallel or in series.

The system has an open architecture so that user written programs can easily be integrated. So far the OASIS–ALADDIN structural optimization system and the SINDBAD fluid mechanics optimization system have been included. The system is designed to handle up to 100 different optimization jobs simultaneously.

Two examples conceived in cooperation with Atlas Copco Tools are presented to illustrate the use of the system.

G. I. N. Rozvany (ed.), Optimization of Large Structural Systems, Vol. II, 1165–1194.
© *1993 Kluwer Academic Publishers.*

OCTOPUS, the multiple discipline optimization concept – an introductory description

Almost every engineering design problem is an optimization problem. A product is going to be designed – for what purpose? – what is its function? – what is its operational environment? – i.e. what constraints are there on the product and what is the objective in the design process?

The normal design process is: the designer has an idea, makes a drawing, he then contacts the analyzer who performs calculations in order to check various behaviors and responses for the product when it is in its operational environment. The results of the calculations are brought back to the designer who will modify the design, and again the design has to be analyzed, etc. The designer and the analyzer have a creative and controlling role, respectively, in the process.

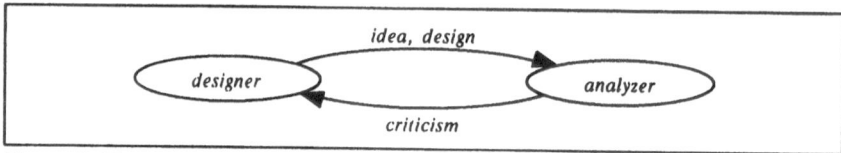

Figure 1. The traditional design process.

The design process is in fact an optimization process but because of tight dead lines, high staff costs etc., it normally stops after very few iterations.

With the structural optimization technique the process is automated and each design step is also made much more efficient.

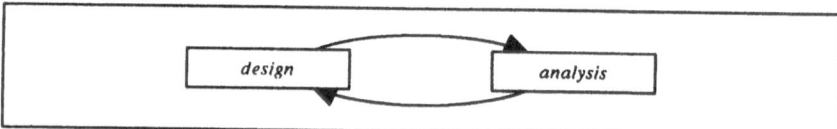

Figure 2. Automatic structural optimization.

Structural optimization is primarily a tool for the designer.

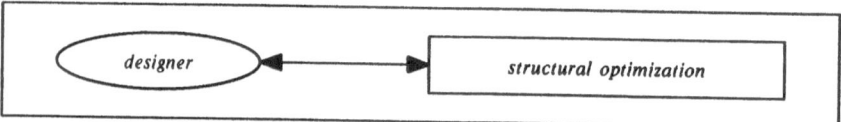

Figure 3. Designer's tool.

Consequences for a company using structural optimization are that, for a specific project, the design process is speeded up and a smaller staff is required to be involved in the design process. Personnel is released and can be used either as creative designers or as controlling analyzers for more complicated and qualified analyses. The efficiency is increased, and costs are reduced.

Most important is to define the problem. Let us look at two examples:

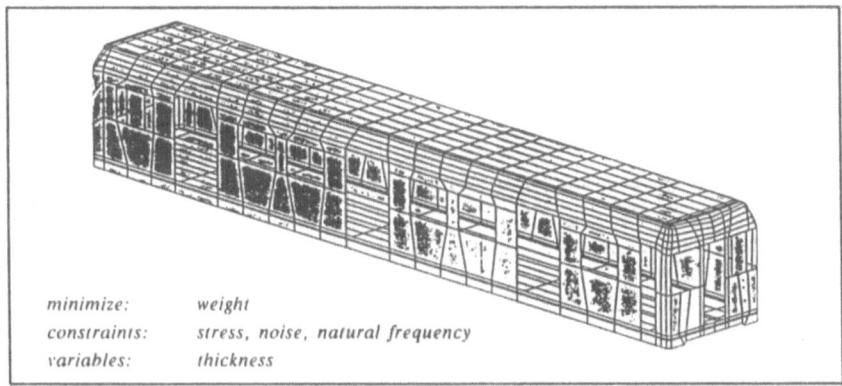

minimize: *weight*
constraints: *stress, noise, natural frequency*
variables: *thickness*

Figure 4. Subway car

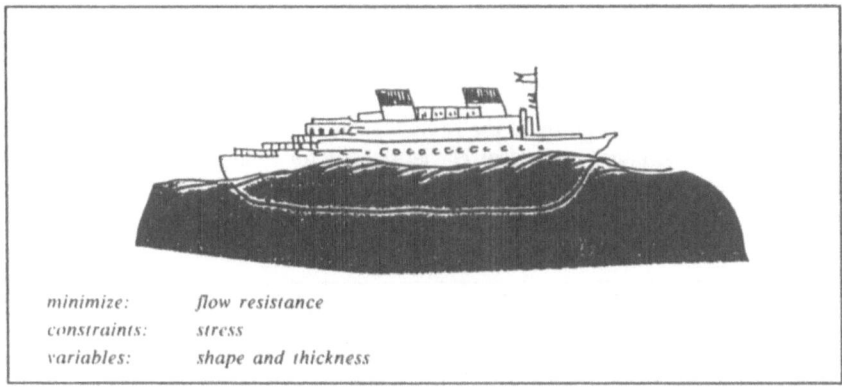

minimize: *flow resistance*
constraints: *stress*
variables: *shape and thickness*

Figure 5. Ship hull

Specific about these problems is that there are several disciplines involved. In the subway car case we have weight, stress, natural frequency and acoustic calculations. In the ship case we have fluid flow and stress analyses. Design of different products require different types of analyses. An efficient optimization tool must be able to handle many types of analyses simultaneously, we have just mentioned a few typical. Others are production costs, thermal, magnetic and electrical response, etc.

A major objective for the producer is to maximize his profit for a certain product:

$$\text{maximize: } profit_p = income_p - costs_p$$

Figure 6. The producers problem.

The costs consist of material costs, tool costs, wear costs, etc.

The producer sells his product to the customer. Also the customer wants to maximize his profit:

1168

$$\textit{maximize:} \quad \textit{profit}_c \;=\; \textit{income}_c \;-\; \textit{costs}_c$$

Figure 7. The customers problem.

The customers cost$_c$ is partly equal to the producers income$_p$. To some extent the producer and customer have interfering problems. They have a good reason to collaborate and set up a common goal:

$$\textit{maximize:} \quad \textit{coeff}_c \;\cdot\; \textit{profit}_c \;+\; \textit{coeff}_p \;\cdot\; \textit{profit}_p$$

Figure 8. Producer–customer collaborative objective.

We have a combined objective function which in turn consist of a number of combined functions. Both the producer and the customer have a great number of constraint functions originating from a large number of disciplines.

In order to solve such complex problems we have developed the OCTOPUS optimization system. It can handle up to one hundred simultaneous parallel and/or sequential analyses.

The individual analyses may all be of the same type or completely different, e.g. FEM for the structural analysis, panel method or Navier–Stokes for fluid dynamics, etc. They can be run on the same computer or on different machines, in parallel or sequentially, according to the user's preferences and requirements, and what the available hardware will allow.

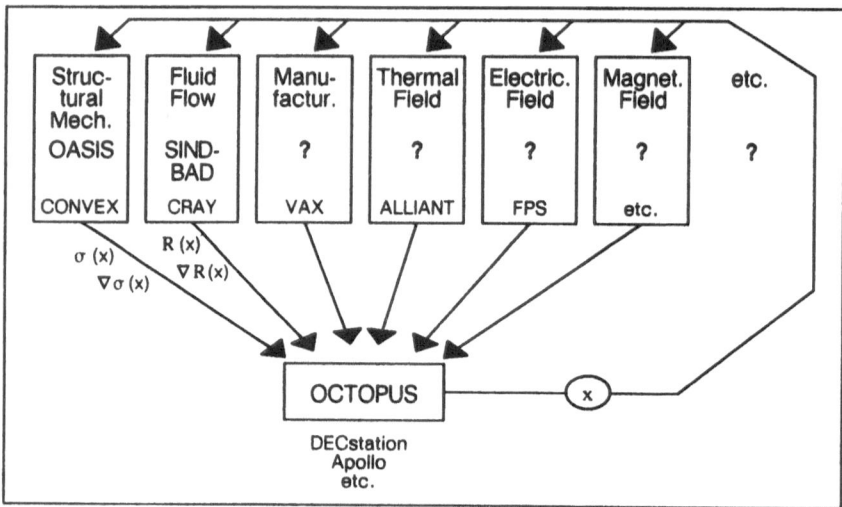

Figure 9. OCTOPUS principal system function.

Let the figure above illustrate the general case. In each iteration the current values of the objective and constraint functions are calculated, as well as their derivatives with respect to the design variables. Each separate type of response is treated on a separate machine with the appropriate type of software. Once per iteration this information is assembled in the optimization module OCTOPUS which calculates new values for the design variables. The new values are then transferred back to the different sub-processes where they constitute start values for the next iteration.

OCTOPUS solves a sequence of approximate subproblems. Each approximation needs information about the function values F_i and their gradients ∇F_i. The system uses the Method of Moving Asymptotes, MMA (ref. 1.), and can handle both continuous and discrete design variables.

OCTOPUS and the analysis programs communicate by small ascii files containing either design variable values x_i, or function values F_i and gradients ∇F_i. Their contents will be discussed later. However, their respective sizes are relatively small and because they are ascii files they can easily be sent between different computers (see below and appendix A).

The OASIS program in Figure 9 is a system for structural optimization (ref. 2.), SINDBAD is a system for fluid flow optimization. Both can be integrated with the OCTOPUS concept.

One particular application is illustrated in Figure 10. A structure is analyzed by the same method in two parallel processes, only with slightly different aims and with different discretizations. Let this concept be illustrated by a railway car body: one analysis concentrates on natural frequencies and utilizes an element mesh appropriate for that; the other analysis deals with stresses and uses a different and partly much finer mesh. However, both analyses are performed with the same method (FEM) and perhaps also with the same program. Consequently, two parts of one larger analysis may be performed in parallel on separate machines instead of in sequence on the same cpu, and also using different and tailored computational meshes.

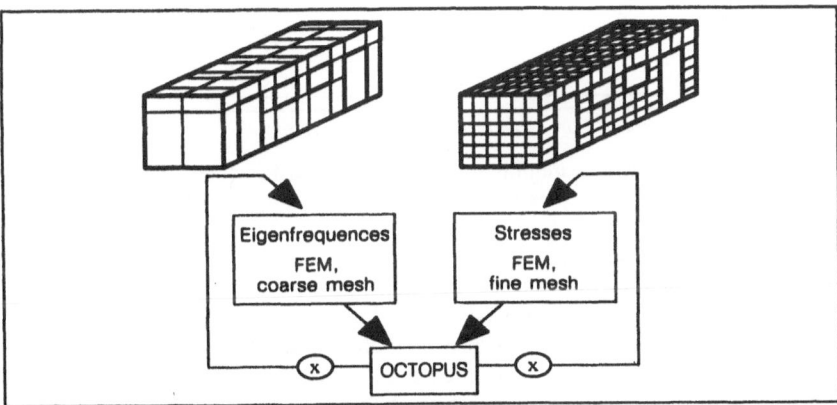

Figure 10. Parallel analyses on separate computers.

A third example of how the system may be utilized is illustrated in Figure 11. Here OCTOPUS brings together the computer resources (as regarding both hard- and software) of a company's different engineering departments. The structural analysis is performed at the structures department using their computer(s) and code (FEM), the flow calculations are performed at the fluid dynamics department, and so on. Detailed modeling can be performed at the appropriate department where the specific competence is found.

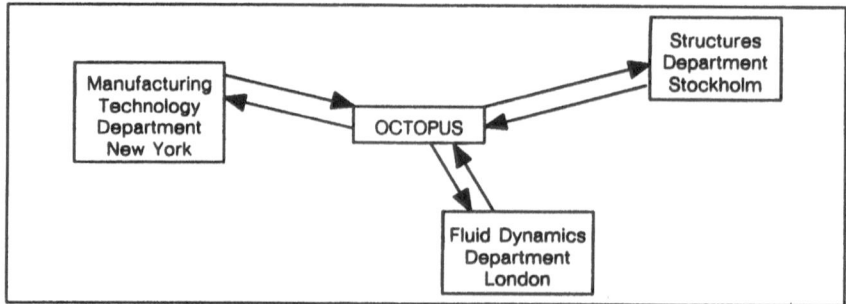

Figure 11. Geographically and/or organizationally separate analyses.

The whole process is automatically controlled by a mother process, and can be monitored and controlled by the project leader who has access to all relevant information on his computer.

A different application is illustrated in Figure 12 below. Due to errors and inaccuracies in the production methods there will always be some deviation from the desired design. If the found optimum is very sharp it means that design deviations may be fatal. However, this can be avoided if the optimization includes consideration of the production errors. This is easily done if a set of similar designs are considered simultaneously. One is perfect, i.e. no production errors. The others are different variations of the first – each with a small deviation. The same set of design variables determines all designs, the only difference is the set of small deviations.

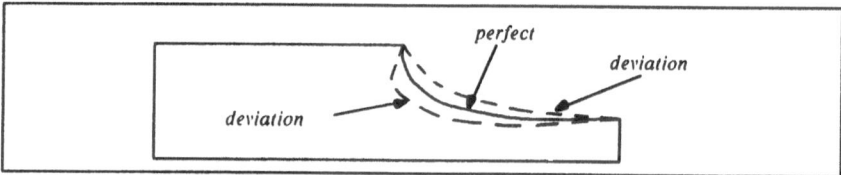

Figure 12. Design optimization considering production errors.

OCTOPUS is equipped with a monitor function, GANDALF, which enables the coordinator (supervisor) of all the analyses to display all participating subjobs, their status, objective and constraint function values, design variable values, etc. OCTOPUS is dynamic meaning that the coordinator can include new subjobs during the optimization process. I.e. the coordinator can, e.g. after a few iterations, give the command INCLUD, followed by the subjob name, to the monitor GANDALF. In the next iteration the new subjob will become part of the process. He can also temporarily REMOVE a job and put it in a waiting mode. This may be the case if a job does not contribute to the objective function and all constraints related to it are far from being infeasible. By giving the ANALYS command to GANDALF a removed job can be forced to perform a pure function analysis, i.e. without gradient evaluation. A subjob can thus be inspected also when it is in the passive waiting mode. Relevant design variable values will be used. With the INCLUD command it can be reincluded. The KILL command kills a specific job, and the KILLAL command will terminate all processes.

The "sub-jobber", i.e. the person responsible for a certain subjob can also inspect his own job as the process is running. The monitor GANDALF can to some extent be used to modify parameters, check constraints and design variable values, etc., specific for that job. If the sub-jobber finds errors in his job he can either give the REMOVE or the KILL command. A message is then automatically sent to OCTOPUS which temporarily halts and issues a message to the coordinator who must decide if he shall accept to REMOVE or KILL the job and continue without it, or if he shall kill all processes (KILLAL).

OCTOPUS also coordinates jobs which are dependent on each other in a sequential way, i.e. output from one job can be input to another. For instance structural mechanics calculations require information about the thermal field which has to be calculated first and then transferred. This has to be done in every iteration because the temperature field will depend on the design variables.

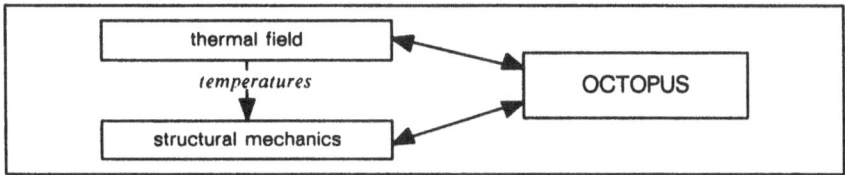

Figure 13. Sequential jobs.

Parallel and sequential job may be combined.

In the following a more detailed description of OCTOPUS' functions and organization will be given.

Organization – parallel jobs

Each subjob is started separately by the 'sub-jobber', as he gives the command OASIS, SINDBAD, USER etc. to 'his' computer. OASIS is used for structural analysis/optimization, SINDBAD for fluid flow, etc. USER is a user written program for some other type of analysis.

To start with the system issues the following questions:

```
The O A S I S module
----------------------

Give input files for:
Structural Analysis  /.INP/ ?    OAS1
Structural Optimization  /OAS1.OPT/ (Optional) ?   <return>
Optimization database  /OAS1.OD1/ ?   <return>
```

Underlined (typewriter font) text is the user's response. In this specific case we start an OASIS structural optimization subjob. The files asked for have the names *OAS1.INP, OAS1.OPT* and *OAS1.OD1*, respectively. In the OASIS or SINDBAD case they are created by the ALADDIN pre/post processor.

ALADDIN is an interactive CAD-oriented program interfaced with the system which creates all the information and files necessary for analysis and optimization. It can be used for all types of analyses/optimizations where a computation mesh is needed. Geometry, loads, constraints and objective functions, design variables, convergence criteria, etc., are

1172

defined in ALADDIN. ALADDIN also creates an optimization job database, in this case the *OAS1.OD1*. Each OASIS/SINDBAD subjob will have its own database. In the example USER (described below) neither input files nor a database is used and none of the questions above will arise. OASIS-ALADDIN and SINDBAD-ALADDIN use the computation oriented database program BAGIS. BAGIS can also be used for user written applications and it has to be used if the sequential concept in OCTOPUS is applied. BAGIS works with datasets given by names. One or several datasets can be transferred from the database of one job to that of another (see sequential jobs).

BAGIS is entirely written in the C programming language. There is also a FORTRAN shell around BAGIS (called BAGCOM). The shell, which has been used in these examples, has exactly the same parameters as corresponding subroutines in the database system MEMCOM (ref. 3.), which is an alternative for the user who wants a complete, general purpose database system.

Once the input files are accepted by the system a program WRJOB (OCTOPUS program) will be executed and issue the following questions:

```
Give job name :  SUB1
Give OCTOPUS computer name :  CONV1
After job (job name) :  <return>
```

The OASIS job is given the name SUB1. Let us say it is executed on the computer/processor with the symbolic name VAX1. The coordinating OCTOPUS program will run on the computer/processor with the symbolic name CONV1. See below and appendix A (Communication) concerning physical correspondences to the symbolic names and the communication between the computers/processors. To the question ' After job:' we just give a blank return to indicate that this job does not depend sequentially on any other job.

In the general case the subjob program consists of two parts – one initialization part and one iterative part. A communication file (*x.COM1*; where *x* is the job name, in this case SUB1) is created at the end of the initialization part. It contains all the information that OCTOPUS needs to coordinate this job with the others, i.e. job name, analysis type, computer name, design variable numbers and types, their initial values, lower and upper limits, the maximum number of constraints that will appear, convergence criteria, etc. (see appendix A).

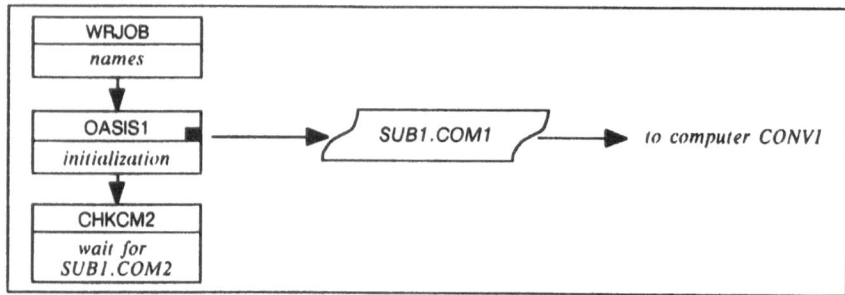

Figure 14. OASIS – initialization

The communication file *x.COM1* (ascii file) will be transferred to the OCTOPUS computer as soon as it is released. After the initialization phase the subjob passes to the iterative part which starts with the program CHKCM2 (OCTOPUS program – it appears also as subroutine CHCOM2). The program CHKCM2 waits for a communication file (*x.COM2*) from OCTOPUS and it will not pass on until that has arrived. Figure 14 illustrates the OASIS example above.

Similarly other jobs which shall be included are started.

Now it is time to start the coordinating OCTOPUS program on the computer with the symbolic name, in our case, CONV1. The coordinator gives the start command OCTO-PUS to that computer. The system will give the following response:

```
The  O C T O P U S  executing module
--------------------------------------
```

Directly a program WRJOB1 will be executed and it will ask the following questions:

```
OCTOPUS job name ?  MAIN
Next job name ?  SUB1
Computer name ?  VAX1
Next job name ?  <return>
```

The OCTOPUS job will be given the name MAIN. A file *MAIN.JOB1* is created and it will contain the names of all subjobs given, and their host computers, in our case the names SUB1 and VAX1 only. Next, the program CHKCM1 is started and it will check if the communication files *x.COM1* exist. Their names are listed in the file *MAIN.JOB1*. In our case there is only one file, *SUB1.COM1*. When all *x.COM1* files are ready OCTOPUS will continue to the next module. OCTOPUS1 which will examine them. The problem size is determined, and a database *MAIN.OD2* is created. Compatibility between the participating subjobs is checked, e.g. that design variables with the same number in different subjobs are of the same type. The section of all lower and upper design variable limits will be used, the last given initial design variable value, the last given convergence tolerance is used, etc. During initialization of the OCTOPUS job MAIN with subjob SUB1 the following happens in OCTOPUS:

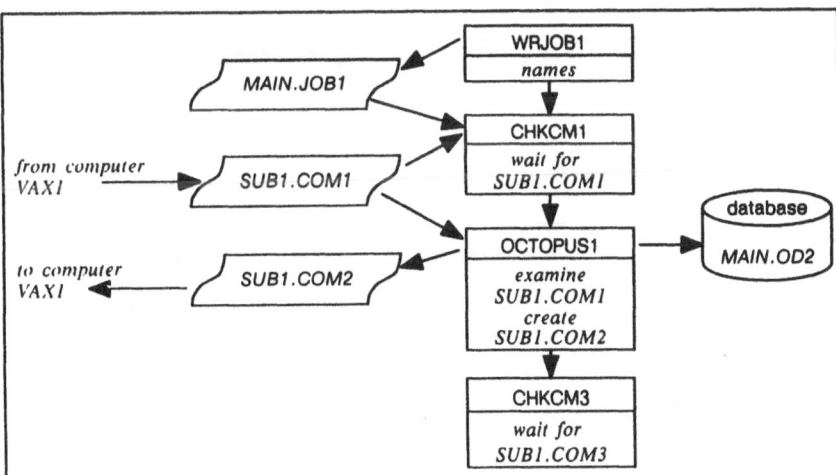

Figure 15. OCTOPUS – initialization

Communication files *x.COM2*, containing the current design variable values, their limits, etc., are created and transferred to the host computers of the different subjobs. Finally the file *MAIN.JOB1* is deleted by OCTOPUS1. The following module CHKCM3 will wait for communication files *x.COM3* which contain function values and gradients for objective function contributions and constraints from all subjobs. The OCTOPUS objective function is assembled of the contributions from the different subjobs.

Over to the subjobs: When files *x.COM2* have arrived at the respective computers, the program CHKCM2 (and subroutine CHCOM2) will pass on to the iterative part of the subjob program. Design variable values etc. are read and the desired function values and gradients are computed and written to the communication files *x.COM3*. Program CHKCM2 is entered again and a new iteration starts. In our special case we have:

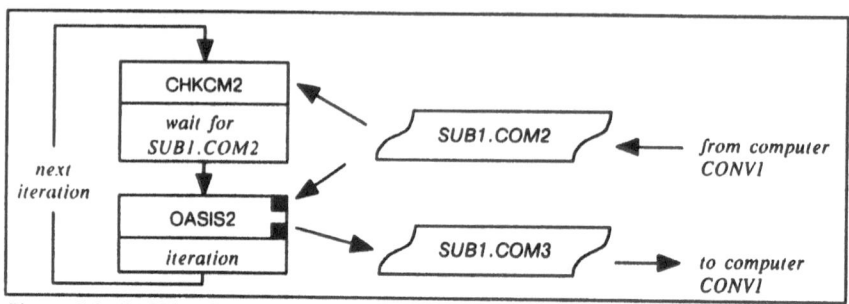

Figure 16. *OASIS – iteration*

▭▬ indicates that subroutines for communication from the OCTOPUS library are used. The files *x.COM2* will be deleted as soon as the information in them has been used.

Now over to OCTOPUS: the files *x.COM3* now exist and the module OCTOPUS2 is entered. It will collect the information from all *x.COM3* files. It checks feasibility and convergence, and creates and solves the subproblem. OCTOPUS then jumps up to module CHKCM1 and a new iteration starts. In our case we get:

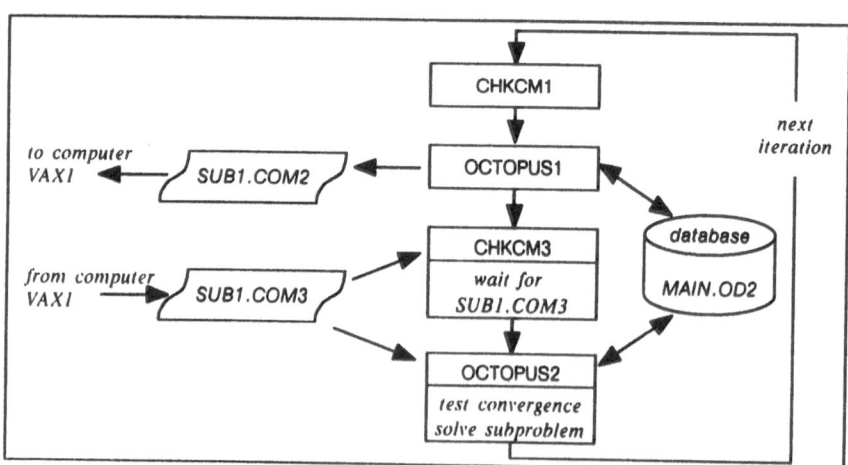

Figure 17. *OCTOPUS – iteration*

In the next iteration CHKCM1 will check if a new file, *MAIN.JOB1*, containing subjob names exist – if not it passes on, otherwise it will wait for files *x.COM1* corresponding to the new names in *MAIN.JOB1*. The files *x.COM3* will be deleted in OCTOPUS2 when the information in them has been used. The interaction between OASIS and OCTOPUS will be:

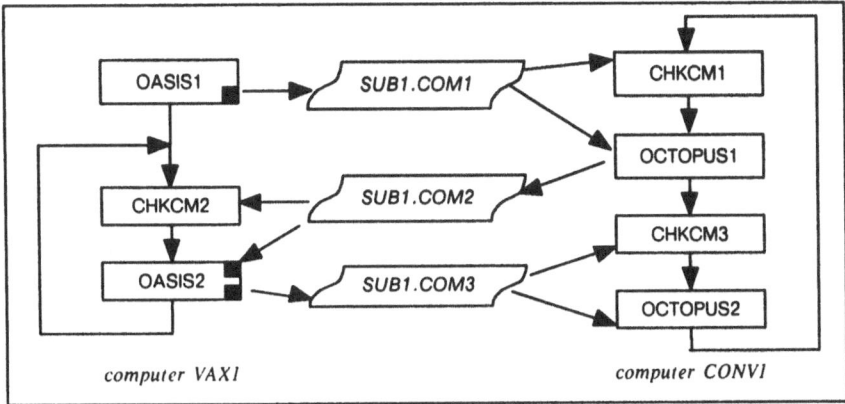

Figure 18. Interaction OASIS – OCTOPUS

Another subjob can be included, let us say for instance a fluid analysis (SINDBAD) job running on a computer called 'DEC1'. Its name (SUB2) must be included in the joblist file *MAIN.JOB1*. The monitor program GANDALF is used to handle this, see below.

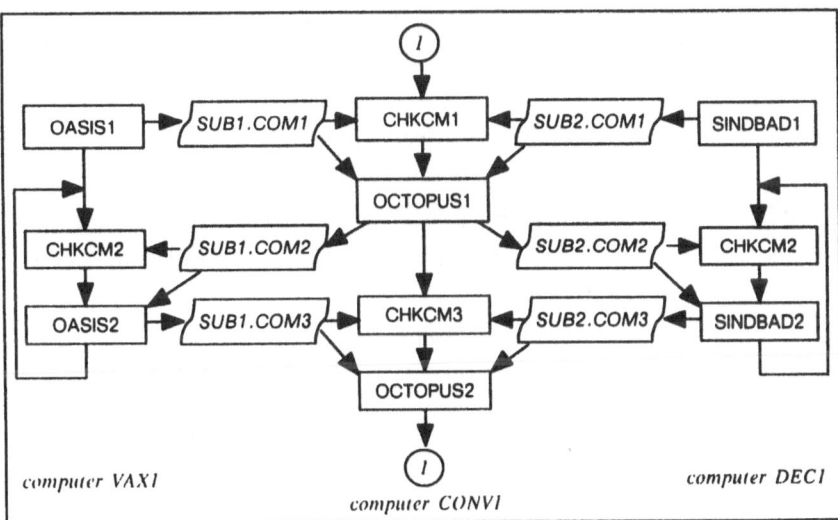

Figure 19. Integration of multiple disciplines

OCTOPUS2 checks if convergence to a feasible solution has been achieved or if the maximum number of iterations has been reached. If that is the case OCTOPUS1 will write a message in files *x.COM2*. The subjobs and the OCTOPUS job will stop the next time after their iteration phases and after OCTOPUS2, respectively.

The GANDALF monitor function

Both OCTOPUS1 and OCTOPUS2 will in each iteration create database files *x.OD4*, containing information about activity levels and feasibility for all subjobs. They also contain convergence histories, design variable values, etc These database files can be used at any moment by the monitor program GANDALF. If we return to our specific case where the OCTOPUS job has the name MAIN the situation is:

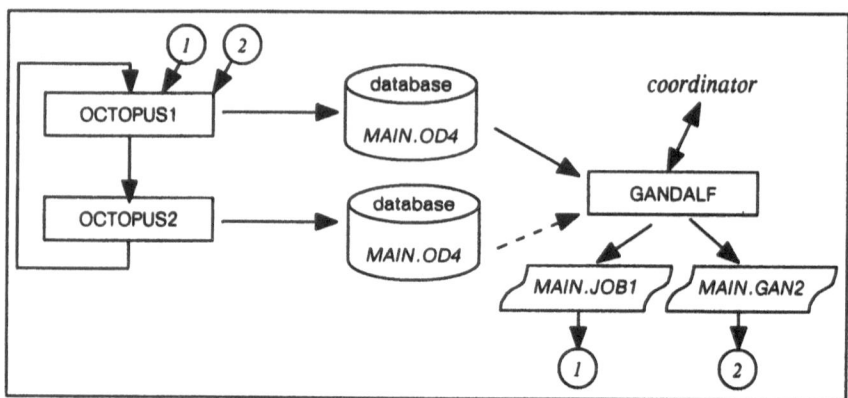

Figure 20. The GANDALF monitor

The commands given in GANDALF will result in the creation of an ascii file, in this case called *MAIN.GAN2*, containing update information for OCTOPUS. If there is any INCLUD command an ascii file *MAIN.JOB1*, containing the names of the subjobs to be included, is also created. Once per iteration OCTOPUS1 will test for the appearance of any update file *MAIN.GAN2* or include file *MAIN.JOB1*.

GANDALF has two modes of operation: a SHOW mode for inspection and a CHANGE mode for modification.

Some of the SHOW commands are listed below:

STATUS – lists the activity levels for all jobs, e.g. INCLUDed, REMOVEd, ANALYSe mode or KILLed

OBF – display objective function values versus iteration number

CON – display values of worst constraints among all INCLUDed subjobs versus iteration number

ALFGAM – display design variable values versus iteration number or versus design variable number

Some of the CHANGE commands are:

INCLUD – include or re-enter a job

REMOVE – remove a job temporarily, i.e. put it in a waiting mode. The job can later be re-entered with the INCLUD command

ANALYS – perform analysis on a job that has been put in the waiting mode, i.e. calculate a function value. No gradients are calculated and the evaluated constraint value is not considered by OCTOPUS (i.e. no $x.COM3$ file is created)

KILL – kill a job. It is then permanently removed from OCTOPUS. The only way to re-enter a KILLed job is to perform a restart

KILLAL – kill all jobs included in the OCTOPUS main job

The commands are transferred to OASIS, SINDBAD, etc., from OCTOPUS via the $x.COM2$ communication files.

Figure 21. Commands via OCTOPUS

GANDALF can also be used to monitor subjobs created by OASIS and SINDBAD. For instance, OASIS will in each iteration create a postprocessing database, $x.OD3$, which can be examined by GANDALF. It contains various postprocessing data specific for this job such as constraint histories, design variable values, etc. Both the SHOW and CHANGE commands can be used. A few CHANGE mode commands are:

REMOVE – as for OCTOPUS above

ANALYS – as for OCTOPUS above

KILL – as for OCTOPUS above

All CHANGE mode instructions are written to the ascii file $x.GAN1$ and then transferred to OASIS which in turn will transmit it to OCTOPUS via the communication file $x.COM3$. In our example we have:

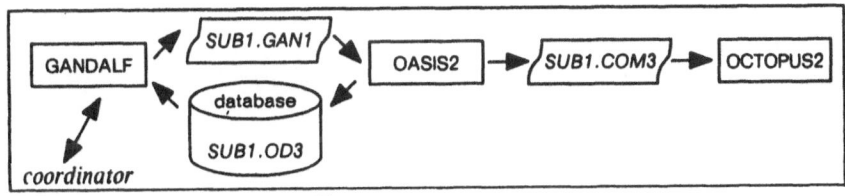

Figure 22. Commands via OASIS

If there is a REMOVE/KILL command in the communication file $x.COM3$, OCTOPUS will go directly to OCTOPUS1 which will send a message to the coordinator that the specific subjob has been removed/killed. Thus no subproblem will be created or solved. The coordinator is requested to use GANDALF on the OCTOPUS job to give his acceptance, i.e send a REMOVE/KILL command. Until OCTOPUS has received this acceptance it will be put in a waiting mode.

Organization – sequential jobs

Let us say that the program PROGX (job SUB4) needs information from OASIS (job SUB3). The information will be stored with the data set names "A" and "B" in the database of job SUB3 and we want them to be transferred to job SUB4 and stored in its database with the same data set names.

First we start program OASIS or PROGX, let us say OASIS on a VAX computer (called by the logical name 'VAX1'), and we get the following questions (we intend to run the OCTOPUS job MAIN2 on the 'CONV1' computer) :

```
Give job name :  SUB3
Give OCTOPUS computer name :  CONV1
After job :  <return>
```

Then PROGX is started on a DECstation (called by the logical name 'DEC1'), and the following questions must be answered:

```
Give job name :  SUB4
Give OCTOPUS computer name :  CONV1
After job :  SUB3
Computer name :  VAX1
Data to be transferred
Next dataset name :  A
Next dataset name :  B
Next dataset name :  <return>
```

It is not necessary that the depending jobs SUB3 and SUB4 run on the same computer. Job SUB4 will now create an ascii file, *SUB3.SET*, which contains the names A and B of the datasets to be transferred. The two jobs are coordinated by the OCTOPUS job MAIN2 (running on the Convex machine 'CONV1'):

```
OCTOPUS job name ?  MAIN2
Next jobname ?  SUB3
Computer name :  VAX1
Next jobname ?  SUB4
Computer name :  DEC1
Next jobname ?  <return>
```

The process has now started and the program and data flow is:

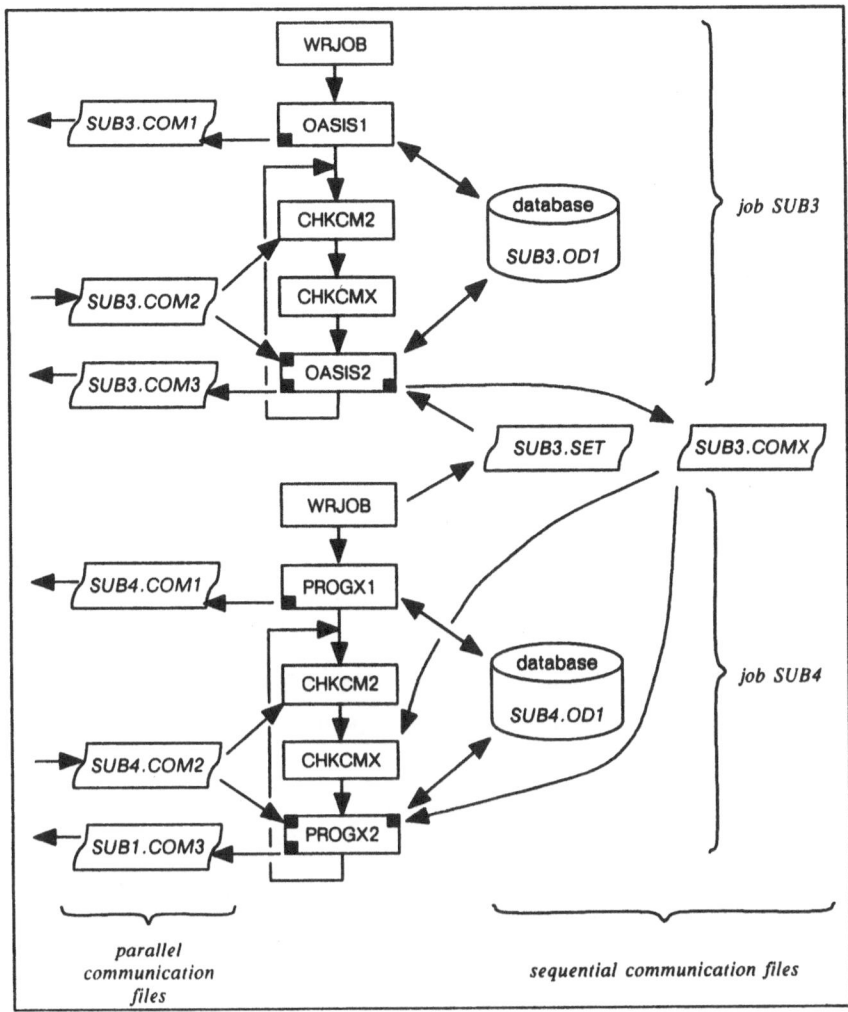

Figure 23. Communication between sequential jobs

Aside from the normal OCTOPUS procedure the following happens. The dataset file *SUB3.SET* is examined at the end of program OASIS2. Datasets found in *SUB3.SET* are read from the database, *SUB3.OD1*, of OASIS and written to ascii file *SUB3.COMX*. The next iteration of job SUB3 can start.

Module CHKCMX in job SUB4 will wait for file *SUB3.COMX* until it arrives. In the beginning of program PROGX2 all datasets in *SUB3.COMX* will be read and written to the database, *SUB4.OD1*, of PROGX. Once *SUB3.COMX* has been used it will be deleted. All use of *x.COMX* files is handled by standard OCTOPUS subroutines which are available in a library. (PROGX2 is the second, iterative part of PROGX.)

Several jobs may come in a sequence, and several sequences may run in parallel. However, no crosswise dependence is possible. At present OCTOPUS can handle up to 100 subjobs – parallel and sequential.

User written programs

The most basic form of a user written program compatible with the OCTOPUS concept follows:

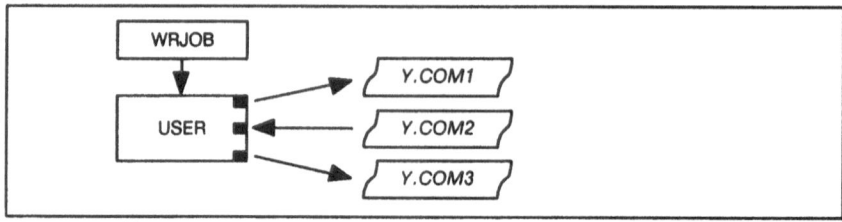

Figure 24. Closed form user written program

This a closed form, i.e. the iteration loop is inside the program. The OCTOPUS concept includes a subroutine library, and in this case four routines have been used: WTCOM1, CHCOM2, RDCOM2 and WTCOM3 (see appendix A for details on the subroutines). These subroutines handle all the communication with OCTOPUS. WTCOM1 writes the *Y.COM1* file. All information that is to be written is given as parameters in the subroutine call. Similar are RDCOM2 which reads *Y.COM2*, and WTCOM3 which writes *Y.COM3*. CHCOM2 holds the program until *Y.COM2* has arrived from OCTOPUS.

In our special case we want to solve the following problem:

$$\min \quad w(x_4, x_{10}) \;=\; x_4 + 2 \cdot x_{10}$$
$$\text{subject to:} \quad g_1(x_4, x_{10}) \;=\; (x_4 - 2)^2 + (x_{10} - 5)^2 \le 9$$
$$g_2(x_{10}) \;=\; x_{10} \ge 3$$
$$-3 \le x_4 \le 3$$
$$0 \le x_{10} \le 10$$

OCTOPUS requires all constraints to be on the form $g(x) \le 1$. g_1 and g_2 are therefore reformulated:

$$g_1(x_4, x_{10}) \;=\; (x_4 - 2)^2 + (x_{10} - 5)^2 - 8 \;\le\; 1$$
$$g_2(x_{10}) \;=\; 4 - x_{10} \;\le\; 1$$

The FORTRAN code for the sample program USER is found in appendix B.

A more advanced, open form, use of OCTOPUS where also the database system BAGIS and the sequential communication possibility is used may have the following configuration:

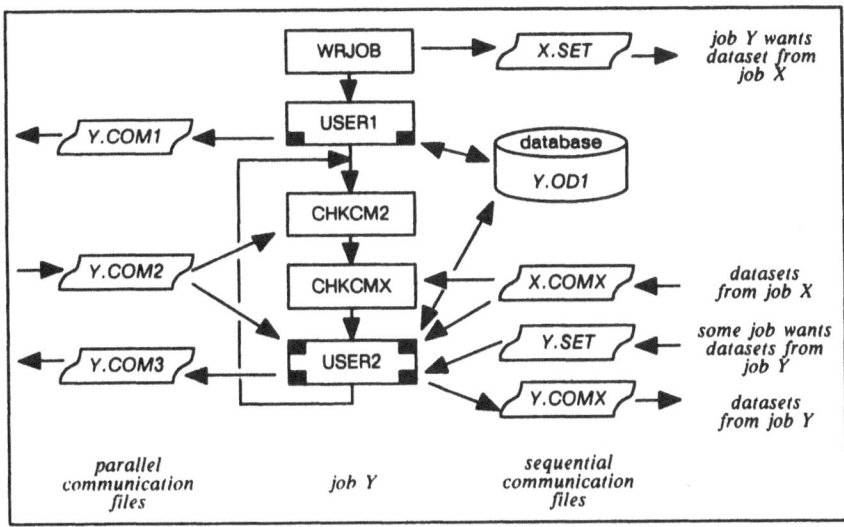

Figure 25. Open form user written program

The modules WRJOB, CHKCM2, CHKCMX and the subroutines in BAGIS are taken from the OCTOPUS library.

Communication between computers

Since OCTOPUS is strongly dependent on file transport between processes possibly running simultaneously on different types of computers, some thought has been put into solving this problem. The target has been to produce a general, easily expandable method that uses already existing systems for ASCII–file transport between computers. The kernel of the OCTOPUS file transport system is a subroutine that reads lines of text from a communication configuration file (*Comconf*), substitutes the file (job) name and extension into those lines, and then sends the lines as command lines to be interpreted by the operative system. The use of the *Comconf* file makes it possible for the user to add a new channel of communication or a new machine simply by editing the file. An entry in the *Comconf* file consists of a logical name to identify the communication channel, commands needed to send a file, a separation line (hyphens), and commands needed to receive a file. A simple example could look like:

```
:Comp1
rcp  $ID$.$EXT$  msh:/tmp/$ID$.$EXT$
-----
mv  $Storage_area/$ID$.$EXT$  $ID$.$EXT$
```

ID is replaced (by OCTOPUS) with the actual job name, and EXT with the file extension. In more complex cases it might be best to let the Comconf command start another process that takes care of the file transport. The reason for the send–receive structure is twofold:

I. It is quite often true that a process running on a machine of type A easily can send files to and receive files from a machine type B but that the reverse is not that simple

II. It may be necessary to use disk space on a third machine as intermediate storage if direct communication between two machines is impossible. This makes it possible to run OCTOPUS on loosely coupled computers since any type of file transfer can be used, like remote copy, Kermit, mailing systems, or even manually transported tapes or floppy disks.

A detailed description of the communication subroutines (programs) is found in appendix A.

Reliability

In a large scale OCTOPUS execution with many subjobs and many computers involved there will always be a significant risk that things go wrong. There are a few safety checks built into the system.

To start, a subjob program may crash with a dump. The job then directly passes on to a "last will" program, WRCOM3, which will send a file *x.COM3* containing a crash message to OCTOPUS. The message will be displayed to the coordinator who then has to decide what to do. The "last will" configuration is:

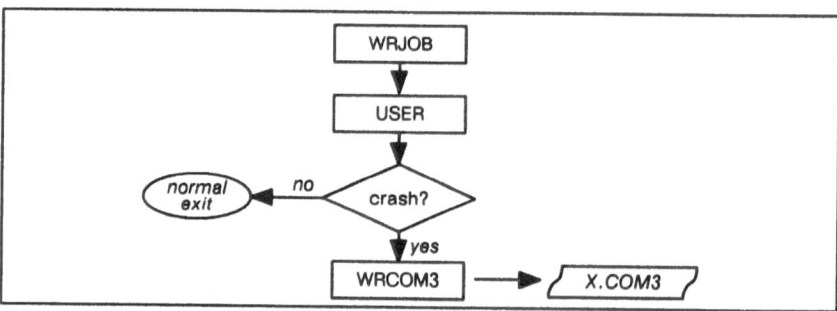

Figure 26. Last will exit

Whatever his decision is he can use the monitoring functions in GANDALF in order to continue. Either he has the the subjob error fixed and restarts the job. He must then use the INCLUD command in GANDALF. Or he simply REMOVEs the failing subjob until it can be fixed, or he KILLs it. Alternatively, he can kill the entire OCTOPUS job with the KILLAL command.

On the other hand if the subjob computer goes down or if its connection to the OCTOPUS main job goes down a "last will" does not help. In order to somewhat assist the coordinator there is a "statistical" check in module CHKCM3. If the elapsed time for a subjob iteration significantly exceeds that of the preceding iteration there is cause to suspect that something has gone wrong, and so a message is sent to the coordinator to alert him to take action.

In many cases a simple way to automatically check system integrity is to try to read a dummy file at certain intervals, or to have the computers ping each other and ask for identification. The best way to do this is of course installation dependent, and it may not always be practical, or even possible.

Example 1: Reaction bar

As a first application example let us study the reaction bar for a pneumatic pistol–grip nutrunner, shown schematically in Figure 27. The function of the reaction bar is to absorb the reaction torque when tightening nuts. The bar is supplied flat to the user who then shapes it to fit a suitable fixed support. It is desired that the reaction bar be as light as possible, but it must of course still satisfy certain strength and stiffness requirements. The manufacturer of course wants to minimize the cost, which in this case is considered to be the sum of net material cost and tooling cost.

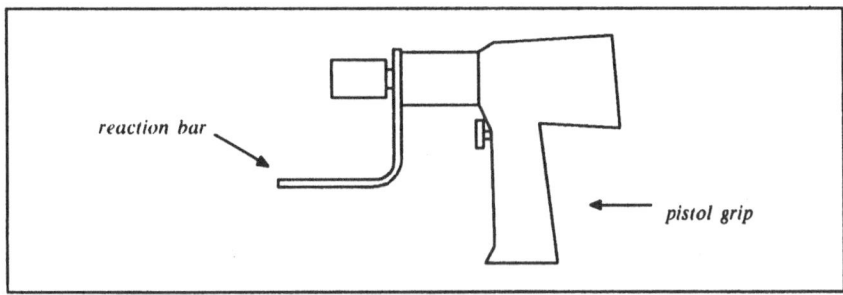

Figure 27. Pistol–grip nutrunner with reaction bar

> *Objective function = minimize(milling cost + net material cost)*
>
> *Constraints : strength and stiffness for two different geometries*

The total OCTOPUS optimization job MHF comprises three OASIS subjobs; MH1F, MH2F and MH3F.

MH1F handles the structural constraints for one geometry, but gives no contribution to the combined objective function. It calculates function values as well as gradients for the constraints.

MH2F handles the structural constraints for the second geometry, as well as the net material cost contribution to the total cost, i.e weight · cost/kg (function values and gradients).

MH3F calculates the milling cost contribution to the total cost by measuring the length of the machine tool path and multiplying it with a cost/unit length (function values and gradients). It gives no contribution to the constraints.

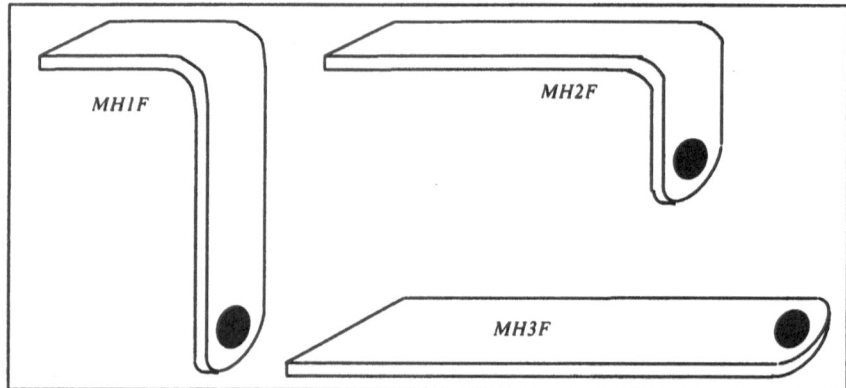

Figure 28. Geometries for the different subjobs

MHF is the main optimization job that assembles all the contributions to form the actual optimization problem. It creates and solves the subproblem, and then sends back the new design variable values to the separate subjobs for the next iteration.

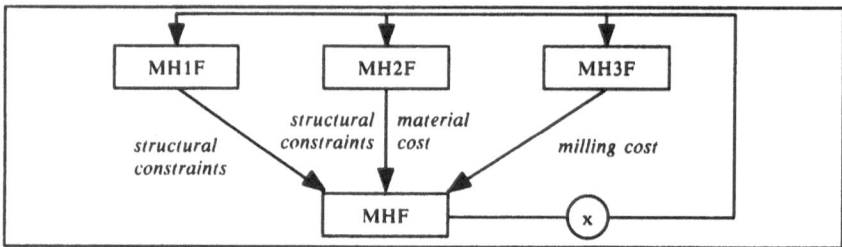

Figure 29. The "MHF" OCTOPUS job with subjobs

Example 2: Reaction bar, extended

As a second example let us expand the MHF job to include the sales price and a manufacturing constraint. This will lead to a 'global' optimization job MHS which includes both parallel and sequential subjobs. The objective in this case will be to maximize the profit.

> *Objective function = minimize (milling cost + material cost – sales price)*
> *i.e. maximize the profit*
> *Constraints : strength and stiffness for two different geometries*
> *milling speed*

The total OCTOPUS optimization job MHS comprises three OASIS subjobs; MH1S, MH2S and MH3S, as well as three subjobs of user written programs; PRIC1, PROD1 and MATC1.

MH1S handles the structural constraints for one geometry, but gives no contribution to the combined objective function, just as MH1F in the first example. It calculates function values as well as gradients.

MH2S handles the structural constraints for the second geometry (function values and gradients), and supplies weight information to the PRIC1 subjob (again, function values as well as gradients).

MH3S measures the length of the machine tool path (the circumference of the machined part; function values and gradients) and supplies that information to the subjob PROD1. It gives no contribution to the constraints.

PRIC1 calculates the market selling price, based on the assumption that the customer is willing to pay a higher price for a lighter tool than for a heavier. The price versus weight curve is shown in Figure 30. PRIC1 takes input (the weight and its gradients) from MH2S (through the dataset 'DWTDAG') and delivers price information (function value and gradients) to MHS (through the dataset 'DWDAG').

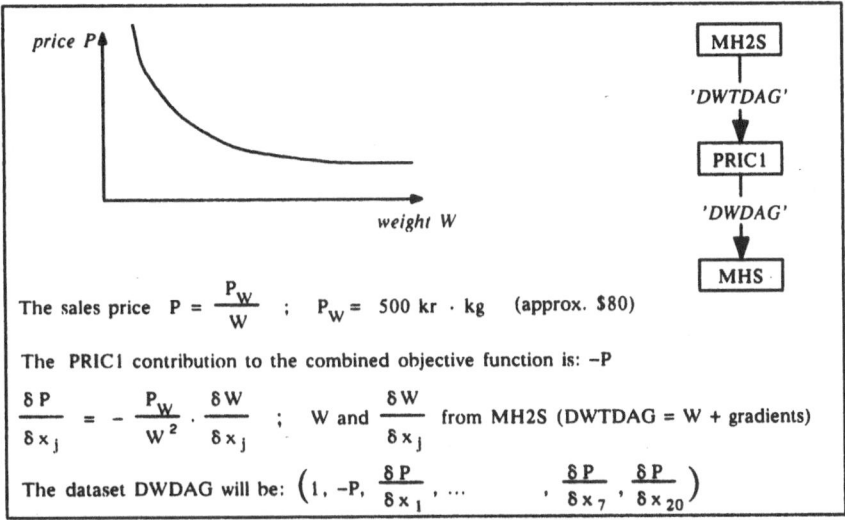

The sales price $P = \dfrac{P_W}{W}$; $P_W = 500 \text{ kr} \cdot \text{kg}$ (approx. $80)

The PRIC1 contribution to the combined objective function is: $-P$

$$\frac{\delta P}{\delta x_j} = -\frac{P_W}{W^2} \cdot \frac{\delta W}{\delta x_j} \quad ; \quad W \text{ and } \frac{\delta W}{\delta x_j} \text{ from MH2S (DWTDAG = W + gradients)}$$

The dataset DWDAG will be: $\left(1, -P, \dfrac{\delta P}{\delta x_1}, \dots, \dfrac{\delta P}{\delta x_7}, \dfrac{\delta P}{\delta x_{20}}\right)$

Figure 30. The PRIC1 subjob

PROD1 takes input (tool path length) from MH3S and calculates the milling cost, and also handles a constraint on the milling speed. It then delivers function values and gradients to MHS.

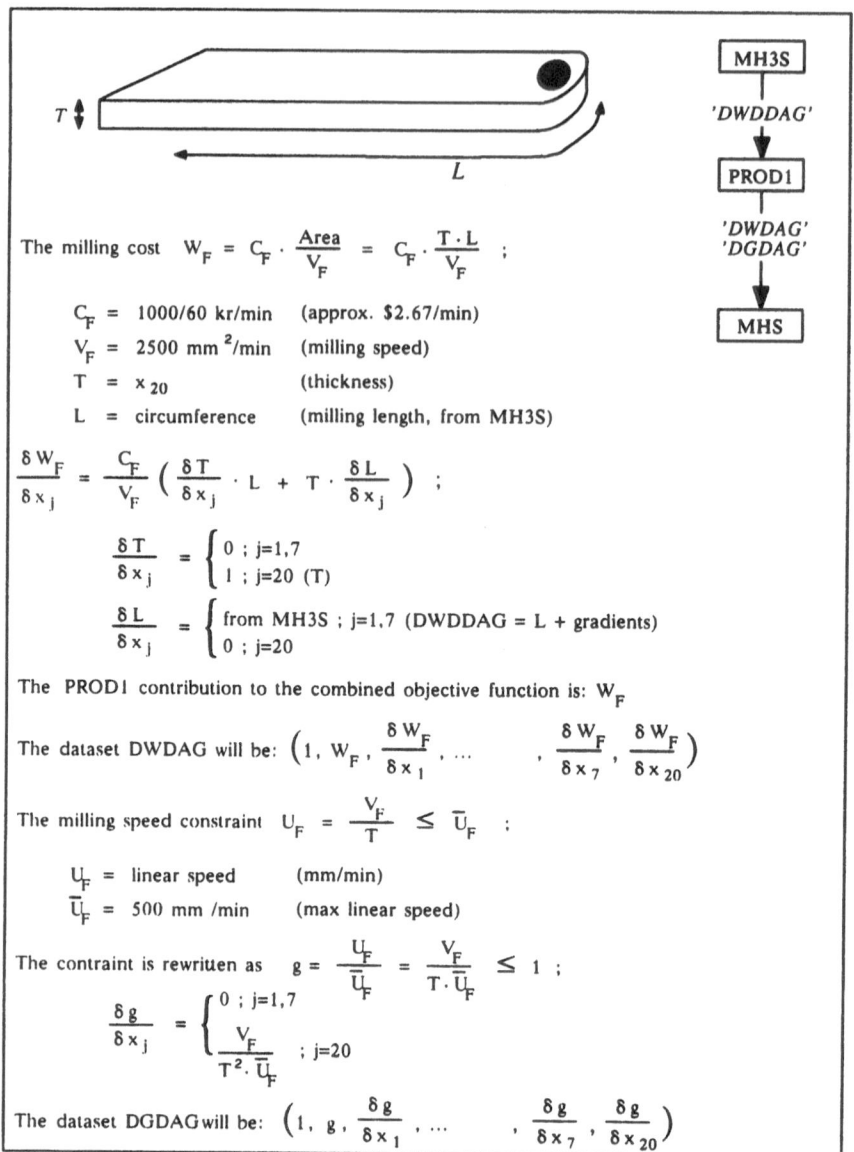

The milling cost $W_F = C_F \cdot \dfrac{Area}{V_F} = C_F \cdot \dfrac{T \cdot L}{V_F}$;

C_F = 1000/60 kr/min   (approx. \$2.67/min)
V_F = 2500 mm^2/min   (milling speed)
T = x_{20} (thickness)
L = circumference   (milling length, from MH3S)

$$\frac{\delta W_F}{\delta x_j} = \frac{C_F}{V_F} \left(\frac{\delta T}{\delta x_j} \cdot L + T \cdot \frac{\delta L}{\delta x_j} \right) ;$$

$$\frac{\delta T}{\delta x_j} = \begin{cases} 0 \ ; \ j=1,7 \\ 1 \ ; \ j=20 \ (T) \end{cases}$$

$$\frac{\delta L}{\delta x_j} = \begin{cases} \text{from MH3S} \ ; \ j=1,7 \ (DWDDAG = L + \text{gradients}) \\ 0 \ ; \ j=20 \end{cases}$$

The PROD1 contribution to the combined objective function is: W_F

The dataset DWDAG will be: $\left(1, \ W_F, \ \dfrac{\delta W_F}{\delta x_1}, \ \dots \qquad , \ \dfrac{\delta W_F}{\delta x_7}, \ \dfrac{\delta W_F}{\delta x_{20}}\right)$

The milling speed constraint $U_F = \dfrac{V_F}{T} \leq \overline{U}_F$;

U_F = linear speed   (mm/min)
$\overline{U}_F$ = 500 mm /min   (max linear speed)

The contraint is rewritten as $g = \dfrac{U_F}{\overline{U}_F} = \dfrac{V_F}{T \cdot \overline{U}_F} \leq 1$;

$$\frac{\delta g}{\delta x_j} = \begin{cases} 0 \ ; \ j=1,7 \\ \dfrac{V_F}{T^2 \cdot \overline{U}_F} \ ; \ j=20 \end{cases}$$

The dataset DGDAG will be: $\left(1, \ g, \ \dfrac{\delta g}{\delta x_1}, \ \dots \qquad , \ \dfrac{\delta g}{\delta x_7}, \ \dfrac{\delta g}{\delta x_{20}}\right)$

Figure 31. The PROD1 subjob

MATC1 calculates the raw material cost from the size of the unmachined material, and delivers function value and gradients to the main job MHS.

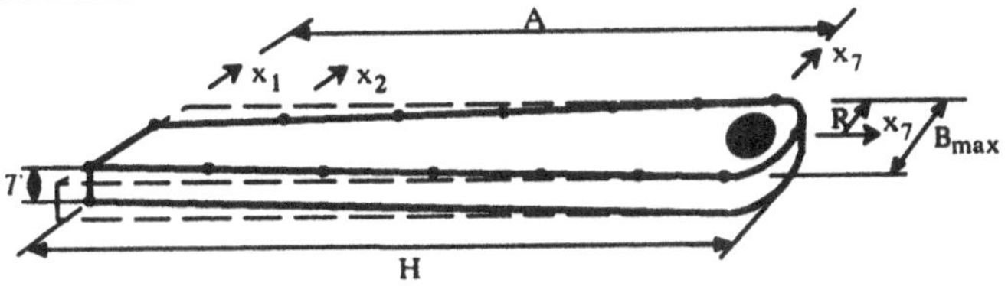

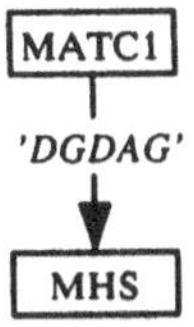

The raw material cost $W_M = B_{max} \cdot H \cdot T \cdot C_W \cdot \rho$;

B_{max} = largest width of the machined part

H = total length

T = x_{20} (thickness)

C_W = 20 kr/kg (approx. \$3.20/kg)

ρ = $7.8 \cdot 10^{-6}$ kg/m^3 (density)

H = $A + R$

R = $35 + x_7$

B_{max} = max(B_j) ; j=1,7

B_j = $70 + 2 \cdot x_j$

W_{Mj} = $B_j \cdot H \cdot T \cdot C_W \cdot \rho$; j=1,7 (candidates for the min–max objective function)

$$\frac{\delta W_{Mj}}{\delta x_k} = \left(\frac{\delta B_j}{\delta x_k} \cdot H \cdot T + B_j \cdot \frac{\delta H}{\delta x_k} \cdot T + B_j \cdot H \cdot \frac{\delta T}{\delta x_k} \right) \cdot C_W \cdot \rho ;$$

$$\frac{\delta B_j}{\delta x_k} = \begin{cases} 2 & ; k=j \\ 0 & ; k \neq j \end{cases}$$

$$\frac{\delta H}{\delta x_k} = \begin{cases} 1 & ; k=7 \\ 0 & ; k \neq 7 \end{cases}$$

$$\frac{\delta T}{\delta x_k} = \begin{cases} 1 & ; k=20 \\ 0 & ; k \neq 20 \end{cases}$$

The MATC1 contribution to the combined objective function is: max(W_{Mj}) ; j=1,7

The dataset DGDAG will be: $\left(1, W_{Mj}, \dfrac{\delta W_{Mj}}{\delta x_1}, \ldots \quad, \dfrac{\delta W_{Mj}}{\delta x_7}, \dfrac{\delta W_{Mj}}{\delta x_{20}} \right)$; j=1,7

There are no constraints in this job.

Figure 32. The MATC1 subjob

MHS assembles all the subjob contributions to form the actual optimization problem. It creates and solves the approximate subproblem, and then sends back the new design variable values to the separate subjobs for the next iteration.

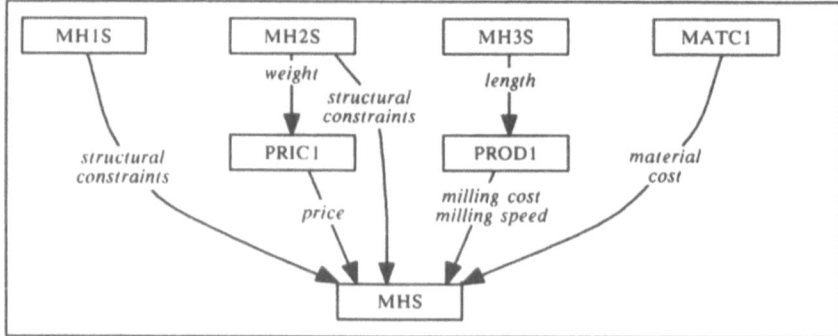

Figure 33. The "MHS" OCTOPUS job with subjobs

Plots from the actual computer runs are found in appendix C, convergence histories are found in appendix D.

The OCTOPUS Club

OCTOPUS has an open and very flexible architecture, and users can easily include their own programs/disciplines. The OCTOPUS Club administrates and distributes integrated programs to members and users. An OCTOPUS Library containing such programs will be implemented and made available to the members, and a list of its contents distributed to all users/members. A licensed OCTOPUS user may become a member by submitting a qualified contribution to the library.

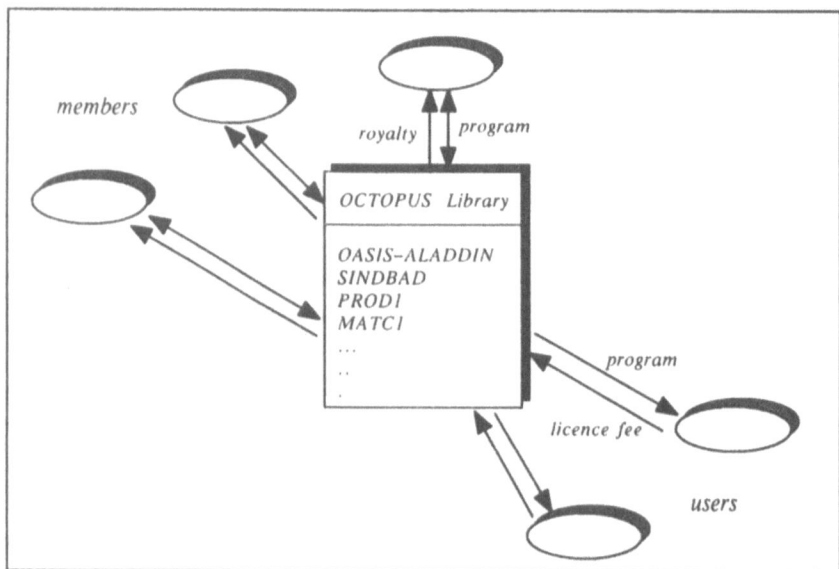

Figure 34. The OCTOPUS Club

Communication between computers

Subroutine WTCOM1

Subroutine WTCOM1 is called by the user program directly after its initialization phase. The FORTRAN call is:

```
call WTCOM1(ANTYP,IFIL,IFIOUT,LDESV,LSEQ,IOCT,OCTR,
            IEXAG,ITYPAG,ALFGAM,AGMINF,AGMAXF,STEP,
            IEXSEQ,NVALUE,VALUE)
```

where the input is:

ANTYP – type of analysis (CHARACTER*15)

IFIL   – file number for x.COM1

IFIOUT – output file number for user information

LDESV = max(1,NDESV)

LSEQ   = max(1,NSEQ)

IOCT(1–100)   – integer array where:
  IOCT(3)   – NCNMAX = maximum number of constraints that will ever occur in this job
  IOCT(10)   – NDESV = number of design variables
  IOCT(12)   – NSEQ = number of design variable sequences for discrete variables
  IOCT(14)   – MAXITO = maximum number of iterations allowed in the optimization process (OCTOPUS will use the largest value from all occurring subjobs)
  IOCT(16)   – IAPPROX = approximation method to be used in OCTOPUS (=6)
  IOCT(17)   – MNNMXX= 0 – combined objective function
                            = 1 – min–max objective function
  IOCT(18–23) – occupied for test purposes

OCTR(1–100)   – real (double precision) array where:
  OCTR(4)   – WTOL = convergence tolerance for OCTOPUS. The processes stop when either the maximum number of iterations (MAXITO) is reached, or convergence to a feasible solution has been achieved. The convergence criterion is:

$$\frac{w^{(k+1)} - w^{(k)}}{w^{(k)}} \leq WTOL$$

where $w^{(k)}$ is the value of the objective function at iteration number k. Normally WTOL = 0.01 – 0.0001. OCTOPUS will use the last value of WTOL given.

IEXAG(J), J=1,NDESV — external, user defined design variable numbers

ITYPAG(J), J=1,NDESV — design variable types:
  1 — sizing variables (thickness etc.)
  3 — material orientation variables
  5 — shape variables
  6 — material variables
  10 — auxiliary
  The design variable type must be the same for all occurrences of a variable with the same external number

ALFGAM(J), J=1,NDESV — initial design variable values. If a value conflicts with already existing values in OCTOPUS it will be neglected (the first value to appear will be used)

AGMINF(J), J=1,NDESV — lower bounds for the variables (the first value to appear will be used)

AGMAXF(J), J=1,NDESV — upper bounds for the variables (the first value to appear will be used)

STEP(J), J=1,NDESV — used for discrete variables:
  > 0 — design variable value is a multiple of STEP(J)
  = 0 — continuous variable
  < 0 — design variable values are picked from the sequence with number no = −STEP(J)

IEXSEQ(J), J=1,NSEQ — external, user defined design variable sequence numbers

NVALUE(J), J=1,NSEQ — number of discrete values in the respective sequence

VALUE(I,J), I=1,20; J=1,NSEQ — discrete design variable values (20) in the respective sequence

and the output is:

IOCT(41) − KBSTAT = a status flag:
  1 — the file x.COM1 has been written ok
  3 — failed in writing x.COM1 − program stops

Subroutine RDCOM2

RDCOM2 is called by the user program in the beginning of each iteration loop. Its FORTRAN call is:

```
call RDCOM2(ANTYP,IFIL,IFIOUT,LENG,LDESV,LOCONS,IOCT,OCTR,
            INTAG,IEXAG,ALFGAM,AGMINF,AGMAXF,RLAMDA,DUMMY)
```

where the input is:

ANTYP, IFIL, IFIOUT, LDESV — as in WTCOM1

LENG — upper limit for external, user defined design variable number. LENG is used to dimension arrays INTAG, IEXAG and DUMMY. Set for instance to 1000.

LOCONS = MAX(1,NOCONS)

IOCT(2)   – NOCONS = number of constraints in preceding iteration

IOCT(10) – NDESV

IOCT(41) – KBSTAT – status flag:
   1   – INCLUDe mode
   2   – REMOVE mode
   3   – KILL mode
   4   – ANALYSe mode
  12 – request to REMOVE
  13 – request to KILL
  14 – request to ANALYSe

INTAG(J), J=1,LENG – user internal design variable numbers for external numbers J

IEXAG(J), J=1,LENG   – workspace

DUMMY(J), J=1,LENG – workspace

and output is:

IOCT(1)   – JOBNO   = subjob number in OCTOPUS

IOCT(13) – KOPTIT   = iteration number

IOCT(14) – MAXITO – see WTCOM1

IOCT(17) – MNNMXX – see WTCOM1

IOCT(19) – LASTOP
   = 0 no convergence – continue
   = 1 convergence – stop

IOCT(20) – IFEAS = 1, will be reset IFEAS = 0 by the user program if the current iteration is not feasible

IOCT(41) – KBSTAT – status flag:
   1, 2, 3, 4   – see input
  11 – command from OCTOPUS to INCLUDe
  12 – command from OCTOPUS to REMOVE
  13 – command from OCTOPUS to KILL
  14 – command from OCTOPUS to ANALYSe

OCTR(5) – WTOT   = initial value for the objective function value for this subjob in this iteration (reset by USER if MNNMXX = 0)

OCTR(7) – CWST   = initial value for the worst constraint (reset by USER)

ALFGAM(J), J=1,NDESV   – current design variable values

AGMINF(J), J=1,NDESV   – checked lower bounds

AGMAXF(J), J=1,NDESV   – checked upper bounds

RLAMDA(J), J=1,NOCONS — Langrangian multipliers for the respective constraints in the preceding iteration (> 0 – active)

Subroutine WTCOM3

WTCOM3 is called by the user program at the end of the iteration loop when all gradients have been calculated. Its FORTRAN call is:

```
call WTCOM3(ANTYP,IFIL,IFIOUT,LDESV,LDESVP,LOCONS,
            IOCT,OCTR,IEXAG,AGMIN,AGMAX,DWDAG,DGDAG,
            IMNMX,ILINAR)
```

where the input is:

ANTYP, IFIL, IFIOUT, LDESV   as before

LDESVP = LDESV + 2

LOCONS = MAX(1,NOCONS)

IOCT(1)   – JOBNO

IOCT(2)   – NOCONS = number of constraints in this iteration

IOCT(10) – NDESV

IOCT(17) – MNNMXX – see WTCOM1

IOCT(19) – LASTOP – see RDCOM2

IOCT(20) – IFEAS =
   0   non feasible solution
   1   feasible

IOCT(41) – KBSTAT =
   1, 2, 3, 4   see RDCOM2
   12 – request from user to REMOVE the job
   13 – request from user to KILL the job
   14 – request from user to ANALYSe the job

OCTR(5) – WTOT = objective function contribution from this subjob (only if MNNMXX = 0)

OCTR(7) – CWST = value of worst constraint

IEXAG(J), J=1,NDESV   –   see WTCOM1

AGMIN(J), J=1,NDESV   –   modified, temporary, lower design variable limits

AGMAX(J), J=1,NDESV   –   modified, temporary, upper design variable limits

DWDAG(J), J=1,NDESV+2 where
   DWDAG(1) = 1
   DWDAG(2) = objective function value (=WTOT)

DWDAG(3) = derivative with respect to first design variable
DWDAG(4) = derivative with respect to second design variable
DWDAG(5) = etc.

DGDAG(J,I), J=1,NDESV+2, I=1,NOCONS constraints where I = constraint number
DGDAG(1,I) = upper limit (=1)
DGDAG(2,I) = constraint value
DGDAG(3,I) = first derivative
DGDAG(4,I) = second derivative
etc.

IMNMX(I), I=1,NOCONS
0 – normal constraint
1 – aspirant in the min–max optimization

ILINAR(I), I=1,NOCONS
0 – non linear constraint
1 – explicit linear constraint

Subroutines RDCOMX and WTCOMX

Both RDCOMX and WTCOMX have the same parameters:

```
call RDCOMX(IFIL,IFIOUT,IFISET,IFILDB,NETT)
```

where the input is:

IFIL   – file number for *x.COMX*

IFIOUT   – output number file for user information

IFISET   – file number for *x.SET*

IFIDB   – file number for data base

NETT   – net identifier (CHARACTER*15) in the data base (normally NETT=' ')

Subroutines CHCOM2 and CHCOMX

Both have the same parameters:

```
call CHCOM2(IFIL,IFIOUT)
```

where the input is:

IFIL   – file number for *x.COM2* or *x.COMX*

IFIOUT – output number file for user information

1194

References

1. K. Svanberg, Method of Moving Asymptots – A New Method for Structural Optimization, Int Journal for Numerical Methods in Engineering, Vol. 24, 359–373, 1987.

2. OASIS–ALADDIN User's Manual, ALFGAM Optimering AB, Stockholm, 1990

3. MEM–COM, SMR Corporation, Bienne, 1985

TITLES OF CONTRIBUTED PAPERS
(in alphabetical order of the first authors)

A.D. Belegundu
Shape Optimization Using Natural Basis Shapes with MSC/NASTRAN

J. Blachut
Externally Pressurized Filament Wound Domes — Scope for Optimization

T. Burczynski
Recent Advances in Boundary Element Sensitivity Analysis and
Optimization of Large Structural Systems

C.-M. Chan
An Optimality Criteria Algorithm for Tall Steel Building Design Using
Commerical Standard Sections

A. Chattopadhyay and T.R. McCarthy
Recent Efforts at Multicriteria Design Optimization of Helicopter Rotor
Blades

J.L. Chen, S.H. Sun, E.Y. Lai and S.B. Lin
Artificial Intelligence Techniques for Optimum Structural Design Process

G. Cheng and L. Song
Computational Aspects of Sensitivity Analysis of Nonlinear Discrete
Structures

A.R. Díaz and N. Kikuchi
A Method for Shape and Topology Optimization for Maximum Eingenvalues

M.E.M. El-Sayed
An Efficient Approach for Considering Durability Schedules in Structural
Optimization

B. Esping and O. Romell
Structural Optimization Using the OASIS-ALADDIN System

P. Fedelinski
Shape Optimization with Respect to Vibration Using the Boundary Element
Method

1196

W. Gollub and G.I.N. Rozvany
Michell Layouts for Complex Boundary Conditions

R.V. Grandhi
Optimum Design of Large Structures with Plate Bending Elements

Y. Gu and G. Cheng
Design Modelling, Sensitivity Analysis, and Decomposition Algorithm of
Structural Optimization

W. Gutkowski, Z. Iwanow and M. Pyrz
Controlled Enumeration Method with Size Preselection in Large
Structure Optimization

M.R. Hansen
An Automated Procedure for Dimensional Synthesis of Mechanisms

E. Hinton, N.V.R. Rao, J. Sienz and M. Özakça
Shape Optimisation and Adaptivity

S. Jendo and W.M. Paczkowski
Multicriteria Discrete Optimization of Large Scale Truss Systems

O. Jørgensen
Optimization of the Flutter Load by Material Orientation

S. Kimmich, R. Reitinger and E. Ramm
Integration of Different Numerical Techniques in Shape Optimization

V.K. Koumousis and P.G. Georgiou
An Expert System for the Optimal Design of Steel Truss Roofs

A.D. Larichev:
Optimization of Thin-Walled Composite Bars in Large Structures

T. Larsson and M. Rönnqvist
A Second Order Approximation Method for Structural Optimization

K.R. Leimbach, P.K. Umesha and D. Hartmann
Parallel Optimization of Large Truss Structures Subjected to Multiple
Loading

T. Lekszycki
Optimization of Vibrating Systems with Damping

U. Lepik
Optimal Design of Rigid Plastic Structures Under Dynamic Loading

R. Levy
Some Results on the Buckling of Optimized Plates

T. Lewiński
On Recent Developments in the Homogenization Theory of Elastic Plates
and Their Application to the Optimal Design

V.P. Malkov
Energy Criterion of Structural Efficiency

W. Marb and H. Suski
Design Sensitivity and Optimization in MSC/NASTRAN

H.P. Mlejnek
Some Aspects in the Genesis of Structures

C.M. Mota Soares, V.M. Franco Correia and J. Herskovits
Semi-Analytical Sensitivity Analysis and Optimal Design of Thin
Composite Shell Type Structures

E. Nikolaidis and J.S. Yang
Further Research on Probabilistic Optimization of Aircraft Wings

A. Osyczka, W. Kuchta and R. Czuła
Computer Aided Multicriterion Optimization System for Computationally
Expensive Functions

T. Petersen and P.S. Frederiksen
Fillet Design in Cold Forging Dies

U.T. Ringertz
Numerical Methods for Optimization of Nonlinear Shell Structures

H.C. Rodrigues
A Mixed Variational Formulation for Shape Optimization of Structural
Components

A.J.G. Schoofs, M.B.M. Klink and D.H. Van Campen
Approximation of Structural Optimization Problems by Means of Designed
Numerical Experiments

S.V. Selyugin
On Optimization of Bar Structures Made of Work-Hardening Elasto-Plastic
Materials

A.P. Seyranian
Sensitivity Analysis and Optimization in Flutter Problems

L.M.C. Simões
Reliability of Portal Frames with Discrete Design Variables

A.V. Soeiro, C.A. Conceição and A.T. Marques
Multilevel Optimization of Laminated Composite Structures

K. Szuwalski
Problems of Optimal Design of Structures against the Ductile Creep
Rupture

V.V. Toropov, A.A. Filatov and A.A. Polynkin
Multiparameter Structural Optimization Using FEM and Multipoint
Explicit Approximations

M. Utku
A Penalty Function Plate Bending Element

B.P. Wang
On "Closed Form" Solutions in Structural Optimization with an
Eigenvalue Constraint

B.P. Wang and D.P. Costin
Optimum Design of Composite Structures with Manufacturing Constraints

M. Weck and T. Nottebaum
Adaptive Meshing — Saving Computational Costs During the
Optimization of Composite Structures

L. Xu
Geometrical Stiffness and Sensitivity Matrices for Optimization of Semi-
Rigid Steel Frameworks

H. Yamakawa and M. Arakawa
Optimizations of Large Structural Systems by Using Qualitative and
Quantitative Transfer Matrix Methods

SUBJECT INDEX

Adjoint
- sensitivity analysis 477-493
- structure 8-9, 65, 1036-1037

Aircraft structures 226, 230-232, 966-971, 1000-1001, 1011-1022

Anisotropic materials
- optimal distribution 667-675
- optimal orientation 649-667, 707-730, 1051-1071
- shape design 676-680

Approximation concepts
- analytical solutions 1115-1133
- general 235-287
- global 241-247
- in layout optimization 131-137
- intermediate response quantities 260-263
- intermediate variables 244-246, 258-259
- local 236-240
- mid-range 248-251
- reduced basis 275-276
- series expansion 273-278

Beams
- Bernoulli 10-11, 14-15, 27-43, 68-69, 82-83, 103, 376-378
- dynamic control 818-819
- hollow square 549-550
- plastic design of 17-21

Bi-linearly elastic materials 683-695

Contact problems 697-705

Composite materials
- aircraft structures 1011-1022
- laminates 623-648, 997-1009
- mechanical response 708-715
- sandwich structures 951-972

Constraints
- behavioural 1
- compliance 81-82
- deflection 10-11, 15, 27-58
- geometrical 1
- global 1
- local 1
- stress 10-11, 15, 44-49

Control problems
- dynamic displacements 815-828
- eigenvalues 829-842
- general 811-842

Decomposition
- by dynamic programming 196-198
- general 193-195
- hierarchic 198-210
- hybrid 217
- non-hierarchic 211-217
- parallel processes 778-781

Dynamic problems 973-985

Elasto-plastic materials 697-705

Finite elements 366-382, 779-781

Frames 69-70, 863-881, 883-896, 897-926

Genetic algorithms 634-636, 851-852

Global optima 579-588, 609-622

Layout optimization
- analytical solutions 80-102
- arch-grids (Prager structures) 100-101

- beam systems (grillages) 78-79, 87-89, 91-98
- discretized solutions 103-116
- perforated plates 101-102
- solid plates 101
- trusses (Michell structures) 99-100, 104-116, 126-136, 139-155

Mathematical programming
- convex linearization 1101-1114
- dual methods 509-530, 873-891
- genetic algorithms 634-636
- goal programming 1135-1153
- integer programming 627-630
- interior point method 589-608
- linear 124-125, 131-132
- method of moving asymptotes (MMA) 555-578
- non-linear 131-132
- penalty functions 630-631
- probabilistic search 631-634
- sequential convex programming 531-553
- simulated annealing 632-634

Motor cars 177-191, 337-342

Multicriteria optimization 793-809, 850-851

Multidisciplinary optimization
- decomposition 193-217
- general 987-996
- software OCTOPUS 1165-1194

Mutual energy 13-14

Neural networks, artificial
- applications 731-745
- self-organization 747-765

Optimality criteria methods
- continuum formulation (COC) 1-120
- continuum formulation, discretized (DCOC) 3, 59-75
- discretized formulation (DOC) 3
- displacement based 139-155
- general 851

- reinforced concrete frames 883-896
- steel frames 863-872

Parallel processes xx, 767-792

Plane stress 43-44

Plastic design, optimal 16-21, 80-81

Plates 101-102, 166- 169, 493-508, 636-641, 661-672

Reinforced concrete structures
- frames 883-896
- members 927-949

Selfweight 22

Sensitivity analysis
- anisotropic materials 657-661
- built-up structures 313-343
- control problems 833-834
- dynamic frequency 329-343
- dynamic response 313-328
- eigenvalues 296-302, 323, 455-476
- error analysis 361-383
- error elimination 385-396
- finite difference 290-292
- finite strain 433-455
- first and second order 477-493
- general 289-310
- in decomposition methods 221-223
- non-conservative problems 345-359
- non-linear 455-476
- parallel processing 782
- plates 493-508
- shape design 345-359, 397-432
- shell structures 1035-1042
- static response 292-296
- transient response 302-305, 317-323

Shape optimization
- generalized 77, 116-120
- sensitivity analysis 345-359, 397-432, 493-508
- shell structures 1023-1049
- stress design 1073-1086

Shells 171-172, 177-191, 1023-1049

Stability 636-642, 951-972, 1087-1099, 1155-1164

Theory and practice 843-862

Steel frameworks 863-881

Topology optimization
• by optimality criteria 80-120, 139-155

• by homogenization 157-191
• design considerations 121-138

Trusses 49-58, 72-75, 83-85, 126-135, 139-155, 282-285, 550-552, 683-695, 737-739, 1141-1151

Variational calculus 14-15